INTEGRATED OPTICAL CIRCUITS AND COMPONENTS

OPTICAL ENGINEERING

Series Editor

Brian J. Thompson

Distinguished University Professor
Professor of Optics
Provost Emeritus
University of Rochester
Rochester, New York

1. Electron and Ion Microscopy and Microanalysis: Principles and Applications, *Lawrence E. Murr*
2. Acousto-Optic Signal Processing: Theory and Implementation, *edited by Norman J. Berg and John N. Lee*
3. Electro-Optic and Acousto-Optic Scanning and Deflection, *Milton Gottlieb, Clive L. M. Ireland, and John Martin Ley*
4. Single-Mode Fiber Optics: Principles and Applications, *Luc B. Jeunhomme*
5. Pulse Code Formats for Fiber Optical Data Communication: Basic Principles and Applications, *David J. Morris*
6. Optical Materials: An Introduction to Selection and Application, *Solomon Musikant*
7. Infrared Methods for Gaseous Measurements: Theory and Practice, *edited by Joda Wormhoudt*
8. Laser Beam Scanning: Opto-Mechanical Devices, Systems, and Data Storage Optics, *edited by Gerald F. Marshall*
9. Opto-Mechanical Systems Design, *Paul R. Yoder, Jr.*
10. Optical Fiber Splices and Connectors: Theory and Methods, *Calvin M. Miller with Stephen C. Mettler and Ian A. White*
11. Laser Spectroscopy and Its Applications, *edited by Leon J. Radziemski, Richard W. Solarz, and Jeffrey A. Paisner*
12. Infrared Optoelectronics: Devices and Applications, *William Nunley and J. Scott Bechtel*
13. Integrated Optical Circuits and Components: Design and Applications, *edited by Lynn D. Hutcheson*
14. Handbook of Molecular Lasers, *edited by Peter K. Cheo*
15. Handbook of Optical Fibers and Cables, *Hiroshi Murata*
16. Acousto-Optics, *Adrian Korpel*

17. Procedures in Applied Optics, *John Strong*
18. Handbook of Solid-State Lasers, *edited by Peter K. Cheo*
19. Optical Computing: Digital and Symbolic, *edited by Raymond Arrathoon*
20. Laser Applications in Physical Chemistry, *edited by D. K. Evans*
21. Laser-Induced Plasmas and Applications, *edited by Leon J. Radziemski and David A. Cremers*
22. Infrared Technology Fundamentals, *Irving J. Spiro and Monroe Schlessinger*
23. Single-Mode Fiber Optics: Principles and Applications, Second Edition, Revised and Expanded, *Luc B. Jeunhomme*
24. Image Analysis Applications, *edited by Rangachar Kasturi and Mohan M. Trivedi*
25. Photoconductivity: Art, Science, and Technology, *N. V. Joshi*
26. Principles of Optical Circuit Engineering, *Mark A. Mentzer*
27. Lens Design, *Milton Laikin*
28. Optical Components, Systems, and Measurement Techniques, *Rajpal S. Sirohi and M. P. Kothiyal*
29. Electron and Ion Microscopy and Microanalysis: Principles and Applications, Second Edition, Revised and Expanded, *Lawrence E. Murr*
30. Handbook of Infrared Optical Materials, *edited by Paul Klocek*
31. Optical Scanning, *edited by Gerald F. Marshall*
32. Polymers for Lightwave and Integrated Optics: Technology and Applications, *edited by Lawrence A. Hornak*
33. Electro-Optical Displays, *edited by Mohammad A. Karim*
34. Mathematical Morphology in Image Processing, *edited by Edward R. Dougherty*
35. Opto-Mechanical Systems Design: Second Edition, Revised and Expanded, *Paul R. Yoder, Jr.*
36. Polarized Light: Fundamentals and Applications, *Edward Collett*
37. Rare Earth Doped Fiber Lasers and Amplifiers, *edited by Michel J. F. Digonnet*
38. Speckle Metrology, *edited by Rajpal S. Sirohi*
39. Organic Photoreceptors for Imaging Systems, *Paul M. Borsenberger and David S. Weiss*
40. Photonic Switching and Interconnects, *edited by Abdellatif Marrakchi*
41. Design and Fabrication of Acousto-Optic Devices, *edited by Akis P. Goutzoulis and Dennis R. Pape*
42. Digital Image Processing Methods, *edited by Edward R. Dougherty*
43. Visual Science and Engineering: Models and Applications, *edited by D. H. Kelly*
44. Handbook of Lens Design, *Daniel Malacara and Zacarias Malacara*
45. Photonic Devices and Systems, *edited by Robert G. Hunsperger*
46. Infrared Technology Fundamentals: Second Edition, Revised and Expanded, *edited by Monroe Schlessinger*
47. Spatial Light Modulator Technology: Materials, Devices, and Applications, *edited by Uzi Efron*
48. Lens Design: Second Edition, Revised and Expanded, *Milton Laikin*

49. Thin Films for Optical Systems, *edited by François R. Flory*
50. Tunable Laser Applications, *edited by F. J. Duarte*
51. Acousto-Optic Signal Processing: Theory and Implementation, Second Edition, *edited by Norman J. Berg and John M. Pellegrino*
52. Handbook of Nonlinear Optics, *Richard L. Sutherland*
53. Handbook of Optical Fibers and Cables: Second Edition, *Hiroshi Murata*
54. Optical Storage and Retrieval: Memory, Neural Networks, and Fractals, *edited by Francis T. S. Yu and Suganda Jutamulia*
55. Devices for Optoelectronics, *Wallace B. Leigh*
56. Practical Design and Production of Optical Thin Films, *Ronald R. Willey*
57. Acousto-Optics: Second Edition, *Adrian Korpel*
58. Diffraction Gratings and Applications, *Erwin G. Loewen and Evgeny Popov*
59. Organic Photoreceptors for Xerography, *Paul M. Borsenberger and David S. Weiss*
60. Characterization Techniques and Tabulations for Organic Nonlinear Optical Materials, *edited by Mark Kuzyk and Carl Dirk*
61. Interferogram Analysis for Optical Testing, *Daniel Malacara, Manuel Servín, and Zacarias Malacara*
62. Computational Modeling of Vision: The Role of Combination, *William R. Uttal, Ramakrishna Kakarala, Sriram Dayanand, Thomas Shepherd, Jagadeesh Kalki, Charles F. Lunskis, Jr., and Ning Liu*
63. Microoptics Technology: Fabrication and Applications of Lens Arrays and Devices, *Nicholas F. Borrelli*
64. Visual Information Representation, Communication, and Image Processing, *Chang Wen Chen and Ya-Qin Zhang*
65. Optical Methods of Measurement: Wholefield Techniques, *Rajpal S. Sirohi and Fook Siong Chau*
66. Integrated Optical Circuits and Components: Design and Applications, *edited by Edmond J. Murphy*

Additional Volumes in Preparation

Computational Methods for Electromagnetic and Optical Systems, *John M. Jarem and Partha P. Banerjee*

Adaptive Optics Engineering Handbook, *edited by Robert K. Tyson*

INTEGRATED OPTICAL CIRCUITS AND COMPONENTS

DESIGN AND APPLICATIONS

EDITED BY

EDMOND J. MURPHY

Uniphase Telecom Products
Bloomfield, Connecticut

MARCEL DEKKER, INC.   NEW YORK • BASEL

ISBN: 0-8247-7577-5

This book is printed on acid-free paper.

Headquarters
Marcel Dekker, Inc.
270 Madison Avenue, New York, NY 10016
tel: 212-696-9000; fax: 212-685-4540

Eastern Hemisphere Distribution
Marcel Dekker AG
Hutgasse 4, Postfach 812, CH-4001 Basel, Switzerland
tel: 41-61-261-8482; fax: 41-61-261-8896

World Wide Web
http://www.dekker.com

The publisher offers discounts on this book when ordered in bulk quantities. For more information, write to Special Sales/Professional Marketing at the headquarters address above.

Copyright © 1999 by Lucent Technologies. All Rights Reserved.

Neither this book nor any part may be reproduced or transmitted in any form or by any means, electronic or mechanical, including photocopying, microfilming, and recording, or by any information storage and retrieval system, without permission in writing from the publisher.

Current printing (last digit):
10 9 8 7 6 5 4 3 2 1

PRINTED IN THE UNITED STATES OF AMERICA

To
Kathy, Ed, Colleen, Dan, Tim, and Sean
for their support and patience
during this endeavor

From the Series Editor

Some topics in Optical Engineering are so important and so dynamic that they need to be revisited. This is certainly the case with the subject *Integrated Optical Circuits and Components: Design and Applications.* Volume 13 of the Optical Engineering Series was published with the above title in 1987 and was edited by Lynn D. Hutcheson, who prepared the introduction (Chapter 1) and contributed two other chapters. As an edited work, twelve authors contributed to the success of the volume's nine chapters.

So here we are some twelve years later with an entirely new volume with the same title under the editorship of Edmond J. Murphy with a total of twenty-two authors contributing twelve chapters. George Stegeman contributed to both volumes. The other notable change is the evidence of the global importance of this field—the authors come from nine different countries.

Integrated optical circuits and components have now become secure technologies that are applied in commercial systems. This current volume certainly reflects that change from the research and development work reported in the original volume.

Topics covered include devices such as lithium niobate modulators; silica, erbium-doped glass, and erbium-doped lithium niobate waveguides. Indium phosphide-based photonic circuits and components, polymeric thermo-optic digital optical switches, and optics devices in communication systems also receive attention. Chapters on some of the current trends include ''hybrid integration of optical devices on silicon,'' ''integrated optics in sensors,'' and ''nonlinear integrated optical devices.'' The volume is completed with a discussion of the design and simulation tools available for integrated optics.

Volume 13, published in 1987, was very successful. Volume 66 presented here is a worthy revision to that earlier volume and will no doubt achieve the same success.

Brian J. Thompson

Preface

The field of integrated optics has evolved dramatically since the earlier incarnation of this book, edited by Lynn Hutcheson, was published in 1987. Integrated optic (IO) devices have moved from an exciting, fast-moving research and development phase to a more exciting, faster-moving commercial deployment phase. This book focuses on technical developments while capturing a flavor of commercial success by incorporating contributions from leaders of commercially successful enterprises. Despite the maturation of the technology in some areas, there is much fertile ground for future research and development work. A view of this work is included in many of the chapters.

The wide and successful deployment of IO devices is a tribute to decades of hard work by research and development teams around the world. A conscious effort was made to identify authors from a cross section of companies and countries to reflect the broad range of workers in this field.

This book is written for scientists and engineers as well as managers and students who desire a deep and up-to-date view of a broad range of IO topics, and it is unique in this regard. To my knowledge, no similar review book dealing with integrated optics has been published in the past few years.

The book represents the joint effort of 22 contributors. The reader thus has the opportunity to learn from a broad range of experts in this field. An extensive set of references supplement each chapter, and the index provides a means of cross-referencing related subjects.

I wish to sincerely thank all the authors for their responsiveness in preparing the chapters. I also wish to thank the staff at Marcel Dekker, Inc., for their professional assistance and their drive to keep the book on schedule. Finally, I want to express my deep thanks to Kathy, Ed, Colleen, Dan, Tim, and Sean for their support and patience during this endeavor.

Edmond J. Murphy

Contents

Contributors

M. R. Amersfoort BBV Design, Enschede, The Netherlands

Gaetano Assanto Terza University of Rome, Rome, Italy

Denis Barbier TEEM Photonics, Meylan, France

J. Bos BBV Software, Enschede, The Netherlands

Peter De Dobbelaere Akzo Nobel Electronic Products Inc., Sunnyvale, California

Mart Diemeer Akzo Nobel Central Research, Arnhem, The Netherlands

Christopher R. Doerr Lucent Technologies, Holmdel, New Jersey

Rien Flipse Akzo Nobel Photonics, Arnhem, The Netherlands

Mats Gustavsson Ericsson Components AB, Stockholm, Sweden

Robert L. Hyde Alp Optics, Laffrey, France

Rino E. Kunz Centre Suisse d'Electronique et de Microtechnique SA (CSEM), Zurich, Switzerland

X. J. M. Leijtens Delft University of Technology, Delft, The Netherlands

Fred J. Leonberger Uniphase Telecom Products, Bloomfield, Connecticut

Edmond J. Murphy* Lucent Technologies, Breinigsville, Pennsylvania

Katsunari Okamoto NTT Photonics Laboratory, Tokai, Japan

Robert G. Peall Nortel Networks, Harlow, England

Anat Sneh Lucent Technologies, Holmdel, New Jersey

Wolfgang Sohler University of Paderborn, Paderborn, Germany

George I. Stegeman University of Central Florida, Orlando, Florida

Hubertus Suche University of Paderborn, Paderborn, Germany

Paul G. Suchoski, Jr. Uniphase Telecom Products, Bloomfield, Connecticut

H. J. van Weerden Twente University of Technology, Enschede, The Netherlands

* *Current affiliation:* Uniphase Telecom Products, Bloomfield, Connecticut

INTEGRATED OPTICAL CIRCUITS AND COMPONENTS

1

Introduction

Edmond J. Murphy*

Lucent Technologies
Breinigsville, Pennsylvania

1 BACKGROUND

Twelve years have passed since an earlier version of this book was published [1]. That volume was devoted to descriptions of integrated optic circuits and components, which at that time were still very much in a research phase. Indeed, in the Preface, Hutcheson referred to the many research organizations pursuing topics in integrated optics and used this as an indication of the potential of the technology. This current book is, in part, a celebration of this potential realized. Interest in and use of this technology has blossomed in the intervening years. In 1987, researchers in integrated optics were just learning how to work with the extremely tight tolerances required for single-mode devices, and the push to a focus on 1.3- and 1.55-micron systems was just beginning. The developments in this field have been fast and furious as the range of applications and the number of technical workers in the field have increased. One of the most significant changes in the field over the past few years has been the successful commercialization of integrated optic devices. In this volume, we attempt to demonstrate the advances in the field by highlighting the commercial success of some devices. It is exciting to see that many of the same people who were involved in technology research 12 years ago have stayed with the technology and seen it through to commercial success. In one of the first articles on integrated optics, P. K. Tien "cautions the reader that the technology involved is difficult because of the smallness and perfection demanded by thin-film optical devices." This book provides a snapshot of the achievement of that perfection.

* *Current affiliation:* Uniphase Telecom Products, Bloomfield, Connecticut

The years since the earlier book have brought forth many significant developments in technology, systems, and commercialization. The technology demonstrations range from materials development for erbium-doped devices to fabrication of very large arrays of integrated devices. Experimental system demonstrations have included the DARPA-funded MONET (Multi-Wavelength Optical Network) program and the RACE-funded Stockholm Gigabit Network. Actual system applications of integrated devices have been seen in wavelength multiplexed transmission networks, cable TV systems, and high-speed (10 Gbit/sec) networks. Not surprisingly, the commercial success of the technology has come in areas related to the system applications. Silica waveguide devices are a key element in the wavelength-division multiplexing (WDM) systems, and lithium niobate modulators are the technology of choice in certain analog and high-speed networks.

These years have also brought impressive achievements in the level of achievable integration. Integrated optic devices are no longer made on small substrates in research fabrication facilities. Rather, they are made in state-of-the-art clean rooms with the same wafer-level processing equipment used in silicon integrated circuit manufacture. The complexity of these devices, as measured by the variety of functions on a chip, the length of devices on a chip, and the number of functional elements, has continued to increase. Examples include array waveguide gratings with waveguides that are controlled well enough to maintain exacting phase relationships over centimeters of length. Devices are fabricated on 100- and 125-mm-diameter substrates. Well-understood and robust fabrication technologies have been developed that allow fabrication of hundreds of individual passive lightguide circuit devices on a single wafer and over 200 active optical switch elements in a single package module. The commercial success of the technology could not have come without the successful move to these mass fabrication techniques and without a strong understanding of the basic device physics and materials science. Without this knowledge, the reproducibility and reliability required for deployment could not be guaranteed. Themes of cost and manufacturability will be seen in many of the chapters—reflecting the advanced stages of these devices.

This success, as measured by use of the technology in general applications, will expand as optical communications systems become more complex and as more and more system-level functions are moved from the electrical domain to the optical domain. Increased numbers of signal channels and expanded optical bands will provide opportunities for use of more integrated optic components. As the number of channels and the complexity of these systems increase, the advantages of integration, miniaturization, and solid-state devices will become increasingly important. The expanding capacity of communications systems will naturally lead to increased optical functionality. Sophisticated integrated devices will be required to provide optical-layer protection, restoration, add/drop, and cross-connect functions. Optical amplifiers provide another example. These devices are becoming increasingly complex as they provide system functionality beyond simple amplification. In today's networks, amplifiers are subsystems that also provide system monitoring, gain equalization, and dispersion compensation. The continued expansion of optical functions performed at the amplifier node will result in more use of active and passive integrated devices within the amplifiers.

The field of integrated optics is a tremendously broad and rich subject area, as evidenced by the range of topics presented herein. Because of the constraints of time and space, it is unfortunate that not all integrated optic areas are covered in this book. The topics we have been able to include are representative but certainly not all-inclusive. However, recently published articles provide excellent summaries of the state of the art of technologies we have not been able to cover here. Interested readers are referred to Ref. 3 for lithium niobate modulators, Ref. 4 for lithium niobate switch arrays, Ref. 5 for integrated laser devices, Ref. 6 for integrated receiver arrays, and Ref. 7 for vertical-cavity surface-emitting lasers (VCSELs).

The book has been organized so that each chapter is an independent discussion of a particular topic. The book represents the joint effort of 22 authors, each a recognized expert in the particular subject area. By compiling articles from leading experts, we offer the reader the advantage of the in-depth knowledge that a person intimately involved in the technology can best describe. Authors have been deliberately chosen from many different companies and countries so that the book represents a broad view of the capabilities of the technology.

Much of the focus is on communication systems—a bias that reflects, in part, the optical fiber communication applications that have driven much of this technology and, in part, the editor's own experience and knowledge base. Hopefully, this focus will not keep the reader from appreciating the applications of the technology in sensors (Chapters 2 and 10) and in signal processing and optical logic (Chapter 11).

The book is intended for scientists and engineers who wish to have an introduction to the many facets of integrated optics, as well as for experienced integrated optics technologists who wish to have a ready reference for reviewing advances in related fields. Students should be able to use the text and the extensive reference list provided with each chapter to obtain a thorough understanding of the field. Hopefully, the chapters that deal with established commercial technologies will provide an added source of motivation to workers in newer integrated optics areas that will enable them to see a path to meaningful applications.

2 Overview of Chapter Content

Chapter 2: Commercialization of Lithium Niobate Modulators. As mentioned earlier, one of the most striking achievements in integrated optics over the past decade has been the large-scale deployment of devices in commercial applications. In this sense, Chapter 2 is a success story. This book would not be complete without a view of the factors involved in moving a technology from research concept to end use. Moving a device from the research phase into a robust manufacturable product is expensive and time consuming and requires a technology appropriate for market conditions. In Chapter 2, Paul G. Suchoski, Jr., and Fred J. Leonberger provide an insightful perspective from the front lines of business development. Despite the demonstration of technological capability for lithium niobate devices in the mid-1980s, large volume applications did not appear until the mid-1990s. Suchoski and Leonberger chronicle the evolution of communication system needs and concepts and the associated evolution of a commercially viable technology. Their discussion illustrates both the need for developing the underlying technology and the need to focus on applications when the market demand materializes.

Chapter 3: Requirements for Integrated Optic Devices in Communication Systems. As the experience related in Chapter 2 clearly demonstrates, system applications imply specific requirements on devices that must be met before a technology or a component will be accepted. For optical communications systems, those requirements revolve around optical bandwidth, insertion loss, polarization sensitivity, and crosstalk. In Chapter 3, Mats Gustavsson takes a theoretical look at these and other parameters and the interactions between device capability and system performance. Although theoretical in nature, the chapter is practical in tone—the discussion focuses on real-world systems issues associated with real-world limits on device performance. After a discussion of a broad range of parameters, Gustavsson deals with the problem of optical crosstalk in depth and finally applies the crosstalk analysis to semiconductor amplifier switch matrices. The chapter provides a solid introduction to the technical issues facing device designers.

Chapter 4: Silica Waveguide Devices. Arrayed waveguide grating multiplexers have been one of the technology drivers in the deployment of dense wavelength-division multiplexing (DWDM) optical communication systems. In Chapter 4, Katsunari Okamoto provides an overview of this important technology. The development of this technology presents unique challenges for fabrication. The large features and the long waveguides must be fabricated with low defect densities and with reproducible phase relationships among the many waveguides. Okamoto briefly describes the fabrication process and then describes the design and performance of the multiplexer. He includes some recent research-level demonstrations of silicon waveguide device capability, including an optical switch matrix that integrates 256 switches on a single wafer. The last section of the chapter contains a short description of the use of silica waveguides and the mechanical features of silicon for hybrid lightguide circuits. This section thus serves as a preview of the more extensive discussion in Chapter 9.

Chapter 5: Erbium-Doped Glass Waveguide Devices. The erbium-doped fiber amplifier has been the most significant technology driver for present-day optical communication systems. The success of this technology has prompted interest in the development of integrated versions of the current fiber devices. There is hope that erbium-doped integrated optic devices will eventually provide a cost, functionality, or performance advantage over fiber amplifiers in some high-volume applications such as fiber distribution networks. Integration of erbium-doped waveguides in glass with passive devices such as power splitters offers the possibility of fabricating ''lossless'' (loss-compensated) integrated devices. In Chapter 5, Denis Barbier and Robert L. Hyde cover the design, fabrication, and performance of erbium-doped waveguide devices in glass.

Chapter 6: Erbium-Doped Lithium Niobate Waveguide Devices. The combination of gain from rare-earth dopants such as erbium with active devices leads to novel integrated devices. In Chapter 6, Wolfgang Sohler and Hubertus Suche provide details on the device physics and applications of such devices. They describe erbium-doped waveguide devices in lithium niobate that make use of the acousto-optic and electro-optic properties of this crystalline material. Device demonstrations include amplifying modulators, tunable wavelength filters, and mode-locked, Q-switched, and tunable lasers. The higher functionality achievable by pairing a gain medium with electrical or acoustical control may lead to a variety of applications in communications, signal processing, and sensing.

Chapter 7: Indium Phosphide–Based Photonic Circuits and Components. One of the true driving thrusts for integrated optic development has always been the goal of optical and electronic integration on a single substrate. For this reason, there has been much research on waveguide devices in semiconductor materials. Many difficult materials and physics issues have been addressed and resolved in order to demonstrate device functions on indium phosphate and gallium arsenide substrates. In Chapter 7, Anat Sneh and Christopher R. Doerr begin with a short description of waveguide design and fabrication on indium phosphide. They then discuss the design and operation of increasingly complex devices, including optical switch arrays, integrated laser sources for wavelength multiplexed systems, dynamic wavelength routing devices, and integrated demultiplexers/receivers. A driving force for integrated components is system simplicity. As the number of wavelength channels increases dramatically in communication systems, it becomes more and more likely that arrays of lasers integrated on indium phosphide with waveguide routers will provide a path to reduce the number of discrete lasers deployed in these systems.

Chapter 8: Polymeric Thermo-Optic Digital Optical Switches. Another driving force for integrated components is the reduction of cost. Polymeric materials offer great hopes for low material and processing cost. However, serious interest in the application of this technology had not developed until recently because of concern for material stability and reliability. In recent years, new materials have been developed, and we have seen demonstrations of high-speed electro-optic modulators as well as arrays of thermo-optic switches. The thermo-optic switches described in Chapter 8 compare quite favorably with competing technologies. Mart Diemeer, Peter De Dobbelaere, and Rien Flipse combine to provide a detailed view of the device design. A novel part of their contribution, which reflects the requirements of a successful commercial technology, is the section on the reliability studies of these components—a study that is essential to the acceptance of any new components. This chapter gives a clear view of the close coupling between the drive of technology capability and the requirements for successful system applications.

Chapter 9: Hybrid Integration of Optical Devices on Silicon. Silicon is, of course, the well-known workhorse of the electronics industry. In Chapter 4, we see how silicon wafers can be used as simple substrates for optical waveguide devices. However, silicon also has crystallographic and thermal properties that make it very useful as a mechanical material for hybrid integration. Cost has been a hindrance to the mass deployment of fiber-optic devices, and packaging operations have been a significant contributor to cost. Optical devices are distinguished from electronic devices in this area. For electronic devices, wafer features are measured in submicron units, while packaging tolerances are measured in microns and tens of microns. For optical devices, the converse is true. The tight alignment tolerances required for efficient alignment between optical devices and transmission fibers drive the cost up. In Chapter 9, Robert G. Peall describes the uses of silicon in the hybrid integration of optical devices. Over the next few years, we can expect to see increased use of this hybrid packaging technology as well as the demonstration of larger, more sophisticated hybrid circuits. These eventually will include the integration of active and passive waveguide devices.

Chapter 10: Integrated Optics in Sensors: Advances Toward Miniaturized Systems for Chemical and Biochemical Sensing. As stated earlier, much of the focus of this book and of the development of integrated optics has been in the area of optical communications. However, the technology has important applications in other areas. In Chapter 10, Rino E. Kunz explores the applications of integrated optics in chemical and biochemical sensors. In the first section, Kunz makes an important point about considering not just the component function but the full system function of integrated optic technology. The chapter describes the underlying properties of chemical sensors and the design of optical waveguides that provide sensitive sensing elements. Attention is given to low-cost fabrication techniques, and many interesting and novel devices are described.

Chapter 11: Nonlinear Integrated Optical Devices. Beyond the current applications that much of this book focuses on lie the frontiers of higher speed and more sophisticated systems with new and novel device needs. In Chapter 11, George I. Stegeman and Gaetano Assanto provide an update to the associated chapter in the earlier version of this book. Integrated waveguides provide an ideal means to take advantage of higher-order effects because of the high optical intensities that can be achieved. This chapter describes the principles of these effects and reviews a set of pertinent devices.

Chapter 12: Design and Simulation Tools for Integrated Optics. The book ends with a chapter that is essential for new device beginnings. The fabrication and testing of new, novel devices is time-consuming and expensive. Rapid and efficient development of new device concepts and designs necessitates sophisticated modeling tools. Along with the commercial market for integrated optic devices has come a market for modeling tools. In Chapter 12, M. R. Amersfoort, J. Bos, X. J. M. Leijtens, and H. J. van Weerden describe the capabilities of optical waveguide modeling tools. These tools have been developed rapidly over the past few years to keep up with the demands of researchers working in many different waveguide materials. The authors describe the impressive software packages that are currently available. These packages integrate waveguide layout tools, passive and active individual device modeling, and graphical output displays. More recent developments include tools for analyzing the performance of multiple components within integrated devices.

REFERENCES

1. Lynn D. Hutcheson, ed. Integrated Optical Circuits and Components. New York: Marcel Dekker, 1987.
2. P. K. Tien. Light waves in thin films and integrated optics. Appl. Optics 10:2395–2413, 1971.
3. F. Heismann, S. K. Korotky, and J. J. Veselka. Lithium niobate integrated optics: selected contemporary devices and systems application. In: I. P. Kaminow and T. L. Koch, eds. Optical Fiber Telecommunications IIIB. San Diego: Academic Press, 1997, ch. 9.
4. E. J. Murphy. Photonic switching. In: I. P. Kaminow and T. L. Koch, eds. Optical Fiber Telecommunications IIIB. San Diego: Academic Press, 1997, ch. 10.
5. T. L. Koch. Laser sources for amplified and WDM lightwave systems. In: I. P. Kaminow and T. L. Koch, eds. Optical Fiber Telecommunications IIIB. San Diego: Academic Press, 1997, ch. 3.
6. L. D. Garret, S. Chandrasakar, J. L. Zyskind, J. W. Sulhoff, A. G. Dentai, C. A. Burrus, L. M. Lunardi, and R. M. Derosier. Performance of eight-channel OEIC *p*-I-*n*/HBT receiver array in 8×2.5 Gbit/s WDM transmission system. J. Lightwave Technol 15:827–832, 1997.
7. L. A. Coldren and B. J. Thibeault. Vertical-Cavity Surface-Emitting Lasers. In: I. P. Kaminow and T. L. Koch, eds. Optical Fiber Telecommunications IIIB. San Diego: Academic Press, 1997, ch. 6.

2

Commercialization of Lithium Niobate Modulators

Paul G. Suchoski, Jr., and Fred J. Leonberger

Uniphase Telecom Products
Bloomfield, Connecticut

1 INTRODUCTION

By the mid-1980s, researchers had demonstrated high-performance lithium niobate integrated optic modulators, switch arrays, and fiber-optic gyro circuits. Yet it would take another decade before there was adequate market demand for lithium niobate devices to support a healthy supplier base. This chapter initially examines issues that inhibited early market acceptance of this technology and then describes the strong market pull that led to volume production of modulators for use in CATV (cable television) signal distribution and dense wavelength division multiplexing (DWDM) telecommunications systems. Today, tens of thousands of modulators are being manufactured each year for use in these applications.

2 EARLY APPLICATIONS ARE TECHNOLOGY DRIVEN AND LACK MARKET PULL

In the early 1980s, research efforts were initiated worldwide at major telecommunications and aerospace companies as well as several government laboratories to develop lithium niobate integrated optical circuits. A significant achievement was realized at AT&T Bell Laboratories in 1981 when researchers demonstrated very low-loss lithium niobate waveguides (less than 1 dB fiber-to-fiber insertion loss at 1300 nm) by locally diffusing titanium into the surface of the material.

By 1984, switches and modulators were demonstrated with efficient fiber coupling, low propagation loss, low drive voltage, and bandwidth in excess of 5 GHz.

Efforts at telecommunications companies concentrated primarily on developing the basic waveguide, device, and packaging technology to realize optical switch arrays. In these structures, light from any one of *m* input fibers can be rapidly switched (<1 nsec) to any one of *n* output fibers by applying a set of control voltages to the device. At several companies, 8 × 8 switch arrays were demonstrated with reasonable insertion loss, drive voltage, and channel isolation. Despite these successful demonstrations, research in this area was largely discontinued by the early 1990s due to a number of technical limitations, including polarization dependent loss and crosstalk, tight manufacturing tolerances, the need for complex control circuitry for driving the switch arrays, and difficulty in scaling the technology to arrays larger than 8 × 8. In addition, advocates for this technology realized that telecommunications network designers would not be interested in switching in the optical domain until the year 2000 or later. Until then, fiber optics would be used strictly for high-capacity point-to-point links while switching would occur in the electrical domain.

Efforts at aerospace companies and government laboratories concentrated largely on developing lithium niobate devices for use in fiber optic gyroscopes (FOGs) and antenna remoting, i.e., transmitting radio frequency (RF) and microwave signals over optical fiber with high fidelity and low RF/microwave loss. In both applications, the integrated optics technology demonstrations were quite successful, resulting in fiber-optic systems that outperformed competing technologies. Unfortunately, widespread availability of low-cost global positioning systems (GPS) receivers largely eliminated the commercial market for fiber optic gyroscopes. Moderate-volume military applications for this technology exist, and companies such as Litton are aggressively establishing FOG manufacturing facilities. However, fiber-optic gyroscopes are competing against well entrenched technologies such as ring laser gyros (RLGs) and spinning mass mechanical gyros which puts significant pricing pressure on fiber-optic gyros except in niche markets such as satellite stabilization. The resulting integrated optics business for FOG (total revenues) is not large enough to support a stand-alone supplier of these components. Similarly, the fiber optic market for antenna remoting is highly fragmented and largely military. The lack of high-volume opportunities tends to keep component cost high, which in turn limits the technology to low-volume niche markets.

Fortunately, these early research efforts resulted in a strong integrated optic technology base that is at the heart of today's commercial lithium niobate modulator products. The switch array research focused on titanium diffusion in z-cut $LiNbO_3$, leading to a thorough understanding of low-loss channel waveguide fabrication for efficient fiber coupling, low-loss waveguide bends, and efficient switching structures based on directional couplers. In addition, device engineers developed elegant solutions for tough packaging problems such as V-groove arrays for multiple fiber attachment. The FOG and antenna remoting research concentrated on x-cut $LiNbO_3$ using both titanium diffusion and annealed proton exchange (APE). Efficient device structures were developed for Mach–Zehnder structures and phase modulators with excellent RF and microwave performance. In addition, robust packaging solutions were demonstrated for minimization of insertion loss, polarization crosstalk, and optical back-reflection over extreme military operating conditions. Much of this technology is currently being used in commercial CATV and OC-48 modulators.

3 CATV SIGNAL DISTRIBUTION: A STRONG MARKET PULL DRIVES COMPONENT MANUFACTURERS

A typical CATV distribution system is depicted in Fig. 1. At a CATV headend, various television signals are received from satellites, terrestrial antennae, and storage media. Each signal is mixed up to the center frequency corresponding to the desired television channel (e.g., channel 2 in the United States has a center frequency of 55.25 MHz) on which it will be broadcast. The various channels

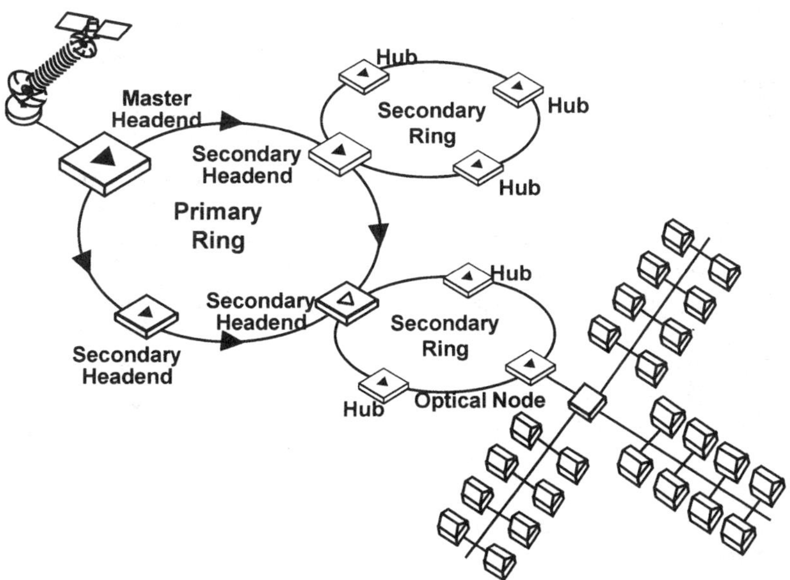

Figure 1 Cable TV distribution system.

are then combined and broadcast out into the network over a coaxial cable and/or fiber-optic network. Traditionally, all channels are analog, each with a bandwidth of 6 MHz in North America.

In the late 1980s, American CATV providers (such as TCI, Cox Cable, and Time Warner) wanted to upgrade the quality of their product offerings to increase revenues and to increase the likelihood that local regulators would reissue their franchise licenses. At that time, most CATV operators were offering a maximum of 30 channels, and novel revenue generating concepts such as pay-per-view and premium movie channels were in their infancy. In addition, CATV customers were generally dissatisfied with their CATV service because of poor signal quality and frequent system outages.

These systems problems were attributed largely to the coaxial cable distribution plants that were in use throughout the CATV networks at that time. Coazial cable has limited bandwidth and such high transmission loss that RF amplifiers are required every 2000 feet to boost the CATV signals. Each amplifier introduces noise and degrades the signal quality because of nonlinearities asociated with the amplifiction process. In addition, most CATV system outages are associated with RF amplifier failures.

Cable TV operators opted in the early 1990s to begin deploying analog fiber-optic technology to transmit CATV signals from the headends to the nodes. Surprisingly, fiber-optic technology is considerably less expensive to deploy than coaxial cable-based systems. In addition, it has significantly higher bandwidth (860 MHz is now commonly available) than coaxial cable, which allows CATV operators to increase their product offerings to at least 80 and in many cases 110 channels. Signal quality and the frequency of system outages also improve with the use of fiber optics. Fiber-optic technology has been so successful in this application that it is now used exclusively for connecting CATV headends to nodes. Cable TV providers were one of the largest consumers of single-mode fiber-optic cable and passive components in the 1990–1995 time frame.

The first analog fiber-optic technology to be deployed was based on directly modulated 1310-nm DFB lasers. This technology was pioneered by Ortel Corporation. A 1310-nm fiber-optic link runs from the headend to the hub, where the optical signal is detected and the resultant RF signal is amplified. A second 1310-nm fiber optic link is then used to connect the hub to one or more nodes. At the nodes, the optical signal is detected, amplified, and then transmitted in the RF domain over coaxial cable to each of the several hundred to several thousand homes served by the node. There has been a general trend to decrease the node size, i.e., the number of homes served by a given node, as the price of fiber optic technology has decreased. Many CATV providers are currently laying out networks that have only one RF amplifier between the node and the home in order to optimize system performance.

1310-nm analog lasers are specially designed DFB laser structures that have low noise levels when modulated and nearly linear optical power-versus-current transfer functions. By varying the laser current linearly with the RF CATV signal, the RF CATV signal is transferred onto the optical output of the laser, which is then carried throughout the CATV distribution plant over single-mode fiber. Higher linearity is achieved by first characterizing each laser for its second- and third-order nonlinearities and then electronically predistorting the laser drive current to compensate for them. Link budgets for this technology typically range from 3 to 13 dB, with fiber loss of 0.4 dB per km at 1310 nm. Thus, unrepeatered transmission over distances up to 30 km are possible. Alternatively, the CATV network designer might opt to split the optical signal and support several closely spaced nodes using a single 1310-nm laser.

Second-generation CATV transmission systems utilize high-power CW lasers and lithium niobate modulators (Fig. 2). Most externally modulated systems operate at 1550 nm to take advantage of erbium-doped fiber-optic amplifiers (EDFAs). High-power (25–60 mW) DFB lasers with extremely low noise (RIN less than-165 dB/Hz) are used as the CW source in these systems. Dual-output Mach–Zehnder interferometric modulators are typically used in CATV applications, since they have predictable sinusoidal transfer functions that make electronic predistortion relatively easy. Using a bias control circuit to operate the modulator

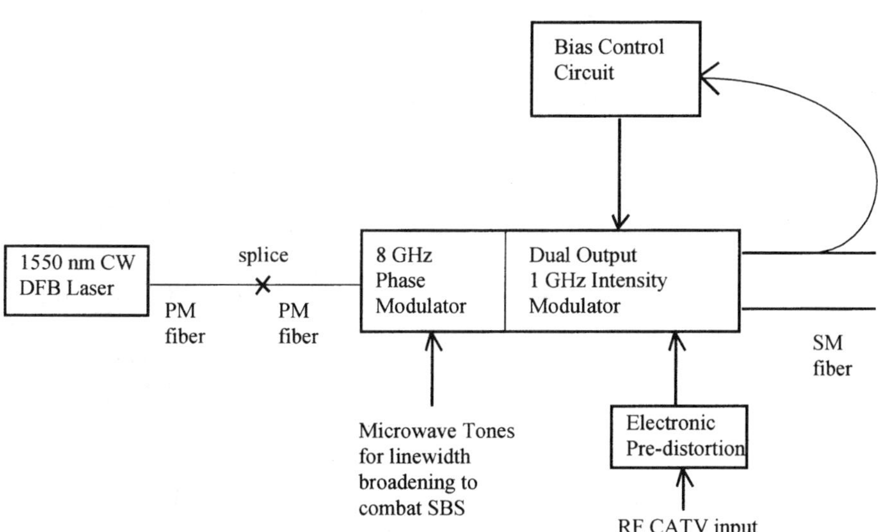

Figure 2 Externally modulated 1550-nm CATV transmitter.

at quadrature (the 3-dB power point, which is the most linear region of the device transfer function), second-order distortion terms are minimized to very low levels. Third-order distortion terms are reduced to acceptable levels by electronically predistorting the RF voltage prior to applying it to the modulator. In these systems, optical power levels up to 17 dBm can be launched into single-mode fibers, which allows for unrepeatered fiber runs up to 85 km (fiber transmission loss is 0.2 dB/km at 1550 nm). An in-line EDFA can increase the fiber span to 150 km.

Integrated optics manufacturers have worked closely with CATV transmitter manufacturers to optimize modulators for this application. Uniphase Telecom Products (UTP) has been the market leader in this lithium niobate modulator market since its inception, drawing heavily from the company's past experience in high-end analog military fiber-optic technology. In terms of optical performance, lithium niobate CATV modulators have less than 4 dB insertion loss, optical return loss more than 55 dB, and two optical outputs that are simultaneously at quadrature. In terms of electrical performance, the frequency response of the modulator is flat to ± 0.4 dB from 45 MHz to 860 MHz, the electrical return loss (S11) is greater than 18 dB over this frequency band, and the half-wave voltage is less than 5.5 V at 100 MHz. In addition, a 6-GHz phase modulator is incorporated into the lithium niobate circuit. High-frequency tones can be applied to this phase modulator to broaden the optical spectrum in order to increase the SBS (stimulated Bruillion scattering) threshold of the fiber optic system to 17 dBm.

A large global market has developed over the past 2 years for this technology. In the United States, 1550-nm externally modulated transmitters are widely used for interconnecting CATV headends and for connecting headends to hubs. In addition, they are being widely deployed in overseas CATV builds where there is no existing CATV plant in place. In these deployments, one 1550-nm transmitter and multiple high-power (up to 22 dBm) fiber amplifiers can provide more than 20,000 homes with CATV service at 40-50% lower cost than can be achieved with directly modulated 1310-nm DFB lasers. Although CATV has largely been a North American product concept in the past, today the overseas market for CATV fiber-optic equipment is approximately twice as large as the North American market. Every major CATV equipment manufacturer now offers externally modulated transmitters and EDFAs. Critical optical components, including high-power 1550-nm CW DFB lasers, lithium niobate modulators, EDFA gain blocks, and optical receivers, are available from multiple suppliers.

4 DWDM SYSTEMS SATISFY THE HUGE DEMAND FOR TELECOMMUNICATIONS CAPACITY

Since 1994, the demand for capacity on North American long-haul fiber-optic networks run by carriers such as AT&T, MCI, Sprint, and WorldCom has more than doubled every 18 months. This voracious demand for bandwidth, which is expected to continue through at least 2005, is being driven by nonvoice applications such as the Internet, video conferencing, e-mail, and fax traffic. In 1998, total data transmission on North America's long-distance networks exceeded total voice traffic. As shown in Fig. 3, by the year 2060 data traffic is expected to exceed voice traffic by at least a factor of 5. The only viable method for satisfying this demand for bandwidth is to utilize state-of-the-art fiber-optic systems throughout the long-distance network.

Fiber-optic networks have been widely used in the North American long-haul market for more than 10 years. First-generation systems were based on directly modulated 1300-nm DFB lasers. Modulation rates of 2.5 gigabits/sec are widely deployed in long-distance networks today. The optical fiber utilized in these systems is 1310-nm single-mode fiber with propagation loss of 0.4 dB/km and minimal dispersion at 1300 nm. As shown in Fig. 4, these systems require 3R (receive and convert to the electrical domain, retime and reshape the electrical signal, and retransmit in the optical domain) stations every 40 km to compensate for fiber transmission loss and dispersion. Typical intercity spans are up to 600 km in length, thus requiring fourteen 3R stations.

Second-generation systems operate over the previously installed 1300-nm fiber and utilize 1550-nm CW DFB lasers, external modulators, and fiber amplifiers. With this architecture, it is possible to go more than 600 km without con-

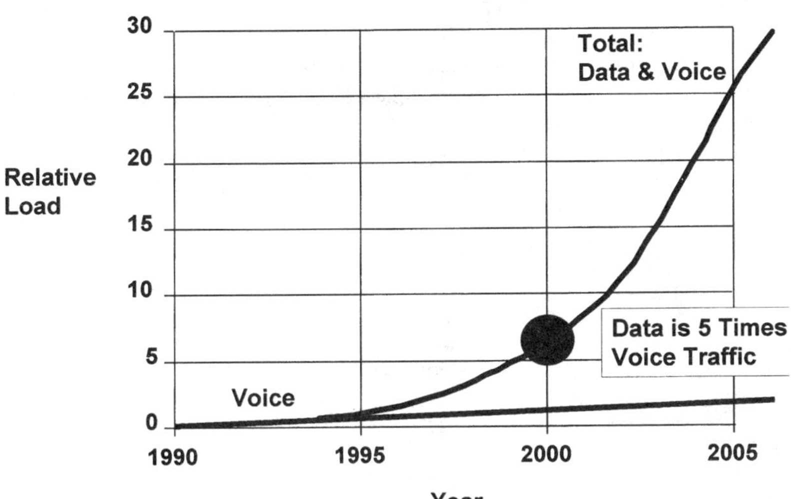

Figure 3 Volume of data traffic on North American long-distance fiber-optic plant will be 10 times larger than volume of voice traffic by the year 2005.

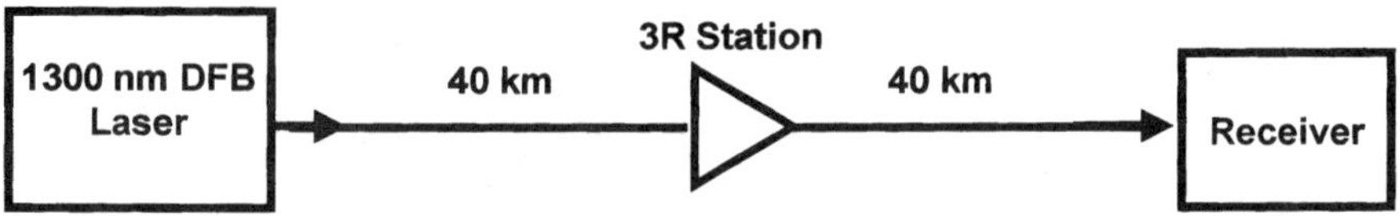

Figure 4 First-generation 2.5 Gbit/sec systems utilized directly modulated DFB lasers and multiple 3R stations.

verting to the electrical domain (Fig. 5). Since fiber transmission loss is only 0.2 dB/km, fiber amplifier spacing can be increased to 120 km. Thus, a 600-km fiber span based on 1550-nm externally modulated technology requires only five fiber amplifiers (each of which is less expensive than one 3R station), compared to 14 3R stations that are required in a 600-km fiber span based on 1300-nm technology. Lithium niobate modulators are the technology of choice for this generation of systems, since they do not introduce wavelength chirp in the modulation process. Maintaining a narrow optical spectrum is critical, since the 1300-nm fiber is highly dispersive at 1550 nm. Directly modulating a 1550-nm DFB laser results in wavelength chirp (a broadening of the laser linewidth), which limits directly modulated lasers to fiber lengths of 150 km with current state-of-the-art lasers.

Third-generation systems also operate at 1550 nm and utilize dense wavelength division multiplexing (DWDM) technology to transmit up to 100 wavelengths, each operating at 2.5 gigabits/sec onto a single fiber (Fig. 6). This technology has been pioneered by Ciena, Lucent, and Pirelli. DWDM products were first introduced in 1996 with 16 channels. By 1998, several companies had announced plans for 80-channel systems. By 2000, 80-128 channel systems are expected to be available, providing a fiber capacity of 200–250 gigabits.

Wavelength-controlled CW DFB lasers, external modulators, wavelength-flattened EDFAs, wavelength multiplexers, and wavelength demultiplexers are

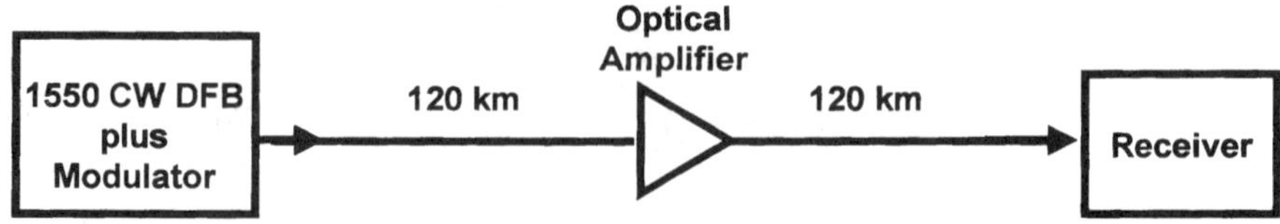

Figure 5 Second-generation 2.5 Gbit/sec systems utilized externally-modulated 1550-nm transmitters and erbium-doped fiber-optic amplifiers (EDFAs) to eliminate the need for 3R stations.

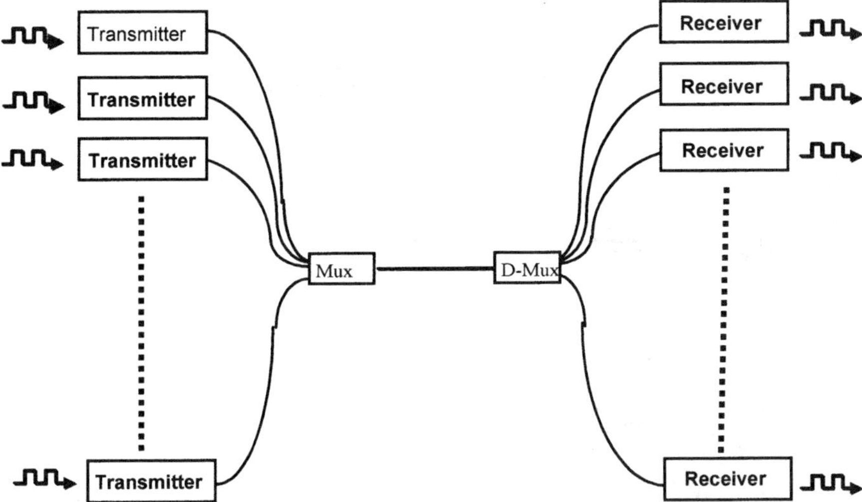

Figure 6 Third-generation 2.5 Gbit/sec systems utilize up to 100 DWDM channels, each operating at 2.5 Gbit/sec to transmit up to 250 Gbit/sec over a single optical fiber.

used in these systems. Lithium niobate modulators are again the technology of choice, for four reasons. First, they operate chirp-free. Second, one modulator operates over the entire wavelength range of the EDFA, greatly simplifying inventory issues. Third, by separating the light-generation function from the modulation function, more precise control of the laser wavelength is possible. Finally, lithium niobate modulators are widely available from several vendors, thus reducing manufacturing risks for the systems providers.

Integrated optics manufacturers have worked closely with systems engineers to develop digital modulators that are optimized for this application. Today, several thousand 2.5 gigabits/sec modulators are manufactured each month. These devices have optical insertion loss less than 4 dB, on/off extinction greater than 20 dB, drive voltage less than 3.2 V, S21 greater than 3 GHz, and S11 less than −10 dB from DC to 3 GHz. The modulators must interface effectively with commercially available driver amplifiers to yield acceptable eye diagrams with more than 16 dB eye opening, less than 125 psec rise/fall times, and less than 50 psec of jitter. Finally, the modulator manufacturer must guarantee that the bias point of the modulator must be very stable over the 25-year lifetime of the part. UTP has addressed this problem by developing a bias-free modulator, as shown in Fig. 7. In this structure, the passive bias point of the modulator is set to quadrature during manufacturing so that the user does not need to apply any DC bias voltage to the device. By carefully packaging the modulator to minimize

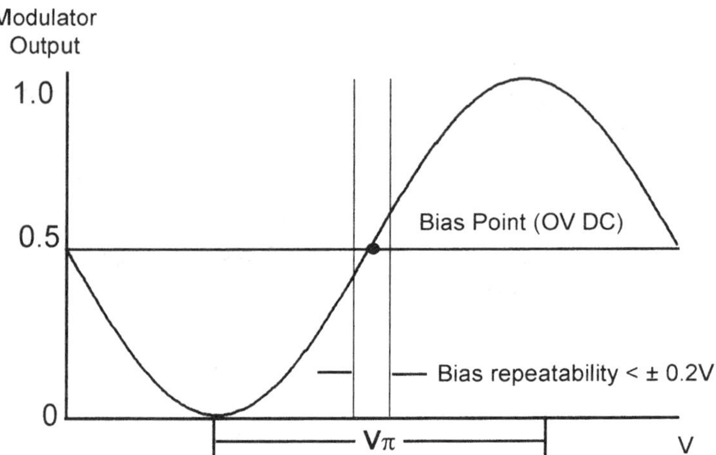

Figure 7 Bias-free 2.5-Gbit/sec modulator pioneered by Uniphase Telecommunications Products (UTP) greatly simplifies the use of lithium niobate modulators in digital systems.

stress on the chip and by eliminating any DC voltage on the device, UTP can guarantee that the passive bias point of the modulator is stable to ±10° optical phase from 0 to 70°C for 25 years. This performance is adequate to guarantee better than a 16-dB eye opening for the lifetime of the system.

Technical and product innovation in the long-haul telecom market is being driven entirely by favorable market dynamics. The long-distance carriers have to invest in increased fiber capacity or risk losing market share to rivals. There is virtually no "dark" (unused) fiber in the existing intercity fiber plant. At the same time, it is extremely expensive to install new fiber optic cables (up to one million per mile) because of the high costs associated with purchasing right-of-ways and burying fiber. Thus, in most cases carriers are choosing to upgrade their terminal equipment by embracing the latest optical component technology rather than by installing new fiber. North American carriers are presently spending several billion dollars per year on this technology and an international market for this technology, is rapidly developing. The deployment of DWDM technology in North American long-haul telecommunications networks is by far the largest photonics program in the world.

CONCLUSIONS

Over the past 15 years, lithium niobate integrated optic devices have been developed for optical switch arrays, fiber-optic gyroscopes, antenna remoting, CATV signal distribution, and long-haul telecommunications. The device technology has been successfully transitioned to high-volume manufacturing in the latter two applications. In all of the applications described in this chapter, the lithium niobate technology was ''customized'' to solve specific systems problems. For this effort to be successful, device engineers must work closely with systems engineers to understand system and end-user design and implementation issues fully.

To become qualified as a supplier of modulators into these demanding systems, manufacturers must produce lithium niobate modulators that meet or exceed specific performance targets, as well as demonstrate that these components will have very low field failure rates over a 25 year lifetime. In addition, the systems manufacturers must be convinced that a component manufacturer can routinely produce high-quality devices in high volume with short lead times.

The effort and cost associated with transitioning a ''new'' technology, such as lithium niobate integrated optical circuits, into high-volume production is immense. Production ramp-up, technology qualification, and customer design-in costs can easily add up to over $10 million and take several years. In order to justify this type of investment, there has to be strong market dynamics that pull the technology or product through to maturity. The five areas of device research mentioned in this chapter (switch arrays, fiber-optic gyro chips, modulators for RF antenna remoting, CATV modulators, and OC-48 modulators) were all fundamentally sound from a technology viewpoint. However, only two of the research areas (CATV modulators and OC-48 modulators) experienced adequate market pull to justify the transition into high-volume production.

3

Requirements for Integrated Optic Devices in Communication Systems

Mats Gustavsson

Ericsson Components AB
Stockholm, Sweden

1 INTRODUCTION

Optical fiber communication has gone through a remarkable development. During a period of less than 25 years, the capacity of point-to-point fiber-optic transmission links has increased by about six orders of magnitude—from around 1 Gb/s · km in the mid 1970s (see, e.g., Ref. 1) to more than 1 Pb/s · km in the late 1990s [2,3]. In recent years, this growth in capacity has been achieved via the development of optically amplified transmission systems employing wavelength-division multiplexing (WDM). Currently, there is also considerable interest in extending the use of fiber optics beyond point-to-point transmission links to create systems with a more complex optical network layer that can benefit from the flexibility that WDM offers. The demand for communication networks with ever-increasing capacity in combination with research in the areas optical network architectures, fiber-optic transmission techniques, and optical device technologies paves the way toward such sophisticated optical network layers.

The subject of the present chapter is requirements for integrated optic devices in communication systems; in this context it is relevant to consider the four concepts of *application, system, device*, and *technology*. An application-driven communication system implies that these four concepts are discussed in terms of a chain of *requirements*. The application imposes requirements on the system used to implement the application, which, in turn, define requirements for device performance, which, finally, translate into technology requirements. On the other hand, for a technology-driven communication system, these four concepts are instead discussed in terms of a chain of *possibilities*, starting from the possibilities associated with current state-of-the-art technology. In general, however, as just indicated, communication systems are both application and technology driven; consequently, requirements and possibilities are in practice considered simultaneously.

Many signal-processing functions required or desired in optical communication systems may adequately be realized using integrated optic technology, providing for physically compact devices of high stability and reliability as well as low power consumption. It is therefore essential to understand what requirements are imposed on integrated optic devices. Like any other type of device, an integrated optic device is characterized by a set of properties, and, depending on the application of the device, more or less strict requirements are associated with each of its properties. The various performance requirements imposed on integrated optic devices employed in a particular communication system are of different kinds and have different origins; the origins of these requirements may, roughly, be any combination of the following (partly interrelated) categories:

1. The requirement to meet capacity and coverage specifications for the communication system
2. A desire to minimize the cost associated with implementing, operating, and maintaining the communication system
3. A desire or requirement to meet standards and recommendations for the communication system

The requirements implied by category 1, which are determined by fundamental physical limitations associated with the transfer of information in the communication system, are related to device (and system) parameters, such as optical bandwidth, insertion loss, dispersion, crosstalk, and temperature sensitivity, that allow pure physical analysis. In addition to design guidelines obtained through the analysis of alternative implementations of a specific communication system, standards and recommendations provide specifications for its construction. Other important device (and system) parameters, such as physical size, fabrication yield, packaging, and manageability, may favorably be analyzed emphasising the point of view of cost minimization or standards and recommendations. Among the series G Recommendations of the Telecommunication Standardization Sector of the International Telecommunication Union (ITU-T) are a number of recommendations for fiber-optic transmission systems, and others, including recommendations for multichannel systems and optical networking, are under development.

The present chapter is directed toward drawing the reader's attention to some integrated optic device properties that must be considered in optical network design. Because the number of device parameters that can be subject to requirements is large within each of the preceding three categories, it would not be appropriate to try to give even a reasonably complete treatment in this chapter. Most of the parameters mentioned will therefore be addressed merely in brief, with the aim of providing a simple and introductory discussion rather than taking into account every possible aspect of the subject. However, following the more general discussion on requirements for integrated optic devices in Sec. 2, the chapter will be more specific in Sec. 3 and turn to its main subject, namely, crosstalk requirements. Optical crosstalk, in particular interferometric crosstalk, has received great attention over the last few years, and requirements for device crosstalk will be treated more thoroughly. In the context of requirements for integrated optic devices, it is of interest to give an indication of what performance can be obtained in practice. Therefore, addressing one type of integrated optic device in Sec. 4, properties of some examples of space switches—of various categories, demonstrated by different laboratories—will then be given, as well as a more detailed discussion of results reported by Ericsson. Finally, in Sec. 5, the chapter will be summarized by way of some concluding remarks.

2 GENERAL DISCUSSION

The specific requirements for integrated optic devices depend on the application (see, e.g., Ref. 4), and relevant parameters should be analyzed simultaneously for each network configuration. In general, of course, integrated optic devices for use in low-cost, high-volume applications, such as in access networks, should allow for simple and efficient implementation, and, at the same time, provide sufficiently good performance without employment of complex external, power-consuming electronic circuitry or feedback loops. Which particular optical devices are of interest for access networks depends on the network architecture chosen and may be exemplified by transceivers, optical power splitters, and optical amplifiers. On the other hand, severe requirements are imposed on integrated optic devices for use in the telecommunications core transport network, in particular, for long-distance transmission. Because many subscribers share such equipment, the device cost can be allowed to be higher here, and more complex electronic feedback systems may be used to extract the best possible performance and to compensate for possible temperature sensitivity. There are numerous optical devices of interest for transport network applications, e.g., integrated laser source and modulator, wavelength multiplexers and demultiplexers, add-drop multiplexers, optical space switches, optical filters, and optical amplifiers (see, e.g., Refs. 5–7).

In this section, a selection of device properties and associated requirements will be briefly discussed. Requirements associated solely with integrated optic light sources or transmitters will not be treated; it might, however, be worthwile to note that depending on the application and the type of light source, there will be requirements imposed on a variety of parameters, such as side-mode suppression ratio, chirp, output power, and wavelength tunability, that translate into detailed design requirements for the transmitter structure (see, e.g., Ref. 8).

2.1 Examples of Requirements Related to Capacity and Coverage

There are in general many parameters that together determine the transport characteristics, i.e., transmission and switching characteristics, of integrated optic devices, e.g., optical bandwidth, insertion loss, polarization dependence, dispersion, linear crosstalk, switching time, number of WDM signals, maximum output power, input power dynamic range, noise generation, and nonlinear effects. It should be emphasized that not every integrated optic device possesses all of the properties just mentioned; for example, it is not relevant to speak about noise generation in the context of nonamplifying devices or about switching time for fixed filters. Because a network connection may involve many concatenated optical network elements, one has to consider accumulation of, e.g., crosstalk and optical noise, as well as the effective end-to-end polarization dependence and optical transmission bandwidth.

This section will touch upon some parameters of importance for transport capabilities.

2.1.1 Optical Bandwidth

The function of integrated optic devices is in general frequency—or, equivalently, wavelength—dependent. There is, however, a general desire to realize, in the frequency range of operation, as flat a device transmission spectrum as possible in order to support WDM, and to allow the cascading of many devices without too quickly reducing the effective end-to-end optical bandwidth. As an example, for optical switches the wavelength dependence may manifest itself in increased insertion loss and increased crosstalk when the signal departs from its nominal wavelength. In the case of optical filters, there is also the requirement of having sufficiently good suppression of all neighboring channels; i.e., the filter shape should be characterized by steep slopes in addition to having flat transfer characteristics in its passband. Notice also that requirements for filter center-wavelength tolerances become more pronounced for transmission links involving a cascade of filters. Systems with filters that have flat passbands become more tolerant to variations in source wavelength and filter center-wavelength misalignments.

Different types of filters exhibit different shapes of the transmission spectrum—for amplitude as well as for group delay transfer characteristics—thus displaying different concatenation properties [9,10]. There are various ways of realizing integrated optic filters with improved flatness (see, e.g., Refs. 11 and 12). Here, it may be of interest to give just a few simple examples of the power transmission requirements for different device transmission spectra. For instance, if the effective full width at half maximum (FWHM) optical frequency bandwidth of a channel must not be less than $\Delta\nu$ when N number of devices with identical Gaussian device transfer functions, each characterized by an FWHM of $\Delta\nu_i$ (and centered at the same wavelength), are cascaded, the requirement for the FWHM of each individual device becomes

$$\Delta\nu_i > \Delta\nu\sqrt{N} \tag{1}$$

The corresponding requirement for devices with Lorentzian transmission spectra is

$$\Delta\nu_i > \Delta\nu(\sqrt[N]{2} - 1)^{-1/2} \approx \Delta\nu\left(\frac{N}{\ln (2)}\right)^{1/2} \tag{2}$$

where the approximative equality applies for large N. As a simple example of a filter device with better concatenation properties than filters with Gaussian or Lorentzian transfer functions, consider the four-mirror Fabry–Perot interferometer, the intensity transmission spectrum $T(\nu)$ of which, near the peak of its passband, can be approximated as [13]

$$T(\nu) \approx \left[1 + \left(2\frac{\nu - \nu_c}{\Delta\nu_i}\right)^6\right]^{-1} \tag{3}$$

where ν denotes optical frequency and ν_c is the center frequency of the filter. This transmission spectrum, which exhibits a relatively flat passband and steep slopes, is obtained for suitably chosen reflectivities and cavity lengths of the four-mirror Fabry–Perot interferometer [13]. The requirement corresponding to Eqs. (1) and (2) becomes

$$\Delta\nu_i > \Delta\nu\ (\sqrt[N]{2} - 1)^{-1/6} \approx \Delta\nu\left(\frac{N}{\ln (2)}\right)^{1/6} \tag{4}$$

where, again, the approximative equality holds for large N. The requirements for Gaussian, Lorentzian, and four-mirror Fabry–Perot filters are illustrated in Fig. 1, which clearly shows the superior concatenation properties of the four-mirror Fabry–Perot interferometer filter; even for as many as 100 cascaded filters, the individual filter bandwidth need not be more than about a factor 2.3 larger than the desired effective end-to-end bandwidth. Notice that none of the three types of filter shapes discussed here implies that the requirement for device bandwidth increases faster than $\sqrt{N}$.

Optical filters can also be used to equalize the transmission spectrum of optical devices, which imposes requirements on accurate, nonflat shapes of their transmission spectra. Such equalizing filters in the form of, e.g., long-period fiber

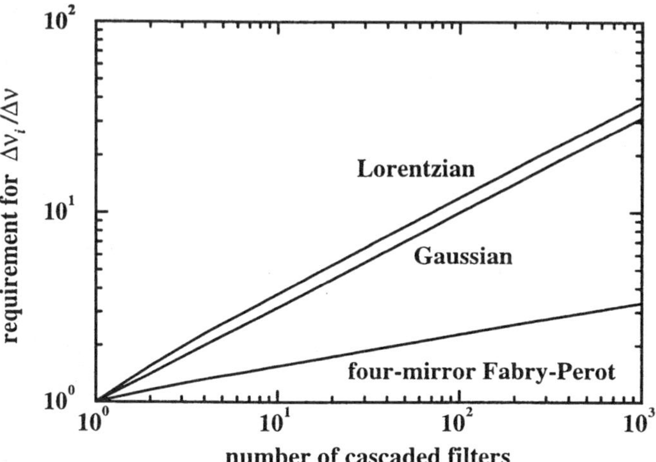

Figure 1 Requirements for device optical bandwidth. The figure shows requirements for device optical bandwidth $\Delta\nu_i$ for cascaded optical devices characterized by identical Gaussian, Lorentzian, or four-mirror Fabry–Perot interferometer transmission spectra, respectively. Requirements for different effective end-to-end optical bandwidths $\Delta\nu$ translate into different requirements for $\Delta\nu_i$ of each individual device.

gratings have been used to obtain wide-bandwidth erbium-doped fiber amplifiers [14].

2.1.2 Transmission Flatness

Ideally, optical devices in communication systems should have flat transmission spectra within their operating window. The term *transmission ripple* describes to what extent the transmission characteristics of an optical device vary across its optical bandwidth of interest, i.e., irregularities superimposed on the main, smooth shape of its transmission spectrum. Fabry–Perot cavity undulations are a good example of transmission ripple. There are generally several reflectivities associated with an integrated optic device, and two or more reflectivities in an optical path form an optical cavity. In general, more than one device is included in an optical transmission link, and the overall, accumulated transmission ripple that can be tolerated in the link determines the requirement for transmission flatness of each individual device employed.

As an example, consider the simple Fabry–Perot cavity: if the worst case, total transmission ripple for N cascaded, noninteracting, identical Fabry–Perot cavities should be less than a factor of Q, then each cavity must meet the following requirement:

$$G_s\sqrt{R_1 R_2} < \frac{\sqrt[2N]{Q} - 1}{\sqrt[2N]{Q} + 1} \approx \frac{\ln(Q)}{4N} \tag{5}$$

where G_s is the single-pass intensity gain of the cavity and R_1 and R_2 are the intensity reflectivities at the cavity ends. (If there is a net single-pass loss in the cavity, then $G_s < 1$.) The approximative equality in Eq. (5)—which indicates a linear increase in the requirement for $G_s\sqrt{R_1R_2}$—applies when $N/\ln(Q)$ is large; the requirement for $G_s\sqrt{R_1R_2}$ based on the exact expression in Eq. (5) is illustrated in Fig. 2. For $N = 1$, i.e., for a single Fabry–Perot cavity, $Q < 1$ dB requires the quantity $G_s\sqrt{R_1R_2}$ to be less than approximately $6 \cdot 10^{-2}$. The requirement becomes more severe as the number of cascaded cavities increases: for example, for a cascade of 10 cavities, $Q < 1$ dB requires $G_s\sqrt{R_1R_2}$ to be less than approximately $6 \cdot 10^{-3}$. The transmission ripple in general changes in magnitude and shifts in wavelength with changes in temperature, optical power, and electron density due to induced changes in the complex index of refraction or the optical path length. In general, the ripple is also polarization sensitive, because of polarization-dependent single-pass gain, waveguide effective index, and reflectivities.

Obviously, if a good transmission flatness is desired, large single-pass gains require low reflectivities at the cavity ends; this has been a main concern for optical amplifiers, in particular, for discrete semiconductor optical amplifiers

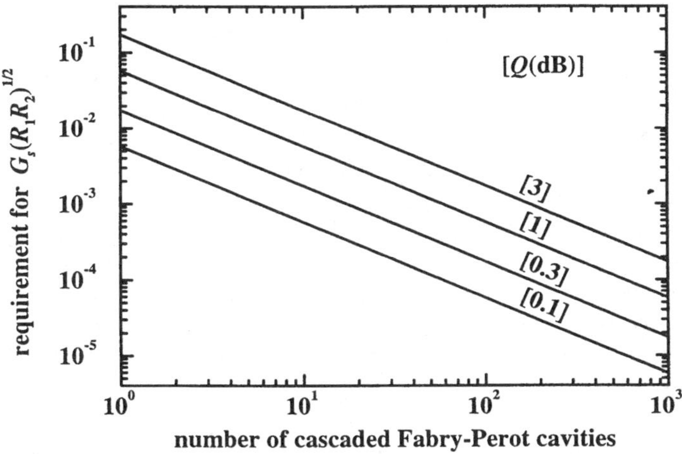

Figure 2 Optical device requirements related to transmission flatness. The figure shows requirements for single-pass gain G_s and reflectivities R_1 and R_2 of cascaded, noninteracting, identical Fabry–Perot cavities. Requirements for different worst-case total transmission ripple Q translate into different requirements for $G_s\sqrt{R_1 R_2}$ of each individual cavity.

(SOAs), due to the high refractive index of gallium arsenide and indium phosphide. There are, however, a number of ways by which the reflectivity experienced by the optical mode can be reduced, as compared with the reflectivities displayed by an amplifier with ordinary cleaved facets; any of the following steps or any combination of them may be taken: evaporation of antireflective coatings to its facets [15]; designing the amplifier with a tilted waveguide [16], with buried facets [17], or with a tapered waveguide [18]. Modal reflectivities down to around 10^{-5} have been attained. More than two reflectivities create complex multicavity transmission characteristics. Because an integrated optic device may comprise several concatenated integrated substructures, such as phase shifters, power dividers and combiners, and amplifiers, it is not unusual that one has to consider many reflectivities within the device, in addition to the reflectivities of output and/or input optical ports.

2.1.3 Insertion Loss and Polarization Sensitivity

In communication systems, the insertion loss of optical devices often has to be compensated for in some way. Although losses can be offset by optical amplification, it is at the expense of increased noise and increased power consumption. In order to minimize the amount of optical amplification required, optical losses should be reduced as much as possible, within the device as well as at the coupling between device and interfacing optical fiber. For guided-wave devices, tapered lensed fibers [19] or optical mode-size transformers [20] may be required in order to avoid unreasonably high coupling losses between standard single-mode fibers and the guided-wave chip. This is often the case for semiconductor integrated optic devices that offer the advantage of tightly confined modes with the ensuing efficient usage of substrate area. Devices with integrated mode-size transformers in the regions near the chip facets are particularly attractive, because, in addition to reducing coupling losses, they provide relaxed fiber alignment tolerances, as a result of the larger mode size.

Integrated optic devices may be polarization sensitive because of asymmetrical optical waveguides and/or employment of waveguide material with polarization-dependent absorption or gain characteristics. The polarization sensitivity of devices for operation over a wide wavelength range manifests itself mainly as polarization-dependent insertion loss or insertion gain, whereas integrated optic filters also may exhibit passbands at different wavelengths for the two modes of polarization. The polarization sensitivity of each optical network element contributes to the total polarization sensitivity of the connection, and the accumulated polarization sensitivity of all the optical network elements determines the maximum polarization-induced signal-power fluctuations at the receiver. These power fluctuations must fall within the dynamic range of the link, which means, e.g., that the power variations must not violate the margin of the system. By way of feedback loops, an intrinsic device polarization sensitivity can be eliminated.

2.1.4 Dispersion

Chromatic dispersion [21] and polarization mode dispersion [22] of the optical fiber incur power penalties in optical transmission systems, and dispersion of integrated optic devices must thus be considered, too. The differential group delay between the polarization modes, i.e., the (deterministic) polarization mode dispersion, of integrated optic devices should be minimized. However, the chromatic dispersion of devices may be used as a part of the dispersion management of the communication system. Therefore, the chromatic dispersion of devices need not necessarily be very low, although its magnitude, which must be well controlled, is important for transmission systems design. The accumulated chromatic dispersion for the transmission link of interest should be sufficiently low as not to incur too large a dispersion-induced power penalty.

2.1.5 Crosstalk

Crosstalk between transmission channels degrades signal quality by contributing to closing the eye at the receiver and may incur substantial power penalties or even actually prohibit transmission. Requirements for the crosstalk properties of optical network elements will be treated in Sec. 3, but a few preliminary comments are given here.

As will be shown in Sec. 3, crosstalk requirements become severe for meshed optical networks, and current state-of-the-art devices do not always meet these requirements. The importance of interferometric crosstalk has been recognized over the last few years, and considerable efforts have been devoted to realizing devices with lower and lower crosstalk. Relaxed requirements for device crosstalk may, however, be achieved with alternative, but also more elaborate, network nodes. For implementation of switch matrices, dilation represents a possibility of obtaining relaxed requirements for the constituent elementary switch units: through the use of dilated switch architectures, crosstalk due to imperfect switch elements can be reduced [23]. Dilated switches thus give better crosstalk performance, although it is at the expense of a larger number of switch elements and larger switch matrix physical size, with the ensuing increase of insertion loss, and deterioration of effective switching bandwidth, i.e., narrowing of the effective shape of optical transmission spectrum. Reduction of crosstalk due to imperfect filtering can straightforwardly be achieved by filter cascading, resulting, however, in the usual disadvantages associated with device concatenation.

2.1.6 Temperature Insensitivity

Requirements for temperature insensitivity are determined partly by whether the application is an indoor or outdoor application. Obviously, outdoor applications require low temperature sensitivity, unless it is reasonable to employ equipment for active temperature control. In applications that require low-cost devices, e.g., in access networks, it is strongly desirable to avoid active cooling equipment;

because integrated optic devices employed in access network applications may have to operate either indoors or outdoors, a sufficiently low temperature sensitivity is of great interest for such devices.

2.1.7 Reconfiguration Time

Different applications, such as dynamic network configuration, network protection, and optical packet switching, imply different requirements for the reconfiguration time of switches. For packet switching, the reconfiguration time should be as short as possible, because it is desirable to minimize guard times between packets, implying reconfiguration times typically in the nanosecond range [24]. Fast switching requires, in turn, high-speed electronic drivers, in conjunction with sophisticated microwave packaging techniques. The reconfiguration time for optical switches used for network protection should be compatible with current telecommunications standards [25]. This indicates significantly more moderate reconfiguration time requirements than those for packet switching; detailed requirements for an optical network layer are, however, presently under discussion within the ITU-T.

Devices for optical packet switching consequently require fast switching mechanisms, such as electro-optically or possibly carrier-injection controlled switches. Switches for network protection, on the other hand, either can rely on these fast switching mechanisms or may employ other, less fast mechanisms, such as thermo-optically controlled structures.

2.1.8 Number of Input and Output Ports and Blocking Properties

For space-division switches, there are requirements for the number of input and output ports and the switch blocking properties, in addition to the requirements already discussed. For wavelength multiplexers and wavelength demultiplexers, requirements are imposed on the number of wavelength channels, i.e., the number of input or output ports. Requirements associated with this type of parameters are, of course, heavily dependent on the network architecture used and the traffic carried by the network. For example, in the case of space switches, there will be different requirements when switches are used in connection with network protection, cross-connection, add-drop multiplexing, or switching in access networks.

It is worth noticing, however, that these parameters may influence transmission performance indirectly; for instance, when the number of input and output ports of an integrated optic space switch increases, crosstalk will in general increase due to, e.g., a larger number of (nonideal) waveguide crossings on the chip.

2.1.9 Additional Remarks

Optical amplifiers (see, e.g., Refs. 26–28), are sometimes used in integrated optic devices. The employment of optical amplifiers makes it necessary to consider optical amplifier parameters, such as saturation output power, linearity, and noise figure of these integrated amplifiers, in addition to the corresponding properties of any other in-line optical amplifiers used in the communication system. In an optical transmission line connection, accumulation of amplified spontaneous emission increases the noise level at the receiver and contributes also to saturating the amplifiers. Furthermore, nonlinear effects associated with optical amplifiers may cause distortion in analog transmission systems and crosstalk in WDM systems.

2.2 Comments on Parameters Related to Fabrication and Operation

The parameters discussed in Sec. 2.1 relate mainly to transmission performance, which has to be considered for all in-line optical devices. In contrast to the requirements related to capacity and coverage of the communication system, it is in many cases difficult to discuss in general terms requirements related to fabrication and operation, because often they are application dependent and/or originate in a desire to minimize cost. Nevertheless, a few brief comments on parameters that are important for device fabrication and operation might be of interest.

The total *power consumption* of devices should be as low as possible. One reason for this is that heat dissipation should be minimized; if possible, active cooling should be avoided. Low *drive voltage* or *current* for dynamic device control, e.g., for rearranging space switches or for filter tuning, is also desirable, in particular, for high-frequency applications, in order to relax implementation difficulties of electronic driver circuits. Furthermore, *packaging* of optical devices is of great importance. If high-frequency electrical connections are required, then advanced packaging techniques are needed. There are also application-dependent requirements for *environmental stability* associated with (packaged) integrated optic devices. From the point of view of cost reduction, integrated optic devices compatible with small packages and efficient assembly are desired. This is facilitated by way of the fabrication of integrated optic chips of small *physical size* with high *fabrication yield*. Small device modules help to reduce consumption of cabinet space, leading to reduced fabrication and operating costs. Devices must also meet requirements for *reliability* in accordance with specifications for the application of interest. As a final comment, it is worthwhile noting that requirements for *manageability* imply, e.g., requirements for adequate interface possibilities with the management system.

3 CROSSTALK REQUIREMENTS

Meshed networks that involve wavelength-selective space-division routing may be limited by accumulated crosstalk due to imperfect optical space-division switching and imperfect optical filtering [29–42]; in such WDM systems, it is likely that the wavelengths have to be well controlled, which can lead to severe problems with interferometric crosstalk effects [30–42] when a signal is mixed with other disturbing signals of essentially the same wavelength. The worst-case crosstalk requirement for each individual crosstalk-generating device is network and connection dependent, but may be in the range below -50 dB.

This section will address crosstalk requirements for optical devices employed in communication systems. The basic model for analysis of requirements imposed on device crosstalk applies to any type of optical network elements that generate crosstalk, not only to integrated optic devices. This crosstalk model [38,39] will be described in some detail; device crosstalk requirements for the case when the crosstalk contributions are incoherent as well as device crosstalk requirements for the case of interferometric crosstalk will be discussed and compared with a statistical analysis of interferometric crosstalk, showing that device crosstalk requirements may be relaxed, as compared with the worst-case requirements.

3.1 Incoherent and Interferometric Crosstalk

In an optical network or in a single-fiber point-to-point transmission line connection, signal leakage between transmission channels will in general occur. This signal leakage gives rise to crosstalk, unless it is completely suppressed before detection. Depending on the relations between the frequencies of the interfering optical signals and the electrical bandwidth of the receiver, signal leakage manifests itself as either incoherent or interferometric crosstalk. These two types of crosstalk, both of which cause closure of the received eye and receiver power penalty, can be explained as follows.

Consider first the upper part of Fig. 3. In an optical WDM communication system, a fraction of a neighboring WDM signal will mix with the main, desired signal at the receiver. This occurs because of nonideal filtering characteristics of the wavelength demultiplexer. Because beating frequencies between the crosstalk and the main signals typically fall outside the receiver bandwidth in this case, it is the relative optical *powers* of the crosstalk and the main signal that determine the magnitude of crosstalk as detected by the receiver. This type of crosstalk can therefore be termed *incoherent* crosstalk.

Consider now an optical network that comprises space switches, as schematically illustrated in the lower part of Fig. 3. The signal incident upon input 1 of the space switch is routed to output 2, and vice versa. However, a fraction of the signal incident upon input 2 will mix with the main signal at output 2 as a result of nonideal switching characteristics. If the space switch routes signals of essentially the same wavelengths, the crosstalk cannot be suppressed by additional filtering, as would be possible in the previous case. The beating frequencies due to mixing of the electromagnetic fields fall within receiver bandwidth; thus, su-

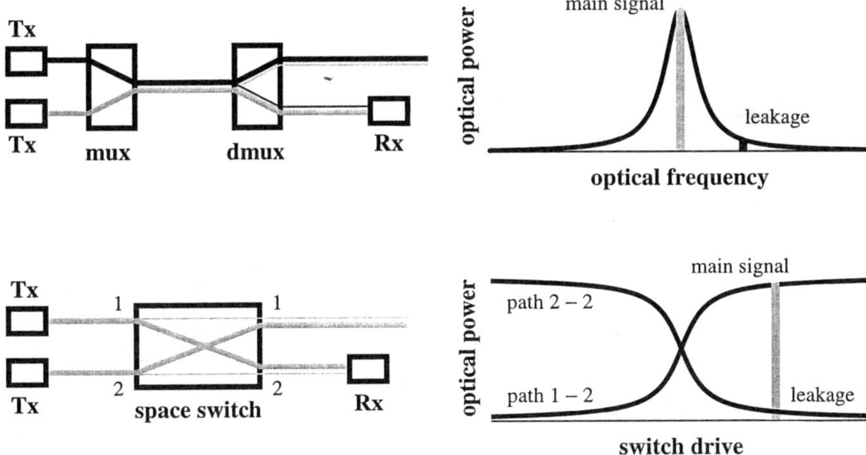

Figure 3 Schematic illustration of crosstalk in optical networks. The *upper part* of the figure shows how *incoherent* crosstalk may arise in a WDM system. The grey signal is the desired, main signal to be detected at the receiver. However, the (nonideal) filter characteristics of the demultiplexer cause a fraction of the black signal to reach the detector; this leakage gives rise to incoherent crosstalk. The *lower part* of the figure shows how *interferometric* crosstalk may arise in an optical network employing space-division switches. The two signals routed through the space-division switch are assumed to have nominally the same wavelength. At the receiver, the thick, grey signal is the desired, main signal. However, the (nonideal) switching characteristics of the space-division switch cause a fraction of the other signal to reach the detector; this leakage gives rise to interferometric crosstalk.

perposition of electromagnetic *fields* here determines the magnitude of crosstalk as detected by the receiver. This phenomenon can thus be referred to as *interferometric* crosstalk.

Both incoherent and interferometric crosstalk may occur in optical network routing nodes. This situation is schematically shown in Fig. 4 for a routing node with two input and two output fibers, each with two wavelength channels. The accumulation of crosstalk can be made clear by following the solid black signal through the node in Fig. 4. As a result of imperfect demultiplexing just after the node inputs, a fraction of the solid grey signal will propagate along with the solid black one; the leakage of the solid grey signal will thus give rise to (incoherent) crosstalk at the receiver, unless it is entirely suppressed by optical filtering before detection. When these two signals—the main, solid black signal and the fraction of the solid grey one—are routed through the space-switch stage, they will, due to imperfect switching, be accompanied by a fraction of the dashed black signal;

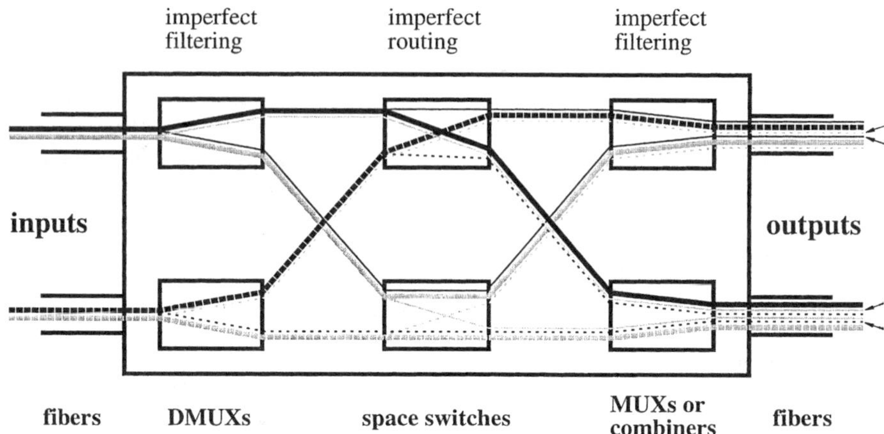

Figure 4 Schematic illustration of linear crosstalk in WDM network routing nodes. Incoherent as well as interferometric crosstalk accumulate in the node. Some of the crosstalk contributions—indicated by arrows at the node output—will be suppressed to second-order crosstalk contributions if wavelength multiplexers are used instead of conventional power combiners just before the node outputs.

because the solid and dashed black signals are of the same nominal wavelength, the leakage of the dashed black signal will inevitably contribute to (interferometric) crosstalk at the receiver. Finally, when this group of signals is combined with the signals coming from the other space-switch stage, additional crosstalk contributions arise. Some of the crosstalk contributions will be suppressed to second-order crosstalk contributions if wavelength multiplexers are used instead of conventional power combiners just before the node outputs; it is especially important that some of the interferometric contributions can be suppressed to second-order crosstalk. A connection established in an optical network will typically involve transmission through a number of concatenated routing nodes, resulting in a large number of crosstalk contributions at the receiver. The number of crosstalk contributions generated in each node increases of course when the size of the nodes grows, in terms of the number of input and output fibers and/or the number of wavelength channels.

A simple mathematical illustration of the mixing of a desired signal with a fraction of another signal may be of interest. Neglect signal modulation, and let the fields of the main signal and the crosstalk at the receiver have the same states of linear polarization and be expressed by

$$F_0(t) = a_0 \cos[\omega_0 t + \phi_0(t)] \tag{6}$$

$$F_1(t) = a_1 \cos[\omega_1 t + \phi_1(t)] \tag{7}$$

where t denotes time, a_0 and a_1 are the signal amplitudes of the main signal and the crosstalk, respectively, ω_0 and ω_1 represent their carrier angular frequencies, and ϕ_0 and ϕ_1 are their carrier phases. Assuming appropriate normalization of the fields, the detected current in a square-law photodetector with responsivity R is given by

$$\begin{aligned} I_{\text{det}}(t) &= RL_p[(F_0(t) + F_1(t))^2] \\ &= R[P_0 + P_1 + a_0a_1 \cos[(\omega_1 - \omega_0)t + \phi_1(t) - \phi_0(t)]] \end{aligned} \tag{8}$$

where the operator L_p denotes low-pass filtering. Thus, as a consequence of square-law detection, there are three terms that form the detected current. The first term constitutes the detected current due to the optical power P_0 of the main signal; the second term is the detected current due to the optical power P_1 of the disturbing signal, thus representing incoherent crosstalk; and the third, mixed term is due to the beating of their electromagnetic fields, thus representing interferometric crosstalk. Equation (8) can be rewritten as

$$I_{\text{det}}(t) = RP_0[1 + X_P + 2\sqrt{X_P} \cos[(\omega_1 - \omega_0)t + \phi_1(t) - \phi_0(t)]] \tag{9}$$

where $X_P = P_1/P_0$ is defined as the incoherent crosstalk. Equation (9) clearly shows the interferometric crosstalk to be particularly detrimental, since it is proportional to $\sqrt{X_P}$. The impact of the mixing term, i.e., the interferometric crosstalk, is significant if the phases $\phi_0(t)$ and $\phi_1(t)$ are correlated over the measurement time and if the optical frequency difference between the main signal and the crosstalk is less than or comparable to the electrical bandwidth of the receiver. This situation may arise in, e.g., WDM networks in which each wavelength channel carries bit streams in the Gb/s range; interferometric crosstalk must thus be considered here because the semiconductor laser sources employed have emission linewidths typically in the megahertz range, and the wavelength tolerances in the network can be tight.

3.2 Statistical crosstalk analysis

The total crosstalk level at a receiver is, in general, composed of many different contributions. Intuitively, the probability that all these contributions would add to the signal in the worst possible way, i.e., exhibit identical carrier phase, polarization, and wavelength at all time instants, should be rather low. It might therefore be of interest to investigate, from probabilistic arguments, whether it would be possible to relax the crosstalk requirements for the crosstalk-generating elements. A possible criterion could be to specify the time fraction during which the system generates unacceptably high bit-error rates (BERs) as a result of crosstalk; a measure of this time fraction could be the probability p of obtaining BERs larger than an acceptable value. This approach [38,39] is motivated partly by the fact that the relative states of polarization and phases among the main and interfering signals typically vary only slowly as compared to the time duration of one bit; once a certain crosstalk level is established, it will remain for a relatively long time.

3.2.1 Theoretical Model

From the description in Sec. 3.1, it is clear that crosstalk accumulates in optical network connections. During its propagation through the network, the signal field desired at the receiver has in general been mixed with a number N of interfering signal fields due to optical leakage from other connections in the network that use essentially the same wavelength. The structure of multiple interferometric crosstalk is therefore simple: a given main signal (described by its amplitude, state of polarization, and optical phase) is mixed with a number of interfering signals (described by their amplitudes, states of polarization, and optical phases). Crosstalk from signals of different wavelengths is assumed to be negligible, owing to sufficient suppression by way of optical filtering. Regarding the polarization states and the optical phases of the interfering signals as mutually independent, uniformly distributed random variables, it is of interest to find the probability density function (pdf) of detected current at the receiver, from which the pdf of total crosstalk follows. The detailed derivation of these pdf's is given elsewhere [38], but a brief summary of the ideas and main results are given here.

The detected current resulting from a main signal mixed with a disturbing one was described in Sec. 3.1. When there are many crosstalk contributions, i.e., many disturbing signals, a generalized discussion is required. In analogy with Eq. (8), the detected current at the receiver in the presence of multiple interference can be expressed as

$$I_{\text{det}}(s_0) = RL_p\left[\left|\mathbf{F}_0(t,\, s_0) + \sum_{i=1}^{N} \mathbf{F}_i(t,\, s_i)\right|^2\right] \tag{10}$$

where the main signal field vector $\mathbf{F}_0$ and each of the interfering signal fields $\mathbf{F}_i$ are characterized by their field amplitudes, carrier phases, and states of elliptical polarization. Further, the quantity s_0 denotes the symbol (mark or space) of the main signal. In the presence of interfering signals, the detected current $I_{\text{det}}(s_0)$ will in general depart from what would be detected without any interference. The total crosstalk level $X(s_0)$ at the receiver is defined as this current excursion normalized by $I_{\text{det},0}(\text{m})$:

$$X(s_0) \equiv \frac{I_{\text{det}}(s_0) - I_{\text{det},0}(s_0)}{I_{\text{det},0}(\text{m})} \tag{11}$$

where the detected current due to the optical power $P_0(s_0)$ of the main signal in the absence of crosstalk is denoted by $I_{\text{det},0}(s_0)$, which can be either the current of a mark, $I_{\text{det},0}(\text{m})$, or the current of a space, $I_{\text{det},0}(\text{sp})$. Without loss of generality, the detected current is expressed in terms of a normalized detected current, $J(s_0) \equiv I_{\text{det}}(s_0)/I_{\text{det},0}(\text{m})$, in terms of which the crosstalk can be written as

$$X(s_0) = J(s_0) - \rho^{-1} \tag{12}$$

The parameter $\rho = \rho(s_0)$ is thus different for marks and spaces: $\rho(\text{m}) = 1$ for a mark and $\rho(\text{sp}) = r_0$ for a space, r_0 being the extinction ratio of the main signal.

Application of the central limit theorem allows the pdf of normalized detected current to be calculated in closed form:

$$f_J(j;\, X_T,\, \rho) = \frac{2\sqrt{\rho j}}{X_T} \exp\left(-\frac{2(\rho j + 1)}{\rho X_T}\right) I_1\left(\frac{4\sqrt{\rho j}}{\rho X_T}\right) \tag{13}$$

which is defined in the interval $[0, \infty]$, and where I_1 is the modified Bessel function of the first kind and first order; further, the sum X_T of the incoherent crosstalk contributions is given by

$$X_T = \sum_{i=1}^{N} X_{P,i} \tag{14}$$

where $X_{P,i}$ is defined as the ratio between the optical power of an undesired signal i and the optical power of the main signal. This model allows crosstalk contributions of considerably different magnitudes, but, as a consequence of the use of the central limit theorem, there must be a significant number of crosstalk contributions around each of these different magnitudes. The mean and variance of J follow by integration using Eq. (13): $E(J) = \rho^{-1} + X_T$, and the variance $V(J) = X_T(\rho^{-1} + X_T/2)$. As can be observed in Fig. 5, the pdf of J is asymmetrical. This is not surprising, because there is a minimum possible detected current, namely, zero, in the presence of interferometric crosstalk, but there is in principle no upper limit for the detected current. It is also observed that the variance of the detected current for a mark in the main signal is larger than the variance for a space in the main signal; this can be explained by the fact that the crosstalk interferes with the main signal, which implies that there is a larger range of possible values of detected current for the high optical power of a mark in the main signal than for the lower optical power of a space in the main signal; cf. Eq. (8).

By observing Eq. (12), one obtains the pdf of total crosstalk,

$$f_X(x;\, X_T,\, \rho) = f_J(x + \rho^{-1};\, X_T,\, \rho) \tag{15}$$

which is defined in the interval $[-\rho^{-1}, \infty]$. Knowing $E(J)$ and $V(J)$, the mean and variance of X follow directly from Eq. (12) as $E(X) = X_T$, as might have been expected, and $V(X) = V(J)$.

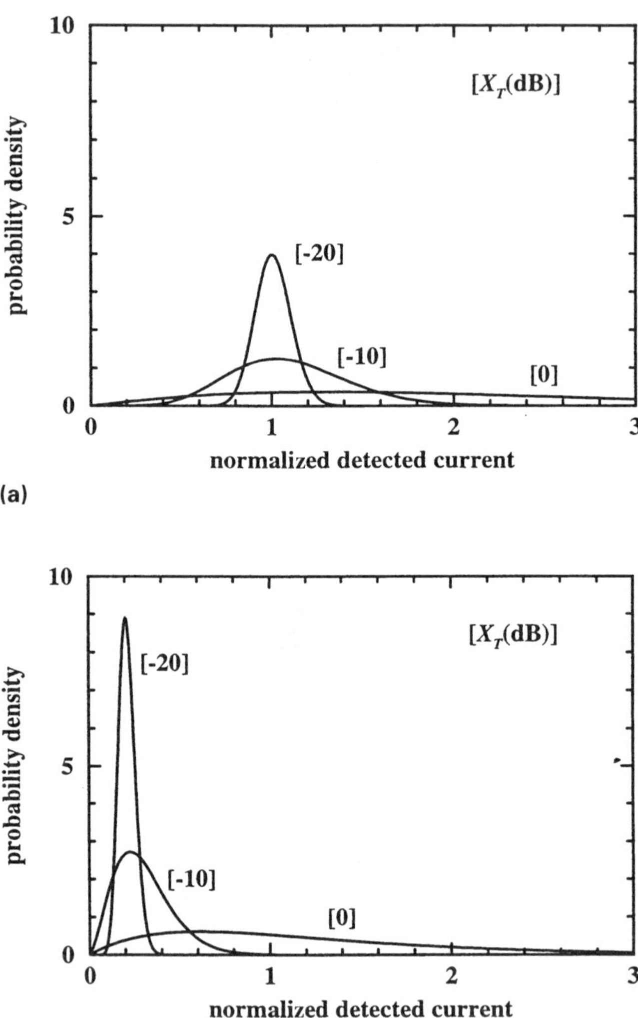

Figure 5 Probability density function of normalized detected current with the total incoherent crosstalk level X_T as parameter. (a) Mark in main signal. (b) Space in main signal; for the purpose of illustration, the extinction ratio is set to a low value of $r_0 = 5$. (From Ref. 38 © 1997 IEEE.)

3.2.2 The Bounds of Detected Current and Total Crosstalk

As a consequence of the use of the central limit theorem, the calculated pdf of J is defined in the normalized current interval $[0,\infty]$. The exact pdf of J for a finite number N of interfering signals does not of course extend to infinite currents and not always even down to zero. The purpose of calculating the bounds of detected current and of the total crosstalk is twofold: first, these bounds are required in order to calculate worst-case total crosstalk levels and, thus, worst-case device crosstalk requirements; second, a better approximation of the true pdf's can be expected if the calculated pdf's of J and X are truncated at these bounds.

Minimum detected current results—typically—if all interfering signals are in counterphase with the main signal (see Ref. 38 for more details); maximum detected current results when all interfering signals are phase aligned with the main signal. Now, introducing

$$X_C = \sum_{i=1}^{N} \sqrt{X_{P,i}} \tag{16}$$

the bounds of normalized detected current, i.e., the minimum and maximum possible values of normalized detected current, $J_{\min}$ and $J_{\max}$, respectively, can be written as:

$$J_{\min}(X_C, \rho) = \begin{cases} \left[\dfrac{1}{\sqrt{\rho}} - X_C\right]^2, & 0 \le X_C \le \dfrac{1}{\sqrt{\rho}} \\ 0, & X_C > \dfrac{1}{\sqrt{\rho}} \end{cases} \tag{17}$$

$$J_{\max}(X_C, \rho) = \left[\frac{1}{\sqrt{\rho}} + X_C\right]^2 \tag{18}$$

The bounds $J_{\min}$ and $J_{\max}$ of normalized detected current give, according to Eq. (12), the minimum and maximum possible total crosstalk levels $X_{\min}$ and $X_{\max}$, respectively:

$$X_{\min}(X_C, \rho) = J_{\min}(X_C, \rho) - \rho^{-1} \tag{19}$$

$$X_{\max}(X_C, \rho) = J_{\max}(X_C, \rho) - \rho^{-1} \tag{20}$$

whereas the total crosstalk, if the contributions were incoherent, would amount to X_T.

3.2.3 Requirements for Device Crosstalk Performance

Crosstalk can contribute to closing the eye at the receiver, but a certain crosstalk level can be accepted for a given power penalty. Let the worst acceptable total crosstalk level at the receiver be $X_A(s_0)$. Given $X_A(s_0)$, the purpose is now to formulate the requirements for the crosstalk of the system; these requirements are found from Eqs. (19) and (20), if the system is plagued by interferometric crosstalk. The discussion on crosstalk requirements is separated for marks and spaces, since for a mark the received signal level decreases for $X < 0$ (the eye closes from the top), and for a space the received signal level increases when $X > 0$ (the eye closes from the bottom), which implies that Eq. (19) determines the system crosstalk requirement for marks, and Eq. (20) for spaces. Three different crosstalk requirements will be discussed here: the incoherent requirement, the worst-case requirement, and a statistical requirement for interferometric crosstalk.

Incoherent Requirement. When all crosstalk contributions are incoherent, the requirement $X_T(s_0)$ for X_T is determined by increase of the signal level for a space in the main signal:

$$X_T(\text{sp}) < X_A(\text{sp}) \tag{21}$$

Because incoherent crosstalk always increases the signal level, irrespective of the symbol of the main signal, the signal level for a mark in the main signal is not decreased because of incoherent crosstalk. For a mark in the main signal, the error probability does therefore not increase in the presence of incoherent crosstalk.

Worst-Case Requirement. For a mark in the main signal, the most stringent requirement $X_C(\text{m})$ for X_C follows from $X_{\min}[X_C(\text{m}), 1]$ in Eq. (19), which describes the maximum possible decrease of the signal level of a mark in the main signal:

$$X_C(\text{m}) < 1 - \sqrt{1 + X_A(\text{m})} \approx -\frac{X_A(\text{m})}{2} \tag{22}$$

where the approximative equality applies when $|X_A(\text{m})| \ll 1$. The requirement for a space, which, typically, becomes less demanding than the requirement for a mark, is obtained analogously [38].

Statistical Requirement. When there is a large number of independent crosstalk contributions, it is unlikely that they combine to produce the maximum possible crosstalk level at the receiver, which motivates formulation of a statistical requirement for the system element crosstalk levels. Such a statistical assessment can be delineated in connection with Fig. 6. The maximum tolerable power penalty corresponds to the normalized detected currents $j_{\text{det},A}(s_0)$. These detected currents correspond, in turn, via Eq. (12), to the maximum acceptable crosstalk levels

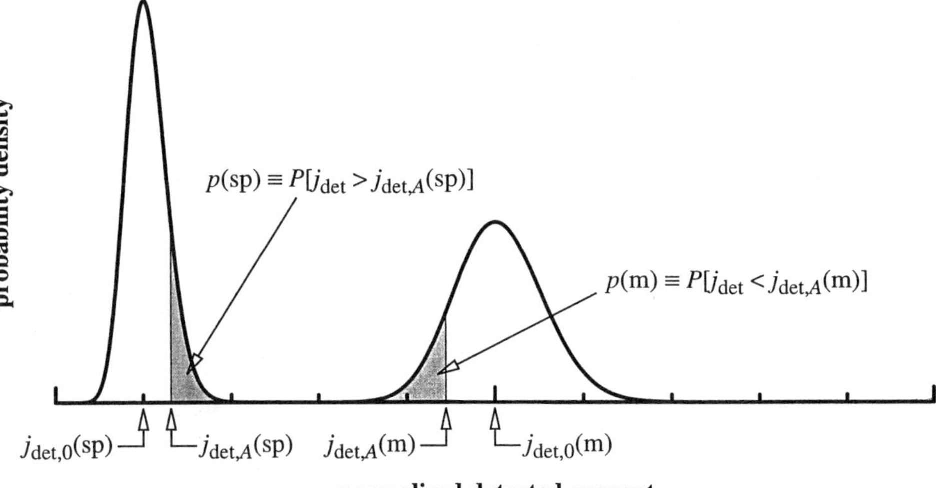

Figure 6 Schematic illustration of the meaning of the probabilities $p(\mathrm{sp})$ and $p(\mathrm{m})$. The maximum tolerable power penalty corresponds to the normalized detected currents $j_{\mathrm{det},A}(\mathrm{sp})$ and $j_{\mathrm{det},A}(\mathrm{m})$. The normalized detected currents $j_{\mathrm{det},0}(\mathrm{sp})$ and $j_{\mathrm{det},0}(\mathrm{m})$ for a main-signal mark and a main-signal space, respectively, in the absence of crosstalk, are also indicated. The shaded areas represent the probability that the crosstalk exceeds what is acceptable, i.e., $p(\mathrm{sp})$ and $p(\mathrm{m})$, respectively.

$X_A(s_0)$, with $X_A(\mathrm{m}) < 0$ and $X_A(\mathrm{sp}) > 0$. According to Eqs. (19) and (20), the total crosstalk at the receiver can assume a range of values, and, with a probability $p(s_0) > 0$, it will exceed the acceptable crosstalk levels $X_A(s_0)$ if the system does not meet worst-case crosstalk requirements. For a mark in the main signal, the statistical crosstalk requirement is defined as the device crosstalk levels $X_{P,i}(\mathrm{m})$ for which $X(\mathrm{m}) < X_A(\mathrm{m})$ with probability $p(\mathrm{m})$. Analogously, for a space in the main signal, the statistical crosstalk requirement is defined as the device crosstalk levels $X_{P,i}(\mathrm{sp})$ for which $X(\mathrm{sp}) > X_A(\mathrm{sp})$ with probability $p(\mathrm{sp})$. It is adequate to set $p(\mathrm{m}) = p(\mathrm{sp}) = p$, which will be the case throughout the rest of the discussion. As already discussed, the total crosstalk varies randomly with time in a communication system. If the total crosstalk is an *ergodic* stochastic process, then the probability p corresponds to relative times; i.e., p is a measure of the time fraction during which the system suffers from unacceptably high crosstalk. Regarding the probability p as an additional parameter for system specification, it is of interest to investigate, from probabilistic arguments, whether it would be possible to relax the requirement for $X_{P,i}$, as compared to Eq. (22). With this end in view, the requirement $X_{P,i}(s_0)$ for each crosstalk-generating element can be calculated using

Eq. (15). However, because X is bounded [see Eqs. (16)–(20)], $f_X(x;\, X_T,\, \rho)$ does not give an accurate description outside these bounds; a better approximation for small p can be expected if the pdf outside the crosstalk bounds is set to zero. For a mark in the main signal, the requirements $X_{P,i}(\mathrm{m})$ are then given by

$$\int_{X_{\min}[X_c(\mathrm{m}),1]}^{X_A(\mathrm{m})} f_X[x;\, X_T(\mathrm{m}),\, 1]\, dx < p \tag{23}$$

The requirement for a space is obtained analogously [38]; however, because $V[X(\mathrm{m})] > V[X(\mathrm{sp})]$, the requirement for a space becomes less demanding than the requirement for a mark.

Numerical Illustrations and Discussion. Assuming the amplitudes of all interfering signals to be equal, i.e., $X_{P,i} = X_P,\ \forall i$, computations of the three different requirements as functions of N are presented in Fig. 7, with p as parameter. For simplicity, $|X_A(\mathrm{m})| = X_A(\mathrm{sp}) = -7$ dB is assumed; for incoherent crosstalk, an

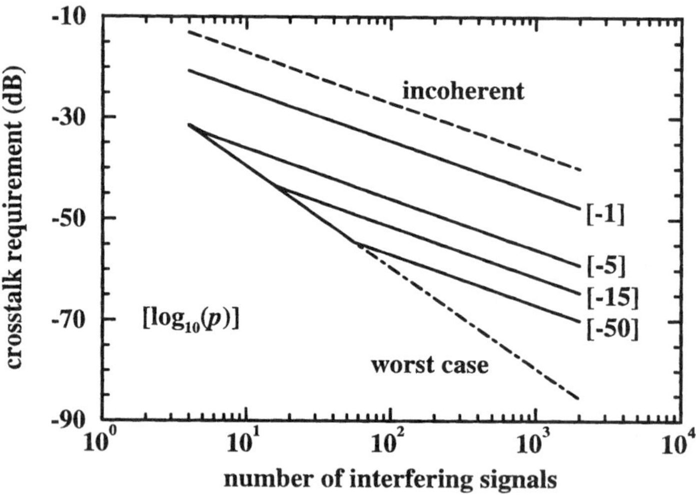

Figure 7 Requirement for the crosstalk X_P of each crosstalk-generating element as function of the number N of crosstalk contributions. *Dashed line*: incoherent crosstalk requirement according to Eq. (21) for $X_A(\mathrm{sp}) = -7$ dB. *Dot-dashed line*: worst case interferometric crosstalk requirement according to Eq. (22) for $|X_A(\mathrm{m})| = -7$ dB. *Solid lines*: statistical crosstalk requirement according to Eq. (23) for $|X_A(\mathrm{m})| = -7$ dB, with $\log_{10}(p)$ as parameter. If the total crosstalk is an ergodic stochastic process, then the probabilities $p = 10^{-1}, 10^{-5}, 10^{-15}, 10^{-50}$ correspond to failure times per year of around 37 days, 5 minutes, 32 ns, and $3 \cdot 10^{-43}$ s, respectively. (From Ref. 38 © 1997 IEEE.)

acceptable crosstalk level of -7 dB corresponds to around 1 dB power penalty, in case receiver noise is independent of incident optical power. For a reasonable number of interfering signals, say, $N = 100$, and a relatively low probability, say, $p = 10^{-5}$, that $X(\mathrm{m}) < X_A(\mathrm{m})$, X_P may be around 13.5 dB larger as compared with the most stringent requirement; however, the crosstalk levels required are still very low, on the order of $X_P < -45$ dB, and it should be emphasized that these crosstalk requirements are strongly dependent on which probability p can be accepted.

A more detailed model for network transmission performance [38,39] is obtained by combining the crosstalk model [38,39] with models for accumulation of amplified spontaneous emission, amplifier saturation, and signal power regulation [43]. The Multi-Wavelength Transport Network (MWTN) [9], which is a meshed optical network, has been analyzed [38,39,42] using this transmission model. The MWTN cross-connect nodes perform wavelength-selective optical routing through a three-stage structure, similar to the one in Fig. 4, i.e., by way of optical demultiplexing followed by space-switching and combining (see Ref. 9 for more details). The number of crosstalk contributions is strongly dependent on whether optical power combiners or wavelength multiplexers are used for the combining stage, and it is also important in what way the WDM signals are routed through the nodes. For the case of a node with four input and output fibers, each carrying four WDM signals, there are—for each WDM signal—six first-order crosstalk contributions of the same nominal wavelength per node, if power combiners are used [38–40,42]; this corresponds to 240 contributions in total for transmission through 40 nodes. Employing wavelength multiplexers at the output as well as a wavelength routing scheme that does not allow signals of the same wavelength to be routed in the same space-switch module, there will be no first-order crosstalk contributions, but only nine second-order crosstalk contributions [38–40,42], corresponding to 360 second-order contributions for 40 nodes. Calculations using the more detailed transmission model indicate that the crosstalk requirement for each crosstalk-generating element, for $p = 10^{-5}$, must be less than about -48 dB to allow transmission through 40 nodes at 10 Gb/s per WDM signal in the first case, and less than about -25 dB in the second case [42]. These figures are close to those that result from the general crosstalk requirements shown in Fig. 7, viz., device crosstalk requirements of about -50 dB for 240 first-order crosstalk contributions and about -26 dB for 360 second-order contributions, respectively, when $p = 10^{-5}$.

4 PROPERTIES OF INTEGRATED OPTIC SWITCH MATRICES

Having discussed a selection of requirements imposed on integrated optic devices in the previous sections of this chapter, it is of interest to give an indication of what performance can be obtained in practice. For this end, this section will address some properties of one type of integrated optic device—the space-division switch.

4.1 General Remarks

Different space-switch technologies represent alternatives in terms of fabrication methods and performance characteristics (see, e.g., Ref. 4); some examples of specific implementations of guided-wave switch matrices, each comprising a number of integrated elementary switch units, are given in Table 1. Table 1 is, of course, not in any way exhaustive, but its purpose is merely to show that guided-wave space-division switches have been implemented in different material systems and that a variety of switching principles have been used.

A major advantage of monolithic, guided-wave switch matrices is a high degree of integration that implies small physical device size, and potentially low cost. Disadvantages are relatively large insertion loss unless optical amplifiers are integrated, a limited wavelength range as determined by the switching principle and material used, and the difficulty of obtaining sufficiently low crosstalk levels. As mentioned previously, dilation constitutes one way of improving crosstalk performance, which has been recently exemplified by demonstration of low-crosstalk lithium niobate switch matrices [55]. Switch matrices using an array of gates in conjunction with fiber-optic couplers [56,57] avoid waveguide crossings, thus eliminating one contribution to crosstalk. The crosstalk problem is completely absent in the case of optomechanical switches that rely on electrically controlled connections of fibers [58,59]; such switches have excellent spectral transfer characteristics, limited only by the properties of fiber-optic connectors. Disadvantages are relatively large physical size and long rearrangement times.

4.2 Semiconductor Optical Amplifier Gate Switch Matrices: An Example

4.2.1 4 × 4 Switch Matrices

To exemplify the transmission properties of one specific type of guided-wave space switch, characteristics of monolithic SOA gate switch matrices reported by Ericsson are discussed (see Ref. 60 for a more thorough review). The switching function of these switches is based on optical power splitting, gating with SOAs, and optical power combining. Low insertion loss or net insertion gain, high switching extinction ratio, low crosstalk, and low polarization sensitivity are achieved for such 4 × 4 InGaAsP/InP switches of small physical size. Although the employment of amplifiers makes the transmission properties complex, polarization-sensitive 4 × 4 SOA gate switches incur merely reasonable power

Table 1 Examples of Guided-wave Space Switches

Laboratory	Year	Ref.	Material	Switch principle	Port size	Architecture	Size (mm^2)	Loss (dB)	Δ (dB)	Crosstalk (dB)	Drive[a]
AT&T	1995	44	Ti:$LiNbO_3$	EO DC	8 × 8	dilated Benes		9	—[f]	−22[a,b]	9.4[c] V
Ericsson	1994	45	Ti:$LiNbO_3$	EO DOS	8 × 8	tree	80 × 15	14.6[b]	1.6[c]	−15.5[b]	±105 V
NTT	1997	46	SiO_2/Si	TO MZI	16 × 16	PI-Loss	100 × 107	6.6[c]		+55[c,d]	1.06[c] W
Akzo Nobel	1996	47	polymer/Si	TO DOS	8 × 8	tree	80 × 65	10.7[c]	0.5	+30.3[c,d]	92 mW
NEC	1993	48	AlGaAs/GaAs	EO DC	4 × 4	simplified tree	17 × 1.4	1.6[e]	—[f]		13.5 V
Alcatel	1993	49	InGaAsP/InP	CI DOS	4 × 4	tree	20 × 2.25	<15		−12[c]	
ETHZ	1996	50	InGaAsP/InP	EO MZI	2 × 2	tree	20 × 5	5	<1	<−15[a,b]	4.5 V
Hitachi	1994	51	InGaAsP/InP	TIR/SOAG	4 × 4	crossbar	11.8 × 2.0	0	2	+54[d]	250 mA
NTT	1992	52	InGaAs/GaAs	SOAG	4 × 4	Benes	1.2 × 1.7	0		<−30	16[c] mA
Ericsson	1992	53	InGaAsP/InP	SOAG	4 × 4	tree	6.7 × 3.0	0	6–12	<−30	50 mA
Ericsson	1995	54	InGaAsP/InP	SOAG	4 × 4	tree	6.7 × 3.0	<2	<1	>+40[d]	32[c] mA

(Notice that crosstalk and extinction of the different switches are indicated in various ways, which makes direct comparison difficult.)

Key to symbols and acronyms: Δ: polarization sensitivity; EO: electro-optic; TO: thermo-optic; CI: carrier-injection; TIR: total internal reflection; DC: directional coupler; DOS: digital optical switch; MZI: Mach–Zehnder interferometer; SOAG: SOA gate; PI-Loss: path-independent loss arrangement.

[a] per switch element
[b] worst case
[c] average
[d] extinction
[e] coupling losses to fibers not included
[f] single-polarization operation

penalty levels for single-wavelength transmission, multiwavelength transmission, and for single-wavelength transmission through cascaded switch paths. The input power dynamic range, which is limited by amplified spontaneous emission and by gain saturation, is one of the important parameters. For single-wavelength transmission at 622 Mb/s, the input power dynamic range for power penalty less than 2 dB is larger than 20 dB for such a 4 ×4 switch [61,62]. Optical feedback may deteriorate switch performance [53,54]; in particular, too high external and/or internal residual reflectivities reduce the dynamic range [61,62], as compared with the foregoing value. An adequate dynamic range should be retained also for WDM transmission. As shown in Fig. 8, simultaneous transmission of two, three, and four WDM channels can be performed at moderate penalty levels of less than 3 dB; for four channels, the dynamic range for power penalty of less than 3 dB is 7 dB per channel. Concatenation of switches is also of central importance; a single-wavelength transmission experiment involving 2 × 80-km fiber of the Stockholm Gigabit Network, erbium-doped fiber amplifiers, and transmission three times through a 4 × 4 switch at 2.488 Gb/s resulted in a moderate power penalty level of less than 5 dB [61,62]. The polarization dependence of SOA gate switch matrices can be virtually eliminated [54], and a higher saturation output power would contribute to an increased input power dynamic range; increased saturation output power may be obtained by using strained quantum well amplifiers (cf. Ref. 63).

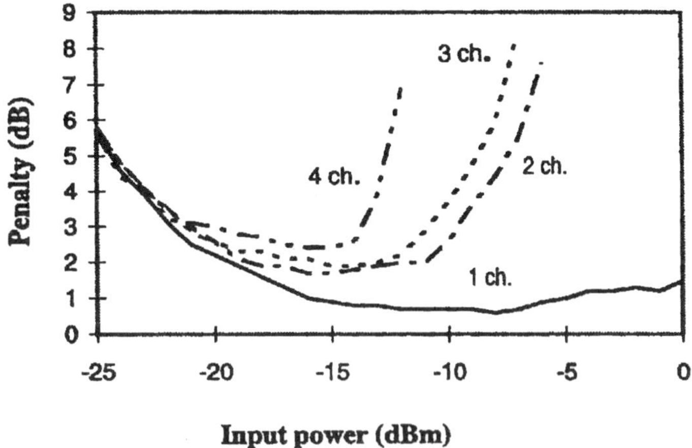

Figure 8 Power penalty at a BER of 10^{-9} as function of input power per WDM channel for transmission at 622 Mb/s. Signal wavelengths are {1548, 1552, 1556, 1560} nm. (From Ref. 61.)

4.2.2 Linear Crosstalk in a 4 × 4 Switch Matrix

The purpose of this section is to illustrate the crosstalk performance of a 4 × 4 switch matrix and, by way of this example, what influence interferometric crosstalk within an integrated optic device may have.

Crosstalk has been measured for a fully loaded, packaged 4 × 4 SOA gate switch matrix, including investigation of the influence of the state of the switch [64,65]. Measurement results are shown in Fig. 9; the mean crosstalk level is −30.7 dB for the fully loaded switch, which implies a mean crosstalk contribution of −35.4 dB from each of the three disturbing signals. There are considerable variations around the mean among the different switch states, due partly to variations in the number of crosstalk contributions within the switch and partly to interferometric effects. Less crosstalk could probably be obtained by an improved design and fabrication process. From a network point of view, worst-case crosstalk should of course be sufficiently low. As already discussed, reasonably large meshed optical networks impose severe requirements on the crosstalk of the constituent network elements.

It is of interest to analyze the crosstalk performance of these SOA gate switch matrices in some detail. Contributions to crosstalk for such switch matrices include incomplete signal absorption in closed gates, crosstalk in waveguide crossings, and scattered light in the chip. For the case of a fully loaded 4 × 4 switch, the crosstalk situation due to contributions from gates and waveguide crossings

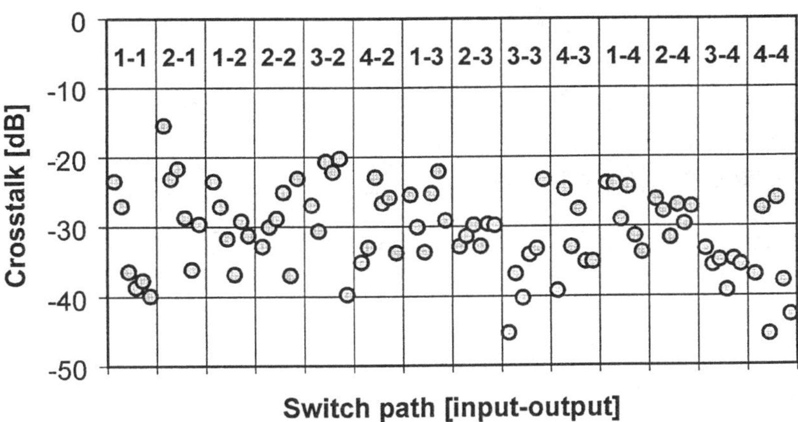

Figure 9 Crosstalk levels for 84 different switch states, measured for a fully loaded 4 × 4 SOA gate switch matrix. For each main signal path (indicated above the circles), there are six different combinations. (From Ref. 64.)

is schematically shown in Fig. 10; this particular switch state involves seven crosstalk contributions for one of the main signals. In the case of mutually uncorrelated signals at the four different inputs of a 4 × 4 switch matrix with uniform gain/loss profiles for all paths, the corresponding total first-order crosstalk for one of the worst-case paths—input 1 to output 4—amounts to

$$\begin{aligned} X = {} & L_\times^{-1}(3 + L_\times^{-3})X_\times + L_\times^{-2}(1 + L_\times^{-1} + L_\times^{-3})K^{-1} \\ & + 2L_\times^{-5/2}X_\times \cos(\Delta\phi_1) + 2L_\times^{-3/2}(K^{-1}X_\times)^{1/2} \\ & \times (\cos(\Delta\phi_2) + L_\times^{-1/2}\cos(\Delta\phi_3) + L_\times^{-3/2}\cos(\Delta\phi_4) + L_\times^{-2}\cos(\Delta\phi_5)) \end{aligned} \tag{24}$$

where X denotes the total power crosstalk at output port 4 of the 4 × 4 switch, $X_\times < 1$ is the power leakage in a waveguide crossing, $L_\times < 1$ is the power loss of a waveguide crossing, $\Delta\phi_i$ are optical phase differences on the 4 × 4 switch chip, and $K = G/X_G$ denotes the SOA gate switch extinction ratio, where G denotes the power amplification in an open SOA gate and $X_G < 1$ is the power leakage through a closed SOA gate. In addition to the incoherent crosstalk contributions, there are a number of interferometric contributions that can substantially change the total crosstalk at the output as a consequence of changes in the optical path length differences, which can be induced by varying temperature, injection current, or optical power levels. Worst-case interferometric crosstalk must be taken into account here, especially because of the distinct coherence that prevails within the small-sized switch chip for any given signal.

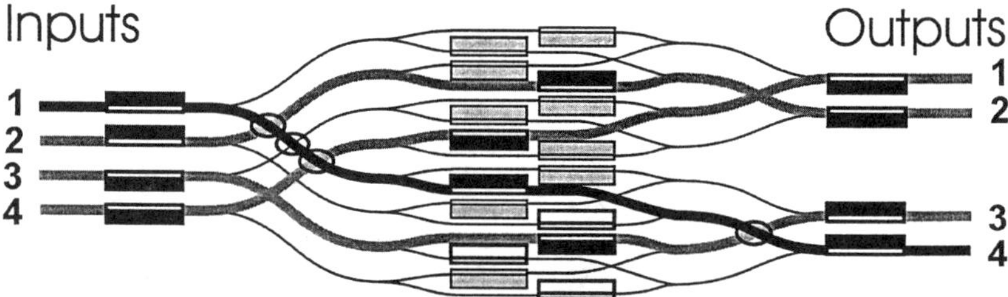

Figure 10 Schematic illustration of crosstalk—for a particular switch state—in a fully loaded, monolithic 4 × 4 SOA gate switch matrix. Black SOAs are open; the others are closed. There are seven crosstalk contributions interfering with the black signal: the four circles indicate leakage at contributing waveguide crossings, and the three white-colored SOAs indicate leakage at contributing closed gates. Notice that, although optical power is divided at each power splitter, only paths of interest for the routing of the four input signals are marked with thick lines. (From Ref. 65 © 1997 IEEE. The drawing is based on the original mask layout of Ref. 53. See Ref. 65 for more details.)

In case there are no losses associated with waveguide crossings, then (neglecting losses due to crosstalk) $L_\times \equiv 1$, and the total crosstalk at output 4 of the 4×4 switch reduces to

$$X = 4X_\times + 3K^{-1} + 2X_\times \cos(\Delta\phi_1) + 2\sqrt{K^{-1}X_\times} \sum_{i=2}^{5} \cos(\Delta\phi_i) \tag{25}$$

In case of uniform gain/loss profiles, it is thus observed that the crosstalk is dependent on only the leakage at waveguide crossings, the gate extinction ratio, and the various phase differences occuring on the chip. Differential losses would contribute to increased effective crosstalk. The losses due to waveguide crossings introduce differential losses, but they can be made uniform for the switch, in the sense that all paths will have the same number of waveguide crossings; this also implies that first-order crosstalk in a first approximation will be insensitive to these losses. It is easy to verify that 24 dummy waveguide crossings should be included in the switch to achieve reduced differential losses and thereby to lower crosstalk for worst-case paths.

As an example, assume that the crosstalk associated with waveguide crossings and gates is the same; i.e., assume that $X_\times = K^{-1} = X_0$. Then

$$X = \left[7 + 2 \sum_{i=1}^{5} \cos(\Delta\phi_i)\right] X_0 \tag{26}$$

which gives $X_{max} = 17X_0$ and $|X|_{min} = 0$. If it is required that the total crosstalk for the 4×4 switch should be less than, say, -40 dB, then X_0 must fulfill the condition $X_0 < -52.3$ dB.

As another example, assume that the crosstalk contributions from the gates are negligible in comparison with the waveguide crossing contributions; i.e., assume that $K^{-1} \ll X_\times$. Then

$$X \approx 4X_\times + 2X_\times \cos(\Delta\phi_1) \tag{27}$$

which gives $X_{max} = 6X_x$ and $X_{min} = 2X_\times$. If it is required that the total crosstalk should be less than -40 dB in this case, then $X_\times$ must fulfill the condition $X_\times < -47.8$ dB.

As a final example, assume that leakage through closed gates is much larger than the leakage at waveguide crossings; i.e., assume that $K^{-1} \gg X_\times$. Then

$$X \approx 3K^{-1} \tag{28}$$

which gives $X_{max} = X_{min} = 3K^{-1}$, since there are no interferometric effects present in this approximation. If it is required that the total crosstalk be less than -40 dB in this case, then the gate extinction ratio must be larger than 44.8 dB.

5 CONCLUSIONS

Integrated optic devices must be implemented in such a way that capacity and coverage specifications for the communication system of interest can be met, meaning, e.g., that they should not impose significant transmission limitations. In particular, it is important to consider the implications of device concatenation, because an optical connection established in a communication system may involve transmission through a significant number of optical devices; the larger the number of cascaded devices, the more severe the individual device requirements related to transport characteristics become. Depending on the application, additional requirements related to fabrication and operation, e.g., physical size, fabrication yield, packaging, and manageability, are imposed.

One of the most pressing current problems is transmission limitations due to interferometric crosstalk. In light of the severe requirements for the crosstalk of each crosstalk-generating network element, networks have to be designed with great care to meet worst-case requirements. Such designs might be possible, but they imply tight fabrication tolerances on constituent network optical devices and/or probably relatively complex network switching nodes. In a meshed optical network there are typically many crosstalk contributions that interfere with a main signal, so the slow variations of states of polarization and optical phases indicate that a statistical treatment of crosstalk may be of interest. A probability p that the total crosstalk level exceeds a certain value can be introduced, and it can be shown that the crosstalk requirement for optical network elements may be significantly relaxed if it can be accepted that it is possible, although most unlikely, that the total crosstalk exceeds a maximum tolerable value. Whether a probability $p > 0$ is acceptable in installed networks is left for further discussion, for instance, within technical committees for telecommunications standards.

REFERENCES

1. P. S. Henry, R. A. Linke, and A. H. Gnauck. Introduction to lightwave systems. In: S. E. Miller and I. P. Kaminow, eds. Optical Fiber Telecommunications II. New York: Academic Press, 1988, pp. 781–831.
2. N. S. Bergano, C. R. Davidson, M. Ma, A. Pilipetskii, S. G. Evangelides, H. D. Kidorf, J. M. Darcie, E. Golovchenko, K. Rottwitt, P. C. Corbett, R. Menges, M. A. Mills, B. Pedersen, D. Peckham, A. A. Abramov, and A. M. Vengsarkar. 320 Gb/s WDM transmission (64 × 5 Gb/s) over 7,200 km using large mode fiber spans and chirped return-to-zero signals. Tech. Dig. Optical Fiber Commun. Conf., postdeadline papers, PD12, San Jose, CA, 1998.
3. H. Taga, N. Edagawa, M. Suzuki, N. Takeda, K. Imai, S. Yamamoto, and S. Akiba. 213Gbit/s (20 × 10.66 Gbit/s), over 9000 km transmission experiment using disper-

sion slope compensator. Tech. Dig. Optical Fiber Commun. Conf., postdeadline papers, PD13, San Jose, CA, 1998.
4. M. Erman. What technology is required for the pan-European network, what is available and what is not. Proc. European Conf. on Optical Commun., vol. 5, Oslo, Norway, 1996, pp. 87–94.
5. M. J. Karol, G. Hill, C. Lin, and K. Nosu, eds. Special issue on broad-band optical networks. J. Lightwave Technol. 11:(5/6), 1993.
6. M. Fujiwara, M. S. Goodman, M. J. O'Mahony, O. K. Tonguz, and A. E. Willner, eds. Special issue on multiwavelength optical technology and networks. J. Lightwave Technol. 14(6):1996.
7. R. L. Cruz, G. R. Hill, A. L. Kellner, R. Ramaswami, G. H. Sasaki, and Y. Yamabayashi, eds. Optical networks. IEEE J. Select. Areas Commun. 14(No. 5):1996.
8. M. Yamaguchi, T. Kato, T. Sasaki, K. Komatsu, and M. Kitamura. Penalty-free (<0.2 dB) 2.5-Gb/s transmission requirements for modulator-integrated DFB LDs. Proc. European Conf. on Optical Commun., Florence, Italy, 1994, pp. 977–980.
9. G. R. Hill, P. J. Chidgey, F. Kaufhold, T. Lynch, O. Sahlén, M. Gustavsson, M. Janson, B. Lagerström, G. Grasso, F. Meli, S. Johansson, J. Ingers, L. Fernandez, S. Rotolo, A. Antonielli, S. Tebaldini, E. Vezzoni, R. Caddedu, N. Caponio, F. Testa, A. Scavennec, M. J. O'Mahony, J. Zhou, A. Yu, W. Sohler, U. Rust, and H. Herrmann. A transport network layer based on optical network elements. J. Lightwave Technol. 11:667–679, 1993.
10. C. Caspar, H.-M. Foisel, C. v. Helmot, B. Strebel, and Y. Sugaya. Comparison of cascadability performance of different types of commercially available wavelength (de)multiplexers. Electron. Lett. 33:1624–1626, 1997.
11. C. Dragone, T. Strasser, G. A. Bogert, L. W. Stulz, and P. Chou. Waveguide grating router with maximally flat passband produced by spatial filtering. Electron. Lett. 33: 1312–1314, 1997.
12. A. Rigny, A. Bruno, and H. Sik. Multigrating method for flattened spectral response wavelength multi/demultiplexer. Electron. Lett. 33:1701–1702, 1997.
13. H. van de Stadt and J. M. Muller. Multimirror Fabry–Perot interferometers. J. Opt. Soc. Am. A 2:1363–1370, 1985.
14. A. M. Vengsarkar, J. R. Pedrazzani, J. B. Judkins, P. J. Lemaire, N. S. Bergano, and C. R. Davidson. Long-period fiber-grating-based gain equalizers. Opt. Lett. 21: 336–338, 1996.
15. M. J. Coupland, K. G. Hambleton, and C. Hilsum. Measurement of amplification in a GaAs injection laser. Phys. Lett. 7:231–232, 1963.
16. C. E. Zah, J. S. Osinski, C. Caneau, S. G. Menocal, L. A. Reith, J. Salzman, F. K. Shokoohi, and T. P. Lee. Fabrication and performance of 1.5 μm GaInAsP traveling-wave laser amplifiers with angled facets. Electron. Lett. 23:990–992, 1987.
17. I. Cha, M. Kitamura, and I. Mito. 1.5 μm band traveling-wave semiconductor optical amplifiers with window facet structure. Electron. Lett. 25:242–243, 1989.
18. C. E. Zah, R. Bhat, S. G. Menocal, N. Andreadakis, F. Favire, C. Caneau, M. A. Koza, and T. P. Lee. 1.5 μm GaInAsP angled-facet flared-waveguide traveling-wave laser amplifiers. IEEE Photon. Technol. Lett. 2:46–47, 1990.

19. I. W. Marshall. Low loss coupling between semiconductor lasers and single-mode fiber using tapered lensed fiber. Br. Telecom Technol. J. 4:114–121, 1986.
20. J. V. Collins, M. Dagenais, and M. Itoh, eds. Alignment tolerant structures for ease of optoelectronic packaging. IEEE J. Select. Topics Quantum Electron. 3(6), 1997.
21. A. F. Elrefaie, R. E. Wagner, D. A. Atlas, and D. G. Daut. Chromatic dispersion limitations in coherent lightwave transmission systems. J. Lightwave Technol. 6: 704–709, 1988.
22. A. Galtarossa, C. G. Someda, F. Matera, and M. Schiano. Polarization mode dispersion in long single-mode-fiber links: a review. Fiber Integrated Optics 13:215–229, 1994.
23. K. Padmanabhan and A. N. Netravali. Dilated networks for photonic switching. IEEE Trans. Commun. 35:1357–1365, 1987.
24. M. Renaud. Keys to optical packet switching: the ACTS KEOPS project. Tech. Dig. Photon. in Switching, paper PWB1, Stockholm, Sweden, 1997.
25. ITU-T Recommendation G.841. Types and characteristics of SDH network protection architectures. 1995.
26. L. J. Andrews, T. Mukai, N. A. Olsson, and D. N. Payne, eds. Special issue on optical amplifiers. J. Lightwave Technol. 9(2), 1991.
27. C. R. Giles, M. Newhouse, J. Wright, and K. Hagimoto, eds. Special issue on system and network applications of optical amplifiers. J. Lightwave Technol. 13(5), 1995.
28. Tech. Dig. Optical Amplifiers and Their Appl., Victoria, B. C., Canada, 1997.
29. J. Zhou, M. J. O'Mahony, and S. D. Walker. Analysis of optical crosstalk effects in multi-wavelength switched networks. IEEE Photon. Technol. Lett. 6:302–305, 1994.
30. E. L. Goldstein, L. Eskildsen, and A. F. Elrefaie. Performance implications of component crosstalk in transparent lightwave networks. IEEE Photon. Technol. Lett. 6: 657–660, 1994.
31. P. J. Legg, D. K. Hunter, I. Andonovic, and P. E. Barnsley. Inter-channel crosstalk phenomena in optical time division multiplexed switching networks. IEEE Photon. Technol. Lett. 6:661–663, 1994.
32. P. J. Chidgey, J. Laws, D. J. Malyon, G. P. Reeve, and P. Swan. Crosstalk in multiwavelength optical networks. IEEE/LEOS Summer Topical Meeting on Optical Networks and their Enabling Technologies, postdeadline papers, Lake Tahoe, NV, 1994.
33. J. Zhou, R. Cadeddu, E. Casaccia, C. Cavazzoni, and M. J. O'Mahony. Crosstalk in multiwavelength optical cross-connect networks. J. Lightwave Technol. 14:1423–1435, 1996.
34. J. J. O'Reilly and C. J. Appleton. System performance implications of homodyne beat noise effects in optical fibre networks. IEE Proc. Optoelectron. 142:143–148, 1995.
35. E. L. Goldstein, L. Eskildsen, Y. Silberberg, and C. Lin. Polarization statistics of crosstalk-induced interferometric noise in transparent lightwave networks. Proc. European Conf. on Optical Commun., Brussels, Belgium, 1995, pp. 689–692.
36. D. J. Blumenthal, P. Granestrand, and L. Thylén. BER floors due to heterodyne coherent crosstalk in space photonic switches for WDM networks. IEEE Photon. Technol. Lett. 8:284–286, 1996.
37. C. P. Larsen, L. Gillner, and M. Gustavsson. Scaling limitations in optical multiwavelength meshed networks and ring networks due to crosstalk. Tech. Dig. Photon. in Switching, Sendai, Japan, 1996, pp. 20–21.

38. M. Gustavsson, L. Gillner, and C. P. Larsen. Statistical analysis of interferometric crosstalk: theory and optical network examples. J. Lightwave Technol. 15:2006–2019, 1997.
39. M. Gustavsson, L. Gillner, and C. P. Larsen. Network requirements on optical switching devices. Proc. 8th European Conf. on Integrated Optics, Stockholm, Sweden, 1997, pp. 10–15; and Tech. Dig. Photon. in Switching, Stockholm, Sweden, 1997, pp. 10–15.
40. C. P. Larsen, L. Gillner, and M. Gustavsson. Linear crosstalk properties of large WDM cross-connects. Tech. Dig. Photon. in Switching, Stockholm, Sweden, 1997, pp. 27–30.
41. D. J. Blumenthal. Coherent crosstalk in photonic switched networks. Tech. Dig. Photon. in Switching, paper PWA1, Stockholm, Sweden, 1997.
42. L. Gillner, C. P. Larsen, and M. Gustavsson. Influence of crosstalk on the scalability of optical multi-wavelength switching networks. Dig. IEEE LEOS Summer Topical Meetings, Technologies for a Global Information Infrastructure, Montreal, Canada, 1997, pp. 32–33.
43. L. Gillner and M. Gustavsson. Scalability of optical multiwavelength switching networks: power budget analysis. IEEE J. Select. Areas Commun. 14:952–961, 1996.
44. E. J. Murphy, C. T. Kemmerer, D. T. Moser, M. R. Serbin, J. E. Watson, and P. L. Stoddard. Uniform 8 × 8 lithium niobate switch arrays. J. Lightwave Technol. 13:967–970, 1995.
45. P. Granestrand, B. Lagerström, P. Svensson, H. Olofsson, J.-E. Falk, and B. Stoltz. Pigtailed tree-structured 8 × 8 $LiNbO_3$ switch matrix with 112 digital optical switches. IEEE Photon. Technol. Lett. 6:71–73, 1994.
46. T. Goh, M. Yasu, K. Hattori, A. Himeno, and Y. Ohmori. Low loss and high extinction ratio 16 × 16 thermo-optic matrix switch using silica-based planar lightwave circuit technology. Proc. Third Asia-Pacific Conf. on Commun., Sydney, Australia, 1997, pp. 232–236.
47. A. Borreman, T. Hoekstra, M. Diemeer, H. Hoekstra, and P. Lambeck. Polymeric 8 × 8 digital optical switch matrix. Proc. European Conf. on Optical Commun., Oslo, Norway, 1996, vol. 5, pp. 59–62.
48. K. Hamamoto, S. Sugou, K. Komatsu, and M. Kitamura. Extremely low loss 4 × 4 GaAs/AlGaAs optical matrix switch. Tech. Dig. Integrated Photon. Research, Palm Springs, CA, 1993, pp. 172–175.
49. J.-F. Vinchant, A. Goutelle, B. Martin, F. Gaborit, P. Pagnod-Rossiaux, J.-L. Peyre, J. Le Bris, and M. Renaud. New compact polarisation insensitive 4 × 4 switch matrix on InP with digital optical switches and integrated mirrors. Proc. European Conf. on Optical Commun., postdeadline paper, Montreux, Switzerland, 1993, pp. 13–16.
50. R. Krähenbühl, R. Kyburz, W. Vogt, M. Bachmann, T. Brenner, E. Gini, and H. Melchior. Low-loss polarization-insensitive InP–InGaAsP optical space switches for fiber optical communication. IEEE Photon. Technol. Lett. 8:632–634, 1996.
51. T. Kirihara, M. Ogawa, H. Inoue, H. Kodera, and K. Ishida. Lossless and low-crosstalk characteristics in an InP-based 4 × 4 optical switch with integrated single-stage optical amplifiers. IEEE Photon. Technol. Lett. 6:218–221, 1994.
52. M. Ikeda, S. Oku, Y. Shibata, T. Suzuki, and M. Okayasu. Loss-less 4 × 4 mono-

lithic LD optical matrix switches. Tech. Dig. International Topical Meeting on Photonic Switching, paper 2B1, Minsk, Belarus, 1992.
53. M. Gustavsson, B. Lagerström, L. Thylén, M. Janson, L. Lundgren, A.-C. Mörner, M. Rask, and B. Stoltz. Monolithically integrated 4 × 4 InGaAsP/InP laser amplifier gate switch arrays. Electron. Lett. 28:2223–2225, 1992.
54. W. van Berlo, M. Janson, L. Lundgren, A.-C. Mörner, J. Terlecki, M. Gustavsson, P. Granestrand, and P. Svensson. Polarization-insensitive, monolithic 4 × 4 InGaAsP–InP laser amplifier gate switch matrix. IEEE Photon. Technol. Lett. 7:1291–1293, 1995.
55. E. J. Murphy, T. O. Murphy, R. W. Irvin, R. Grencavich, G. W. Davis, and G. W. Richards. Enhanced performance switch arrays for optical switching networks. Proc. 8th European Conf. on Integrated Optics, Stockholm, Sweden, 1997, pp. 563–566.
56. E. Eichen, W. J. Miniscalco, J. McCabe, and T. Wei. Lossless, 2 × 2, all-fiber optical routing switch. Tech. Dig. Optical Fiber Commun. Conf., postdeadline papers, PD20, San Francisco, 1990.
57. L. R. McAdams, A. M. Gerrish, R. F. Kalman, and J. W. Goodman. Optical crossbar switch with semiconductor optical amplifiers. Tech. Dig. Photon. in Switching, Palm Springs, CA, 1993, pp. 60–63.
58. M. Tachikura, T. Katagiri, and H. Kobayashi. Strictly nonblocking 512 × 512 optical fiber matrix switch based on three-stage Clos network. IEEE Photon. Technol. Lett. 6:764–766, 1994.
59. S. Sjölinder. Mechanical optical fibre cross connect. Tech. Dig. Photon. in Switching, Salt Lake City, UT, 1995, pp. 115–117.
60. M. Gustavsson. Technologies and applications for space-switching in multi-wavelength networks. International Workshop on Photonic Networks and Technologies, Lerici, Italy, 1996; published in G. Prati, ed. Photonic Networks, London: Springer-Verlag, 1997, pp. 157–171.
61. C. P. Larsen, E. Almström, W. van Berlo, J.-E. Falk, F. Testa, L. Gillner, and M. Gustavsson. Transmission experiments on fully packaged 4 × 4 semiconductor optical amplifier gate switch matrix. Tech. Dig. Photon. in Switching, Salt Lake City, UT, 1995, pp. 95–97.
62. E. Almström, C. P. Larsen, L. Gillner, W. H. van Berlo, M. Gustavsson, and E. Berglind. Experimental and analytical evaluation of packaged 4 × 4 InGaAsP/InP semiconductor optical amplifier gate switch matrices for optical networks. J. Lightwave Technol. 14:996–1004, 1996.
63. L. F. Tiemeijer, P. J. A. Thijs, T. van Dongen, J. J. M. Binsma, E. J. Jansen, and S. Walczyk. 33 dB fiber to fiber gain +13 dBm fiber saturation power polarization independent 1310 nm MQW laser amplifiers. Tech. Dig. Optical Amplifiers and Their Appl., postdeadline papers, PD1, Davos, Switzerland, 1995.
64. C. P. Larsen and M. Gustavsson. Comprehensive investigation of linear crosstalk in 4 × 4 semiconductor optical amplifier gate switch matrix. Tech. Dig. Optical Fiber Commun. Conf., Dallas, TX, 1997, pp. 6–7.
65. C. P. Larsen and M. Gustavsson. Linear crosstalk in 4 × 4 semiconductor optical amplifier gate switch matrix. J. Lightwave Technol. 15:1865–1870, 1997.

4

Silica Waveguide Devices

Katsunari Okamoto

NTT Photonics Laboratory
Tokai, Japan

1 INTRODUCTION

The most prominent feature of silica waveguides is their simple and well-defined waveguide structures [1]. This allows us to fabricate multibeam or multistage interference devices such as arrayed-waveguide gratings and lattice-form programmable dispersion equalizers. A variety of passive planar lightwave circuits (PLCs), such as $N \times N$ star couplers, $N \times N$ arrayed-waveguide grating (AWG) multiplexers, and thermo-optic matrix switches have been developed [2,3]. Hybrid opto-electronics integration based on the terraced-silicon platform technologies are also important, both to the fiber-to-the-home (FTTH) applications and to high-speed signal-processing devices [4,5]. Synthesis theory of the lattice-form programmable optical filters has been developed [6] and has been implemented in the fabrication of variable group-delay dispersion equalizers [7]. This chapter briefly reviews the recent progress in planar lightwave circuit devices for optical wavelength-division multiplexing (WDM) systems and subscriber networks, with particular emphasis on $N \times N$ arrayed-waveguide grating multiplexers and related devices, such as optical add/drop multiplexers (ADMs).

2 WAVEGUIDE FABRICATION

Planar lightwave circuits using silica-based optical waveguides are fabricated on silicon or silica substrate by a combination of flame hydrolysis deposition (FHD)

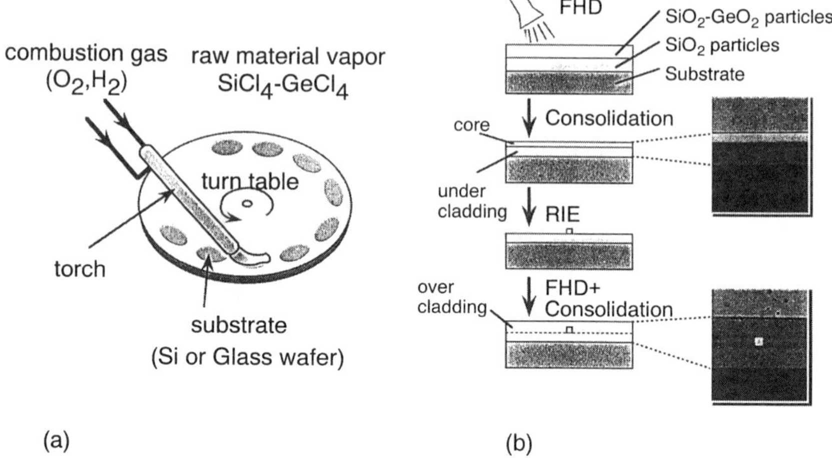

Figure 1 Planar waveguide fabrication technique. (a) Flame hydrolysis deposition; (b) fabrication process.

and reactive ion etching (RIE). Figure 1 shows the planar waveguide fabrication technique. Fine glass particles are produced in the oxyhydrogen flame and deposited on substrates. After undercladding and core glass layers are deposited, the wafer is heated to high temperature for consolidation. The circuit pattern is fabricated by means of photolithography and reactive ion etching. Then core ridge structures are covered with an overcladding layer and consolidated again.

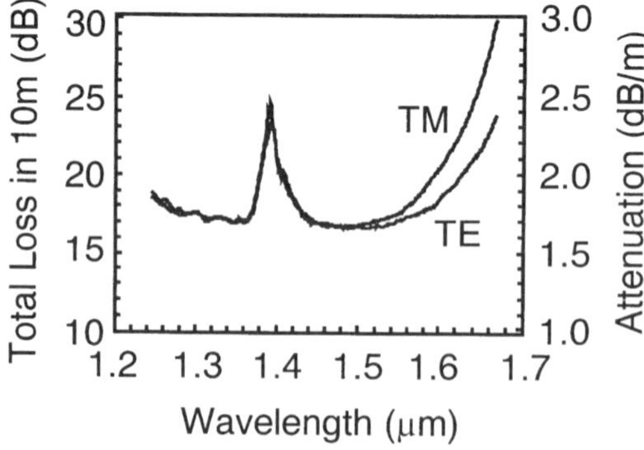

Figure 2 Loss spectra of 10-m-long waveguide.

Table 1 Waveguide Parameters and Propagation Charteristics of Four Kinds of Waveguides

Characteristic	Waveguide			
	Low Δ	Medium Δ	High Δ	Superhigh Δ
Index difference (%)	0.3	0.45	0.75	1.5–2.0
Core size (μ m)	8 × 8	7 × 7	6 × 6	4.5 × 4.5–3 × 3
Loss (dB/cm)	<0.01	0.02	0.04	0.07
Coupling loss (dB/point)	<0.1	0.1	0.4	2.0
Bending radius (mm)	25	15	5	2

Since the typical bending radius R of a silica waveguide is around 2–25 mm, the chip size of the large-scale integrated circuit becomes several centimeters square. Therefore, propagation loss reduction and the uniformity of refractive indices and core geometries throughout the wafer are strongly required. Propagation loss of 0.1 dB/cm was obtained in a 2-m-long waveguide with $\Delta = 2\%$ index difference (R = 2mm) [8] and loss of 0.035 dB/cm was obtained in a 1.6-m-long waveguide with $\Delta = 0.75\%$ index difference (R = 5mm) [9], respectively. Further loss reduction down to 0.017 dB/cm has been achieved (Fig. 2) in a 10-m-long waveguide with $\Delta = 0.45\%$ index difference (R = 15mm) [10]. The higher loss for transverse magnetic (TM) mode (electric field vector is perpendicular to the waveguide plane) may be due to the waveguide wall roughness caused by the RIE process. However, the conversion from transverse electric (TE) to TM mode, or vice versa, was less than −20 dB in 10-m-long waveguides. Various kinds of waveguides are utilized, depending on the circuit configurations. Table I summarizes the waveguide parameters and propagation characteristics of four kinds of waveguides. The propagation losses of low-Δ and medium-Δ waveguides are about 0.01 dB/cm, and those of high-Δ and superhigh-Δ waveguides are about 0.04–0.07 dB/cm, respectively. The low-Δ waveguides are superior to the high-Δ waveguides in terms of fiber coupling losses with the standard single-mode fibers. On the other hand, the minimum bending radii for high-Δ waveguides are much smaller than those for low-Δ waveguides. Therefore, high-Δ waveguides are indispensable in constructing highly integrated and large-scale optical circuits such as $N \times N$ star couplers, arrayed-waveguide grating multiplexers, and dispersion equalizers.

3 *N* × *N* STAR COUPLER

$N \times N$ star couplers are quite useful in high-speed, multiple-access optical networks, since they evenly distribute the input signal among many receivers and make possible the interconnection between them. The free-space type of integrated optic star coupler, in which a slab waveguide region is located between the fan-shaped input and output channel waveguide arrays, is quite advantageous when constructing large-scale $N \times N$ star couplers [11,12]. Figure 3 shows the schematic configuration of the $N \times N$ star coupler. The input power, from any one of the N-channel waveguides in the input array, is radiated to the slab (free space) region, and it is received by the output array. In order to get the uniform power distribution into N output waveguides, the radiation pattern at the output side of the slab–array interface should be uniform over a sector of N waveguides. Since the radiation pattern is the Fraunhofer pattern (Fourier transform) of the field profile at the input side of the slab–array interface, proper side lobes must be produced by the mode coupling from the excited input waveguide to neighboring guides. The star coupler parameters, such as the aperture angle, the radius of slab region, and others, were optimized by using a beam propagation method so as to get the maximum output and good splitting uniformity. Figure 4 shows the splitting loss histogram of 64×64 star coupler measured at $\lambda = 1.55$ μm [12]. The essential splitting loss when light is evenly distributed into 64 output waveguides is 18.1 dB. Therefore, the average of the excess losses is 2.5 dB, and the standard deviation of the splitting losses is 0.8 dB, respectively. Among the 2.5-dB additional loss, the inevitable imperfection (theoretical) loss is about 1.5 dB, the propagation loss is about 0.6 dB, and the coupling loss with single-mode fiber is 0.2 dB/facet, respectively. We have fabricated a series of star couplers ranging from 8×8 to 256×256.

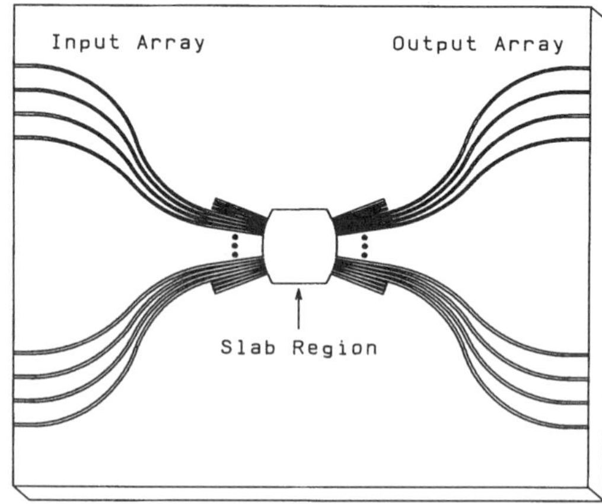

Figure 3 Schematic configuration of $N \times N$ star coupler.

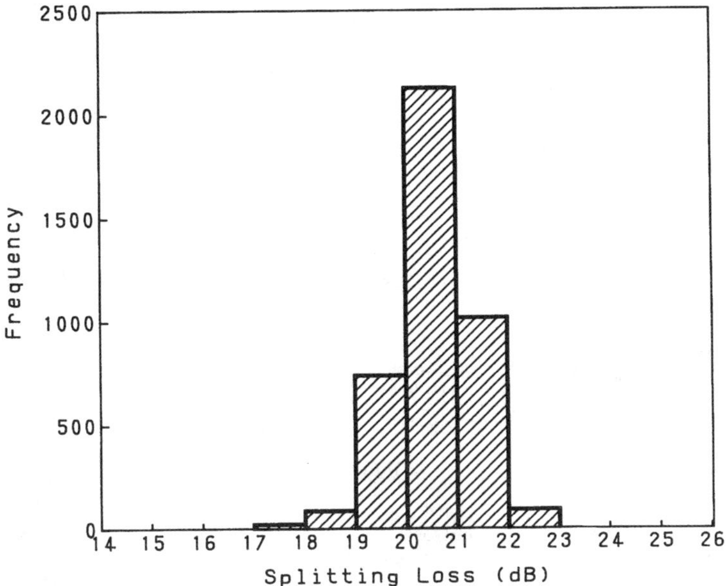

Figure 4 Splitting loss histogram of 64 × 64 star coupler.

4 ARRAYED-WAVEGUIDE GRATINGS

4.1 Principle of Operation and Fundamental Characteristics

An $N \times N$ arrayed-waveguide grating (AWG) multiplexer is very attractive in optical WDM networks, since it is capable of increasing the aggregated transmission capacity of single-strand optical fiber [13,14]. The arrayed-waveguide grating consists of input/output waveguides, two focusing slab regions, and phase array of multiple-channel waveguides with the constant path length difference ΔL between neighboring waveguides (Fig. 5). The input light is radiated to the first slab and then excites the arrayed channel waveguides. After traveling through the arrayed waveguides, the light beams constructively interfere into one focal point in the second slab. The location of this focal point depends on the signal wavelength, since the relative phase delay in each waveguide is given by $\Delta L/\lambda$. The dispersion of the focal position x with respect to the wavelength λ is given by

$$\frac{\Delta x}{\Delta \lambda} = -\frac{N_c f \Delta L}{n_s d \lambda_0} \tag{1}$$

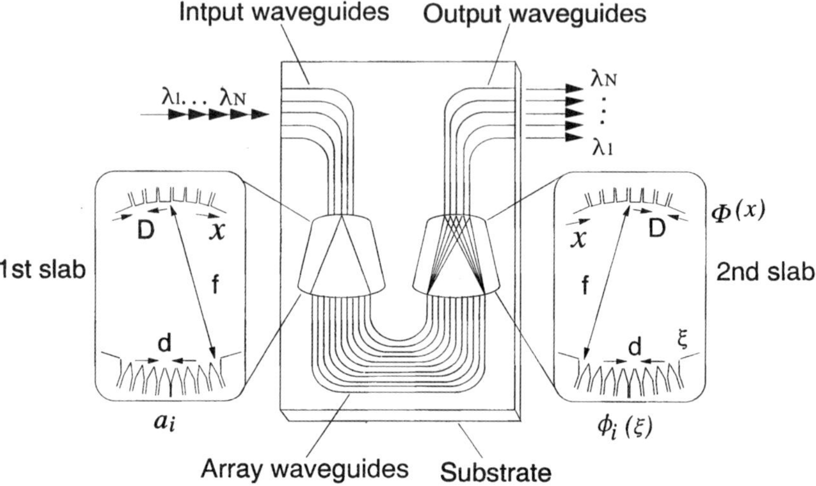

Figure 5 Schematic configuration of arrayed-waveguide grating multiplexer.

where f is the focal length of the converging slab, n_s is the effective index in the slab region, N_c is the group index of the effective index n_c of the array waveguide ($N_c = n_c - \lambda dn_c/d\lambda$), d is the pitch of the array waveguides at the slab–array interface, and λ_0 is the center wavelength of WDM signals, respectively. The diffraction order m is given by $m = n_c \, \Delta L/\lambda_0$. The spatial separation of the mth and $(m + 1)$th focused beams for the same wavelength is expressed as

$$X_{\mathrm{FSR}} = x_m - x_{m+1} = \frac{\lambda_0 f}{n_s d} \tag{2}$$

X_{FSR} represents the free spatial range of the AWG. The number of available wavelength channels N_{ch} is given by dividing X_{FSR}, with the output waveguide separation D as

$$N_{\mathrm{ch}} = \frac{X_{\mathrm{FSR}}}{D} = \frac{\lambda_0 f}{n_s d D} \tag{3}$$

When the center wavelength λ_0 and the number of channels N_{ch} in a WDM system are given, the arcradius f of the slab region is determined by Eq. (3), where waveguide parameters n_s, d, and D are already known. The separation of the focal positions should be $|\Delta x| = D$ for the WDM signals with wavelength spacing $\Delta\lambda$. Therefore, path length difference ΔL is obtained from Eq. (1) as

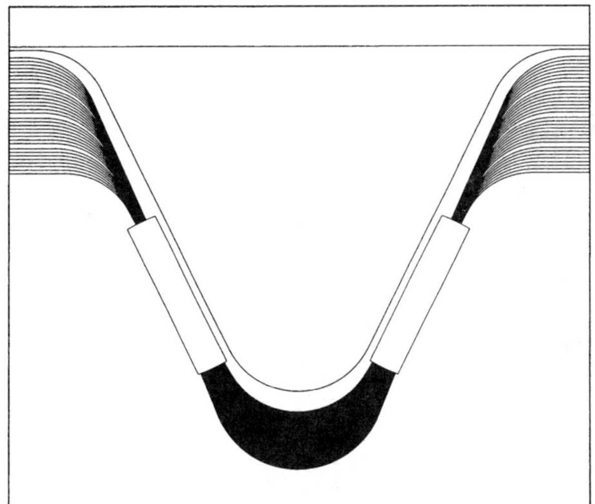

Figure 6 Waveguide layout of a 32-ch 100-GHz-spacing AWG.

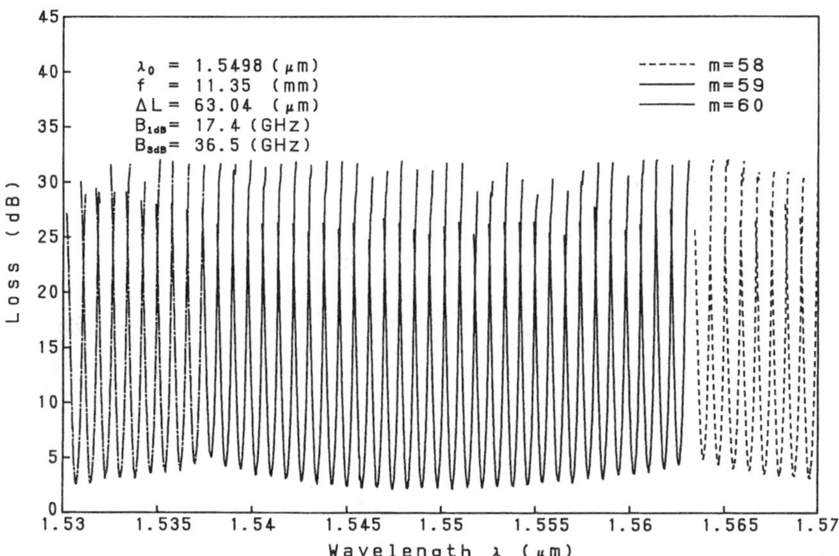

Figure 7 Demultiplexing properties of a 32-ch 100-GHz-spacing AWG.

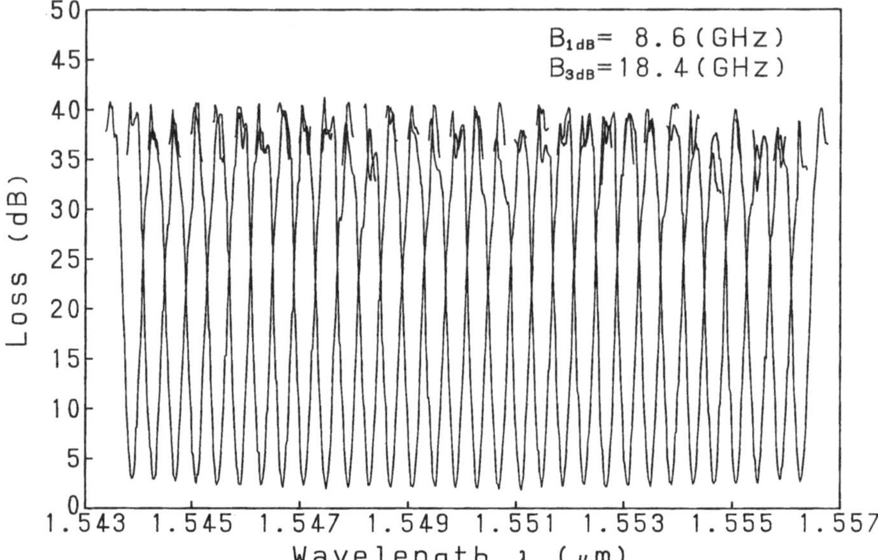

Figure 8 Demultiplexing properties of the central 32 channels in a 64-ch 50-GHz-spacing AWG.

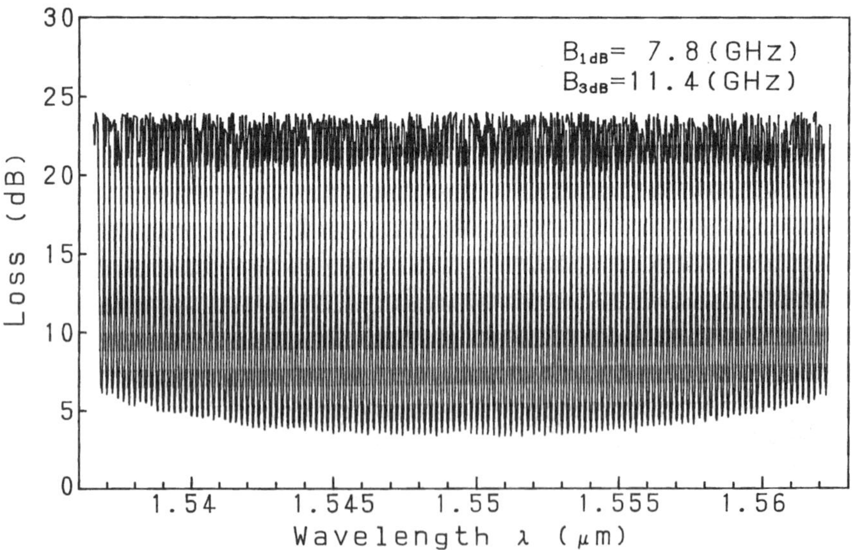

Figure 9 Demultiplexing properties of a 128-ch 25-GHz-spacing AWG.

Table 2 Experimental Performances of Fabricated Multiplexers

Parameter	Experimental and theoretical results				
Number of channels N	8	16	32	64	128
Center wavelength λ_0	1.299 μm	1.5521 μm	1.5498 μm	1.5496 μm	1.5494 μm
(designed value)	(1.30 μm)	(1.552 μm)	(1.550 μm)	(1.550 μm)	(1.550 μm)
Channel spacing $\Delta\lambda$	50 nm	2 nm	0.8 nm = 100 GHz	0.4 nm = 50 GHz	0.2 nm = 25 GHz
Path difference ΔL	2.8 μm	50.3 μm	63 μm	63 μm	63 μm
Slab arc length f	3.58 mm	5.68 mm	11.35 mm	24.2 mm	36.3 mm
Diffraction order m	3	47	59	59	59
Number of arrayed-waveguides	28	60	100	160	388
On-chip loss for λ_0	2.2 dB	2.3 dB	2.1 dB	3.1 dB	3.5 dB
3-dB Bandwidth	27.4 nm	0.74 nm	40 GHz	19 GHz	11 GHz
(BPM simulation)	(28.6 nm)	(0.75 nm)	(37 GHz)	(21 GHz)	(9.5 GHz)
Channel crosstalk	< -26 dB	< -29 dB	< -28 dB	< -27 dB	< -16 dB

(□) = Designed values by beam propagation method.

$$\Delta L = \frac{n_s\, dD\, \lambda_0}{N_c f\, \Delta\lambda} \tag{4}$$

The electric field profile $\Phi(x)$ at the output plane of the AWG (Fig. 5) is the summation of the farfield patterns of ϕ_i's from each array waveguide. Therefore, $\Phi(x)$ is the summation of the spatial Fourier transforms of ϕ_i's. We can exchange the summation and the Fourier transformation in the linear system. Then it is shown that the focused electric field profile $\Phi(x)$ at the output is the Fourier transform of the entire electric field profile $\phi(\xi)$ at slab-array interface.

Figure 6 shows the waveguide layout of a 32-ch 100-GHz-spacing AWG. It consists of 32 input/output waveguides, slab regions with arc length of 11.35 mm, and waveguide array of 100-channel waveguides with the constant path length difference ΔL between neighboring waveguides. The path length difference ΔL is 63 μm; the corresponding grating order at $\lambda_0 = 1.55$ μm is $m = 59$, which gives a free spectral range of 25.6 nm (3.2 THz) and a channel spacing of 0.8 nm (100 GHz). Each arm in waveguide array consists of two straight waveguides of variable length on both sides, and they are smoothly connected to a nonconcentric waveguide bend. The core size and refractive-index difference of the channel waveguides are 7 μm × 7μm and 0.75%, respectively. The bending radius in the array varies from 5 mm to 6.3 mm, and the minimum waveguide separation is 30 μm. The total device size is 30 μm × 26 μm.

We have fabricated various kinds of multiplexers, ranging from 50-nm-spacing 8-channel AWGs to 25-GHz-spacing 128-channel AWGs (Figs. 7–9) [15,16]. Table 2 summarizes experimental and theoretical performances of several kinds of AWG multiplexers. In all of these AWGs except the 128-ch AWG, crosstalks of about −30 dB have been obtained. Numbers in parentheses are the designed values (λ_0's) and the calculated values (3-dB bandwidths) by the beam propagation method (BPM). Experimental results agree well with the theoretical values. The crosstalk in the 128-ch AWG is determined mainly by the subpeaks near the main pass bands. If these subpeaks are eliminated by the improvement of fabrication techniques and/or phase-error compensation techniques [17] (described later), crosstalks can be improved by at least 10 dB down to −25 dB levels.

4.2 Flat Spectral Response Arrayed-Waveguide Gratings

4.2.1 Parabola-Type Arrayed-Waveguide Gratings

Since the dispersion of the focal position x with respect to the wavelength λ is almost constant, the transmission loss of the normal AWG monotonically increases around the center wavelength of each channel. This places tight restrictions on the wavelength tolerance of laser diodes and requires accurate temperature control for both AWGs and laser diodes. Moreover, since optical signals are transmitted through several filters in the WDM ring/bus networks, the cumulative passband width of each channel becomes much narrower than that of the single-stage AWG filter. Therefore, flat and broadened spectral responses are required for AWG multiplexers. Several approaches have been proposed to flatten the pass bands of AWGs [18–20]. We proposed a novel flat-response AWG multiplexer having parabolic waveguide horns in the input waveguides [21]. Figure 10 shows the enlarged view of the interface between (a) the input waveguides and the first slab and (b) the second slab and the output waveguides. The width of the parabolic horn along the propagation direction z is given by [22]

$$W(z) = \sqrt{2\alpha\,\lambda_g z + (2a)^2} \tag{5}$$

where α is a constant less than unity, λ_g is the wavelength in the guide ($\lambda_g = \lambda/n_{\mathrm{eff}}$), and $2a$ is the core width of the channel waveguide. At the proper horn length $z = 800$ μm, a slightly double-peaked flat intensity distribution can be obtained with the horn parameters of $\alpha = 1$ and $2a = 7$ μm. The broadened and double-peaked field is imaged onto the entrance of an output waveguide having normal core width. The overlap integral of the refocused field with the local normal mode of the output waveguide gives the flattened spectral response of the AWG. Figure 11 shows the demultiplexing properties of a 16-ch 100-GHz-spacing AWG having parabolic horns with $W = 40$ μm and $z = 800$ μm. The crosstalks to the neighboring channels are less than -35 dB, and the on-chip loss is about 7.0 dB. The average 1-dB, 3-dB, and 20-dB bandwidths are 86.4 GHz, 100.6 GHz, and 143.3 GHz, respectively.

Figure 10 Enlarged view of the interface between (a) the input waveguides and the first slab and (b) the second slab and the output waveguides in a flat-response 8-ch AWG.

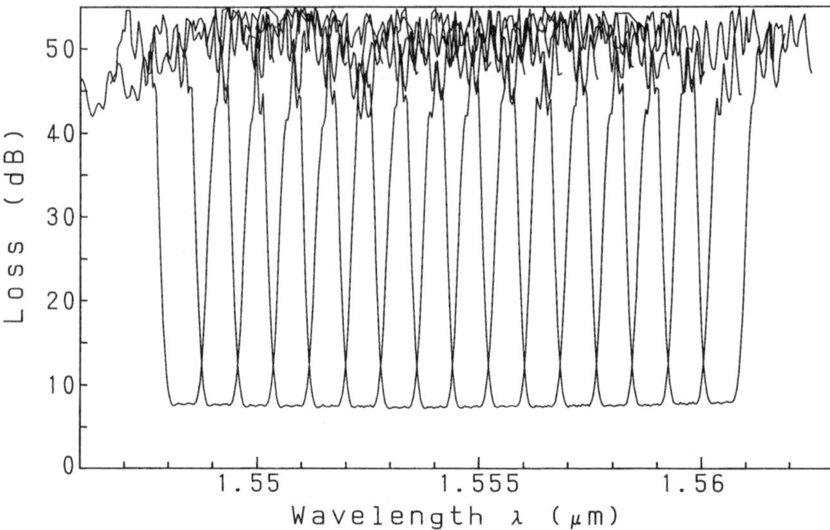

Figure 11 Demultiplexing properties of a 16-ch 100-GHz-spacing AWG having parabolic horns with $W = 40$ μm and $z = 800$ μm.

4.2.2 Sinc-Type Arrayed-Waveguide Gratings

It has been confirmed that in order to obtain a flat spectral response, it is necessary to produce a rectangular electric field profile at the focal plane (interface between the second slab and the output waveguides). Since the electric field profile in the focal plane is the Fourier transform of the field in the array output aperture (interface between the array waveguide and the second slab), such a rectangular field profile could be generated when the electric field at the array output aperture obeys a sin $(\xi)/\xi$ distribution, where ξ is measured along the array output aperture [18]. Figure 12 shows the electric field amplitude and relative phase delays (excess phase value added to $i \times \Delta L$, where i denotes the ith array waveguide) in the sinc-type flat response AWG measured by low-coherence Fourier transform spectroscopy [23]. Sinc-shaped electric field amplitude distribution was realized by introducing an additional loss to each array waveguide. The excess path length differences of π for the negative sinc values were realized by the additional path length $\delta\ell = \lambda/2n_c$ to the corresponding array waveguides. The crosstalk and flat passband characteristics are almost the same as those of parabola-type AWGs.

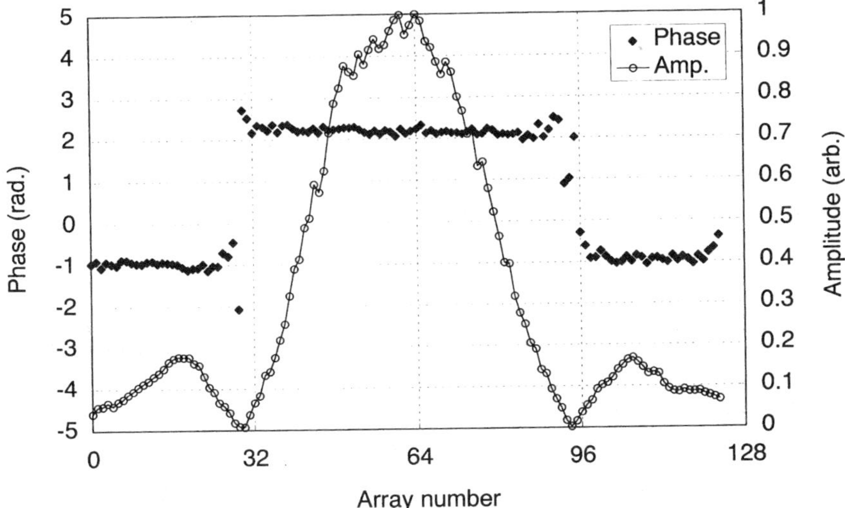

Figure 12 Electric field amplitude and relative phase delays in the sinc-type flat-response AWG.

4.3 Uniform-Loss and Cyclic-Frequency (ULCF) Arrayed-Waveguide Gratings

It is well recognized that $N \times N$ signal interconnection in AWGs can be achieved when the free spectral range (FSR) of the AWG is N times the channel spacing. Here FSR is given by

$$\text{FSR} = \frac{n_c \nu_0}{N_c m} \tag{6}$$

where ν_0 is the center frequency of the WDM signals. Light beams with three different diffraction orders of $m - 1$, m, and $m + 1$ are utilized to achieve $N \times N$ interconnections [24]. The cyclic property provides an important additional functionality, as compared to simple multiplexers or demultiplexers, and plays a key role in more complex devices, such as add/drop multiplexers and wavelength switches. Figure 13 illustrates its functionality. Wavelength routers have N input and N output ports. Each of the N input ports can carry N different wavelengths. The N wavelengths coupled into, for example, input port 3 are distributed among output ports 1–N. The N wavelengths carried by other input ports are distributed in the same way, but cyclically rotated. In this way each output port receives N

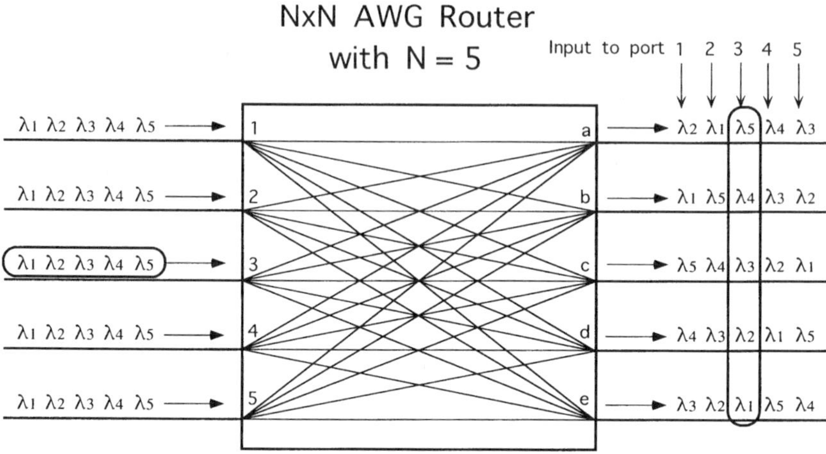

Figure 13 Schematic diagram illustrating the operation of a wavelength router.

different wavelengths, one from each input port. To realize such an interconnectivity scheme in a strictly nonblocking way using a single wavelength, a huge number of switches would be required. Using a cyclic property of AWGs, this functionality can be achieved with only one AWG.

However, such interconnectivity cannot always be realized with conventional AWGs. The typical diffraction order of the 32-channel AWG with 100-GHz channel spacing is $m = 60$, as shown in Table 2. Since FSR is inversely proportional to m, substantial pass frequency mismatch is brought by the difference between three FSRs. Also, insertion losses of the AWG for peripheral input and output ports are 2–3 dB higher than those for central ports, as shown in Fig. 7. These noncyclic frequency characteristics and loss nonuniformity in conventional AWGs are the main obstacles that prevent the development of practical $N \times N$ routing networks. We have fabricated 32×32 arrayed-waveguide gratings having uniform-loss and cyclic-frequency (ULCF) characteristics. Figure 14 shows the schematic configuration of the uniform-loss and cyclic-frequency arrayed-waveguide grating [25]. It consists of an 80-channel AWG multiplexer with 100-GHz spacing and 32 optical combiners that are connected to 72 output waveguides of the multiplexer. The arc length of slab is $f = 24.55$ mm, and the number of array waveguides is 300, having the constant path length difference $\Delta L = 24.6$ μm between neighboring waveguides. The diffraction order is $m = 23$, which gives a free spectral range of FSR = 8 THz. In the input side, 32 waveguides ranging from #25 to #56 are used for the input waveguides so as to secure the uniform loss characteristics. In the output side, two waveguides ($i +$

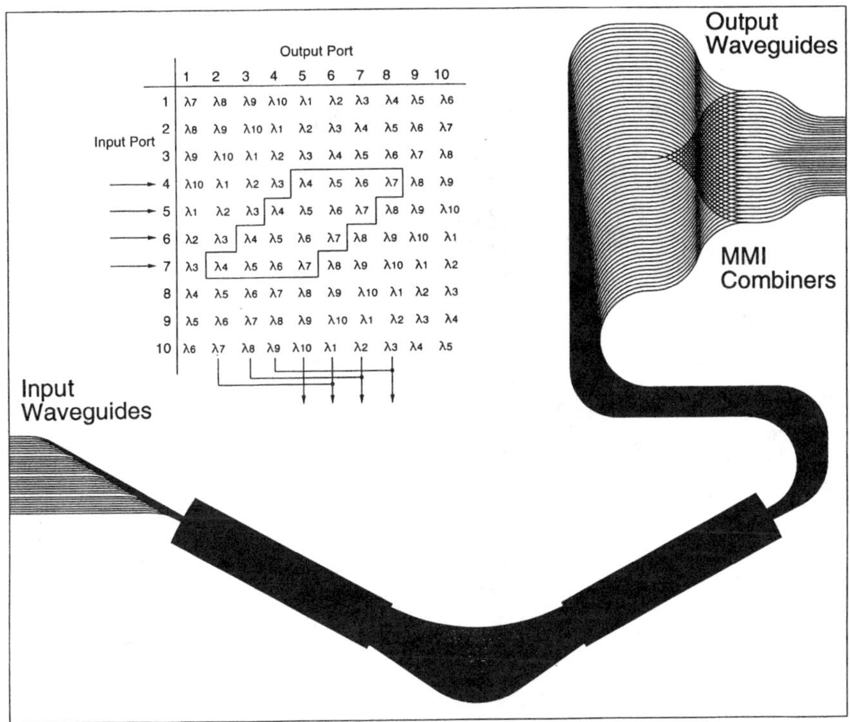

Figure 14 Schematic configuration of a uniform-loss and cyclic-frequency AWG.

8)th and $(i + 40)$th waveguide for $i = 1$–32 are combined through a waveguide intersection and multimode interference (MMI) coupler to make one output port. Since the peripheral output ports are not used, uniform loss characteristics are obtained. The inset of Fig. 14 shows the principle of how the ULCF arrayed-waveguide grating is constructed. In this example a 4 × 4 ULCF AWG is fabricated from a 10 × 10 original AWG. Multiplexed signals with wavelength λ_4, λ_5, λ_6, λ_7 that are coupled to input port #4 are demultiplexed into output waveguides from #5 to #8. When signals λ_4, λ_5, λ_6, λ_7 are coupled to input port #5, signal component λ_4 is folded back into output waveguide #8 through the optical combiner. This is the operational principle of the ULCF AWG. Though we could improve the loss characteristics, we should pay at least a 3-dB loss budget in single-mode optical combiners to achieve uniform-loss and cyclic-frequency characteristics. We first show the light-splitting properties of the conventional 32-channel AWG with 100-GHz spacing. Figure 15(a) shows measured insertion losses for entire input/output combinations in the conventional AWG. The peak-

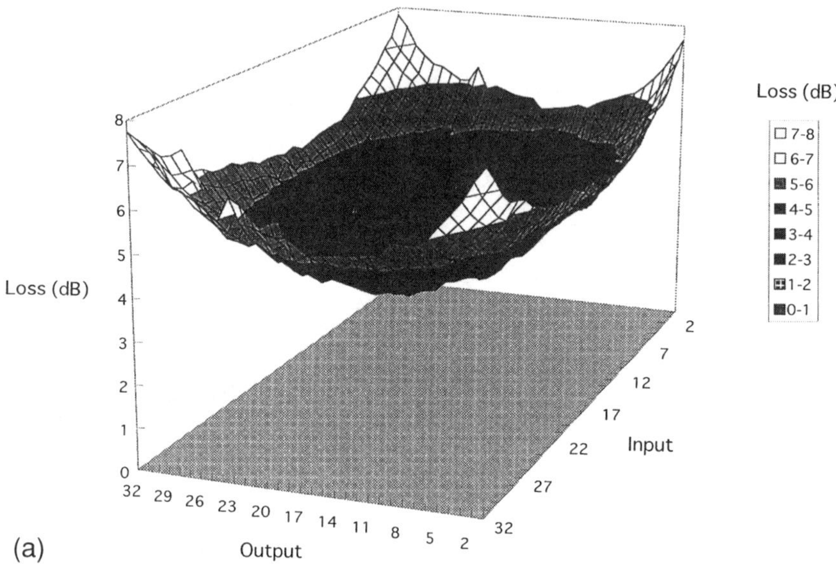

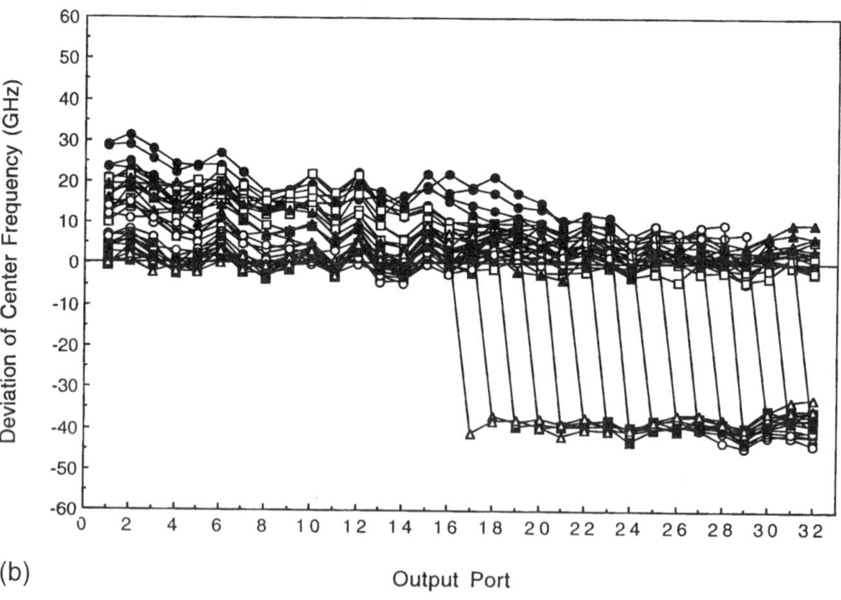

Figure 15 (a) Measured insertion losses for entire 32 × 32 input/output combinations in the conventional AWG. (b) Measured channel center frequency deviations for entire 32 × 32 input/output combinations in the conventional AWG.

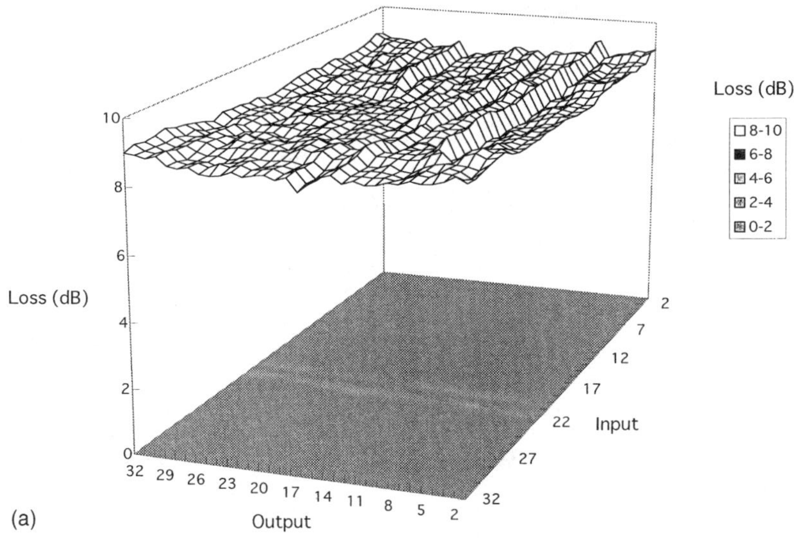

(a)

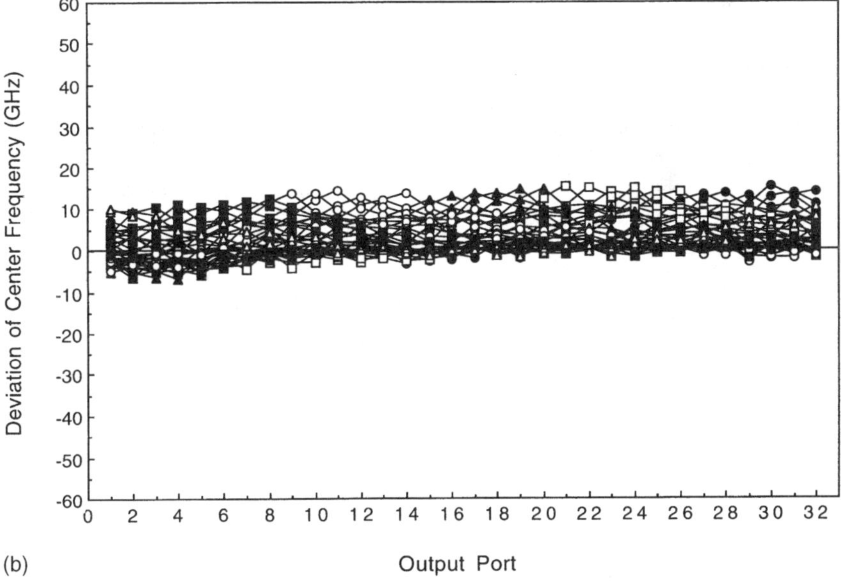

(b)

Figure 16 (a) Measured insertion losses for entire 32 × 32 input/output combinations in the uniform-loss and cyclic-frequency AWG. (b) Measured channel center frequency deviations for entire 32 × 32 input/output combinations in the uniform-loss and cyclic-frequency AWG.

to-peak loss variation is 4.7 dB (minimum loss α_{min} = 3.1 dB, maximum loss α_{max} = 7.8 dB), and the standard deviation is σ_{loss} = 1.0 dB. Figure 15(b) shows deviations of the channel pass frequency from the prescribed grid frequencies. The peak-to-peak variation is 75.6 GHz ($\delta\nu_{min} = -44.5$ GHz, $\delta\nu_{max} = 31.1$ GHz), and the standard deviation is σ_{freq} = 16.7 GHz. The major reason for pass frequency mismatch is the difference of FSR in neighboring diffraction orders. Figures 16(a) and (b) show insertion losses and channel pass frequency deviations for 32-input/32-output combinations in the uniform-loss and cyclic-frequency AWG. The peak-to-peak loss variation is reduced to 1.2 dB (α_{min} = 8.1 dB, α_{max} = 9.3 dB) and the standard deviation is σ_{loss} = 0.2 dB. The peak-to-peak frequency variation is also reduced to 22.3 GHz ($\delta\nu_{min} = -6.8$ GHz, $\delta\nu_{max}$ = 15.5 GHz), and the standard deviation is σ_{freq} = 4.4 GHz. The crosstalk of the AWG is about −26 dB.

4.4 Athermal (Temperature-Insensitive) Arrayed-Waveguide Gratings

The temperature sensitivity of the pass wavelength (frequency) in the silica-based AWG is about $d\lambda/dT = 1.2 \times 10^{-2}$ (nm/deg) ($d\nu/dT = -1.5$ (GHz/deg)), which is determined mainly by the temperature dependence of silica glass itself ($dn_c/dT = 1.1 \times 10^{-5}$ (1/deg)). The AWG multiplexer should be temperature controlled with a heater or a Peltier cooler to stabilize the channel wavelengths. This requires the constant power consumption of a few watts and a lot of equipment for the temperature control. We have succeeded in fabricating an athermal (temperature-insensitive) AWG operating in the 0–85°C temperature range [26]. Figure 17 shows a schematic configuration of an athermal AWG. The temperature-dependent optical path difference in silica waveguides is compensated with a triangular groove filled with silicone adhesive that has a negative thermal coeffi-

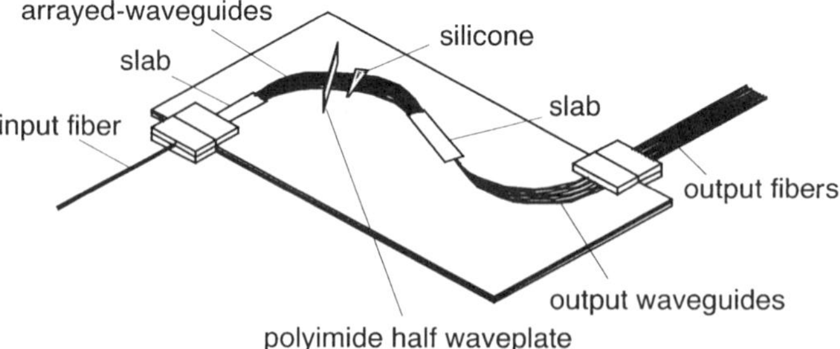

Figure 17 Schematic configuration of an athermal AWG.

cient. Since the pass wavelength is given by $\lambda_0 = n_c \, \Delta L/m$, the optical path length difference $n_c/\Delta L$ should be made insensitive to temperature. Therefore the groove is designed to satisfy the following conditions:

$$n_c \, \Delta L = n_c \, \Delta \ell + \hat{n}_c \, \Delta \hat{\ell} \tag{7}$$

$$\frac{d(n_c \, \Delta L)}{dT} = \frac{d\hat{n}_c}{dT} \Delta \ell + \frac{d\hat{n}_c}{dT} \Delta \hat{\ell} = 0 \tag{8}$$

where $\hat{n}_c$ is the refractive index of silicone and $\Delta \ell$ and $\Delta \hat{\ell}$ are the path length differences of silica waveguides and silicone region, respectively. Equation (7) is a condition to satisfy the AWG specifications, and Eq. (8) is the athermal condition. The temperature sensitivity of silicone is $d\hat{n}_c/dT = -37 \times 10^{-5}$ (1/deg). Therefore the path length difference of silicone is $\Delta \hat{\ell} \cong \Delta \ell/37$. Figure 18 shows the temperature dependencies of pass wavelengths in conventional and athermal AWGs. The temperature-dependent wavelength change has been reduced from 0.95 nm to 0.05 nm in the 0–85°C range. The excess loss caused by the groove is about 2 dB, which is mainly a diffraction loss in the groove. Figure 19 shows the loss change at 1552.52 nm during heat cycles from −40 to 85°C. The loss change in smaller than 0.2 dB. Furthermore, the channel wavelength change was less than 0.02 nm in a long-term test over 5000 hours at 75°C and 90% relative humidity.

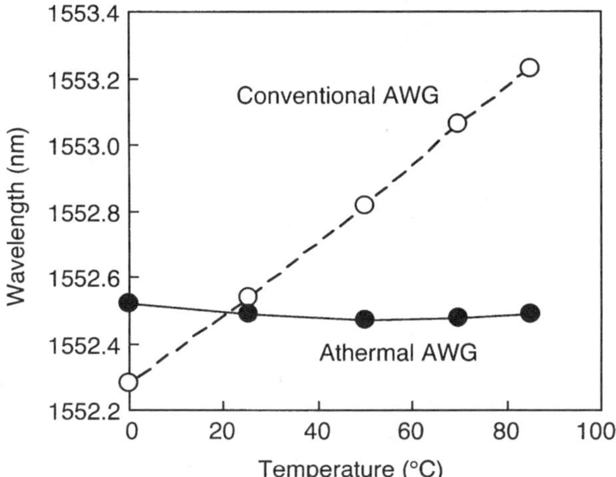

Figure 18 Temperature dependencies of pass wavelengths.

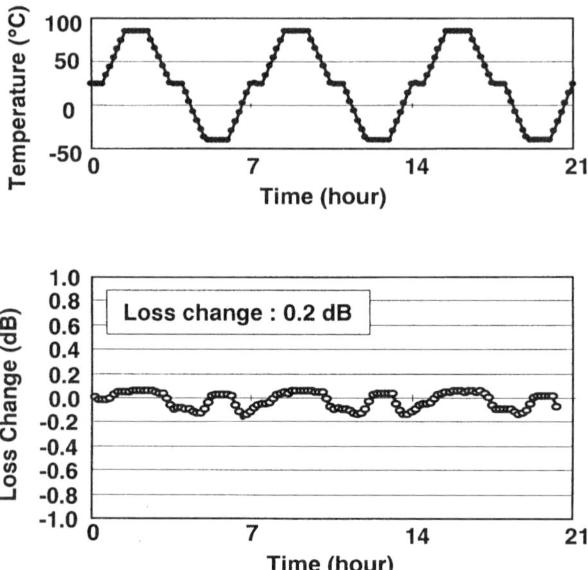

Figure 19 Loss variation under heat cycle test from −40 to 85° C.

4.5 Phase Error Compensation of Arrayed-Waveguide Gratings

Crosstalk improvement is the major concern for the AWG multiplexers, especially for $N \times N$ routing applications. Crosstalks to other channels are caused by the side lobes of the focused beam in the second slab region. These side lobes are attributed mainly to the phase fluctuations of the total electric field profile at the output side of the array–slab interface, since the focused beam profile is the Fourier transform of the electric field in the array waveguides (Fig. 5). The phase errors are caused by the nonuniformity of effective-index and/or core geometry in the arrayed-waveguide region. In order to improve the crosstalk characteristics of AWGs, we carried out the phase error compensation experiment using a 16-ch 10-GHz-spacing ultranarrow AWG filter [27]. Figure 20 shows the configuration of the phase-compensated AWG using an a-Si film for phase trimming. The path length difference of AWG is $\Delta L = 1271$ μm in 64 array waveguides. Since the array waveguide region occupies a large area, accumulated phase errors become quite large. Therefore the crosstalk of the AWG without phase error compensation was about −8 dB. An a-Si stress-applying film was deposited on top of the overcladding of each array waveguide, and the amount of stress-optic effect was adjusted by the Ar ion laser trimming (evaporation) of the a-Si film length.

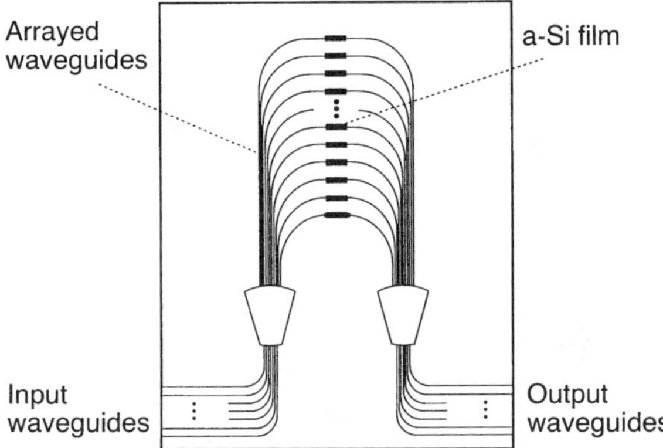

Figure 20 Configuration of a phase-compensated AWG with a-Si film.

Figure 21 shows the phase error distributions before and after phase compensation. The peak-to-peak fluctuation is reduced to about $\pm 2°$ over the 60-mm average path length. The effective-index fluctuations are reduced to within 1.2×10^{-7}. Figure 22 shows the measured demultiplexing properties of the phase-compensated 16-ch 10-GHz-spacing AWG. Crosstalks to the neighboring and all other channels are less than -31 dB.

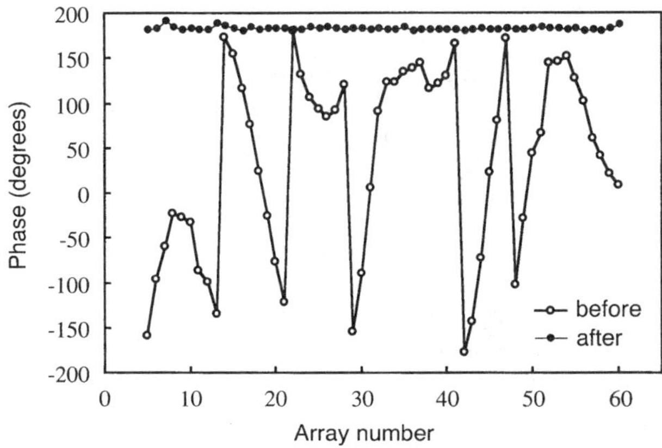

Figure 21 Phase error distributions for the TE mode before and after phase compensation.

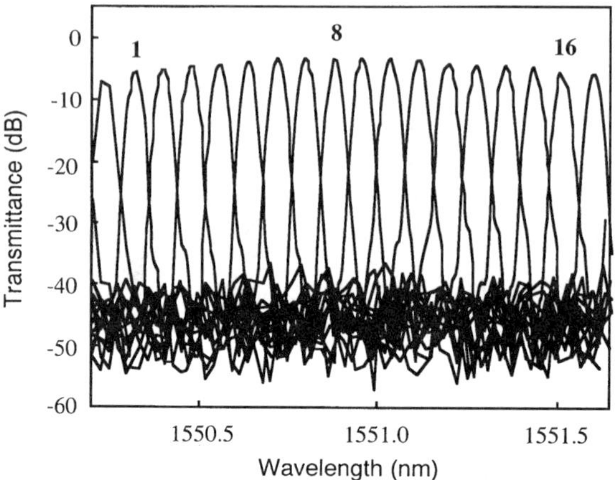

Figure 22 Transmittance of a phase-compensated 16-ch 10-GHz-spacing AWG.

5 OPTICAL ADD/DROP MULTIPLEXER

An optical add/drop multiplexer (ADM) is a device that gives simultaneous access to all wavelength channels in a WDM communication systems. We proposed a novel optical ADM and demonstrated the basic functions of individually routing 16 different wavelength channels with 100-GHz channel spacing [28]. The waveguide configuration of a 16-ch optical ADM is shown in Fig. 23. It consists of four arrayed-waveguide gratings and 16 double-gate thermooptic (TO) switches. Four AWGs are allocated with their slab regions crossing each other. These AWGs have the same grating parameters: channel spacing of 100 GHz and free spectral range of 3300 GHz (26.4 nm) at the 1.55-μm region. Equally spaced WDM signals, $\lambda_1, \lambda_2, \ldots, \lambda_{16}$, which are coupled to the main input port (add port) in Fig. 23, are first demultiplexed by the AWG_1 (AWG_2), and then 16 signals are introduced into the left-hand-side arms (right-hand-side arms) of double-gate TO switches. The crossangles of the intersecting waveguides are designed to be larger than 30° so as to make the crosstalk and insertion loss negligible. Any optical signal coupled into the double-gate TO switch passes through the cross port of one of the four Mach–Zehnder interferometers (MZIs) before reaching the output port. In the single-stage MZI, the light-extinction characteristic of the crossport is much better than that of the throughport even when the coupling ratio of directional coupler is deviated from 3 dB. Therefore, the crosstalk of the double-gate switch becomes substantially improved over that of the conventional single-stage TO switch. We designate here the "off" state of the double-gate switch as the switching condition where the signal from the left input port (right input port)

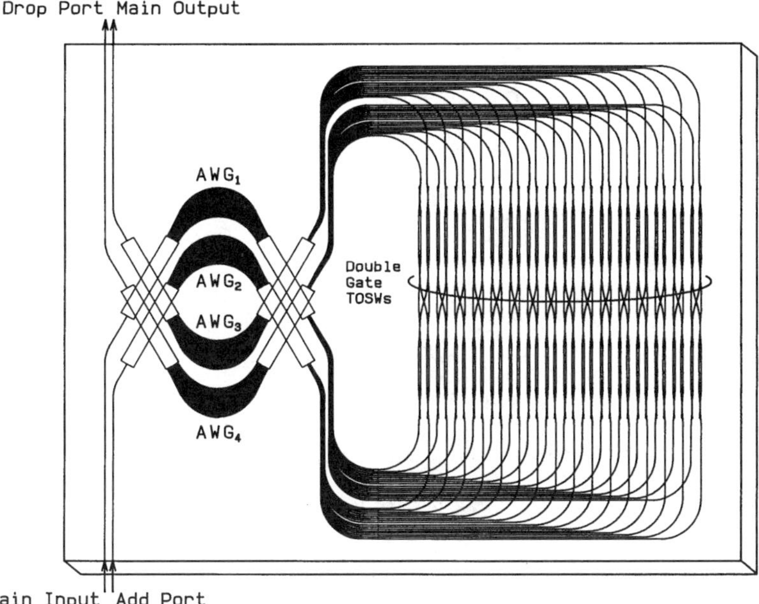

Figure 23 Waveguide configuration of a 16-ch optical ADM with double-gate TO switches.

goes to the right output port (left output port). The "on" state is then designated as the condition where the signal from the left input port (right input port) goes to the left output port (right output port). When the double-gate switch is "off," the demultiplexed light by AWG_1 (AWG_2) goes to the crossarm and multiplexed again by the AWG_3 (AWG_4). On the other hand, if the double-gate switch in is "on" state, the demultiplexed light by AWG_1 (AWG_2) goes to the througharm and is multiplexed by the AWG_4 (AWG_3). Therefore, any specific wavelength signal can be extracted from the main output port and led to the drop port by changing the corresponding switch condition. A signal at the same wavelength as that of the dropped component can be added to the main output port when it is coupled into add port in Fig. 23. When TO switches SW_2, SW_4, SW_6, SW_7, SW_9, SW_{12}, SW_{13}, and SW_{15}, for example, are turned to "on," the selected signals λ_2, λ_4, λ_6, λ_7, λ_9, λ_{12}, λ_{13}, and λ_{15} are extracted from main output port (solid line) and led to the drop port (dotted line), as shown in Fig. 24. The on–off crosstalk is smaller than -30 dB, with on-chip losses of 8–10 dB. Since optical signals pass through both AWG_3 and AWG_4 the crosstalk level here is determined by the crosstalk in the arrayed waveguides. The present optical ADM can trans-

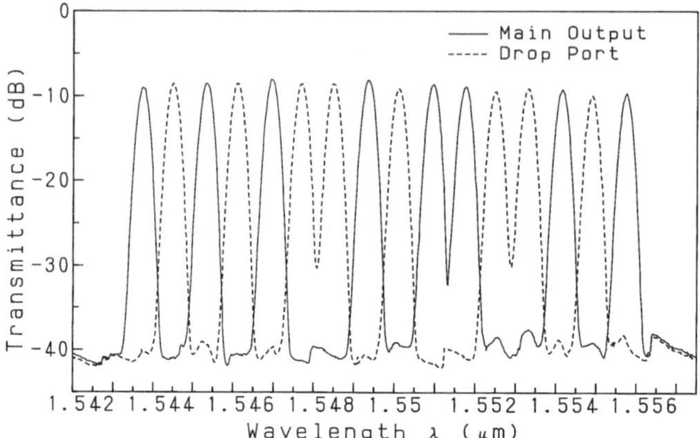

Figure 24 Transmission spectra from main input port to main output port and drop port when TO switches SW_2, SW_4, SW_6, SW_7, SW_9, SW_{12}, SW_{13}, and SW_{15} are "on."

port all input signals to the succeeding stages without inherent power losses. Therefore, these ADMs are very attractive for all optical WDM routing systems and allow the network to be transparent to signal formats and bit rates.

6 $N \times N$ MATRIX SWITCHES

Space-division optical switches are one of the indispensable optical devices for the reconfigurable interconnects in cross-connect systems, fiber-optic subscriber line connectors, and photonic intermodule connectors [3,29,30]. Figure 25 shows the logical arrangement of a 16 × 16 strictly nonblocking matrix switch with a path-independent insertion loss (PI-loss) configuration [31]. This arrangement is

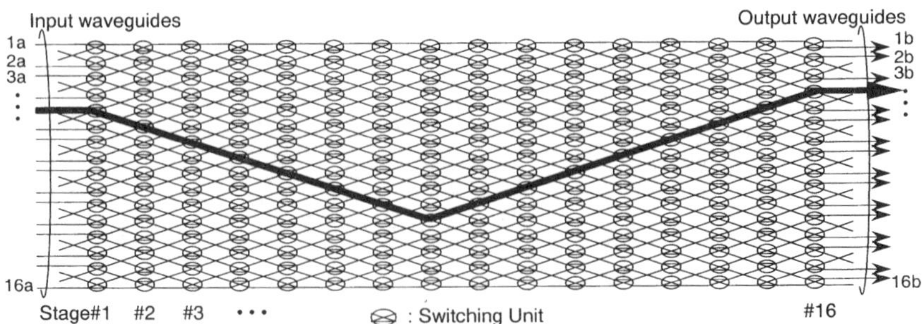

Figure 25 Logical arrangement of a 16 × 16 matrix switch.

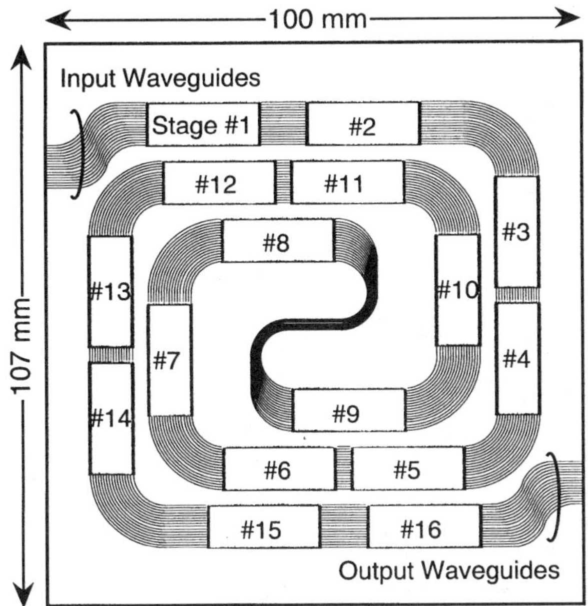

Figure 26 Circuit layout of the 16 × 16 matrix switch.

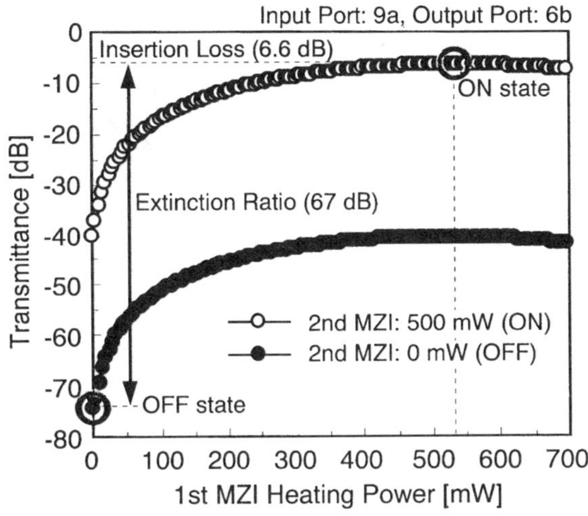

Figure 27 Switching characteristics for the typical input–output combination.

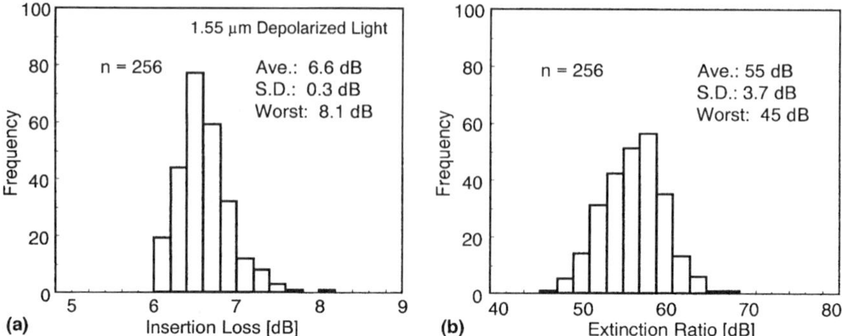

Figure 28 Measured extinction ratios and insertion losses for all 256 input and output connection patterns.

quite advantageous for reducing total circuit length, since it requires only N switching stages to construct an $N \times N$ switch. The switching unit consists of a double-gate switch, which was described in Sec. 5. The circuit layout of the 16×16 matrix switch is shown in Fig. 26. Sixteen switching stages are allocated along the serpentine waveguides. There are 16 switching units in one switching stage. The total circuit length is 66 cm. The fabricated switching unit requires bias power even when it is in the ''off'' state due to the waveguide phase (optical path length) errors. The phase error in each switching unit is permanently eliminated by using a phase-trimming technique, in which high-temperature local heat treatment produces permanent refractive-index change for either arm of the Mach–Zehnder interferometers. Figure 27 shows switching characteristics for one input–output combination when heater power to the second MZI are fixed at 0 mW and 500 mW, respectively. The extinction ratio when first and second MZIs are activated simultaneously becomes double that of the single MZI. Therefore, an extremely high extinction ratio of 67 dB has been successfully achieved with 6.6-dB insertion loss. Figure 28 shows the measured extinction ratios and insertion losses for all 256 input and output connection patterns. The extinction ratio ranges from 45 dB to 67 dB, with an average of 55 dB. The insertion loss ranges from 6.0 dB to 8.0 dB, with an average of 6.6 dB. The total electric power for operating the 16×16 matrix switch is about 17 W (1.06 W for each switching unit).

7 LATTICE-FORM PROGRAMMABLE FILTERS

The transmission distance in optical fiber communications has been greatly increased by the development of erbium-doped fiber amplifiers. In consequence, the main factor limiting the maximum repeater span is now the fiber chromatic dispersion. Several techniques have been reported to compensate for the delay distortion in the optical stage [32–37]. An advantage of the PLC optical delay equalizer [36,37] is that variable group-delay characteristics can be achieved by the phase control of silica waveguides. The basic configuration of the PLC delay equalizer is shown in Fig. 29. It consists of N (=8) asymmetrical Mach–Zehnder interferometers and $N + 1$ (=9) tunable couplers, which are cascaded alternately in series. The crossport transfer function of the optical circuit is expressed by a Fourier series as

$$H(z) = \sum_{k=0}^{N} a_k z^{-k+N/2} \tag{9}$$

where z denotes $\exp(j2\pi\nu\,\Delta t)$ (ν is the optical frequency; $\Delta t = n_c\,\Delta L/c$ is the unit delay time difference in asymmetrical the MZI) and a_k is the complex expansion coefficient. The circuit design procedures are as follows. First the equalizer transfer function to be realized is expressed by the analytical function. Then coefficient a_k's are determined by expanding the analytical function into a Fourier series. Finally, the coupling ratio (ϕ_i) and phase shift value (θ_i) in each stage of the lattice filter are determined by the filter synthesis method [6].

In ultrahigh-speed optical fiber transmission systems (>100 Gbit/s), the effect of the higher-order dispersion (third-order dispersion or dispersion slope) in the dispersion-shifted fiber (DSF) is one of the major factors limiting the transmission distance [38]. Programmable dispersion equalizers can be designed so as to com-

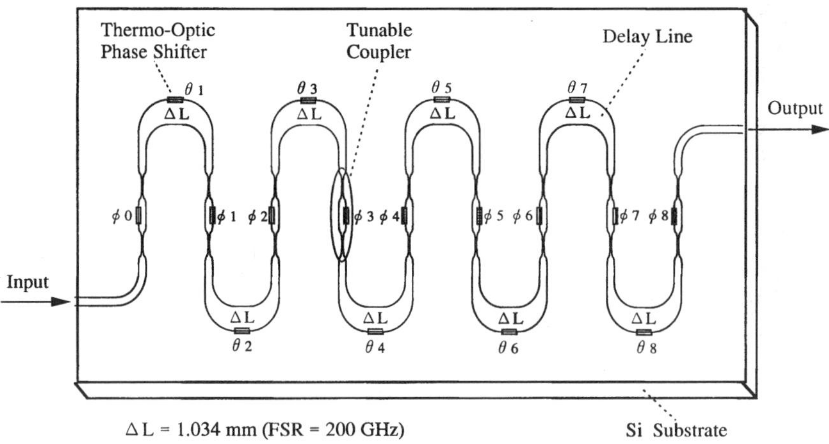

Figure 29 Basic configuration of the PLC equalizer.

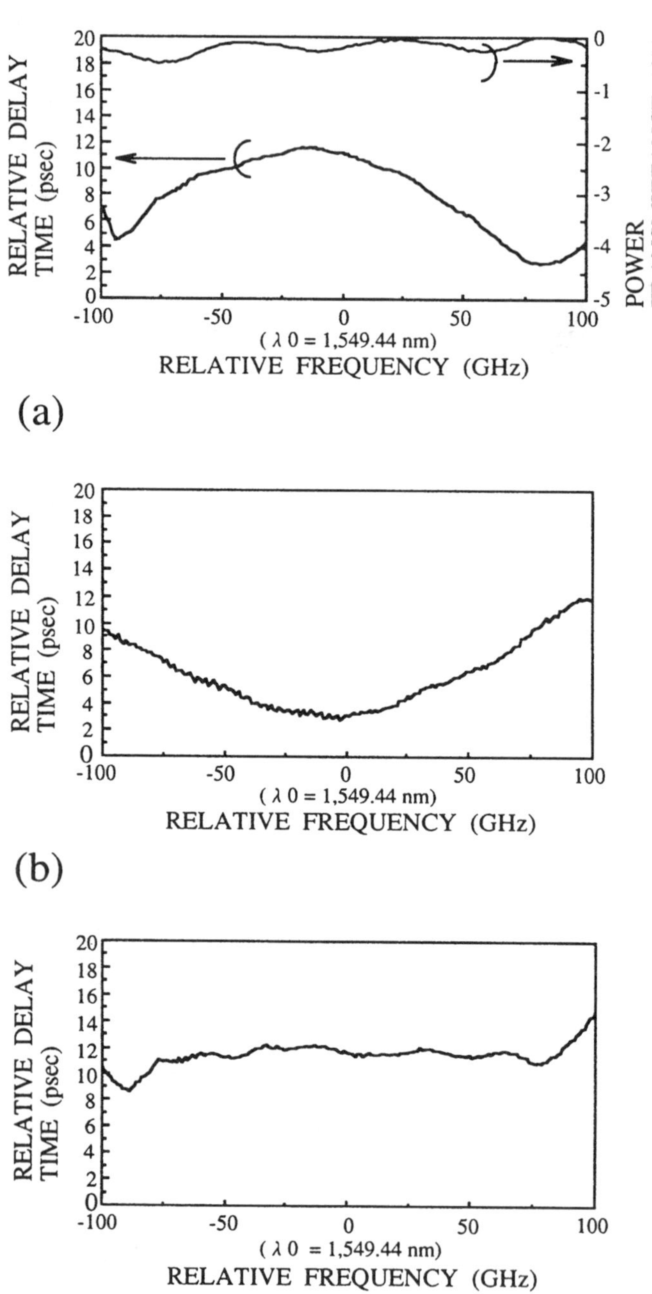

Figure 30 Relative delay times of (a) PLC higher-order dispersion equalizer, (b) 300-km DSF, and (c) 300-km DSF + equalizer.

pensate for the higher-order dispersion of DSFs. Figure 30(a) shows the measured power transmittance and relative delay time of the PLC higher-order dispersion equalizer [39]. The dispersion slope of the equalizer is calculated to be −15.8 ps/nm^2. Figure 30(b) shows the relative delay of the 300-km DSF. The dispersion slope of the DSF is 0.05–0.06 ps/nm^2/km. Therefore, the equalizer can compensate the higher-order dispersion of ~300-km of the DSF. Figure 30(c) shows the relative delay time of the 300-km DSF cascaded with the equalizer. The positive dispersion slope of the DSF is almost completely compensated by the PLC equalizer. A 200-Gbit/s time-division-multiplexed transmission experiment using a dispersion slope equalizer has been carried out over 100-km fiber length [40]. The pulse distortion caused by the dispersion slope was almost completely recovered, and the power penalty was improved by more than 4 dB.

8 HYBRID INTEGRATION TECHNOLOGY USING PLC PLATFORMS

It is widely recognized that optical hybrid integration is potentially a key technology for fabricating advanced integrated optical devices [41]. A silica-based waveguide on a Si substrate is a promising candidate for the hybrid integration platform, since high-performance PLCs have already been fabricated using silica-based waveguides and Si has highly stable mechanical and thermal properties that make it suitable as an optical bench. Planar lightwave circuit platform technology has been utilized in the fabrication of a hybrid integrated external cavity laser [42]. Figure 31 shows the configuration of a multiwavelength external cavity laser with a UV-written grating [43]. Bragg gratings with a 2-nm wavelength interval are written into each waveguide by ArF excimar laser irradiation through

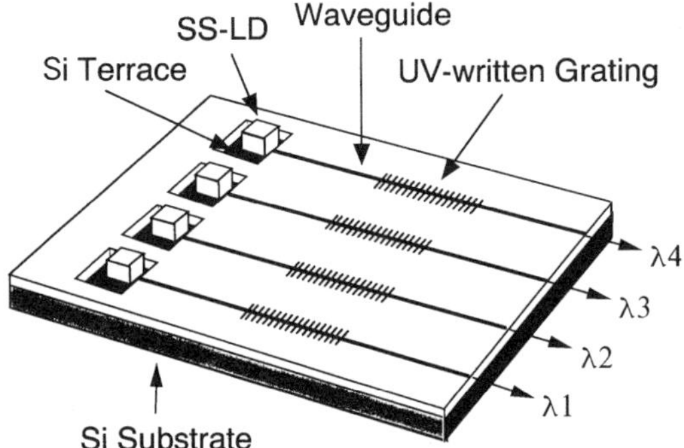

Figure 31 Configuration of a hybrid integrated multiwavelength external cavity laser.

phase masks. Figure 32 shows the measured output spectra. Each laser operates in a single longitudinal mode, with a side-mode suppression of 40 dB. The temperature sensitivity of the oscillation frequency is −1.7 GHz/deg, which is one-eight that of the distributed feedback (DFB) lasers. A four-channel simultaneous-modulation experiment has been successfully carried out at 2.5 Gbit/s [44]. Temperature-stable multiwavelength sources will play an important role in WDM transmission and access network systems.

Semiconductor optical amplifier (SOA) gate switches having spot-size converters on both facets have been successfully hybrid integrated on PLC platforms to construct high-speed wavelength channel selectors and 4 × 4 optical matrix switches [45–46]. Figure 33 shows the configuration of an 8-channel optical wavelength selector module. It consists of two AWG chips with 75-GHz channel spacing and a hybrid integrated SOA gate array chip. It selects and picks up any wavelength channel from multiplexed signals by activating the corresponding SOA gate switch. Three PLC chips are attached directly to each other using UV-curable adhesive. The length of the SOA gate switch is 1200 μm, and their separation is 400 μm. The coupling loss between the SOA and the PLC waveguide ranges from 3.9 dB to 4.9 dB. Figure 34 shows the optical transmission spectra of the wavelength channel selector when only one SOA gate switch is activated successively. The semiconductor optical amplifier injection current is 50 mA for all SOAs. The peak transmittances have 1–3-dB gains; they are, 16–19-dB total chip losses and fiber coupling losses are compensated by SOA gains. The crosstalk is less than −50 dB, and the polarization-dependent loss is smaller than 1.4 dB. However, the crosstalk becomes about −30 dB when two or more SOA gate switches are activated simultaneously. The crosstalk in multigate operation is

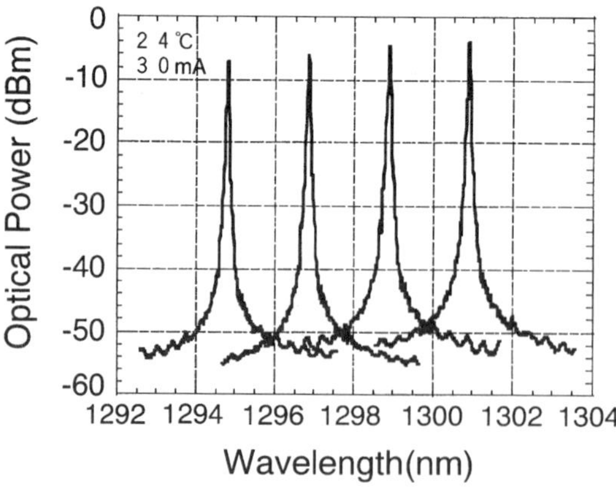

Figure 32 Oscillation spectra of a four-wavelength laser.

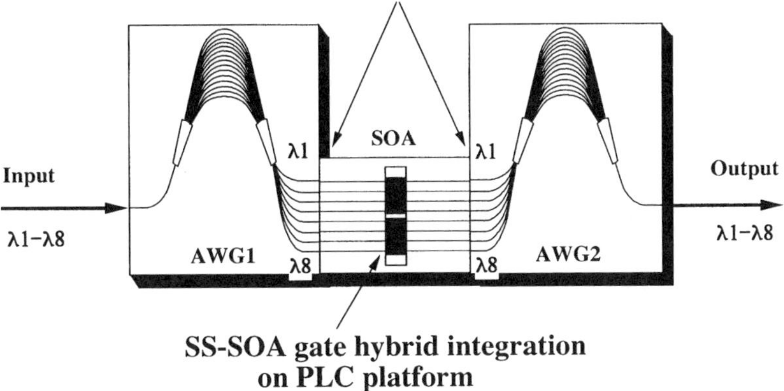

Figure 33 Configuration of an 8-channel optical wavelength selector module.

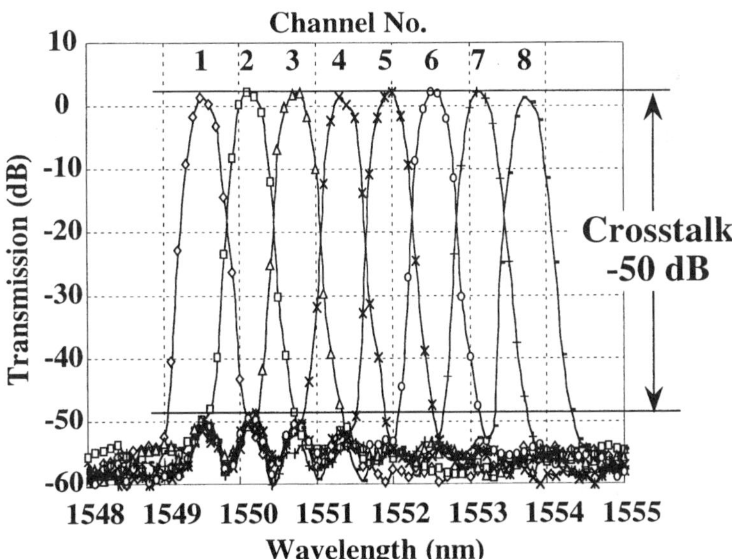

Figure 34 Optical transmission spectra of the wavelength channel selector when only one SOA gate switch is activated successively.

determined by the crosstalk of the AWGs. In the high-speed switching experiments, the rise-and-fall time is confirmed to be less than 1 nsec.

9 CONCLUSIONS

Recent progress in planar lightwave circuits has been reviewed. Although silica-based waveguides are simple circuit elements, various functional devices are fabricated by utilizing spatial multibeam or temporal multistage interference effects such as arrayed-waveguide grating multiplexers and lattice-form programmable filters. Hybrid integration technologies will enable us to realize much more functional and high-speed devices. The PLC technologies supported by continuous improvements in waveguide fabrication, circuit design, and device packaging will further lead to a higher level of integration of optics and electronics, aiming at the next generation of telecommunication systems.

ACKNOWLEDGMENTS

The author would like to thank all members in the Okamoto Research Laboratory and Photonic Component Laboratory for their contribution to this work.

REFERENCES

1. M. Kawachi. Silica waveguide on silicon and their application to integrated-optic components. Opt. Quantum Electron. 22:391–416, 1990.
2. K. Okamoto. Planar lightwave circuits (PLCs). In: Photonic Networks. New York: Springer-Verlag, 1997, pp. 118–132.
3. A. Himeno et al. System applications of large-scale optical switch matrices using silica-based planar lightwave circuits. In: Photonic Networks. New York: Springer-Verlag, 1997, pp. 172–182.
4. Y. Yamada et al. Application of planar lightwave circuit platform to hybrid integrated optical WDM transmitter/receiver module. Electron. Lett. 31:1366–1367, 1995.
5. Y. Akahori et al. A hybrid high-speed silica-based planar lightwave circuit platform integrating a laser diode and a driver IC. Proc. IOOC/ECOC '97, Edinburgh UK, 2:359–362, 1997.
6. K. Jinguji et al. Synthesis of coherent two-port lattice-form optical delay-line circuit. IEEE J. Lightwave Tech. 13:73–82, 1995.
7. K. Takiguchi et al. Variable group-delay dispersion equalizer based on a lattice-form programmable optical filter. Electron. Lett. 31:1240–1241, 1995.
8. S. Suzuki et al. Large-scale and high-density planar lightwave circuits with high-Δ GeO_2-doped silica waveguides. Electron. Lett. 28:1863–1864, 1992.

9. Y. Hibino et al. Propagation loss characteristics of long silica-based optical waveguides on 5-inch Si wafers. Electron. Lett. 29:1847–1848, 1993.
10. Y. Hida et al. 10-m long silica-based waveguide with a loss of 1.7 dB/m. Dana Point, CA: IPR '95, 1995.
11. C. Dragone et al. Efficient multichannel integrated optics star coupler on silicon. IEEE Photonics Tech. Lett. 1:241–243, 1989.
12. K. Okamoto et al. Fabrication of large-scale integrated-optic $N \times N$ star couplers. IEEE Photonics Tech. Lett. 4:1032–1035, 1992.
13. M. K. Smit. New focusing and dispersive planar component based on an optical phased array. Electron. Lett. 24:385–386, 1988.
14. H. Takahashi et al. Arrayed-waveguide grating for wavelength division multi/demultiplexer with nanometer resolution. Electron. Lett. 26:87–88, 1990.
15. K. Okamoto et al. Fabrication of 64×64 arrayed-waveguide grating multiplexer on silicon. Electron. Lett. 31:184–185, 1995.
16. K. Okamoto et al. Fabrication of 128-channel arrayed-waveguide grating multiplexer with a 25-GHz channel spacing. Electron. Lett. 32:1474–1476, 1996.
17. H. Yamada et al. Low-crosstalk arrated-waveguide grating multi/demultiplexer with phase compensating plate. Electron. Lett. 33:1698–1699, 1997.
18. K. Okamoto et al. Arrayed-waveguide grating multiplexer with flat spectral response. Opt. Lett. 20:43–45, 1995.
19. M. R. Amersfoort et al. Passband broadening of integrated arrayed waveguide filters using multimode interference couplers. Electron. Lett. 32:449–451, 1996.
20. D. Trouchet et al. Passband flattening of PHASAR WDM using input and output star couplers designed with two focal points. Proc. OFC '97 ThM7, Dallas, TX, 1997.
21. K. Okamoto et al. Flat spectral response arrayed-waveguide grating multiplexer with parabolic waveguide horns. Electron. Lett. 32:1661–1662, 1996.
22. W. K. Burns et al. Optical waveguide parabolic coupling horns. Appl. Phys. Lett. 30:28–30, 1977.
23. K. Takada et al. Measurement of phase error distributions in silica-based arrayed-waveguide grating multiplexers by using Fourier transform spectroscopy. Electron. Lett. 30:1671–1672, 1994.
24. C. Dragone et al. Integrated optics $N \times N$ multiplexer on silicon. Photon. Tech. Lett. 3:896–899, 1991.
25. K. Okamoto et al. 32×32 arrayed-waveguide grating multiplexer with uniform loss and cyclic frequency characteristics. Electron. Lett. 33:1865–1866, 1997.
26. Y. Inoue et al. Athermal silica-based arrayed-waveguide grating multiplexer. Electron. Lett. 33:1945–1946, 1997.
27. H. Yamada et al. Statically-phase-compensated 10 GHz-spaced arrayed-waveguide grating. Electron. Lett. 32:1580–1582, 1996.
28. K. Okamoto et al. 16-channel optical Add/Drop multiplexer consisting of arrayed-waveguide gratings and double-gate switches. Electron. Lett. 32:1471–1472, 1996.
29. T. Matsunaga et al. Large-scale space-division switching system using silica-based 8×8 matrix switches. Proc. OEC '92, Makuhari, Japan, 1992, pp. 256–257.

30. T. Ito et al. Photonic inter-module connector using silica-based optical switches. Proc. GLOBECOM '92, Orlando, FL, 1992, pp. 187–191.
31. T. Nishi et al. A polarization-controlled free-space photonic switch based on a PI-LOSS switch. IEEE Photonics Tech. Lett. 9:1104–1106, 1993.
32. K. Okamoto et al. Guided-wave optical equalizer with α-power chirped grating. IEEE J. Lightwave Tech. 11:1325–1330, 1993.
33. M. Ashish et al. Highly efficient single-mode fiber for broadband dispersion compensation. OFC '93 Postdeadline paper PD13, San Jose, CA, 1993.
34. A. H. Gnauck et al. 10-Gb/s 360-km transmission over dispersive fiber using midsystem spectral inversion. IEEE Photonics Tech. Lett. 5:663–666, 1993.
35. K. O. Hill et al. Chirped in-fiber Bragg grating dispersion compensators; linearlization of dispersion characteristics and demonstration of dispersion compensation in 100 km, 10 Gbit/s optical fiber link. Electron. Lett. 30:1755–1756, 1994.
36. K. Takiguchi et al. Planar lightwave circuit optical dispersion equalizer. IEEE Photonics Tech. Lett. 6:86–88, 1994.
37. K. Takiguchi et al. Dispersion compensation using a variable group-delay dispersion equalizer. Electron. Lett. 31:2192–2193, 1995.
38. S. Kawanishi et al. 200 Gbit/s, 100 km time-division-multiplexed optical transmission using supercontinuum pulses with prescaled PLL timing extraction and all-optical demultiplexing. Electron. Lett. 31: 816–817, 1995.
39. K. Takiguchi et al. Higher order dispersion equalizer of dispersion shifted fiber using a lattice-form programmable optical filter. Electron. Lett. 32:755–757, 1996.
40. K. Takiguchi et al. Dispersion slope equalizing experiment using planar lightwave circuit for 200 Gbit/s time-division-multiplexed transmission. Electron. Lett. 32: 2083–2084, 1996.
41. Y. Yamada et al. Silica-based optical waveguide on terraced silicon substrate as hybrid integration platform. Electron. Lett. 29:444–445, 1993.
42. T. Tanaka et al. Integrated external cavity laser composed of spot-size converted LD and UV written grating in silica waveguide on Si. Electron. Lett. 32:1202–1203, 1996.
43. T. Tanaka et al. Fabrication of hybrid integrated 4-wavelength laser composed of UV written waveguide gratings and laser diodes. OECC '97 10D3-3, Seoul, Korea, 1997.
44. H. Takahashi et al. A 2.5 Gb/s, 4-channel multiwavelength light source composed of UV written waveguide gratings and laser diodes integrated on Si IOOC-ECOC '97, Edinburgh, UK, 1997, pp. 355–358.
45. I. Ogawa et al. Loss-less hybrid integrated 8-ch optical wavelength selector module using PLC platform and PLC-PLC direct attachment techniques. OFC '98 postdeadline paper PDP4, San Jose, CA, 1998.
46. T. Kato et al. 10 Gb/s photonic cell switching with hybrid 4 × 4 optical matrix switch module on silica based planar waveguide platform. OFC '98 postdeadline paper PDP3, San Jose, CA, 1998.

5

Erbium-Doped Glass Waveguide Devices

Denis Barbier

TEEM Photonics
Meylan, France

Robert L. Hyde

Alp Optics
Laffrey, France

1 INTRODUCTION

1.1 Overview

The evolution of telecommunication optical fiber networks from long-haul networks to metropolitan networks and finally to access networks will greatly increase the demands for optical components. The number of needed components will be larger, and, because they will be shared by a smaller number of customers, their cost will have to be small. Furthermore, the complexity and the interconnections of the networks will require subsystems incorporating many optical devices, which will then have to be as compact as possible. For these reasons—large quantity, low cost, and small size—integrated optic technologies are flourishing and offer more and more components with new functionalities. One of these new functions is that of amplification.

Integrated optical amplifiers were first demonstrated by T. Kitagawa in 1992 [1], and this work was followed by a number of other realizations, most of them based on erbium-doped glass materials.

In this chapter we will explain why erbium doped glasses are used to make integrated optics amplifiers, and what is the state of the art in this field. As this subject is quickly evolving, we will give simple explanations and keys, to help the waveguide amplifier users to appreciate the advantages and limitations of these components.

An erbium-doped waveguide amplifier (EDWA), as well as being an active device in its own right, is also an essential element in lossless passive components and integrated erbium laser oscillators. The amplification compensates for absorption, connection, and background losses inherent in these components.

For now, consider the EDWA as a "black box" two-port device. The input will carry both the optical signals and the optical power necessary to pump or power the amplifier. The output delivers the amplified signals. Section 2 gives an explanation of the workings of this "black box."

1.2 General Considerations

1.2.1 Small Signal Gain

The erbium optical amplifier small-signal gain coefficient, as it evolves with pump power, may be simulated, in a first approximation, by the function

$$g = \frac{\chi - 1}{\chi + 1} \tag{1}$$

where χ is proportional to the pump rate. Figure 1 shows a plot of this function that demonstrates the principal features of a ground-terminated (3-level) amplifier. When $\chi = 0$, the guide is totally absorbing. When $\chi = 1$, the guide is transparent, all absorption being reduced to zero by pumping, and as χ approaches infinity, the gain approaches its maximum. χ is the pump rate rationalized by the photoluminescent decay rate, and the normalized gain is logarithmic.

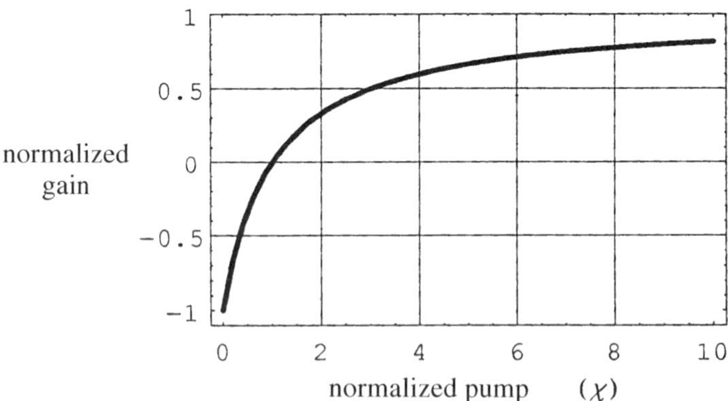

Figure 1 Simulated gain characteristic of a ground-terminated amplifier system. The intrinsic absorption of the erbium ion at 1.55 μm is fully compensated (gain = 0) when the normalized pump power, χ, is equal to 1.

1.2.2 Noise Figure

Optical amplifier noise is characterized by the noise figure (NF), which for 3-level systems at gains much much greater than 1 has the form

$$\frac{2\eta}{2\eta - 1} \tag{2}$$

where η is the metastable-level population. As the metastable level is filled by pumping, the value of the noise figure approaches 2, which is the theoretical limit, usually expressed in decibels. In practical working conditions the NF of an EDWA is between 3 and 5 dB. (See Fig. 2.)

1.2.3 Saturation Behavior

The amplifiers referred to previously are small-signal amplifiers, that is, amplifiers intended to recuperate lost signal at the 0.1–100-μW level. Another application of the amplifier is as a power amplifier or booster. This is operated at the 1–10-mW input level. One may require 10–100 mW of output power, necessitating a gain of 10, but at a signal level comparable with the pump power. Because we are extracting signal levels comparable to the pump power, some saturation is inevitable. Figure 3 shows a characteristic evolution of the normalized gain with signal ξ, simulated by

$$\frac{1}{1 + \xi} \tag{3}$$

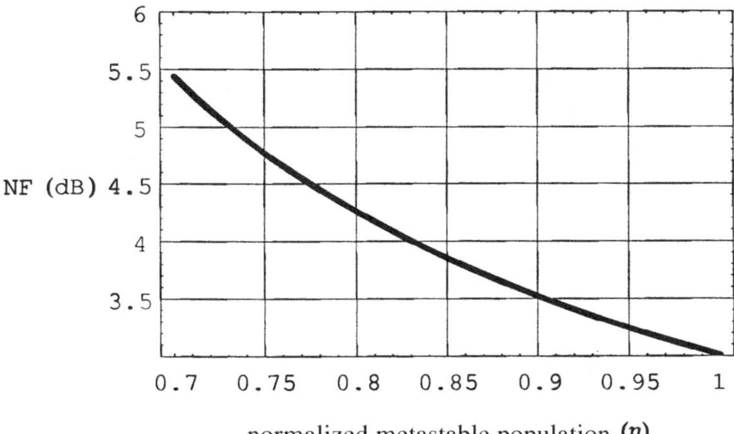

Figure 2 Noise figure of an ideal amplifier as a function of the metastable population. For complete inversion ($\eta = 1$), the noise is at its minimum, 3 dB.

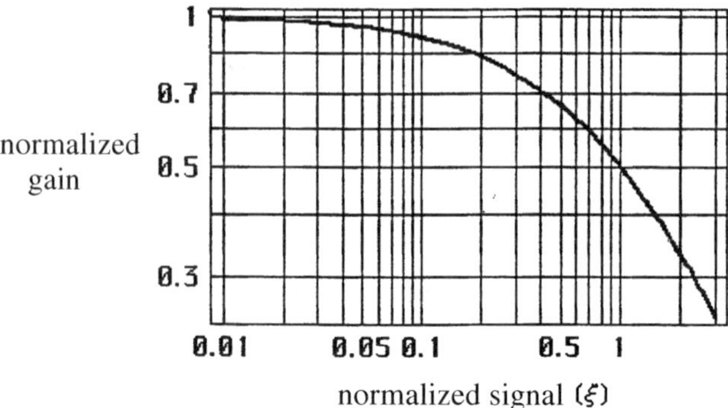

Figure 3 Small-signal gain drops off as the signal increases.

1.2.4 Amplifying Integrated Optic Components

The evident applications of waveguide amplifiers are as integrated preamplifiers, in-line, or as booster amplifiers. But the waveguide amplifiers can also be integrated with lossy components, such as power splitters, as well as with phased arrayed waveguides, add and drop multiplexers, modulators, switches, optical crossconnects, etc. A waveguide laser can also be made by use of the waveguide amplifier. The laser oscillator is in effect an amplifier in an optical feedback cavity.

In the following pages we explain the fundamentals of the EDWA and the different fabrication techniques, and we report on the performance of some of the just-listed components.

2 ERBIUM-DOPED WAVEGUIDE AMPLIFIERS: DESCRIPTION AND SIMPLE THEORY

The erbium-doped waveguide amplifier, an amplifier of optical radiation in the 1.55-μm waveband, is a major element in the different integrated optical devices described in Sec. 1. The EDWA itself consists of an erbium-doped waveguide embedded in, or integrated with, an amorphous substrate, the input and output of the waveguide accessed by pigtailed optical fibers. Both pump and signal beams are input via these optical fibers; that is, the waveguide is end-pumped.

A description of the operation of an EDWA follows, with calculations of the limits of its performance.

2.1 Optogeometrical Properties

The geometrical particularity of a guided device is that both the signal and pump optical beams propagate in a waveguide of small diameter, and consequently the power densities are very high, and the pump and signal optical modes overlap. This greatly reduces the amount of pump power required to reach the full population inversion.

2.2 Spectroscopical Properties

The principal active dopant in an EDWA is the rare-earth element erbium, one of the lanthanide family. The specific spectroscopic feature of the erbium ion is an optical transition between an upper state and the ground state, which has an energy difference corresponding to the energy of optical photons at wavelengths from 1520 to 1620 nm. The upper state is metastable; i.e., excited ions decay radiatively and spontaneously from the metastable level to the ground state at a relatively slow rate, about 100 transitions per second—that is, a photoluminescent decay time of about 10 msec.

In addition to the metastable state, there are higher absorption levels or excited states that may serve as pump transitions, isolated from the metastable state, thereby facilitating a 3-level pumping system (see Fig. 4). The erbium transition, centered at 1535 nm, corresponds to a waveband of the telecommunication spectrum, a waveband at which optical radiation has minimum losses when propagating in silica optical transmission fibers.

2.3 The Rate Equations

The most practical and efficient erbium pumping system is the 3-level system. An ion in the ground state absorbs an incoming pump photon and is stimulated to a higher energy level, from where it rapidly drops nonradiatively to the metastable level. The ion then decays down to its ground state by one of two radiative mechanisms. It may decay spontaneously, emitting an incoherent photon and so

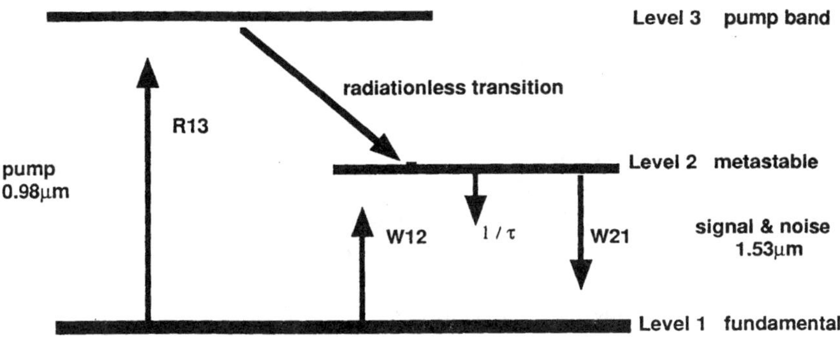

Figure 4 Erbium energy levels.

contributing to the noise in the system, or it may be stimulated by a signal photon of the same energy to drop instantaneously to the ground state, emitting a photon, coherent with the stimulator, and in effect producing a gain of 2. There is also a probability that an incoming signal photon will be absorbed directly, stimulating a ground state ion directly to the metastable state. Finally, one of several lossy processes may be involved, to be addressed in detail in Sec. 2.6.

These emission and absorption processes may be written as a rate equation. The rate of change, with time, of the ion population in the metastable state is given by the sum of the rates of the pumping process (level 1 to level 3), the spontaneous decay process (level 2 to level 1), the stimulated emission (level 2 to level 1), and signal absorption (level 1 to level 2):

$$\frac{dn_2}{dt} = -\frac{dn_1}{dt} = R_{13}n_1 - \frac{n_2}{\tau} - W_{21}n_{21} + W_{12}n_1 \tag{4}$$

where R_{13} is the pump rate from level 1 to level 3, W_{21} and W_{12} are the signal emission and absorption rates, respectively, and $1/\tau$ is the spontaneous decay rate.

In the first approximation, the population of level 3 is negligible, because its nonradiative decay time is much faster than that of the metastable state (a few microseconds). That is, all ions are considered to be in either the ground state or the metastable state, and $n_1 + n_2 = 1$, where $n_{1,2}$ are the normalized level populations, $n_{1,2} = N_{1,2}/N_{Er}$, and N_{Er} is the erbium concentration in the guide.

In the steady-state condition, the rate of change of each level is zero, and thereby one can obtain an expression for the metastable population and the ground state population:

$$n_2 = \frac{R_{13} + W_{12}}{R_{13} + W_{12} + W_{21} + \frac{1}{\tau}} \tag{5a}$$

But for very small signals, W_{12} and $W_{21} \ll R_{13}$ and n_2 becomes

$$n_2 = \frac{R_{13}}{R_{13} + \frac{1}{\tau}} \tag{5b}$$

Our description of an erbium system up to this point serves as an example of the principal radiative processes only. In practice there are other mechanisms, radiative and nonradiative, that intervene and must be included in the rate equations, in particular, parasitic effects (see Sec. 2.6) and the role of ytterbium in a codoped system.

The ytterbium ion has a singular absorption band peaking at 980 nm. This band, which overlaps in energy with that of an erbium absorption band but is stronger, is used as a pump band in short waveguide systems. The process de-

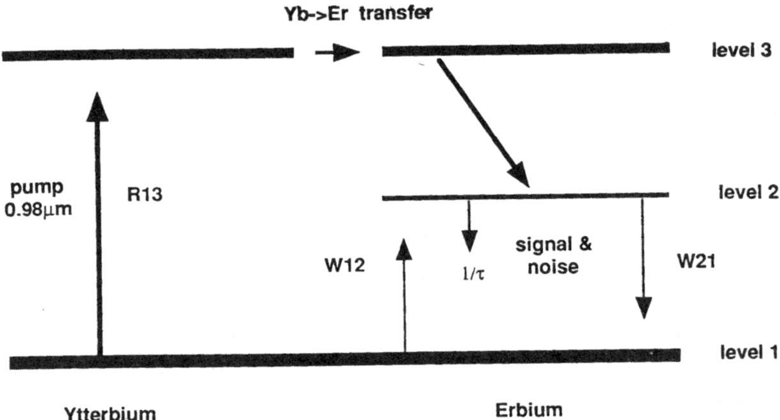

Figure 5 Erbium/ytterbium energy levels.

pends on an efficient transfer of energy from the ytterbium system to the erbium system, which is the case for phosphate glasses (see Fig. 5).

2.4 The Propagation Equation

An optical beam propagating in direction z in an active medium is described by the following differential equation:

$$\frac{dI(z)}{dz} = I(z)\, N_{\text{total}}\, [\sigma_e n_2(z) - \sigma_a n_1(z)] \tag{6}$$

where $I(z)$ is an optical intensity evolving with z.

The metastable state and the ground state are in fact not single energy levels but broad manifolds, and this is modeled by differentiating between a cross section for emission and one for absorption, σ_e and σ_a, respectively.

The propagation equation is a general equation for each of the optical beams propagating in the system, that is, the pump, signal, and noise propagation in the waveguide. Clearly the values of the cross sections are in general different for pump and signal. The cross sections also vary with wavelength in their respective bands.

The beam propagation in a waveguide is transversely confined in the guide, and the ion populations will have a transverse profile f describing the optical mode profile. For example:

$$I(z,r,\theta) \rightarrow I(z)\, f_m(r,\theta)$$

We must also account for possible profiling of the dopant:

$$N_{\text{total}}(r) \rightarrow N_{\text{total}}\, f_d(r)$$

These transverse spatial functions (f) will depend upon which modes are being considered or, in the case of the dopant profile, which technological process is used. For instance, the ion exchange process builds a guide in a ready-doped glass, there is no profiling, and N_{total} is spatially constant; but in a guided-layer technology, channel guides may be made by etching the guiding layer, thereby creating a specific dopant profile.

The beam intensities along a waveguide are found by solving the set of coupled propagation equations that model the optical beams, the energy level populations being determined from the rate equations. Initial pump and signal values are needed for the numerical solutions. A separate solution is found for each pump and signal wavelength.

The form of each coupled differential equation is:

$$\frac{dP(z,r,\theta)}{dz} = I(z) \cdot \int_{\theta=0}^{\theta=2\pi} \int_{r=0}^{\infty} f_m(r,\theta)\{N_{tErl} f_d(r)[\sigma_{e\sigma} n_2(z,r\ ,\theta) - \sigma_a n_1(z,r,\theta)] - \alpha\}\, r\, dr\, d\theta \tag{7}$$

The distributed background losses in the guide are modeled by the α term, in units of inverse length. A typical solution of the coupled equations giving the gain versus guide length is shown in Fig. 6.

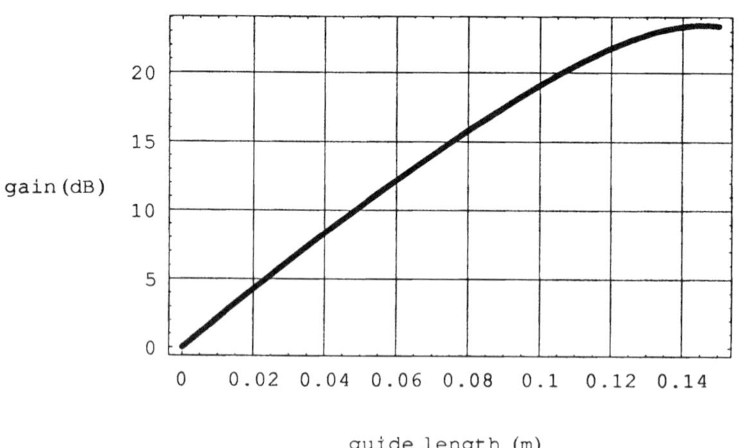

Figure 6 Small-signal gain evolution with guide length, for pump power of 50 mW. The optimum guide length can be seen to be 15 cm for a gain of 24 dB.

2.5 The Full Picture

A complete set of propagation equations for an optical amplifier assembled to calculate the signal, noise, and pump power evolution with guide length must take the following into account:

1. The profile of the intensity modes propagating in the guide
2. The profile of the dopant(s)
3. Small-signal and large-signal performance
4. Amplified spontaneous emission (ASE)
5. Codoping
6. Cooperative energy transfer (CET)

A set of propagation equations taking these factors into account becomes complex and non-intuitive, and the equations quoted earlier, although simplistic, have the advantage of demonstrating the mechanism of the processes involved.

Nevertheless items 1, 2, 5, and 6 above are particularly important in integrated amplifier analysis, and they embody the performance differences to be found in an integrated amplifier (EDWA) and a fiber amplifier (EDFA).

2.6 The Effects of a High Concentration of Erbium

The standard fiber amplifier has a very low erbium concentration, necessitating meters of fiber to obtain sufficient pump absorption and gain.

Integrated waveguides are by definition small devices, with a waveguide length L_g of less than 50 cm, and this reduction entails an increase in the concentration of the erbium ions to achieve practical gains, as is seen from the following expression for the maximum theoretical gain of an amplifier: $N_{\text{total}}\sigma_e L_g$. This higher concentration of erbium, of the order of 10^{26} atoms per cubic meter brings into play the behavior of ions in homogeneous or inhomogeneous (clustered) groups with short interion separations. When ions are in close proximity, cooperative energy transfer can occur and so modify the amplifier operation.

Cooperative Energy Transfer Mechanisms. Note: The energy transfer mechanisms in all of the following cases are concentration dependent and therefore more important in integrated devices.

1. *Sensitized Photoluminescence.* This is the effect of a second ion species (for example, ytterbium) that absorbs pump radiation in a stronger waveband than its equivalent found in erbium itself and efficiently transfers the absorbed energy to the erbium system.
2. *Photoluminescence Quenching.* Energy is transferred from a donor to an acceptor ion, the latter relaxing nonradiatively.
3. *Cross-Relaxation; Self-Quenching.* The donor ion transfers its energy to an acceptor ion in an excited state. The donor is now in the ground state,

and the acceptor moves up to a higher state and quickly relaxes to its metastable state. Cooperative frequency relaxation is a similar mechanism but the final decay path is not to the metastable state but directly to the ground state, emitting a high-frequency photon. This effect is observed in strongly pumped waveguides as a blue/green emission, but other wavelengths are possible at 660 nm, 800 nm, and 980 nm. For an amplifier at 1.53 μm, this emission is clearly unwanted.

4. *Resonant Migration of Energy.* Relaxation of a donor ion to the ground state promotes a neighboring acceptor ion to the excited state. This acceptor in turn becomes a donor, and the energy ''walks'' through the host material.

In Summary, the concentration quenching effects (2, 3, and 4) are detrimental to the coherent amplifying properties. The metastable lifetime is reduced as the concentration increases. A proportion of the erbium ions may be inactive, due to clustering, in the gain mechanism; and because they are for the most part moribund in the ground state, they absorb the signal. These phenomena are introduced in the rate equations as lossy terms, as explained later.

2.6.1 Homogeneous Up-Conversion (HUC)

The disposition of ions in concentration is also a source of loss. Ions pumped to the metastable state may be pumped further by another pump photon to a still higher level. This effect is called *cooperative frequency up-conversion*. This homogeneous up-conversion is due to an overlap of the 1.5-μm erbium absorption band and an excited state band peaking at 1.7 μm [5].

Homogeneous up-conversion has been modeled as a lossy term in the rate equations proportional to the square of the metastable population $C_{\mathrm{up}}n_2^2$, where typically $C_{\mathrm{up}} \approx 10^{-24}\ m^3 \cdot \mathrm{sec}^{-1}$. This process has recently been investigated further [6]. We can therefore describe this HUC effect as:

$$\frac{1}{\tau} = W_{\mathrm{radiative}} + W_{\mathrm{intrinsic}} + W_{OH} + W_{HUC} \tag{8}$$

where $W_{HUC} = C_{\mathrm{up}}n_2$ is dependent upon both pump power and erbium concentration [7].

2.6.2 Inhomogeneous Up-Conversion (IUC)

The introduction of high concentrations of erbium may produce a nonuniform solution of ions, with some ions existing in clusters. The ion-to-ion distances in this case are so small as to cause rapid energy transfer, effectively removing the fraction of ions in clusters from the amplifying process altogether. That is, a fraction k of the erbium ions stay in the ground state as an uninvertable absorbing ion [8]. This effect, when present, is catastrophic but may in principle be ameliorated by material processing. On the other hand, homogeneous effects are concentration based and represent a limiting performance.

2.7 The Effects of OH Impurities in the Glass Matrix

Among the most important parasitic phenomena is that due to impurities in the host medium. The main culprit has been identified as the OH radical, which absorbs in the region 2.7–4 μm, with harmonics in the region of 1.5 μm. The closeness of an excited erbium ion to that of a group of OH radicals rapidly transfers the erbium ion, nonradiatively, to the ground state [2,3].

The spontaneous lifetime is reduced from its theoretical ''radiative lifetime'' to some lower level, due to the erbium and impurity interactions, and can be expressed by the following relationship [3]:

$$\frac{1}{\tau} = W_{\text{radiative}} + W_{\text{intrinsic}} + 8\pi C_{\text{Er–Er}} N_{\text{impurity}} N_{\text{er}} \tag{9}$$

$W_{\text{radiative}}$, the assumed decay rate of an isolated erbium ion, is about 40 per second. The value of $C_{\text{Er–Er}}$ has been determined by experiment to be of the order of $10^{-51} m^6 \cdot \text{sec}^{-1}$. $W_{\text{intrinsic}}$ is a nonradiative internal decay rate independent of impurities. The energy is dissipated as vibrational quanta of the host material and will depend on that material. Its value may vary from zero [3] to 40 per second [4] for certain glasses.

2.8 Output Parameters and Limits of Performance of Erbium-Doped Waveguide Amplifiers

Erbium-doped waveguide amplifier devices are now at a preproduction phase but the performance is still improving. In order not to give values that may be quickly out of date, we will present the limits of performance assuming unlimited pump power but taking into account dopant profiling and predominant cluster effects.

The laser diodes available and suitable for pumping an EDWA are at wavelengths of 800 nm, 980 nm, and 1480 nm. Pumping at 1480 nm is useful for a single doped erbium device; but being a quasi-2-level pumping mechanism, it is less efficient than a 3-level one. Pumping at 800 nm is a 3-level pumping scheme, but strong excited-state absorption (ESA) is an additional parasitic effect.

By codoping with ytterbium, which has a simple spectroscopic structure and a strong absorption at 980 nm, one can pump the erbium indirectly (Fig. 5). This is the system we shall assume next.

2.8.1 Maximum Gain per Unit Length of Guide

From the master differential equation we can derive the local gain when all the erbium ions (except for a fraction *k* due to clustering) are inverted; that is, we assume an infinite pump rate.

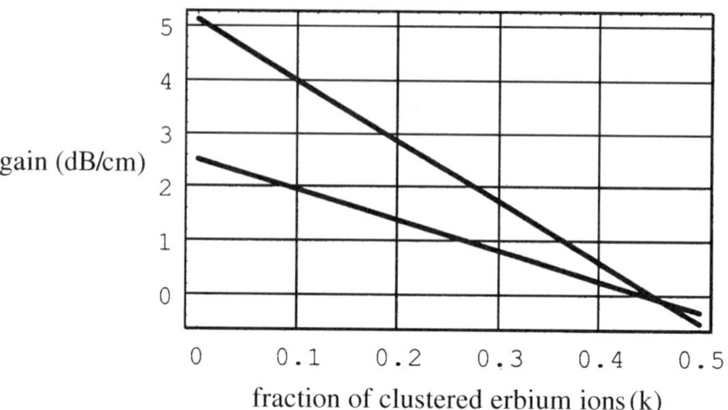

Figure 7 Reduction of the gain with the fraction of clusters k. The lower plot is for a guide with a Gaussian dopant profile. The upper plot is for a guide with no profiling. Complete inversion is assumed.

Assuming that the mode profile is gaussian and that there is no dopant profile, then

$$\gamma = N_{\mathrm{Er}}\,[\sigma_e - \kappa(\sigma_e + \sigma_a)] - \alpha \tag{10}$$

In the case where the dopant concentration is profiled to that of the mode profile, the local gain is given by

$$\gamma_p = \frac{N_{\mathrm{Er}}\,[\sigma_e - \kappa(\sigma_e + \sigma_a)] - 2\alpha}{2} \tag{11}$$

Figure 7 is a plot of the local gains in the two cases, with the fraction of clustered ions, k. Clearly, when nearly half of the ions are quasi-moribund in the ground state, transparency cannot be achieved.

2.8.2 Evolution of Gain with Pump Power

The gain evolution with pump power is shown in Fig. 8. Three plots are shown for $k = 0$ to $k = 0.3$, a severely clustered guide. The pump power axis has units of a normalized power parameter. One can see from this idealized plot the strong effect of the clustered fraction of erbium ions.

2.8.3 Noise Figure for Unlimited Pump Power

The *noise figure* is defined as the ratio of signal to noise at the output and input of the amplifier. In the quantum noise limit, the noise figure for a 3-level system (NF) is defined as the ratio of the metastable state population to the inversion.

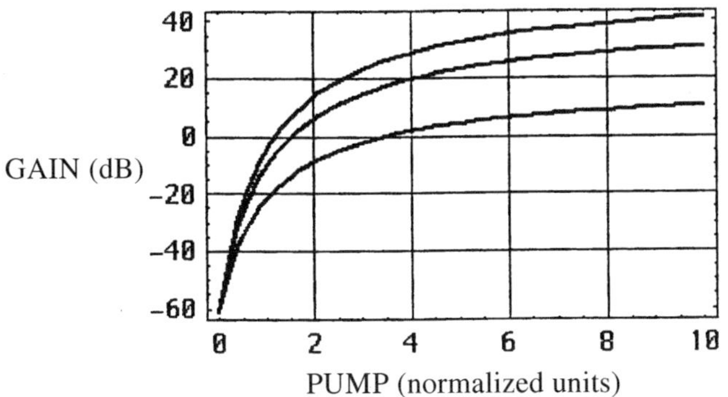

Figure 8 Gain evolution of an optical amplifier showing the effect of clustering. The lower plot ($k = 0.3$) clearly indicates a serious fall-off of gain compared to the other cases ($k = 0.2$ and $k = 0$).

It is usually quoted in decibels (dB). For an unlimited pump power with k moribund ions we have

$$\text{NF (dB)} = 10 \log_{10}\left[2\frac{(1-\kappa)\sigma_e}{(1-\kappa)\sigma_e - \kappa\sigma_a}\right] \tag{12}$$

In the limit for maximum pump power, NF approaches 3 dB for $k = 0$. Figure 9 shows this tendency. As k increases, the noise figure increases rapidly.

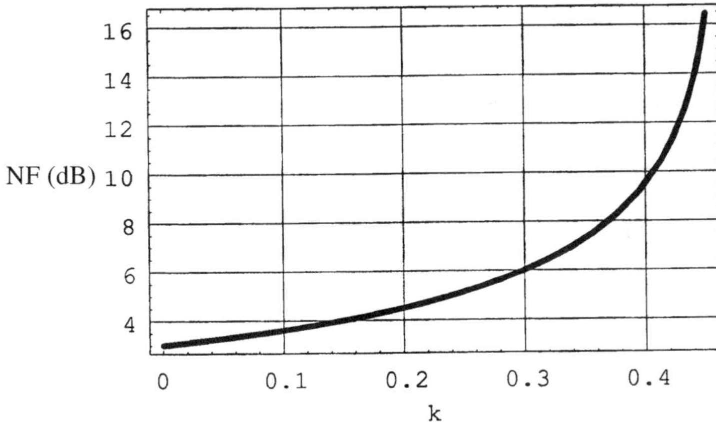

Figure 9 Ideal noise figure as the fraction of moribund ions (k) increases.

2.9 Summary

We have described an optical amplifier theoretically, specifically an EDWA, and have estimated its limitations. The metastable-level decay constant (τ) is an important parameter, because the pump rate necessary for a given gain is inversely proportional to τ. This photoluminescent decay is reduced from its radiative value (25 ms) by impurities and CET mechanisms to less than 10 ms. In the extreme case of a cluster of ions, their time constant is negligibly small (50 ns), and they become moribund in the ground state. The maximum-performance parameters have been estimated for an unlimited pump rate, taking into account the moribund ion fraction k.

3 ADVANTAGES AND LIMITATIONS OF ERBIUM-DOPED WAVEGUIDE AMPLIFIERS

Most of the advantages of erbium-doped waveguide amplifiers (EDWAs) arise from their ability to provide high gain in short optical paths. This can only be obtained in specific erbium-doped materials that are processed with the appropriate technology, which limits the potential performance and manufacturability of these integrated optical amplifiers. We highlight in Table 1 the advantages and limitations in the performance of an EDWA, introduced by the materials and the technological environment used to fabricate them.

The first requirement of a waveguide amplifier is that it provides sufficient gain and output power with the shortest possible guide length. This can be achieved only with a high erbium doping level, and there is an autolimitation here, as explained in Sec. 2. The more erbium ions we add to the glass, the closer they are one to the other; when excited, they may start to exchange energy, thereby reducing their efficiency. Furthermore, at very high doping levels (above 10^{25} atoms/m^3) the ions can get even closer and start to form pairs or clusters,* whereby they become completely inefficient for the amplification process. A compromise has to be found between the high erbium doping level, which helps to create large gain per unit length, and the parasitic effects due to very high doping levels, which hinders the ability of erbium ions to provide gain.

Not all optical materials are able to accept a high doping level of erbium, which is why certain glass materials have been chosen to make waveguide ampli-

* What we call a pair, or a cluster, as opposed to homogenously dispersed ions, are two or more erbium ions close enough to exchange energy via dipole–dipole or dipole–quadripole interactions. The transfer of energy between two close erbium ions is very fast. Piotr Myslinski has measured fluorescence lifetime as short as 50 ns for these pairs [8]. Our extended X-ray absorption edge fine structure (EXAFS) measurements have shown that pairs, or clusters, are made up of neither erbium–erbium nor erbium–oxygen–erbium molecules, which indicates that the distance between the closest erbium ions is larger than 4 Å.

Table 1 Advantages and Limitations of EDWAs According to Materials and Technologies Used

Material and technology	Advantages	limitations
High Er doping level Yb codoping	Short amplifiers (up to 4 dB/cm is feasible) Short amplifiers due to high pump absorption cross section 970–1080-nm pumping band Limitation of the clustering problem	Gain quenching due to clustering of Er ions 1480-nm pumpong is not enhanced ±5-nm pump wavelength selection required Pump lost in Er to Yb transfer mechanism Pump lost in Yb spontaneous emission mechanism
Multicomponent glass materials	High Er doping level achievable Gain spectrum and flatness can be adjusted by varying the glass composition Low-phonon-energy glass matrices can be investigated Adjustable physical and optical properties	Compatibility with the waveguide technology OH contamination has to be lower than 100 ppm Strong know-how is required for glass fabrication Still limited understanding of the cluster formation process
Integrated optics technology	High mode confinement is achievable Integration of amplifiers with other active or passive functions Mass production Low cost, small size High reliablility Compatibility with Bragg grating technology Array of amplifiers	PDL problems with technologies Dopant profiling difficult in some cases High propagation losses Fiber-to-waveguide connections IO process can depreciate Er spectroscopic properties Integration of isolators is difficult

fiers. Not all glass materials are useful; for example, the rare-earth solubility is very low in silica. Multicomponent phosphate or silicate glasses accept much higher erbium doping levels (up to 10^{27} atoms/m^3 for some glasses [9]). Most of the EDWA made uses these two families of glasses (see Sec. 5 for more details).

The large range of compositions achievable with these multicomponent glasses brings an important advantage to waveguide amplifiers. Indeed, by modifying the glass composition we can adjust the gain spectrum toward shorter or longer wavelengths, and we can adjust the gain flatness [10]. We can also reduce the glass phonon energy in order to increase the amplifiers' quantum efficiency. Furthermore, multicomponent glasses can offer other physical properties (photosensitivity, photochromism, optical nonlinearity, etc.), which may be very useful in creating new integrated optical functions, functions to be combined with the basic amplification property.

The range of useful glass compositions may be limited by the technology used to fabricate the waveguides. Indeed the ion exchange, the sputtering, or the etching, as well as the sol-gel processes, are dependent on glass composition and may require time-consuming and precise adjustment from one glass to another. The plasma-enhanced chemical vapor deposition (PECVD) and flame hydrolysis deposition (FHD) processes can be also very restrictive on achievable glass compositions.

Another very important parameter affecting amplifier behavior is the glass OH contamination. Hydroxide creates phonon energy levels that help a nonradiative deexcitation of erbium. The low OH doping level required, less than 100 ppm [11], increases the level of know-how necessary in glass preparation, reducing the availability of good erbium-doped glasses.

If a high erbium doping level is necessary to reach large gains per unit length (a figure of merit, in dB/cm), strong pump absorption is also required to achieve high population inversion. Ytterbium (Yb) has a large absorption cross section around 980 nm, about six times larger than erbium's 980-nm absorption cross section, and Yb transfers the absorbed energy to the erbium metastable level, with an efficiency of more than 80% in some glasses. Ytterbium is therefore used as a codopant to enhance the 980-nm pump absorption in short devices.

There are two other advantages in using ytterbium. One is the large absorption peak, which makes possible the use of 1060-nm lasers to pump Er/Yb codoped glass waveguides for high-power applications. The second is a reduction of the clustering problems, as some of the Er–Er "pairs" are replaced by Er–Yb "pairs," which does not affect the figure of merit.

One drawback to the Yb pump absorption is the reduction in the global pump efficiency. The energy transfer efficiency from Yb to Er is less than 100%, due to phonon deexcitation of Yb ions, to Yb spontaneous emission, and to Yb–Yb energy transfer. Another drawback is the narrow absorption peak of ytterbium, which requires a pump wavelength selection of ± 5 nm around the maximum of the Yb absorption spectrum (see Fig. 11, later).

The integrated optic techniques used to fabricate the EDWA contribute to a wide range of advantages. The most obvious are their compatibility with mass production, which will greatly reduce the fabrication cost of small-size EDWAs, the small size being achieved for waveguides with a highly confined propagating mode [52]. The mode confinement is also required to improve the pump efficiency, but may create some loss of polarization dependence due to birefringence or stress built up in the guiding layer during the deposition process. The polarization problems are highly dependent upon the process used. For example, the waveguide made by the ion exchange technique is generally free from birefringence.

But the main advantage of these techniques is their integration capabilities. Indeed a waveguide amplifier needs other optical components, such as pump multiplexers, filters, Bragg gratings, and mode adapters, to become a ''ready-to-use'' optical amplifier device. The integration of all these components on a single chip enables durability and size and cost reduction of the final amplifier device. One difficulty in this integration procedure is that, for maximum amplifier performance (gain, noise figure, output power, pump efficiency, etc.), the functions other than amplification have to be made on undoped material. The monolithic integration of doped and undoped waveguides has been demonstrated for both PECVD and FHD techniques, and is solved for the ion exchange process by preparing a two-section glass wafer, one undoped and the other doped with rare-earths ions, the passive components being processed on the undoped part of the wafer. This integration approach is also a great advantage in terms of reliability, for there are no splices between the different components, since they are all made on a monolithic piece of glass material, which greatly reduces the potential failure rate.

One drawback in this integration approach, which may reduce the performance of the final device, is that it is difficult to tune the specific performance of each function to its optimum when all functions are made simultaneously. Some other limitations arise from the different technologies; for example, ion exchange makes dopant profiling very difficult to achieve. Even more drastic are the modifications that some steps of the process can bring to the spectroscopic properties of the Er-doped glass material; for example, sputtering [26] and reactive ion etching (RIE) [12] have an influence on the spectroscopic properties.

Another limitation of integrated optic technologies is their incompatibility with isolator fabrication, given that isolators are generally used in optical amplifier assembly to protect the system against parasitic oscillations, against back emission of amplified spontaneous emission, or against Rayleigh back-scattering. This problem is encountered by hybridizing a micro-optics isolator in between the waveguide amplifier and the pigtailed fiber.

In Table 2 we compare the typical performance of Er-doped waveguide amplifiers and fiber amplifiers. It is clear that EDFAs offer better performance than do EDWAs, but fiber amplifiers have benefited from much more important devel-

Table 2 Comparison of EDWA and EDFA Performance

Parameters	Typical performance of EDWA	Typical performance of EDFA
Gain	30 dB demonstrated [13], no limitations	50 dB demonstrated [53], no limitations except back-reflections
Noise figure	3.1 dB measured [10], close to fundamental limit	3.1 dB observed [53]
Output power	14 dBm measured [14], limited by parasitic effects due to high doping level	31 dBm [17], limited by quantum efficiency and available pump power
Gain per unit length	4.2 dB/cm measured [10], limited by Er doping level, ratio of Er in clusters, and emission cross section	0.05 dB/cm, limited by Er doping level and emission cross section
Pump wavelengths used	980 and 1480 nm	800, 980, 1060 and 1480 nm
dB/mW	0.1 observed [15], limited by mode confinement and dopant profiling	5 typical, already optimized
Propagation loss	0.1 dB/cm measured [14]	1 dB/km
Coupling loss with fiber	0.2 dB [14], mode converter generally required	<0.2 dB, expanded-core fiber required
Gain spectrum	1534–1549 nm [16], depends on glass material used	1527–1611nm [18], depends on fiber composition
Gain flatness	Not measured yet	±0.2 dB/32 nm [19]

opment efforts. And despite the limitations brought to waveguide amplifier performance by the spectroscopic behaviors, as explained in Sec. 2, further development should minimize the difference in performance. Indeed, our modeling, based on the rate equations and propagation equations of Sec. 2, predicts a gain higher than 30 dB and a noise figure lower than 3.5 dB for a 25-cm-long waveguide pumped by less than 100 mW of a 980-nm semiconductor laser. For this example, the expected output power of 15–16 dBm is slightly lower than that of the EDFA, due mainly to the cooperative energy transfer phenomena taking place in the highly Er-doped materials used.

Broad gain spectrum as well as gain flatness are achievable with waveguide technology, using similar filtering and multistage schemes to those used in fiber technology. But due to the limited flexibility of integrated optic techniques, especially when many optical functions have to be integrated together, it will be difficult for EDWAs to reach the impressive performance of fiber in terms of gain spectrum and flatness.

In future applications, EDWAs will be able to replace the standard fiber amplifiers but probably not the highly specialized ones. We will find waveguide amplifiers where regular performance together with compact size and low cost are requested. We will find them also in applications where arrays of amplifiers are needed.

4 ERBIUM-DOPED WAVEGUIDE AMPLIFIERS: MATERIALS AND TECHNOLOGIES INVOLVED

4.1 Spectroscopic Properties of Erbium-Doped Glasses

The spectroscopic properties of the rare-earth ions (in their host) are essential basic data for the modeling and understanding of the operation of an optical amplifier. In this section we will present the absorption spectrum of erbium (and ytterbium) in a glass host. We can determine the absorption cross sections and eventually the emission cross sections from this spectrum. The metastable-state temporal decay spectra will be described for determining the photoluminescent rates. Finally, we must address important side effects, for example, rare-earth inhomogeneity and impurities.

In Sec. 2 we saw how the propagation of the signal, the noise, and the pump beam may be described by an equation of the following form:

$$\frac{dI(z)}{dz} = I(z)N_{\text{total}}[\sigma_e n_2(z) - \sigma_a n_1(z)] \tag{13}$$

where N_{total} is the ytterbium or erbium ion concentration (m^{-3}) and $\sigma_{e,a}$ are the cross sections for emission and absorption, respectively. The cross sections are a convenient and practical parameter embodying the emission or absorption spectra and the radiative decay rates. They vary considerably with wavelength and of course from element to element.

The small-signal absorption, α (m^{-1}), is defined by the following propagation equation:

$$\frac{dI(z)}{dz} = -I(z)\alpha(\lambda) \tag{14}$$

where $\alpha(\lambda) \equiv N \cdot \sigma_a(\lambda)$ ($n_1 \to 1$ and $n_2 \to 0$). The solution of this equation gives the optical transmission for a sample thickness L, and is referred to as Beers law, viz:

$$\frac{I(L)}{I(0)} = e^{-\alpha(\lambda)L} \tag{15}$$

Consequently, measuring the absolute transmission and the thickness of a sample over a range of wavelengths will enable the value of $\alpha(\lambda)$ to be obtained. Then, given the value of the dopant concentration (N), the absorption cross section can be determined.

One needs to know the cross-sectional spectra for both the signal and the pump bands. Examples for erbium and ytterbium are shown in Figs. 10 and 11. Each absorption band corresponds to an energy band of the ion in its host, and the overall spectrum is shown in Fig. 12. Given an absolute absorption spectrum, the major source of error in the absorption cross section is in determining N.

The determination of the emission cross sections, $\sigma_e(\lambda)$, is not so direct. There are various methods [20,21], but a description of these will not be given here. The emission cross section can be calculated directly from the Ladenburg–Fuchtbauer equation, knowing the emission spectrum and the *radiative lifetime*. The radiative

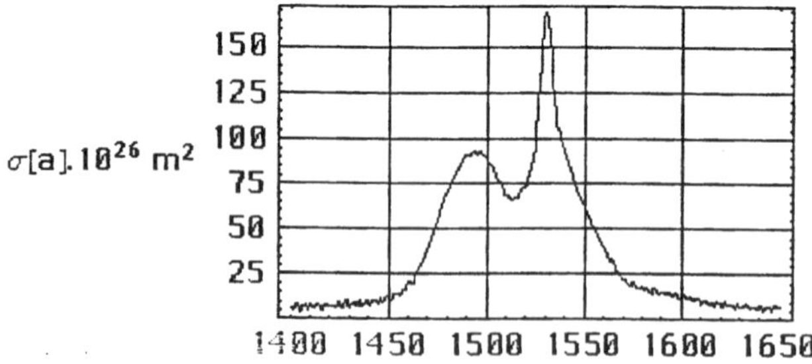

Figure 10 Erbium phosphate glass-absorption cross-sectional spectrum for the signal bandwidth.

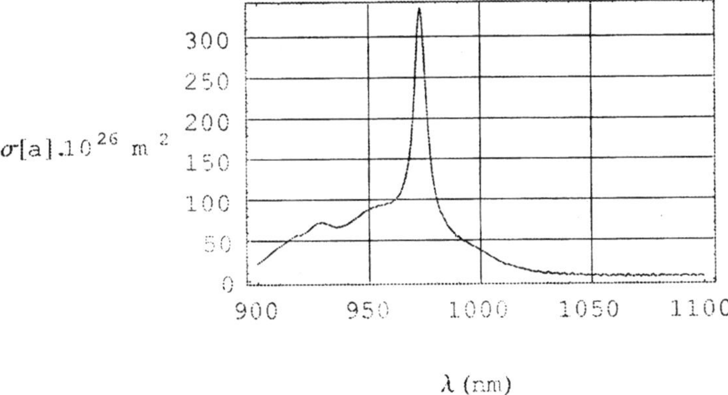

Figure 11 Absorption cross-sectional spectrum for the pump band of a codoped erbium/ytterbium phosphate glass.

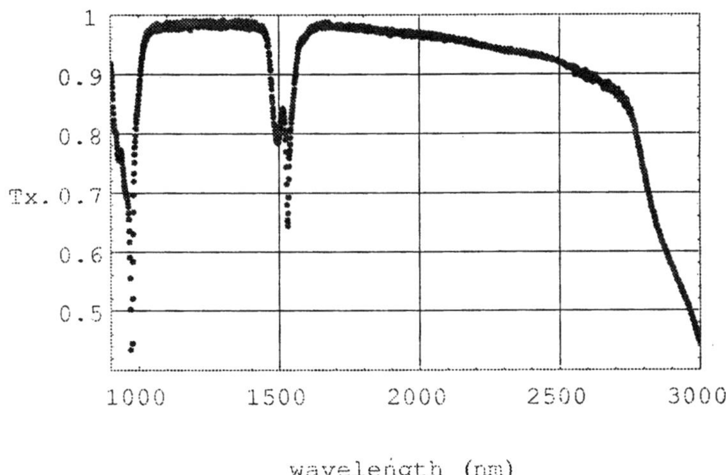

Figure 12 Typical transmission spectrum of an erbium/ytterbium phosphate glass, showing the pump band at 0.98 μm, the signal band at 1.53 μm, and the long-wavelength cutoff, which includes OH absorption at 2.8 μm.

lifetime, in turn, may be determined from an examination of the complete absorption spectrum using Judd–Ofelt theory [22]. We determined the radiative lifetime of erbium ions in a phosphate glass host by the Judd–Ofelt technique and, for a 2 wt % concentration of erbium we found 12 ms.

Metastable Lifetimes. The *photoluminescent lifetime* (τ) of the erbium ion is obtained via direct experimental method by irradiating a thin slice of the erbium-doped glass with low-intensity pump pulses. This determines the ion metastable-state decay time in its given host, as distinct from that of an isolated ion. It is an average value for what in practice will be a range of lifetimes depending on each individual ion's environment. For instance, the lifetime of an ion in a cluster has been measured as low as 50 ns, whereas the lifetime of an isolated ion is about 25 ms. An average value of 7 ms has been determined for the metastable decay time of erbium of concentration 2 wt %, in a phosphate-glass host.

Impurities. The glass should be free of impurity (metal) ions and OH ions. The quality control of a potential glass is carried out by observing the absorption spectrum. The OH ion has an absorption spectrum in the mid-infrared, near to total cutoff of most common glasses. Metal ions absorb in the near ultraviolet and may be traced by spectroscopic analysis [23]. A spectrum of the infrared cutoff of a phosphate glass is shown in Fig. 13.

Inhomogeneity of Erbium Ion Concentration. The determination of the fraction of clustered ions, *k*, has been obtained indirectly by several groups. One such investigation, which involved measuring the residual absorption in a fiber sample and assuming that the "clusters" are ion pairs, is presented in Ref. 24.

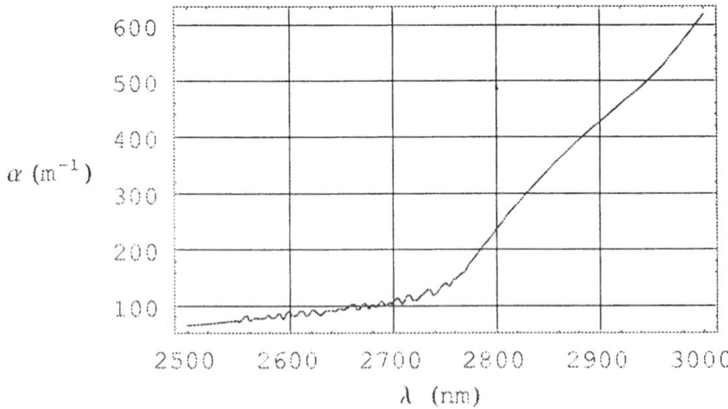

Figure 13 Infrared cutoff of an erbium/ytterbium phosphate glass. The OH absorption lies in the region 2.8–2.9 μm.

Dynamic methods have been pioneered by the Institute for Information Technology, NRC, Canada [8].

An experimental problem remains in determining k in a short waveguide, and this has been temporarily bypassed by invoking the concept of the quasi-moribund ion. We assume that there is a fraction k of ions that remain in the ground state regardless of pump power, except for rapid excursions to upper states. They do not take part in the amplifying process; but because they are almost always in the ground state, they absorb the signal. The variable k is introduced into the rate equations as a free parameter, and values of $k = 0.10–0.20$ have been determined from comparisons between experimental and theoretical gain modeling. This indicates that up to 20% of the erbium ions are inefficient in providing gain.

4.2 Technologies

Various methods are used to produce erbium-doped glass waveguides, the first step being the preparation of the doped material, the second being its processing to form a waveguide.

4.2.1 Bulk Glasses

The glass block is prepared using standard melting techniques, with erbium being added as erbium oxide powder at an early stage of preparation. Two families of oxide glasses are currently used to realize waveguide amplifiers: the alumino-silicate and the alumino-phosphate. In these multicomponent glasses we can find oxide molecules such as SiO_2, P_2O_5, Al_2O_3, Na_2O, CaO, BaO, MgO, and La_2O_3. The main advantage of these glasses is that they accept a very high doping level of erbium (less than 10^{26} atoms/m^3), which is at least a hundred times higher than in silica glass. During the preparation of these glasses, special care is brought to the elimination of the OH impurities, which create strong nonradiative de-excitation of pumped erbium ions.

Erbium-doped fluoride glasses are also examined as a low-phonon-energy matrix [25]. In this case the glass is prepared under controlled atmosphere, and erbium is added as erbium fluoride powder.

Sputtering. The bulk glasses, used as a sputtering target, can form an Er-doped thin-film glass layer on a silica wafer or on a silica-buffered silicon wafer. After an etching process, the thin film is transformed in channel waveguides, which can propagate highly confined modes with moderate propagation losses when protected by a clading layer. The sputtered material may have different spectroscopic properties than the starting one. Lucent Technologies [10], TNO [26], and LETI [12] are investigating this technique.

Ion Exchange. Channel waveguides can be made in the oxide bulk glasses using the ion exchange technique. The waveguides, which benefit from the same spectroscopic properties as the starting material, are buried a few microns below the glass surface; they propagate moderately confined modes with low propagation losses and very good compatibility with optical fibers. Despite the fact that dopant profiling is difficult with this technique, GeeO [27], Corning [28], and the University of Michigan [29] are using it. The ion exchange technique is well established, and it is used to manufacture commercial passive components as well as integrated amplifiers based on the EDWA.

4.2.2 Gas Precursors

The techniques of PECVD and FHD use gas precursors to form silica-based glass materials, with the erbium ions being added as vaporized Er chelate dissolved in butyl acetate [30] (for the PECVD technique) or as Er chloride alcoholic solution (in the case of the FHD technique). The Er-doped layer, deposited on top of a silica-buffered silicon wafer, is the high-index guiding layer, which is etched and covered by a silica cladding layer to form the channel waveguide. These waveguides can handle well-confined modes showing good compatibility with fibers, but may present excessive propagation losses due to the etching process. The silica-based material cannot accept very high doping levels of erbium.

The PECVD and FHD technologies are already used at a production level to manufacture a wide range of passive optical components. Hybrid components, incorporating passive and semiconductor optical chips, are also manufactured.

4.2.3 Liquid Precursors

In the sol-gel technique, organometallic liquid precursors are mixed together and hydrolized to form a solution, which is deposited as a thin layer by spin-coating or dip-coating onto a silica-buffered silicon wafer. Erbium is introduced in the solution as Er alkoxide or Er nitrate. After gelation, the thin film is transformed into dense glass material by an annealing step at high temperature. Channel waveguides are made by etching the guiding layer; they can support highly confined modes but present high propagation losses. No usable gain has been demonstrated yet with this technique [32]. This very young technique is not yet used in integrated optics production lines.

4.2.4 Ion Implantation

Ion implantation can be used to dope a glass thin-film material with erbium. The advantage of this technique is its abilities to form a locally Er-doped waveguide on an undoped wafer and to adjust the dopant profile. Gain has been demonstrated with this technique by FOM in Amsterdam [33].

4.2.5 Summary

The various technologies just described are used to make waveguide amplifiers in different glass materials. The most promising results are presented next, in Sec. 5.

5 ERBIUM-DOPED WAVEGUIDE AMPLIFIER PERFORMANCE AND FABRICATION: STATE OF THE ART

Of the various integrated optics techniques listed in Sec. 4, the first to prove its ability to make a laser was the flame hydrolysis deposition (FHD) of Er-doped silica, work done at NTT in 1991 [31]. The same NTT team demonstrated, one year later, amplification in a silica waveguide made by the same technology [1]. Later, other technological approaches reached maturity, and new demonstrations of Er-doped integrated lasers and amplifiers were made at Corning Europe (Avon, France) with ion exchange in silicate glasses [28], at FOM (Amsterdam) with Er ion implantation in alumina [33]; at GeeO (Grenoble, France) with ion exchange in phosphate glass [27]; at Lucent Technologies (Murray Hill, NJ) with sputtering of soda-lime glass [7]; at MESA (Twente, the Netherlands) with sputtering of Y_2O_3 [34]; at ORC (Southampton, England) with ion exchange in silicate glass [35]; at TNO (Eindhoven, the Netherlands) with sputtering of phosphate glass [26]; and at the University of Michigan with ion exchange in silicate glass [29]. More recently, other groups have measured gain in Er-doped waveguides: the University of Pittsburgh with sputtering of soda-lime glass [36], and the University of Alberta with sputtering of silicate [37].

In addition to these practical gain or lasing demonstrations, there is a lot of work in progress in laboratories around the world on a variety of Er-doped materials combined with various integrated optics technologies. Among them we can mention the new approach of Er-doped SiO_2 deposition by a PECVD technique at MIC (Copenhagen) [30] and the sol-gel technique developed by a European consortium, "CAPITAL," supported by the European program ACTS [32]. There is also a Canadian consortium working on Er-doped sol-gel materials [38].

We should also mention that amplifiers and lasers have been made in Er-doped $LiNbO_3$ [39]. Despite the impressive results obtained and the extensive work done in this field, we will concentrate our analysis on glass materials, where we have most of our expertise.

In order to highlight the main advantages of each technology, we summarize, in Table 3, the measured characteristics of amplifiers made with these various technologies. It appears, on considering the gain per unit length (the gain figure of merit), that the sputtering technique gives the best results, either for phosphate or silicate glasses. These values were obtained with highly confined modes compared to those values measured in ion exchange waveguides. Indeed, the larger mode area of the ion exchange waveguides makes them far from being infinitely pumped, as confirmed partially by the 4-dB noise figure. Better confinement would probably enable ion exchange guides to reach a higher figure of merit, with the remaining difference being attributed to the moribund erbium ions, which seem to be present in a smaller percentage in sputtered glass than in bulk glass.

Table 3 Comparison of Measured Performances of Waveguide Amplifiers

Technology	Host material (doping level 10^{20} at/cm^3)	dB/cm	Net gain (120 mW at 980 nm)	Mode area (μm^2)	Noise figure (dB)	Loss dB/cm	FPC*	Laboratory
Sputtering	Soda-lime (4.3)	**4.2**	10	11	3.1	1	Yes	Lucent [10]
	Phosphate	4.1	>4.1 (25 mW)	5		0.9	No	TNO [26]
	Y_2O_3 (1.3)	1.3	>5.7 (10 mW)	4		0.9	No	MESA [34]
Ion exchange	Phosphate (1.98)	2.9	16	50	4	**0.1**	Yes	GeeO [14]
	Silicate glass (2.4)	2.3	9	26		0.15	Yes	Corning [28]
FHD	$P_2O_5:SiO_2$ (0.64)	0.68	**18.9**	44	3.8	0.15	Yes	NTT [40]
PECVD	$P_2O_5:SiO_2$	0.67	2.5	69		0.17	Yes	NTT [41]
Ion implant	Al_2O_3 (2.7)	0.75	>3 (10mW)	2		0.35	No	FOM [33]

* FPC: Fiber pigtailed component

Table 4 Comparison of Measured Performances of Waveguide Lasers (cw Operation)

Technology	Host material	Slope efficiency (%)	Threshold (mW) (980 pump)	Length (cm)	FPC*	Laboratory
Ion exchange	Phosphate	**27**	66	4.3	Yes	GeeO [42]
	Silicate	5.5	**14.8**	**0.86**	No	U. Michigan [29]
	Silicate	0.15	15	3.4	No	ORC [35]
FHD	P_2O_5; SiO_2	0.81	49	4.5	Yes	NTT [31]

* FPC: Fiber pigtailed component

Both phosphate and silicate glasses were used with the ion-exchange and sputtering techniques and no significant gain-per-unit-length difference was found between these two types of glass. The FHD or PECVD techniques present a lower figure of merit, but the waveguides used had a larger mode area and a lower erbium doping level than those made with the previous techniques. Despite this, the NTT group was able to reach a high gain by making a low-loss long waveguide.

From an examination of Table 3 a question arises: Is it better to make a short, highly doped waveguide amplifier or a lower-doped, slightly longer waveguide? It seems, from NTT's results, that the second option is better. But we believe that the first option is also viable. Indeed, if we analyze the very promising results of the sputtered phosphate glass or sputtered Y_2O_3, as well as the results obtained with the ion-implanted Al_2O_3, we see that high gain can be reached at a very low pump power even with short, highly doped waveguide amplifiers.

Turning now to lasers, both ion exchange and FHD technologies have shown their ability to make Er-doped glass waveguide lasers. We summarize in Table 4 the work done in this field. It is difficult to compare the performances of these lasers, for very different optical cavities were used for each of them. But still, this table shows that Er-doped waveguides are good candidates as amplifying media to realize cw lasers operating in the 1.55-μm band. Indeed, lasers with a 27% slope efficiency have already been demonstrated, as well as lasers as short as 8.6 mm, still with a good slope efficiency.

6 TELECOMMUNICATION APPLICATIONS

6.1 System Evaluation of Erbium-Doped Waveguide Amplifiers

The gain (up to 30 dB) and noise figure (as small as 3.1 dB) performances, as well as the available output power (up to 14 dBm), make Er-doped waveguide amplifiers immediately attractive for telecommunication system applications. Erbium-doped waveguide amplifier prototypes have been evaluated in multigigabits-per-second transmission systems, and they have proven very promising. These first experiments were made under a single wavelength, or a few wavelengths, operation, because the gain spectrum of EDWA is relatively narrow.

Erbium-doped waveguide amplifiers have been used in 10-Gb/s fiber transmission system experiments as preamplifiers, in-line amplifiers, and power boosters [13,43]. An excellent sensitivity of −35.4 dBm has been reported for waveguide preamplifiers, and stable long-term bit-error rate (BER) measurements have been achieved with waveguide in-line amplifiers, through 186 km of fiber dispersion without showing any floor or performance degradations. With the waveguide amplifier used as a booster, no degradation at the receiver was observed down to a BER of 10^{-12}, even after transmission through 72.5 km of dispersion-shifted fiber. These experiments have revealed that waveguide amplifiers have the performances required to be efficient boosters, in-line amplifiers, or preamplifiers in digital telecommunication applications. The compatibility of waveguide amplifiers with analog transmission (AM-CATV) has also been demonstrated [44].

6.2 Gain Spectrum

Another very important point for system applications is the gain spectrum of the amplifier, which determines its compatibility with WDM transmission systems. Indeed, the broader the gain spectrum, the higher the number of wavelengths that can potentially be amplified. This gain spectrum is very dependent on the glass matrix used, as shown by J. Shmulovitch [10]. He compared the amplified spontaneous emission (ASE) spectra of three different glass waveguides: a soda-lime glass-based waveguide with a potential 17-nm gain bandwidth, a Ge/Al codoped silica glass waveguide with 40 nm of gain bandwidth, and an alumino-silicate glass showing an impressive bandwidth of more than 50 nm. Silica-titania glass waveguides, elaborated by the sol-gel technique, have also shown very broad and flat ASE spectra: 1-dB power variation over a 40-nm range [32]. These measurements show that broad gain spectra are potentially achievable with EDWAs by selecting a proper glass matrix.

Gain flattening, which is also an issue for WDM applications, will be obtained by combining filters, with appropriate spectra, with the waveguide amplifier. Pump power level and gain saturation, due to high input power signal, will also affect the gain flatness.

6.3 Lossless Splitters and Wavelengths Combiners

Passive integrated optical components often require an amplifier to compensate for their loss, for example, a waveguide that branches to produce two output channels. This is known as a *power splitter*, and in general a $1 \times N$ splitter. Clearly at each bifurcation there is a loss of power of about a factor 2. Integrate a splitter with a compensating amplifier and the power per output channel may be made equal to or greater than the input power, a zero-loss power splitter.

The power splitter is characterized by the number of its output ports (2, 4, 8, . . . , $N = 2^n$) and the degree of equality of the output power. Clearly, for a $1 \times N$ splitter we require a compensating gain greater than N. Excess losses in the system—junction, connection, and distributive losses—will also require compensating gain. Given connection losses of c dB per connection and y dB per junction and distributed losses of α dB, the total gain required for a $1 \times N$ ports system is:

$$\text{Loss (dB)} = \alpha + yn + 2c + 10_n \log 2 \tag{16}$$

Typical values for a 1×8 ($n = 3$) splitter made by ion exchange technology are $\alpha = 0.3$ dB, $y = 0.1$ dB, and $c = 0.2$ dB, which requires a compensatory gain of 10 dB [45].

One advantage of waveguide amplifiers is that they can be combined with other optical functions in a very compact, lossless assembly. These integrated optical functions can be made either with the same technology as the one used for the amplifier or with alternative technologies. 1×2, 1×4, and 1×8 lossless splitters have been realized [16,45,46] with packaged dimensions smaller than $8 \times 8 \times 120$ mm^3. The amplifier/splitter assembly in Fig. 14 presents a lossless operation in a 1530–1545-nm wavelength range, as shown in Fig. 15. Excess gain of up to 5 dB at the peak wavelength and a noise figure varying from 4.5 to 5.5 dB make this type of component desirable for distribution purposes in metropolitan or access networks.

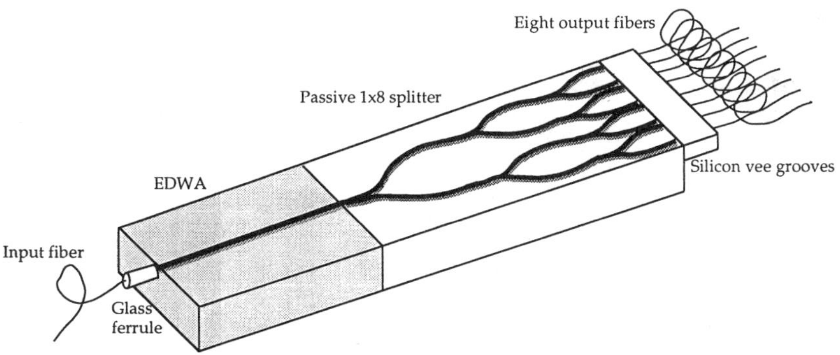

Figure 14 1×8 lossless splitter assembly, combining an Er-doped waveguide and a passive splitter.

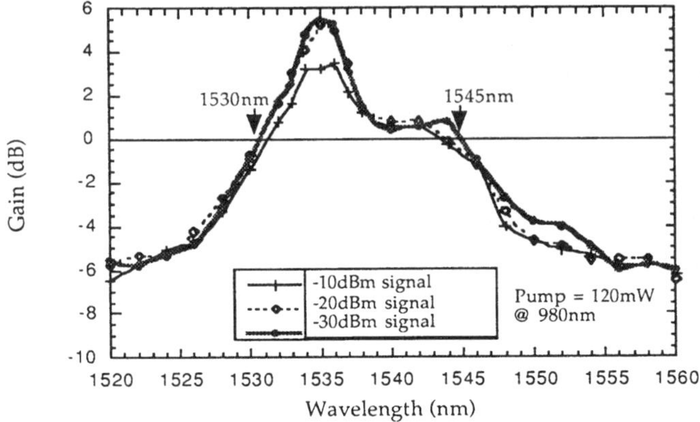

Figure 15 Gain spectrum of the 1 × 8 lossless splitter, for different input signal levels.

The same component can be used as a lossless eight-wavelength combiner, with each of the eight channels being launched in one of the eight ports of the splitter. Figure 16 shows the output spectrum of such a component when eight-18-dBm channels are combined. All channels present a signal-to-noise ratio higher than 32 dB, showing the potential compatibility of this combiner with 10-Gb/s WDM transmission systems.

Used as boosters, such lossless wavelength combiners have been evaluated in an 8 × 2.5-Gb/s transmission system experiment, without inducing any floor or degradation in the BER measurement; on the contrary, its use improves the transmission budget. The use of a 4 × 1 amplifying wavelength combiner has helped to evidence four-wave mixing in an 8 × 2.5-Gb/s bidirectional transmission experiment [47].

6.4 Multiplexer Integration

A duplexer or wavelength coupler is a three-port device. One port contains two beams of differing wavelength. The other ports carry single-wavelength signals. The input beam has been divided in terms of wavelength, or vice versa. This passive device will be used for coupling a pump and a signal wavelength into an integrated amplifier.

The extension of this function is a multiwavelength device, called the WDM, for wavelength-division multiplexer. This is a multiport device; therefore an integrated amplifier may be employed to compensate for the intrinsic loss and connection losses. Because the spectrum may be large (100 nm), the amplifiers will

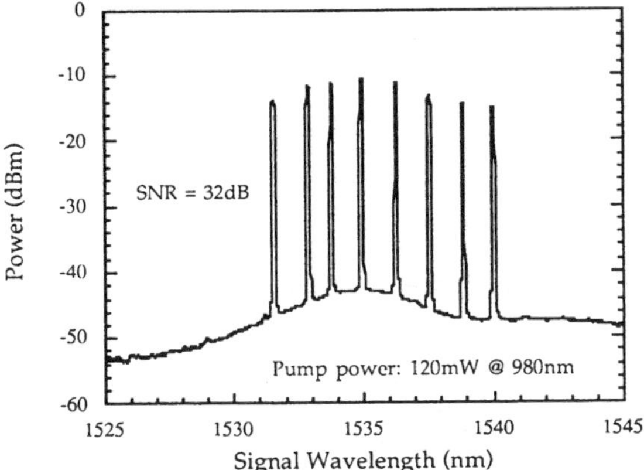

Figure 16 Output spectrum of the amplifying wavelengths combiner, when eight channels are combined. All channels present a signal-to-noise ratio of 32 dB after pre-equalization.

have to encompass this bandwidth with equal gain. Some gain flattening of the amplifier may be necessary.

6.5 Other Applications

Erbium-doped waveguide amplifiers can be combined with any lossy components, such as phased arrayed waveguides, add and drop multiplexers, modulators, switches, or optical cross-connects—in order to make very compact lossless devices. They can also be used as variable attenuators/amplifiers to adjust the signal level in WDM add/drop transmission systems. They can replace optical switches for routing purposes. And matrices of EDWAs can easily be realized for parallel data transmission or processing.

Furthermore, short EDWAs can be used as amplifying medium for solid-state 1.55-μm laser fabrication. Kitagawa et al. have demonstrated the feasibility of the single-longitudinal-mode distributed Bragg reflector (DBR) glass laser for WDM applications [48], with the Bragg mirrors being integrated directly with the amplifying waveguide by UV photoinscription, and GeeO has shown the tunability of DBR lasers, also for WDM applications [49]. The erbium-doped waveguide lasers were also used by Delavaux et al. to make 3-W peak-power Q-switched lasers for sensors and instrumentation applications [50] and by Jones et al. to make a 116-fs soliton source [51].

6.5 Summary

Erbium-doped waveguide amplifiers have proved their efficiency in system applications and will be found where moderate amplifier performances are acceptable but where compactness and/or cost is an important issue.

7 PIGTAILING AND PACKAGING OF ERBIUM-DOPED WAVEGUIDE AMPLIFIERS

Because they are based on planar technology, EDWAs have to be pigtailed with single-mode fibers to be inserted in the optical fiber network. And this induces new constraints on the planar amplifier. Indeed, the coupling loss, between the waveguide and the single-mode optical fiber used in the network, has to be lower than 0.3 dB, especially at the input of the amplifier, where any losses directly increase the noise figure.

Generally, waveguide amplifiers propagate highly confined optical modes, in order to maximize the pump efficiency, but these modes are much smaller in diameter than those of standard telecommunication fibers, which induces unacceptable coupling losses. Therefore there is a need for an adiabatic mode converter, or taper, integrated at the end of the waveguide and able to transform the confined mode of the waveguide into a mode as similar as possible to that of the fiber. Various methods are used to achieve this taper function. In the thin-film planar-waveguide–based technologies, the waveguide ends can be etched as a fan, which gives only a two-dimensional taper, or they can be etched as a needle, in which the modes enlarge as the guiding properties are reduced. For ion-exchanged waveguides, a localized rediffusion at the end of the guide will enlarge the mode. Very efficient tapers are made with this technology. Indeed, fiber-to-waveguide coupling losses lower than 0.2 dB were observed, with the losses in the taper itself being lower than 0.1 dB [14].

The input and output fibers may be different for an EDWA. The input fiber carrying the pump and the signal has to be monomode at both wavelengths, whereas the output fiber carries only the signal. The fibers are generally inserted in a glass ferrule or a silicon V-groove at the end of the polished guide before being aligned and glued via UV curing or with epoxy glue. A simple plastic package is then used to protect the assembly against shock and dust. The packaged component, with a remote pump laser, is then tested under given temperature-cycling conditions and humidity environments. So far the pigtailed waveguide amplifiers have shown excellent behaviors under such conditions, similar to that of passive components. Minor changes of the amplifier performance with temperature are due to known spectroscopic properties of erbium.

8 CHARACTERIZATION OF ERBIUM-DOPED WAVEGUIDE AMPLIFIERS

8.1 Gain and Noise Figure Measurements

At a development stage, as well as at the production stage, EDWAs have to be characterized for their gain, noise figure, losses, gain saturation, gain spectrum, etc. All these are measured with a setup similar to that represented in Fig. 17. An external cavity laser (ECL) delivers a signal at various wavelengths, and this signal is then amplified and launched into the EDWA to be tested, after being combined with the 980-nm pump. With an optical spectrum analyzer (OSA) we measure the signal and noise at the output of the waveguide, and by comparison to the input signal measured with the optical power-meter we calculate the gain and noise figure of our EDWA. For the small-signal gain and NF characterization we reduce the input signal level with the variable attenuator. This can be done at various wavelengths by adjusting the ECL. To analyze the saturation behavior (gain and NF evolutions for large input signals), we progressively increase the input signal power by adjusting the variable attenuator.

The gain is calculated with the following formula:

$$G\ (\text{dB}) = 10\log\frac{P_{\text{out}B} - P_{ASE}}{P_{\text{in}A}} \tag{17}$$

P_{ASE} is the power of the noise (amplified spontaneous emission) measured at a wavelength displacement of 0.5 nm from the signal. This measurement is correct for small signals, but it underestimates the gain and overestimates the noise figure for large signals. The exact noise value should be measured exactly at signal wavelength, but this requires the use of the "polarization nulling technique."

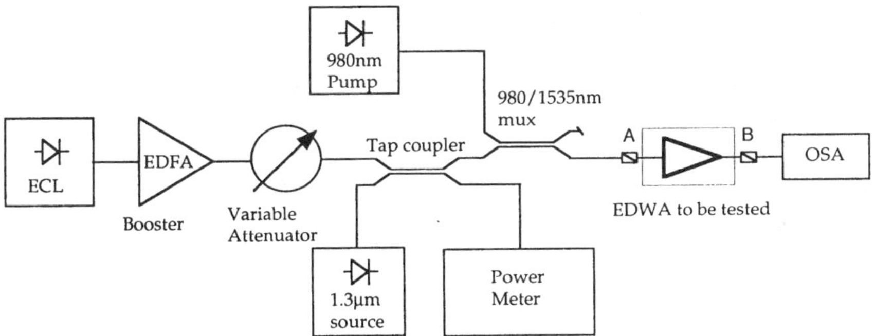

Figure 17 Characterization setup for EDWA gain and noise figure measurements.

The signal is polarized and the noise is not, so we can measure the noise at the signal wavelength, while the signal is present in the amplifier, by using two cross-polarizers. The measured noise value then has to be multiplied by 2, since the polarizer eliminates half of the noise power.

To calculate the noise figure we use the following expression:

$$\mathrm{NF\ (dB)} = 10 \log \frac{1}{10^{G/10}} \left(\frac{1 \cdot 677 \cdot 10^{-5}\, \lambda^3 P_{ASE}}{\Delta\lambda} + 1 \right) \tag{18}$$

G is the gain, in dB, P_{ASE} is the noise, in mW, λ is the signal wavelength, in nm, and $\Delta\lambda$ is the optical bandwidth, in nm, selected on the OSA.

8.2 Gain Spectrum

To analyze the spectral behavior of the EDWA we saturate it with one or more high-power input signals and examine the gain and noise figure with the tunable laser (ECL). The gain flatness may vary with the number and position of the saturating signals and also with the level of pump power launched into the amplifier. This spectral analysis is particularly important for WDM applications.

8.3 Pump Wavelength Dependence

The amplifier performance will vary with the detuning of the pump wavelength, especially for erbium/ytterbium codoped waveguides. The preceding characterization, reproduced for different pump wavelengths, will give the impact of this parameter on amplifier performance.

9 CONCLUSIONS

We have addressed the subject of erbium-doped waveguide amplifier components, which are rapidly developing into marketable devices. As we write, new performance data are being published. Consequently we have summarized their current performance and explained their advantages and limitations, in order to help readers understand and use these components.

To date, integrated lasers and amplifiers have been made with good performance, with some of them being already fully pigtailed and packaged and available as functional prototypes for system evaluation. But it is clear that their performance, as well as their integration with other optical functions, can and will be further improved, with such improvements being driven by the requirements of the rapidly changing fiber telecommunication industry.

The main advances will come from the integration effort. Other optical functions will be added to the amplifying function itself. For example, pump/signal multiplexers and tap couplers will soon be integrated, on a single chip, with the waveguide amplifier, in order to offer a complete amplifier device with limited performance for simple applications. Another very important evolution toward self-sufficient integrated amplifiers will be the direct coupling of the pump semiconductor laser with the pump input waveguide. This hybridization will greatly improve the compactness of the final device. This will require a large technological effort, but we believe it to be a necessary step for the generalization of the use of the waveguide optical amplifier.

On the material side, the erbium clusters limit the performance of the EDWA. A significant improvement of this performance is expected with more advanced glass fabrication techniques.

ACKNOWLEDGMENTS

The authors would like to thank all their colleagues at GeeO and TEEM Photonics, as well as Jean-Marc Delavaux of Lucent Technologies, who have contributed to the design, fabrication, and characterization of the EDWA. We also thank Mireille Lazarevitch and Marie-Anne Barbier for their support.

REFERENCES

1. T. Kitagawa, K. Hattori, K. Shuto, M. Yasu, M. Kobayashi, and M. Horiguchi. Amplification in erbium doped silica based planar lightwave circuit. Electron. Lett. 28: 1818–1819, 1992.
2. A. J. Faber, D. R. Simons, Y. Yan, and H. deWaal. Photoluminescence quenching by OH in Er- and Pr-doped glasses for 1.5 μm and 1.3 μm optical amplifiers. SPIE 80 2290:80, 1996.
3. E. Snoeks, P. G. Kik, and A. Polman. Concentration quenching in erbium implanted alkali silicate glasses. Optical Mater. 5:159–167, 1996.
4. François Auzel. Rare earth doped active films: basic problems to solve and hints for solutions. Proceedings of the International Conference on New Laser Technologies and Applications, Olympia, Greece, 1–4 June 1997.
5. J. E. Román, M. Hempstead, C. Ye, S. Nouh, P. Camy, P. Laborde, and C. Lerminiaux. 1.7μm Excited state absorbtion measurement in erbium-doped glasses. Appl. Phys. Lett. 67:470–472, 1995.
6. J. L. Philipsen and A. Bjarklev. Monte Carlo simulations of homogeneous upconversion in erbium-doped silica glasses. IEEE J. Quantum Electron. 33:845–854, 1997.
7. J. Shmulovitch, Y. H. Wong, G. Nykolak, P. C. Becker, R. Adar, A. J. Bruce, D. J. Muehlner, G. Adams, and M. Fishteyn. 15 dB net gain demonstration in Er glass waveguide amplifier on silicon. OFC '93 Postdeadline Paper Proceedings, PD-3, San Jose, CA, 1993.
8. P. Myslinski, D. Nguyen, and J. Chrostowski. Effects of concentration and clusters in EDFA. OFC '95 Technical Digest. San Diego, CA, February 26–March 3, 1995.

9. F. Auzel. Contribution à l'étude spectroscopique de verres dopés avec Er pour obtenir l'effet laser. Annales des Telecommunications 24:199–229, 1969.
10. J. Shmulovitch. Er-doped glass waveguide amplifier on silicon. Proceedings of Photonics West Conference, San Jose, CA, February 14–17, 1997.
11. Y. Yan, A. J. Faber, and H. De Waal. Luminescence quenching by OH groups in highly Er-doped phosphate glasses. J. Noncrystalline Solids 181:283–290, 1995.
12. Y. Leluyer. Mise au point d'une tecnologie silice dopée erbium en optique intégrée pour l'amplification optique vers 1.5 μm. Ph.D. dissertation, INPG, Grenoble, France, November 27, 1997.
13. J. M. P. Delavaux, Y. K. Park, E. Murphy, S. Grandlund, O. Mizuhara, D. Barbier, M. Rattay, G. Clauss, A. Kevorkian, and J. A. Nagel. High performance Er/Yb planar waveguide amplifiers as in-line and pre-amplifiers in 10 Gb/s fiber system experiments. ECOC '96 Proceedings, Oslo, Norway, September 1996, paper ThB3.6
14. D. Barbier, P. Bruno, C. Cassagnettes, M. Trouillon, R. L. Hyde, A. Kevorkian, and J. M. P. Delavaux. Net gain of 27 dB with a 8.6 cm long Er/Yb doped glass planar amplifier. OFC Technical Digest, San Jose CA, February 22–27, 1998.
15. T. Kitagawa. Rare-earth doped planar waveguide amplifiers. OAA '93 Proceedings, Yokohama, Japan, July 4–6, 1993, pp. 136–139.
16. P. Camy, J. E. Roman, F. W. Willems, M. Hempstead, J. C. van der Plaats, C. Prel, A. Beguin, A. M. J. Koonen, J. S. Wilkinson, and C. Lerminiaux. Ion-exchanged planar lossless splitter at 1.5 μm. Electron. Lett. 32:321–323, 1996.
17. Y. Tashiro, S. Koyanagi, K. Aiso, and S. Namiki. 1.5 W erbium doped fiber amplifier pumped by the wavelength division multiplexed 1480 nm laser diodes with fiber Bragg grating. OAA Technical Digest, Vail, Colorado, July 27–29, 1998.
18. A. K. Srivastava et al. 1 Tb/s transmission of 100 WDM 10 Gb/s channels over 400 km of TrueWave fiber. OFC Proceedings, PD 10, San Jose, CA, February 22–27, 1998, Postdeadline Paper.
19. J. H. Lee and N. Park. Temperature dependent distortion of multichannel gain flatness for silica and ZBLAN based erbium amplifiers. OFC Technical Digest, San Jose, CA, February 22–27, 1998.
20. D. E. McCumber. Einstein relations connecting broadband emission and absorption spectra. Physical Rev. 136:A954–A957, 1964.
21. W. J. Miniscalco and R. S. Quimby. General procedure for the analysis of erbium cross-sections. Optics Lett. 18:258–260, 1991.
22. E. Snoeks, G. N. van den Hoven, A. Polman, B. Hendrikssen, M. B. J. Diemeer, and F. Priolo. Cooperative upconversion in erbium-implanted soda-lime silicate glass optical waveguides. J. Optical Soc. B 12:1468, 1995.
23. J. A. Duffy. Bonding Energy Levels and Bands in Inorganic Solids. New York: Longman Scientific and Technical, 1990.
24. E. Delevaque, T. Georges, M. Monerie, P. Lamouler, and J. F. Bayon. Modeling of pair-induced quenching in erbium-doped silicate fibers. I.E.E.E. Photon. Technol. Lett. 5:73–75, 1993.
25. B. Jacquier, E. Lebrasseur; E. Josse, J. Lucas, B. Boulard, C. Jacoboni, J. E. Broquin, and R. Rimet. Optical amplification in rare earth doped fluoride channel waveguide. CIMTEC proceedings, Florence, Italy, 14–19 June, 1998, p. 254.

26. Y. C. Yan, A. J. Faber, H. de Waal, P. G. Kik, and A. Polman. Net optical gain at 1.53 μm in an Er-doped phosphate glass waveguide on silicon. Appl. Physics Lett. 71:P 2922, 1997.
27. D. Barbier, J. M. Delavaux, A. Kevorkian, P. Gastaldo, and J. M. Jouanno. Yb/Er integrated optics amplifiers on phosphate glass in single and double pass configurations. OFC '95 Postdeadline Papers, San Diego, CA, February 26–March 3, 1995.
28. P. Camy, J. E. Roman, M. Hempstead, P. Laborde, and C. Lerminiaux. Ion-exchanged waveguide amplifier in erbium doped glass for broadband communications. OAA '95 Proceedings, Davos, Switzerland, 15–17 June 1995, pp. 181–184.
29. G. L. Vossler, C. J. Brooks, and K. A. Winick. Planar Er:Yb glass ion exchanged waveguide laser. Electron. Lett. 31:1162–1163, 1995.
30. B. Pedersen, T. Feuchter, M. Poulsen, J. E. Pedersen, M. Kristensen, and R. Kromann. High concentration erbium doped silica on silicon grown by plasma enhanced CVD. ECIO '95 proceedings, Delft, the Netherlands, April 1995, pp. 411–414.
31. T. Kitagawa, K. Hattori, M. Shimizu, Y. Ohmori, and M. Kobayashi. Guided wave laser based on erbium doped silica planar lightwave circuit. Electron. Lett. 27:334–335, 1991.
32. X. Orignac and D. Barbier. Potential for fabrication of sol-gel derived integrated optical amplifiers. SPIE 2997: Integrated optics devices, potential for commercialization 1997, pp. 277–283.
33. G. N. van der Hoven, R. J. I. M. Koper, A. Polman, C. van Dam, J. W. M. van Uffelen, and M. K. Smit. Net optical gain at 1.53 μm in Er doped Al_2O_3 waveguides on silicon. Appl. Physics Lett. 68:1886–1888, 1996.
34. H. J. van Werden, T. H. Hoekstra, P. V. Lambeck, and Th. J. A. Popma. Low threshold amplification at 1.5 μm in Er:Y_2O_3 IO-amplifiers. ECIO '97 Proceedings, Stockholm, Sweeden, April 2–4, 1997, pp. 169–172.
35. J. E. Roman, M. Hempstead, W. S. Brocklesby, S. Nouh, J. S. Wilkinson, P. Camy, C. Lerminiaux, and A. Beguin. Diode pumped ion-exchanged Er/Yb waveguide laser at 1.5 μm in phosphorus-free silicate glass. ECIO '95 Postdeadline Papers Proceedings, Delft, the Netherlands, April 1995, pp. 13–16.
36. C. C. Li, H. K. Kim, and M. Migliuolo. Er doped glass ridge waveguide amplifiers fabricated with a collimated sputter deposition technique. IEEE Photon. Technol. Lett. 9:1223–1225, 1997.
37. M. Krishnaswamy, J. N. McMullin, B. P. Keyworth, and J. S. Hayden. Optical properties of strip loaded Er doped waveguides. Optical Mater. 6:287–292, 1996.
38. T. Touam, G. Milova, Z. Saddiki, M. A. Fardad, M. P. Andrews, S. Juma, J. Chrostowski, and S. I. Najafi. Organoaluminophosphate sol-gel silica glass thin films for integrated optics. SPIE 2997:79–84, 1997.
39. R. Brinkmann, I. Baumann, M. Dinand, W. Sohler, and H. Suche. Er-doped single and double pass Ti:$LiNbO_3$ waveguide amplifiers. J. Quantum Electron. 30:2356–2360, 1994.
40. T. Kitagawa, K. Hattori, K. Shuto, M. Oguma, J. Temmyo, S. Suzuki, and M. Horiguchi. Er-doped silica based planar amplifier module pumped by laser diodes. Proceedings of ECOC '93, Montreux, Switzerland, 1993, postdeadline paper ThC 12.11.
41. K. Shuto, K. Hattori, T. Kitagawa, Y. Ohmori, and M. Horiguchi. Er-doped phospho-

silicate glass waveguide amplifier fabricated by PECVD. Electron. Lett. 29:139–141, 1993.

42. D. Barbier, J. M. P. Delavaux, T. A. Strasser, M. Rattay, R. L. Hyde, P. Gastaldo, and A. Kevorkian. Sub-centimeter length ion-exchanged waveguide lasers in Er/Yb doped phosphate glass. Submitted to ECOC '97, Edinburgh, Scotland, September 22–25, 1997.
43. J. M. P. Delavaux, S. Grandlund, O. Mizuhara, L. D. Tzeng, D. Barbier, M. Rattay, F. Saint André, and A. Kevorkian. Integrated optics Er/Yb amplifier system in 10 Gb/s fiber transmission experiment. IEEE Photon. Technol. Lett. 9:247–249, 1997.
44. J. C. Van Der Plaats, F. W. Willems, W. Muys, A. M. J. Koonen, P. Camy, C. Prel, A. Beguin, M. Prassas, C. Lerminiaux, J. E. Roman, M. Hempstead, and J. Wilkinson. Ion-exchanged planar lossless splitter for analog CATV distribution systems at 1.5 μm. OAA Proceedings, Monterey, CA, July 11–13, 1996, pp. 250–253.
45. D. Barbier, M. Rattay, N. Krebs, F. Saint André, G. Clauss, and J. M. P. Delavaux. Lossless 1 × 8 splitter integrated on Er/Yb doped phosphate glass and silicate glass. ECIO '97 Proceedings, Stockholm, Sweden, April 2–4, 1997, pp. 161–164.
46. D. Barbier, M. Rattay, F. Saint André, G. Clauss, M. Trouillon, A. Kevorkian, J. M. P. Delavaux, and E. Murphy. Amplifying four wavelength combiner, based on Er/Yb doped waveguide amplifiers and integrated splitters. IEEE Photon. Technol. Lett. 9:315–317, 1997.
47. Y. Jaouen, J. M. P. Delavaux, and D. Barbier. Repeaterless bidirectional 8 × 2.5 Gb/s WDM fiber transmission experiment. Optical Fiber Technol. 3: 1997.
48. T. Kitagawa, F. Bilodeau, B. Malo, S. Theriault, J. Albert, D. C. Jihnson, K. O. Hill, K. Hattori, and Y. Hibino. Single frequency Er-doped silica based planar waveguide laser with integrated photo-imprinted Bragg reflectors. Electronics Lett. 30:1311–1312, 1994.
49. D. Barbier, J. M. Delavaux, R. L. Hyde, J. M. Jouanno, A. Kevorkian, and P. Gastaldo. Tunability of Yb/Er Integrated optical lasers in phosphate glass. OAA '95, Davos, Switzerland, June 15–17, 1995, postdeadline paper PD3.
50. J. M. P. Delavaux, A. Yeniay, J. Toulouse, T. A. Strasser, R. Pedrazanni, and D. Barbier. Q-Switched Er/Yb co-doped channel waveguide laser. ECIO '97 Proceedings, Stockholm, Sweden, April 2–4, 1997, pp. 173–176.
51. D. J. Jones, S. Namiki, D. Barbier, E. P. Ippen, and H. A. Haus. 116 fs soliton source based on an Er/Yb co-doped waveguide amplifier. IEEE Photon. Technol. Lett. 10: 666–668, May 1998.
52. Jean-Marc Jouanno. Amplification optique dans des guides d'ondes realises per echange d'ions dans des verres phosphates dopes par des terres rares. Ph.D. dissertation, Institut National Polytechnique de Grenoble, France, January 19, 1995.
53. E. Desurvire. Erbium-Doped Fiber Amplifiers—Principles and Applications. New York: Wiley, 1994.

6

Erbium-Doped Lithium Niobate Waveguide Devices

Wolfgang Sohler and Hubertus Suche

University of Paderborn
Paderborn, Germany

1 INTRODUCTION

In recent years there has been a growing interest in rare-earth-doped optically pumped amplifier and laser devices in $LiNbO_3$ [1–7]. In particular, Er-doped devices operating in the third telecommunication window around 1.55-μm wavelength have attracted much attention [8]. The combination of the amplifying properties of the dopant erbium with the excellent electro-optical and acousto-optical properties of the waveguide substrate $LiNbO_3$ allows the development of a whole class of new waveguide devices of higher functionality. Loss compensated or even amplifying devices have been fabricated by doping the corresponding waveguide structures [9,10]. Very simple but efficient waveguide lasers have been developed [11]. By intracavity integration of modulators and wavelength filters, mode-locked [12,13], Q-switched, and tunable lasers [14] have been realized. Moreover, the different types of lasers and amplifiers can be combined with other active and passive devices on the same substrate [15] to form integrated optical circuits (IOCs) for a variety of applications in optical communications, sensing, signal processing, and measurement techniques.

It is the aim of this chapter to review the state of the art of Er-doped integrated optical devices in $LiNbO_3$. In Sec. 2, the technology of fabricating Er-diffusion-doped waveguides in $LiNbO_3$ substrates is briefly discussed. If optically pumped, such doped waveguides can be used as optical amplifiers (Sec. 3). Examples of loss-compensated and amplifying devices are given in Sec. 4. In Sec. 5, the different types of lasers demonstrated up to now are presented. A laser/modulator transmitter unit is described in Sec. 6 as an attractive example of a monolithically integrated optical circuit in $LiNbO_3$. Finally, the chapter ends with some concluding remarks.

2 ERBIUM-DIFFUSION-DOPED WAVEGUIDES

In principle, there are two ways to fabricate Er-doped waveguides in $LiNbO_3$: homogeneously doped or surface-doped crystalline wafers can be used as waveguide substrate. Homogenous doping can be achieved during crystal growth from an Er-doped melt [16]. However, it is difficult to achieve high-quality striation-free crystals of high doping concentration and large size. The introduction of rare-earth ions tends to increase the number of domains in the crystal [17]. On the other hand, erbium doping of a surface layer as an alternative can be achieved by implantation and annealing [18] or by in-diffusion of vacuum-deposited erbium layers [19,20]. For these techniques, commercially available $LiNbO_3$ wafers of high optical quality and large diameter can be used. The latter method is easy and is ideally suited for a selective doping of the surface using photolithographic patterning of the evaporated erbium or Er_2O_3 layers, respectively. Selective doping is a prerequisite for the monolithic integration of active (optically pumped, Er-doped) and passive (unpumped) devices on the same substrate to avoid reabsorption in unpumped Er-doped waveguides. The process of erbium in-diffusion has been characterized in detail using secondary ion mass spectroscopy (SIMS), secondary neutral mass spectroscopy (SNMS), Rutherford back-scattering spectroscopy (RBS), and atomic force microscopy (AFM) [21]. The site of the Er^{3+} ions in the $LiNbO_3$ lattice has been determined using X-ray standing wave spectroscopy (XSW) [22] and optical site-selective spectroscopy (OSSS) [23].

The diffusion of erbium into $LiNbO_3$ can be described by Fick's laws with a concentration-independent diffusion coefficient and a temperature-dependent maximum solubility [21]. Typical Er-doping profiles before and after the fabrication of a planar titanium in-diffused waveguide in a $\hat{Z}$-cut $LiNbO_3$ substrate are shown in Fig. 1 together with the titanium profile.

Due to the anisotropic crystal structure of $LiNbO_3$, the diffusion coefficient of Er in $LiNbO_3$ depends on the crystal cut with the highest diffusivity along the crystal $\hat{Z}$-axis. This can be clearly seen in Fig. 2, where the temperature-dependent diffusion coefficients derived from measured Er concentration profiles have been plotted as Arrhenius diagrams [21]. From the slope and offset of the graphs, the cut-dependent diffusion constants D_0 and activation energies E_a have been determined, respectively. The diffusion coefficients are about two orders of magnitude smaller than those of titanium. Therefore, the Er diffusion doping has to be done prior to the waveguide fabrication. To achieve a good overlap of doping profile and the optical waveguide modes, the Er diffusion has to be performed at a temperature as close as possible to the Curie temperature of ferroelectric

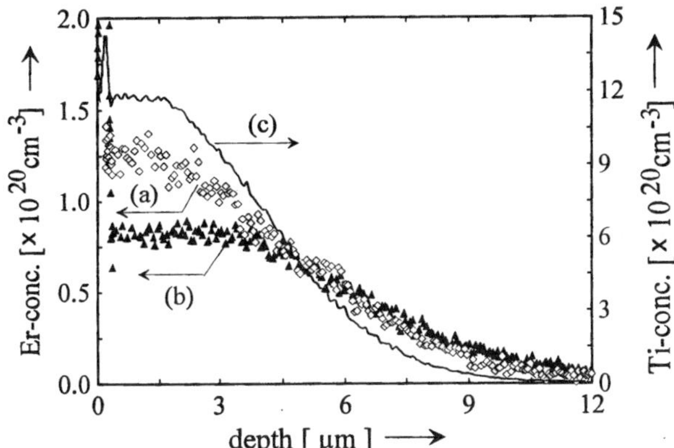

Figure 1 Erbium and titanium concentrations versus depth in a Ẑ-cut $LiNbO_3$ wafer measured by SNMS (Er) and SIMS (Ti), respectively. Erbium profile after in-diffusion of a 22-nm-thick erbium layer at 1130°C within 100 h (a). Erbium profile after additional in-diffusion of a 95-nm-thick titanium layer at 1030°C within 9 h (b). Corresponding titanium distribution (c). (From Ref. 51, © 1996 IEEE.)

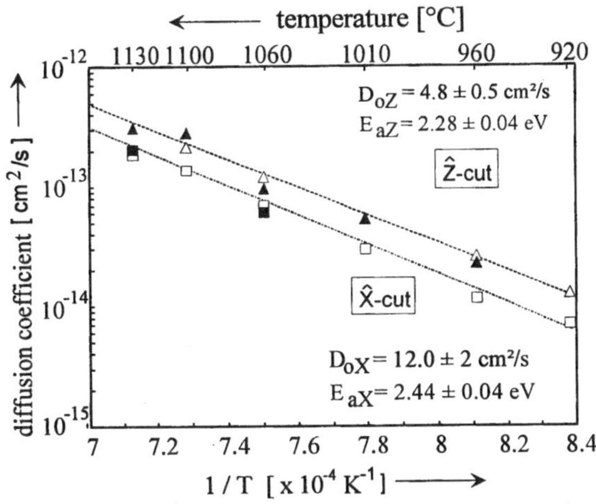

Figure 2 Arrhenius plots of the diffusion coefficients of erbium into $LiNbO_3$ for diffusion parallel to the *c*-axis—Ẑ-cut—(triangles) and perpendicular to the *c*-axis—X̂-cut—(squares) of the $LiNbO_3$ crystal derived from the concentration profiles determined by SIMS (white) and SNMS (black) measurements (see Fig. 1). (From Ref. 51, © 1996 IEEE.)

$LiNbO_3$ (1142°C for congruent melting material [24]) for at least 100 h (see also Fig. 1). The solubility of erbium (about 1.7×10^{20} cm^{-3} at 1100°C) is about one order of magnitude smaller than that of titanium; it grows exponentially with the temperature [21].

During in-diffusion of a vacuum-deposited Er layer, the surface concentration of the dopant corresponds to the solubility level as long as the diffusion reservoir is not exhausted. Afterwards the concentration drops according to Fick's laws. To prevent the formation of precipitates of Er_xNb_y oxide compounds during waveguide fabrication, the Er surface concentration should be at least slightly below the solid solubility level at the waveguide fabrication temperature.

Erbium is incorporated into $LiNbO_3$ on a vacant Li site or it replaces lithium [22]. The exact position compared with the Li site is shifted by 0.46 Å in $(-c)$-direction. Absorption measurements and site-selective spectroscopy found four slightly different erbium sites [23]. This has been attributed to perturbations of the local crystal field due to variations of the arrangement of charge-compensating defects.

Up to now, Er-doped waveguides of very low scattering losses could be fabricated only by Ti in-diffusion into the Er-doped $LiNbO_3$ surface; the proton exchange technique gave waveguides of lower quality and drastically reduced fluorescence yield. To be specific, scattering losses below 0.1 dB/cm have been measured in Ti-doped waveguides with an Er surface concentration up to $\approx 5 \times 10^{19}$ cm^{-3}.

3 WAVEGUIDE AMPLIFIERS

Erbium is incorporated into the $LiNbO_3$ lattice as Er^{3+} ions, preferentially at Li^+ sites [22]. Their $4f^{11}$ electron configuration leads to the ground state $^4I_{15/2}$; it is split into $J + 1/2 = 8$ doubly degenerate levels by the Stark effect in the $LiNbO_3$ crystal field of axial symmetry C_{3v}. The first excited state $^4I_{13/2}$ is split into seven Stark sublevels. Transitions between both manifolds determine absorption and optical amplification by stimulated emission in the wavelength range $1.44\ \mu m < \lambda < 1.64\ \mu m$.

Figure 3 shows the full energy level scheme of Er:$LiNbO_3$ up to a wavenumber (energy) of about 23,000 cm^{-1} as measured by Gabrielyan et al. [25]. The upward-pointing (straight-line) arrows represent possible transitions for optical pumping to populate (after a fast relaxation) the metastable $^4I_{13/2}$ level. If population inversion with respect to the ground state is achieved, optical amplification of a signal wave sets in. The preferred absorption band for optical pumping of Er:$LiNbO_3$ is around $\lambda_p \approx 1480$ nm. Wavelengths in the visible ($\lambda_p \sim 530$–~ 660 nm) and near infrared ($\lambda_p \sim 808$ nm and $\lambda_p \sim 980$ nm) lead to excited-state absorption (ESA), indicated by the upward-pointing (dotted) arrows in Fig. 3. Several (radiative and nonradiative) decay channels are possible to the ground

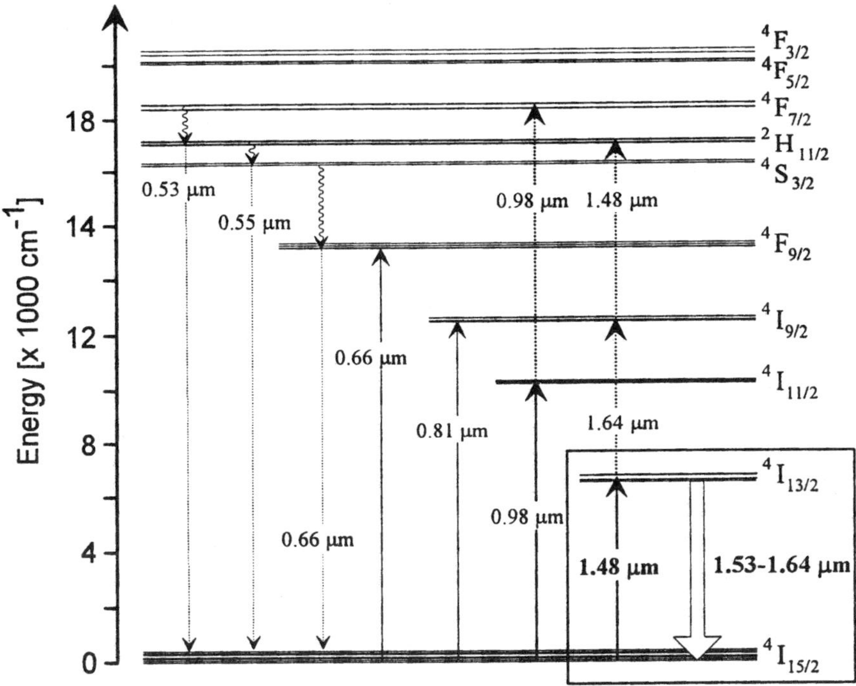

Figure 3 Energy level scheme of Er:$LiNbO_3$. (From Ref. 25.)

state (see, e.g., the downward-pointing (dotted) arrows). These processes, which are weak for $\lambda_p \approx$ 1480-nm pumping, diminish the optical gain by reducing the population density of the $^4I_{13/2}$ level. There are further reasons to prefer $\lambda_p \approx$ 1480-nm pumping: the optical channel guides can be fabricated as single-mode waveguides at both pump and signal wavelengths. This guarantees a good mode overlap and, moreover, a good overlap with the erbium doping profile, sufficient depth provided. Furthermore, high-power laser diodes as pump sources are commercially available in that wavelength range.

Optical amplification in the wavelength range 1530 nm $< \lambda <$ 1620 nm has been intensively investigated for different Er doping levels, and in two different configurations: single pass and double pass [26]. In the latter case, a highly reflecting broadband mirror has been deposited on the rear endface of the waveguide to double the interaction length of pump and signal modes. In this way the pump absorption efficiency is improved and the signal gain is increased.

To excite the Er ions, a laser diode of $\lambda_p \approx$ 1480-nm emission wavelength has been used. In Fig. 4 the results of the pump power-dependent signal transmis-

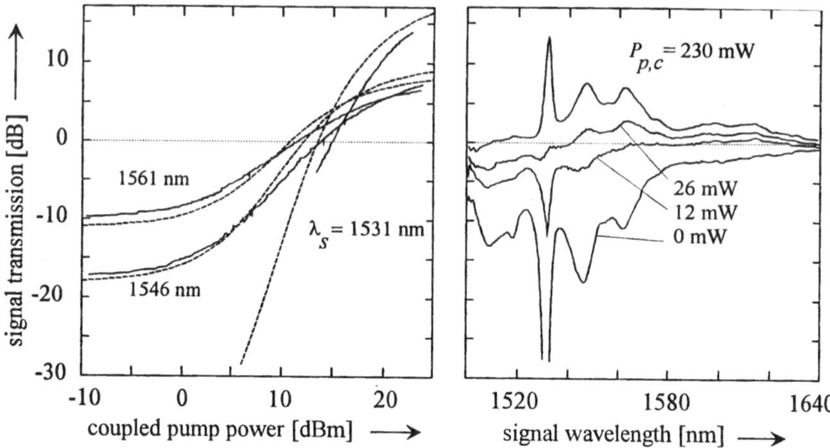

Figure 4 Signal transmission (net absorption/gain) of a 70-mm-long optically pumped Ti:Er:$LiNbO_3$ ($\hat{Z}$-cut) waveguide amplifier versus signal wavelength at different levels of coupled pump power (right) and versus coupled pump power (left) (λ_p = 1480 nm) at the signal wavelengths of gain maxima. Doping and waveguide fabrication correpond to the data of Fig. 1. Both signal and pump polarization are TE (σ). Solid lines: experimental; dashed lines: calculated data. (From Ref. 51, © 1996 IEEE.)

sion of a 7-cm-long single-pass amplifier are shown as an example [26]. The investigated waveguide has been fabricated by in-diffusion at 1030°C for 9 h of a 6-μm-wide, 95-nm-thick Ti stripe into a $\hat{Z}$-cut $LiNbO_3$ substrate Er-doped near the surface by in-diffusion of a 22-nm-thick evaporated Er layer at 1130°C during 130 h. Both pump and signal polarization have been adjusted perpendicular (σ) to the crystal *c*-axis, exciting TE modes in the waveguide. The right-hand diagram gives the transmission spectra for different levels of the coupled pump power. With increasing pump power, the absorption is bleached and gradually converted into gain, first at longer wavelengths. At the highest pump power, the maximum gain peak of 13.8 dB is at 1531-nm wavelength. In the diagram on the left, the measured (solid lines) and calculated dependence (dotted lines) of three gain maxima on the coupled pump power are plotted. It should be emphasized that the length of the amplifying waveguide is far below the optimum length of full pump exploitation.

To increase the effective amplifier length, the Er-doped waveguide devices have been coated with an Au mirror at one endface, resulting in a double pass of signal and pump radiation, simultaneously. The results of the absorption/gain measurements are shown in Fig. 5 for a 48-mm-long $\hat{Z}$-cut sample. On the left-hand side, the dependence of absorption/gain on the coupled pump power $P_{p,c}$

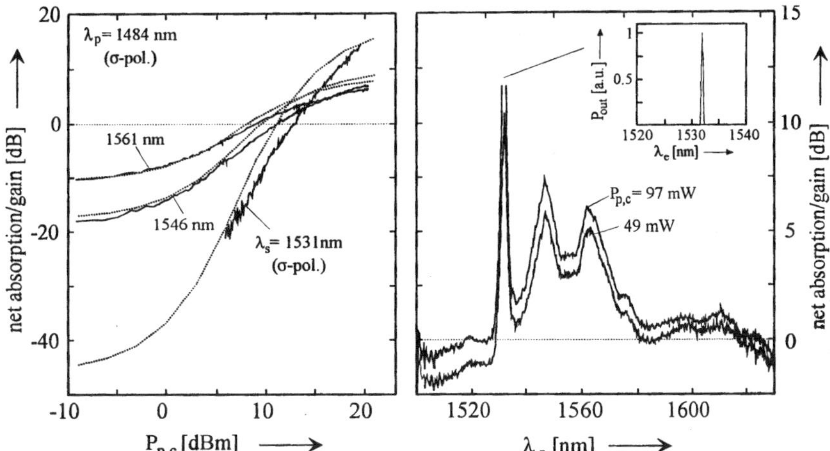

Figure 5 Small-signal net absorption/gain in a 48-mm-long Er-doped $\hat{Z}$-cut Ti: $LiNbO_3$ double-pass waveguide amplifier versus coupled pump power $P_{p,c}$ for different signal wavelengths λ_s (left) and versus λ_s for different levels of $P_{p,c}$ (right). Dotted lines: calculated results using the model presented in Ref. 27. Inset: laser emission at $\lambda_e = 1531$ nm with $P_{p,c} > 90$ mW. (From Ref. 26, © 1994 IEEE.)

is shown for the same signal wavelengths λ_s as earlier for the single-pass amplifier. With only 90 mW of coupled pump power, a maximum gain of 14.7 dB was obtained for $\lambda_s = 1531$ nm. At a pump power level of 50 mW, already more than 10 dB gain was achieved.

On the right-hand side, the spectral dependence of the gain is shown for two different levels of coupled pump power $P_{p,c}$. At $P_{p,c} = 97$ mW, the gain exceeded 3.8 dB in the wavelength range 1542 nm $< \lambda_s <$ 1569 nm, with maxima of 7.3 dB at $\lambda_s = 1546$ nm and 6.2 dB at $\lambda_s = 1561$ nm.

At $\lambda_s = 1531$ nm, the gain for $P_{p,c} = 90$ mW was already high enough to compensate for the output coupling losses of the resonator formed by the Au mirror ($R \approx 96\%$) on the amplifier endface and the endface of the input fiber ($R \approx 3.4\%$) almost in contact with the AR-coated waveguide; lasing set in (see inset in Fig. 5). The losses of this resonator can be calculated to be approximately -14.9 dB, which agrees well with the measured amplification.

To illustrate the potential for further improvements, Fig. 6 presents the predicted net gain versus waveguide length (lines) together with selected experimental results (marks) for single-pass (left) and double-pass amplifiers (right). With a double-pass device fabricated in a commercially available $LiNbO_3$ wafer of 4-inch-diameter, net gain up to 40 dB seems to be possible at a signal wavelength of

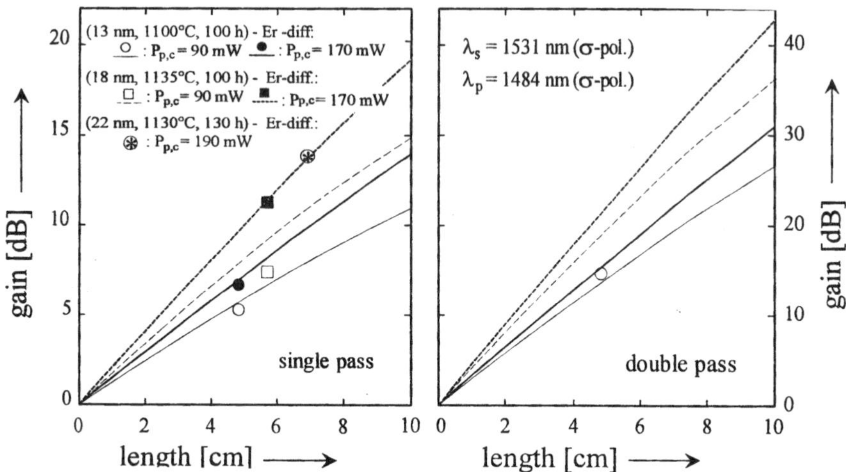

Figure 6 Calculated gain (lines) and measured gain (marks) versus length of Ti:Er:$LiNbO_3$ waveguide amplifiers for single-pass (left) and double-pass configurations (right). Thickness of in-diffused Er layers, diffusion temperatures, diffusion time, and coupled pump power levels are given as inset. (From Ref. 51, © 1996 IEEE.)

λ_s = 1531 nm; however, up-conversion effects, which could reduce the amplifier efficiency at high signal levels, have not been taken into account in the calculations.

4 LOSS-COMPENSATED AND AMPLIFYING DEVICES

In $LiNbO_3$, a variety of passive and active devices of excellent performance, e.g., filters, modulators, and switches, have been developed in the past for which coupling and internal loss (over)-compensation would be highly desirable (''0 dB- and amplifying devices''). Erbium doping can be an attractive means to provide the necessary gain by optical pumping. In the following, two specific examples of such devices are presented.

4.1 Loss-Compensated Electro-Optical Spectrum Analyzers

Finesse enhancement in waveguide cavities by internal gain was first demonstrated in Er-doped fiber- and silica-based integrated optical ring cavities [28,29]. However, these materials only allow a slow thermo-optical or mechanical scanning of the cavity resonances. For fast spectral analysis, an electrooptically tunable Ti:$LiNbO_3$ waveguide cavity is attractive as an integrated optical spectrum analyzer (IOSA); the scanning speed is limited only by the finite cavity build-up time. We have utilized an Er-diffusion-doped intracavity amplifier for loss compensation, leading to finesse enhancement and in this way to a strong increase of the spectral resolution [9]. The 48-mm-long device is an Er-doped Ti:$LiNbO_3$ waveguide resonator with monolithically integrated phase modulator and dielectric mirrors on both endfaces. It has been fabricated in $\hat{X}$-cut $LiNbO_3$, doped near the surface by in-diffusion of a 9.2-nm-thick erbium layer at 1100°C during 100 h. The cavity mirrors have been deposited using ion-beam-assisted evaporation of SiO_2/TiO_2 layers. To allow endfire pumping of the active IOSA, the pump coupler has been made dichroitic with a high transmittance at the pump wavelength ($T \simeq 0.65$ at $\lambda_p = 1479$ nm) and simultaneously a high reflectance ($R \simeq 0.97$) at 1546 nm, the wavelength of the investigated signal source [distributed feedback (DFB) laser diode]. The output coupler has a reflectance of about 0.98 at both pump and signal wavelengths, allowing a double pass of the pump in the cavity. The active IOSA has been diode-pumped at $\lambda_p = 1479$ nm ($E_p \parallel c$). Linear scanning of the cavity resonances has been achieved by driving the intracavity electro-optic phase modulator ($U_\pi \simeq 10.4$ V) with a sawtooth voltage. The results of such a measurement are shown in Fig. 7 with the transmitted spectral power density of the investigated signal source (DFB laser) plotted against time. The change of the sawtooth drive voltage by V_π within $\simeq$ 300 μs corresponds to one free spectral range FSR = 1.423 GHz of the cavity. The parameter of the set of graphs is the launched pump power. A finesse up to 205 has been achieved, corresponding to a spectral resolution of about 6.9 MHz, which allowed measurement of the spectral width (16 MHz) of the DFB laser diode investigated. This improved resolution is accompanied by an increase in the cavity transmittance

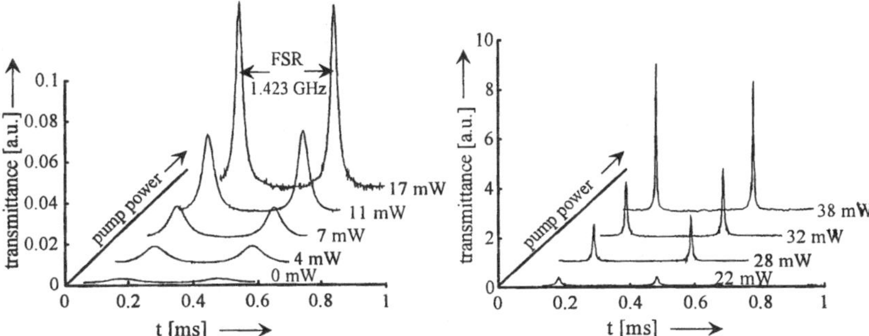

Figure 7 Transmittance of an electro-optically scanned Ti:Er:$LiNbO_3$ waveguide resonator versus time, with the coupled pump power as parameter. The sawtooth drive voltage changes by V_π within $\simeq$300 μs, corresponding to one free spectral range FSR = 1.423 GHz of the cavity. (From Ref. 51, © 1996 IEEE.)

by three orders of magnitude. In this way the spectral investigation of low-power sources is facilitated.

4.2 Amplifying Acousto-Optically Tunable Wavelength Filters

Acousto-optically tunable wavelength filters are very attractive devices for wavelength-division multiplex applications in optical communications due to their unique multiwavelength filtering capability [30]. However, filters of low crosstalk require double-stage conversion concepts [31], inducing increased insertion losses. Therefore, loss compensation or even overall gain is desirable.

As a first step toward a fully integrated, amplifying, double-stage filter, a single-stage acousto-optically tunable TE-TM-mode converter has been fabricated in an Er-doped $LiNbO_3$ substrate. In combination with an external polarizer, tunable filtering over a wavelength range of 70 nm accompanied by amplification has been demonstrated [10]. The structure of the mode converter is shown, together with the filter response at different pump power levels, in Fig. 8. The device has been fabricated in an Er-diffusion-doped $\hat{X}$-cut $LiNbO_3$ substrate. It consists of a combined acoustical and optical waveguide structure, which is defined by titanium in-diffused regions forming the claddings (b) of the acoustical waveguide of 100-μm width and by a 7-μm-wide Ti-diffused stripe forming the core of the optical channel guide (c). The interdigital transducer (a) for the excitation of the guided surface acoustical wave (SAW) has been photolithographically defined. An acoustical absorber (d) terminates the interaction length of SAW and optical modes.

To investigate the performance of the device, a fiber-pigtailed luminescence diode has been used as signal source and a tunable color center laser (CCL) as the pump. The input polarization of both signal and pump have been adjusted to TM (σ). To block the nonpolarization-converted signal at the output, an external polarizer has been used. The signal was then fed into a monochromator of 0.5-nm resolution for spectral analysis.

In the lower part of Fig. 8 the signal transmission through the acousto-optical filter is shown versus wavelength for different levels of coupled pump power (λ_p = 1484 nm). The acoustical frequency, which determines the wavelength of phase-matched acousto-optical mode conversion (here, λ_s = 1531 nm), was 174.94 MHz; the RF power level has been adjusted for complete conversion. At coupled pump power levels higher than about 30 mW, the signal transmission is larger than unity. In this case the filter acts as a narrow-band tunable amplifier. A maximum amplification of 4.8 dB with a coupled pump power of 22 dBm ($\simeq$160 mW) has been obtained at λ_s = 1531 nm. With a coupled pump power level of 140 mW, the gain is higher than 1.6 dB in the whole wavelength band 1540 nm $< \lambda_s <$ 1565 nm, with maxima of 2.9 dB at 1546 nm and 2.6 dB at

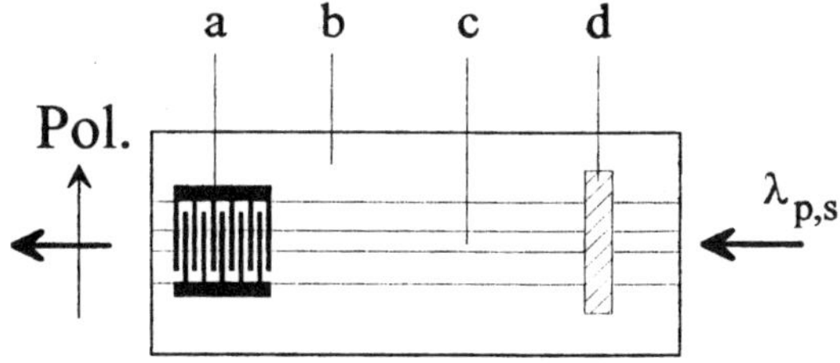

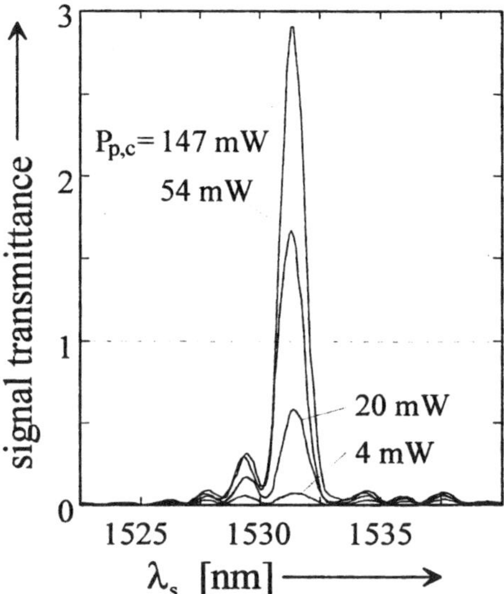

Figure 8 Upper part: Scheme of the amplifying single-stage Ti:Er:$LiNbO_3$ mode converter with external polarizer (Pol.). a: interdigital transducer; b: Ti-diffused claddings of the acoustic waveguide; c: Ti-diffused optical waveguide; d: acoustical absorber. Lower part: transmittance of the filter versus wavelength for different levels of coupled pump power; incident signal, and pump polarization: σ (TM); pump wavelength: 1484 nm; acoustical frequency: 174.94 MHz; rf-drive power: 17 dBm. (From Ref. 51, © 1996 IEEE.)

1561 nm. If a flat gain is required, the transmission of the device can be adjusted by appropriate levels of driving RF power.

These results are encouraging. However, the narrow-band amplifier has a poor side-lobe suppression of the filter characteristics, the frequency of the converted wave is shifted by the acoustical frequency, the required polarizer is not integrated, and the device does not operate polarization independent. Therefore, fully integrated versions of a doped double-stage filter with integrated polarization splitters and a second frequency-shift-compensating mode converter have been realized. One specific structure is presented in Sec. 5.2 as a key element of a tunable laser.

5 WAVEGUIDE LASERS

Waveguide amplifiers are the basic devices to develop integrated lasers. Incorporated in an optical cavity to achieve the necessary feedback, laser oscillation can be obtained by sufficient optical pumping exceeding the threshold level. The simplest laser is a free-running (without intracavity control devices) Fabry–Perot laser. It has the lowest intracavity losses and, therefore, the potential for a high power conversion efficiency. Such a device of optimized efficiency will be presented in Sec. 5.1. More advanced lasers can be developed by integrating a modulator or (and) a wavelength filter in the waveguide resonator. In this way an amplifying acousto-optical filter leads to tunable laser operation; a corresponding device is presented in Sec. 5.2. An intracavity phase or amplitude modulator allows one to obtain mode-locked operation and therefore the generation of ultrashort pulses. This is discussed in Sec. 5.3. Moreover, the intracavity amplitude modulator offers the possibility for Q-switched operation of the laser and in this way for the generation of short pulses of high power. Such a Q-switched laser is presented in Sec. 5.4. All these different types of lasers have dielectric cavity mirrors, deposited on the polished waveguide endfaces. The rear mirror (which is in most cases the output coupler) has a high reflectivity not only at the signal but also at the pump wavelength to provide double-pass pumping and, in this way, an improved pump absorption efficiency.

One (or even both) of the dielectric mirrors can be replaced by a Bragg grating, etched into the surface of the waveguide amplifier, or by a photorefractive grating defined in the waveguide core. Its narrow-band reflectivity precisely determines the wavelength of the laser emission. Moreover, such distributed Bragg reflector (DBR) lasers, which are described in Sec. 5.5, facilitate the monolithic integration of the laser with further devices on the same substrate.

5.1 Free-Running Fabry–Perot Lasers

In a simple Fabry–Perot-type laser, such as shown in Fig. 9, without any wavelength-controlling intracavity device, the emission wavelength is determined by the spectral properties of the cavity (wavelength-dependent mirror reflectivities, amplifier gain spectrum, and waveguide scattering losses) and therefore by the pump power level as well. By a proper choice of the output coupler, laser emission at the different maxima of the Er:$LiNbO_3$ gain spectrum can be achieved.

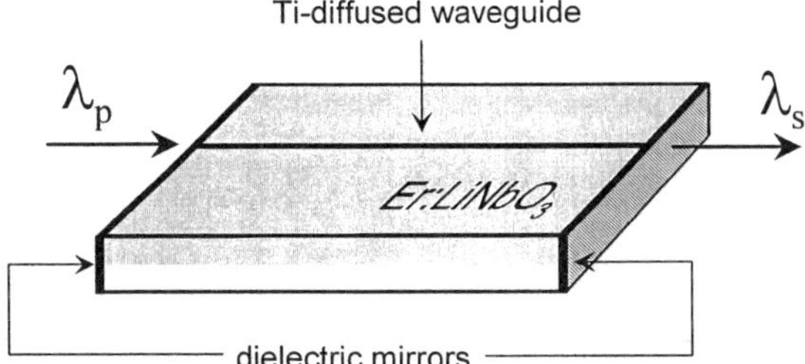

Figure 9 Schematic structure of a free-running Fabry–Perot waveguide laser with dielectric mirrors deposited directly on the waveguide endfaces.

Up to now, lasers of six different wavelengths have been fabricated: λ_s = 1531 nm, 1546 nm, 1562 nm, 1576 nm, 1602 nm, and 1611 nm [11–13,32,33].

It is a problem to design the cavity in such a way that maximum quantum efficiency is obtained. In any case, the waveguide amplifier should have low scattering losses and a high absorption efficiency; both can be achieved by an optimized Er doping profile, an optimized waveguide length, and a double-pass pumping scheme. All the waveguide and amplifier parameters then determine the optimum output coupler. Following these guidelines, a laser of optimized efficiency was recently demonstrated [11]. It has a heavily Er-doped straight Ti-diffused channel guide of 7 cm length with a small-signal gain of up to 13.8 dB achieved with a coupled pump power of 190 mW (see Sec. 3). Using numerical simulations based on the theoretical model for Er-doped waveguide amplifiers [27], the optimum output coupler was determined to have a maximum reflectivity of 30% at λ_s = 1562 nm, corresponding to a 70% output coupling efficiency. This mirror guarantees that laser oscillation sets in at λ_s = 1562 nm with highest quantum efficiency. A maximum output power of 63 mW at an incident pump power level of 210 mW (CCL pumping) has been obtained (see Fig. 10). The lasing threshold was only 24 mW. A slope efficiency up to 37% has been observed. These experimental results are in good agreement with the theoretical predictions. The waveguide laser has been pigtailed, packaged, and characterized again using diode pumping at about 1480-nm wavelength; the result is shown in the lower graph of Fig. 10. Only a slight increase in the threshold and a slight decrease in the output power was caused by the pigtailing. A maximum output power of 14 mW with strongly reduced noise was achieved at a pump power level of 95 mW. Moreover, spectral noise measurements showed that at frequencies above 50 MHz, the laser output is shot-noise limited, while at low frequencies around 350-kHz residual relaxation oscillations are observed. They can be drastically suppressed by a feedback-controlled pumping.

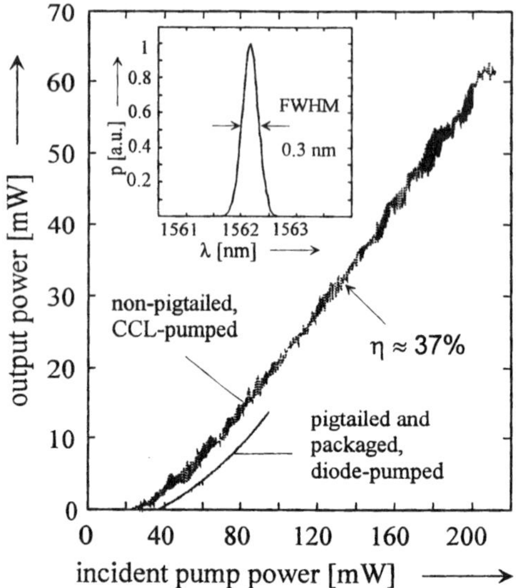

Figure 10 Output power of a laser-diode-pumped pigtailed and packaged Ti:Er:$LiNbO_3$ waveguide laser of optimized efficiency versus incident pump power. Both pump and signal polarization are TE (σ). For comparison, the power characteristics of the color-center-laser- (CCL-) pumped waveguide laser without pigtails is given. Inset: emission spectrum of the diode-pumped (95-mW) waveguide laser. (From Ref. 51, © 1996 IEEE.)

A further improvement of the laser efficiency seems possible. Theoretical calculations predict that a reduction of the scattering losses from 0.18 dB/cm (actual sample) down to 0.08 dB/cm would increase the maximum slope efficiency by more than 10% [11]. A higher Er doping level and improved overlap of the doping profile and the waveguide modes would increase the output power further.

5.2 Acousto-Optically Tunable Lasers

By incorporating an acousto-optical filter in the cavity of a Fabry–Perot-type laser, tunability of its emission wavelength can be achieved. However, due to the frequency shift induced by the acousto-optical interaction, a double-stage filter must be used, with a second stage compensating for the frequency shift of the first one. The design of such a filter is presented in Fig. 11; it can be operated as (nearly) polarization independent or as a polarization-dependent device as sketched in the figure. The routing for both TE-polarized signal and TM-polarized pump is indicated. Only the phase-matched part of the signal is converted by the first acousto-optical mode converter to TM and passes—together with the pump

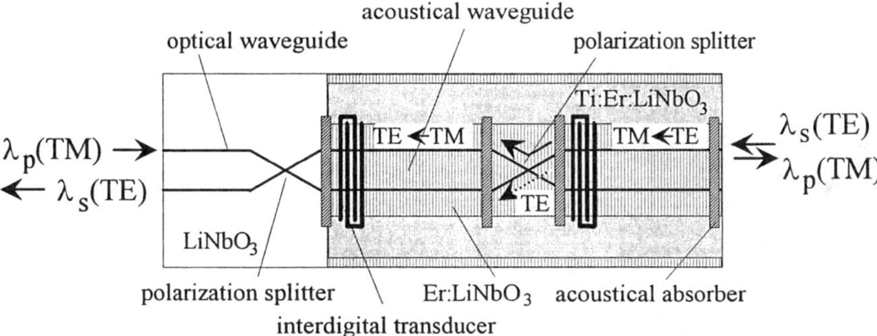

Figure 11 Schematical sketch of an amplifying, double-stage acoustooptical wavelength filter with integrated polarizers (polarization splitters) for polarization-dependent (see diagram) or -independent operation (see text). The routing of the converted and unconverted signal light is indicated for polarization dependent operation. (From Ref. 51, © 1996 IEEE.)

mode—the right polarization splitter in the bar state. The unconverted part of the TE-polarized signal is directed cross to the unpumped lower branch of the left mode converter and hence absorbed. The already once-filtered TM-polarized signal light is converted again to TE in the left (second) mode converter and routed by the second polarization splitter to the lower left output.

To fabricate a laser cavity, both endfaces of the signal waveguide have been coated with mirrors. In the following, a specific example is described [34]. On the left-hand side is an evaporated Au mirror covering the signal arm only; the upper arm of the polarization splitter serves as pump coupler. The output coupler for the laser emission on the right of the 63-mm-long cavity is a dielectric mirror of about 98% reflectance. Low scattering losses of the waveguides—0.1 dB/cm for TE-polarized light and 0.05 dB/cm for TM-polarization—and moderate insertion losses of the other intracavity components resulted in a total cavity round-trip loss of $\simeq$4.8dB.

The laser has been fiber-pigtailed, packaged, and characterized using diode pumping (λ_p = 1484 nm) (see Fig. 12). The minimum threshold of the packaged device was about 50 mW at $\lambda_s \approx$ 1561 nm (see power characteristics as inset in Fig. 13). With about 110 mW of coupled pump power in TE polarization, up to 320 μW of TM-polarized output power has been measured. The nonlinearity of the power characteristics is attributed to a change of the spectrum of the pump laser diode during the sweep of the injection current. The overall tuning behavior for diode pumping of the packaged device is shown in Fig. 13. The tuning slope is −8 nm/MHz. In the gaps of the tuning characteristics the internal gain was not sufficient to overcome the round-trip losses in the laser cavity. However, by

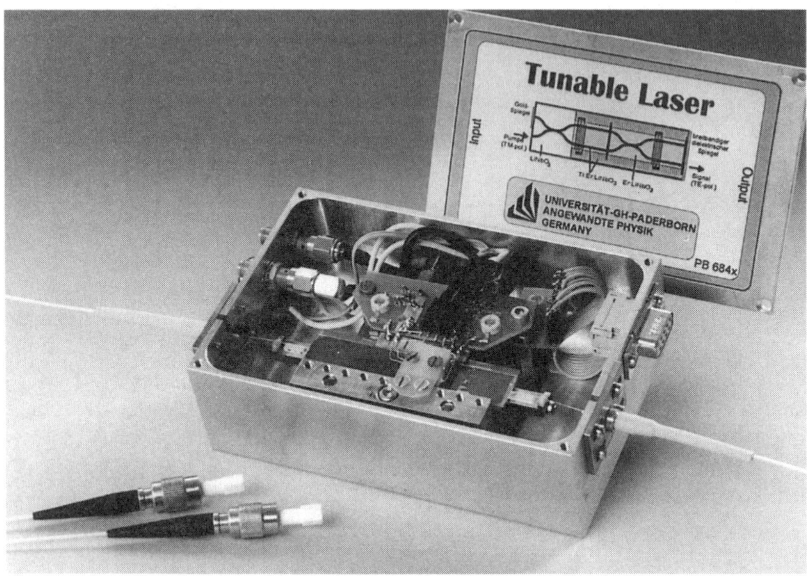

Figure 12 Photograph of the fully packaged acoustically tunable Ti:Er:$LiNbO_3$ waveguide laser. (From Ref. 34, © 1997 IEEE.)

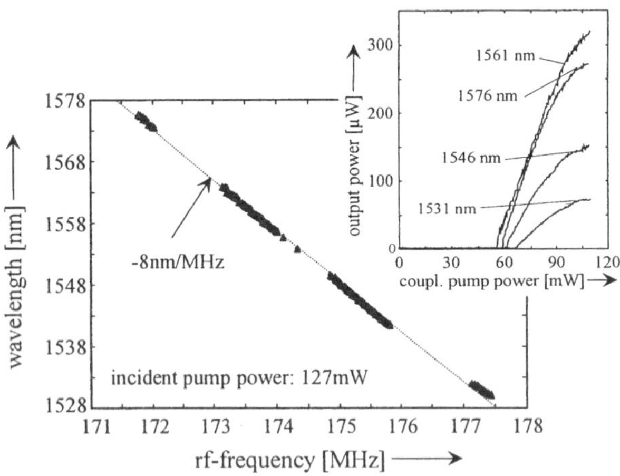

Figure 13 Emission wavelength versus acoustical frequency of pigtailed, packaged, and diode-pumped acousto-optically tunable Ti:Er:$LiNbO_3$ waveguide laser for 127 mW of incident pump power. Inset: power characteristics of the laser for selected wavelengths. (From Ref. 51, © 1996 IEEE.)

using a color center laser as pump source of higher power level, continuous tuning from 1540 nm to 1568 nm was possible. By optical isolation of the laser output and feedback-controlled pumping, fairly stable laser operation with a relative intensity noise (RIN) below -125 dB/Hz for frequencies above 50 MHz has been achieved at an output power level of 108 μW.

5.3 Mode-Locked Lasers

Ti:Er:$LiNbO_3$ waveguides allow the monolithic integration of an active mode-locker by incorporation of an electro-optical phase (or amplitude) modulator in the cavity of a simple Fabry–Perot-type waveguide laser. By modulation synchronous with the fundamental [12] or with harmonics [13] of the axial mode frequency spacing of the laser cavity, a comb of axial modes is phase locked, leading to a train of short optical pulses in the time domain. Harmonic mode-locking in a long cavity allows one to combine efficient pump absorption with the generation of pulses of high repetition frequency. Moreover, pulses can be shortened in comparison with fundamental mode-locking [35].

Mode-locking has been demonstrated with lasers in both $\hat{X}$- and $\hat{Z}$-cut $LiNbO_3$ substrates. Devices in $\hat{Z}$-cut material profit from the slightly lower waveguide attenuation and a higher gain, but require an insulating buffer layer below the modulator electrodes to prevent excess absorption losses.

The design of such a laser in $\hat{Z}$-cut $LiNbO_3$ is shown schematically in Fig. 14. To allow efficient phase modulation at different harmonics of the axial mode frequency spacing, a broadband traveling wave modulator has been used with thick electroplated Au electrodes (details of the fabrication of the waveguide and the laser cavity can be found in Ref. 13). Via the largest electro-optical coefficient, the extraordinary index of refraction is modulated, leading to the coupling of the longitudinal TM modes of the laser. It is pumped by a pigtailed laser diode in TM polarization ($\lambda_p \simeq 1480$ nm). The mode-locked Ti:Er:$LiNbO_3$ waveguide laser has been characterized in terms of power characteristics, time-bandwidth

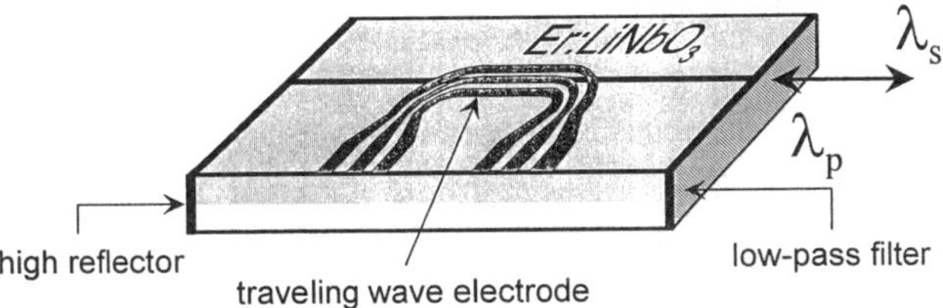

Figure 14 Structure (schematic) of an FM-mode-locked Ti:Er:$LiNbO_3$ waveguide laser with symmetrical coplanar wave (CPW) traveling wave phase modulator as mode-locker.

product of the output pulses, and detuning behavior. By adjustment of the output coupling strength of the laser cavity, mode-locked operation has been achieved at a number of different wavelengths: 1531 nm (π-polarized), 1545 nm (π), 1562 nm (σ), 1575 nm (π), 1602 nm (π), 1611 nm (σ). Threshold figures as low as 9 mW and an average output power of 1.1 mW (at 75-mW pump power) have been achieved at $\lambda = 1602$ nm (slope efficiency 1.6%). The highest output power (up to 12 mW average power) and slope efficiency (14.4%) have been observed at 1575-nm wavelength [36]. To determine the width of the pulses in the time domain and the frequency domain, an optical autocorrelator and a spectrometer of about 0.1-nm spectral resolution have been used. With fundamental mode-locking at 1602-nm wavelength and 1.281-GHz repetition frequency, pulses of 8.6-ps width (FWHM) and 650-mW peak power have been observed.

With harmonic mode-locking, the pulse width could be further reduced. As an example, at the third harmonic (3.843 GHz), pulses of 3.8-ps width (FWHM) and 630-mW peak power could be generated. The corresponding autocorrelation trace and the spectrum are shown in Fig. 15. Pulse width (determined by deconvolution from the correlation trace) and spectral width result in a time-bandwidth product of $\approx$0.42, very close to the transform-limited figure for Gaussian pulses. This low value has been achieved by a slight negative detuning of the mode-lock frequency from the exact fundamental respectively harmonics of the axial mode frequency spacing according to the theory of Kuizenga and Siegman [35].

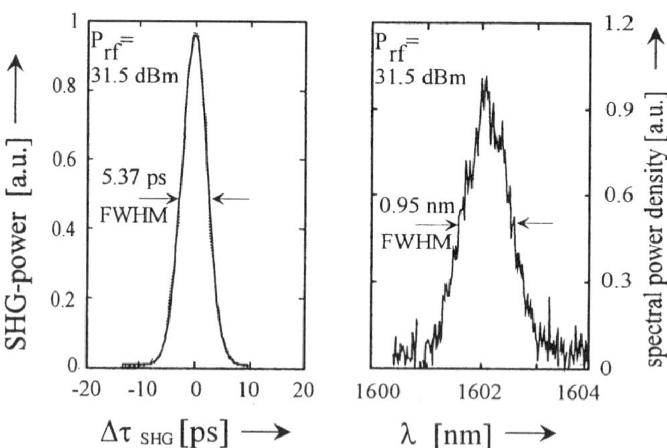

Figure 15 Autocorrelation trace (left) and optical spectrum (right) of an FM-mode-locked Ti:Er:$LiNbO_3$ waveguide laser for third harmonic mode-locking. (From Ref. 51, © 1996 IEEE.)

Harmonic mode-locking has been observed up to the tenth harmonic (10-GHz pulse repetition frequency), limited only by our present RF-signal generator.

In recent years a major effort has been made to stabilize the performance of the integrated mode-locked lasers, for two phenomena can seriously degrade the amplitude stability of the mode-locked laser: low-frequency noise around several hundred kilohertz arises from relaxation oscillations [37]. This issue, which is common to fundamentally as well as harmonically mode-locked lasers, can almost be eliminated by feedback-controlled pumping of the Er laser [36].

High-frequency-amplitude noise can be a serious issue for harmonic mode-locking [38]. During harmonic mode-locking, axial modes are coupled that are an integer multiple of the axial mode spacing apart. According to the harmonic order, the laser can oscillate on a number of different combs of axial modes, so-called *supermodes*. The uncorrelated superposition of supermodes leads to pulse-amplitude fluctuations. To eliminate this problem, the laser has to be stabilized on a single supermode. Intracavity push–pull phase modulation (see Fig. 16) at low frequencies (about 5 MHz) can reduce the supermode beat noise [36]. However, stabilization on a single supermode over a long term has not been achieved in this way.

Much more successful has been the coupling of the laser cavity to a passive reference cavity, as shown in the left part of Fig. 16. The cavity is comprised of a broadband high reflector on the rear side and a glued passive waveguide Fabry–Perot of 4.985 GHz free spectral range on the pump input side with mirror reflectivities of 30% (dielectric low-pass filter) and 4% ($LiNbO_3$-fiber transition), respectively. The free spectral range of the reference cavity almost coincides with the mode-locking frequency. The resulting modulated effective reflectivity of the

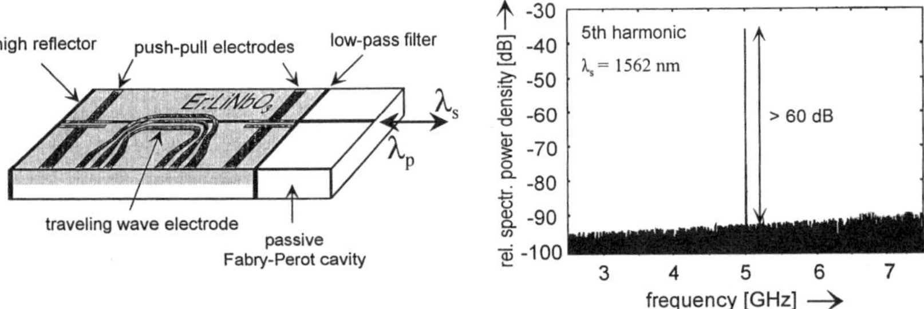

Figure 16 Schematic structure of a coupled cavity FM-mode-locked Ti:Er:$LiNbO_3$ waveguide laser (left) and electronic spectrum of the detector signal (right) within the range of possible supermode beat frequencies for 5th harmonic mode-locking at 4.998 GHz. (From Ref. 39, © 1998 IEEE.)

output coupler mirror of the active laser cavity favors the laser oscillation on one supermode and suppresses the others. Using this concept, stable single supermode operation (side-mode suppression ratio >60dB) over a long term (hours) has been achieved [39] (see right diagram of Fig. 16). Due to the large free spectral range of the integrated laser the supermode stabilization has been achieved with a low-finesse ($F \simeq 2.3$) reference cavity, in contrast to mode-locked fiber lasers, where the small axial mode spacing requires a high-finesse ($F \geq 50$) intracavity filter [40]. With push–pull phase modulation via the additional lumped electrodes on both ends of the active laser cavity and with controlled pumping, the relative intensity noise of the laser reaches almost the shot noise limit for frequencies above 40 MHz. Mode-locked lasers are potential sources for high-bit-rate digital optical transmission. Higher peak power levels of mode-locked pulses, compared to externally modulated continuous wave lasers, are especially attractive for nonlinear optical transmission in the third telecommunications window, to upgrade existing, highly dispersive fiber links.

A diode-pumped, fully packaged harmonically mode-locked Ti:Er:$LiNbO_3$ waveguide laser has been tested for soliton-type data transmission at 10 Gbit/s over highly dispersive (≈17 ps/nm km) standard single-mode (SI) fiber. The laser has been mode-locked at the fifth harmonic (4.998 GHz), and orthogonal polarization switching has been applied to minimize soliton interaction in the line [41]. Without any in-line filtering and dispersion control, a bit-error ratio (BER) of about 10^{-9} has been achieved with five amplified fiber spans of 40 km each.

5.4 Q-Switched Lasers

By monolithically integrating an intensity modulator in the waveguide laser cavity, Q-switched operation can be achieved. Erbium-doped lasers are of particular interest: due to the long fluorescence lifetime of the Er ions incorporated at high concentration levels in the $LiNbO_3$ host, a high-energy storage capability is guaranteed. Moreover, by taking advantage of modulators of high extinction ratio, efficient Q-switched lasers can be designed. Peak power levels in the kilowatt range have been predicted [42], and very recently achieved experimentally [43], with a device schematically shown in Fig. 17.

The actual device has a length of about 75 mm. Half (with respect to the X-direction) of the Z-cut (Y-propagation) $LiNbO_3$ substrate has been doped over the complete length near the surface by in-diffusion of 30 nm of vacuum-deposited Er at 1130°C during 150 h. Subsequently, the photolithographically delineated 7-μm-wide and 100-nm-thick structure of Ti stripes has been in-diffused at 1060°C during 7.5 h to form the waveguide channels. In the undoped region, waveguide scattering losses of 0.03 dB/cm have been measured. The splitter is a Y-junction with X-sine shaped bends located in the center of the cavity. Its excess loss is

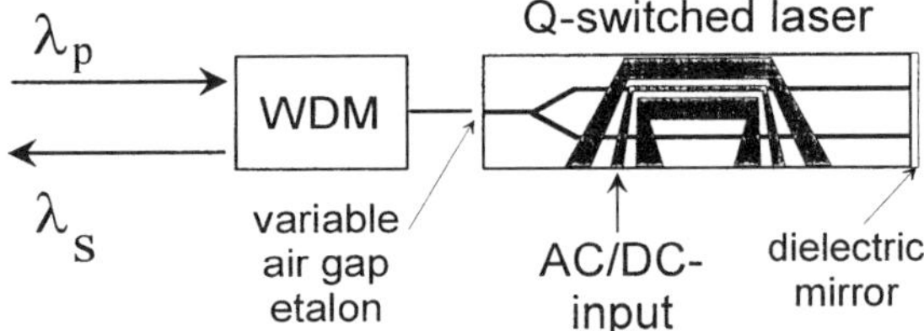

Figure 17 Schematic sketch of the laser structure and the experimental setup for the investigation of Q-switched laser operation. (From Ref. 43, © 1998 IEEE.)

about 0.15 dB, and its deviation from a symmetrical power splitting is below 0.2 dB, leading to an estimated modulator extinction ratio of better than −25dB. On the waveguide structure, an 0.6-μm-thick insulating SiO_2 buffer was vacuum deposited prior to the electrode fabrication.

The electrode structure of the intracavity modulator (Q-switch) is a 25-mm-long symmetrical coplanar microstrip line fabricated by liftoff of a sandwich of sputtered Ti/Au. The modulator was operated as a lumped device without low resistance termination due to the relaxed bandwidth requirements. The halfwave voltage for the laser polarization TE (σ) is about 28 V due to the smaller electro-optical coefficient available in this polarization. The laser has a Fabry–Perot cavity comprised of a dielectric mirror vacuum deposited on the waveguide endfaces and a variable etalon with air gap. The dielectric mirror has a high reflectance (98%) at both emission wavelength ($\lambda_s \approx$ 1562nm, σ-polarized or $\lambda_s \approx$ 1575 nm, π-polarized) and pump wavelength ($\lambda_p \approx$ 1480 nm). In this way, double-pass pumping is provided, allowing for an improved pump absorption efficiency. On the other side, a variable output/pump coupler mirror has been realized by an adjustable, piezoelectrically driven air gap etalon formed by the endfaces of the pump input/signal output fiber (common branch) of the WDM and the polished Ti:Er:$LiNbO_3$ waveguide endface. The effective reflectance of this mirror can be adjusted in the range $0.03 < R < 0.3$.

Using a pigtailed laser diode ($\lambda_p \approx$ 1480nm) of up to 145 mW of output power as the pump source, a threshold of about 90 mW (σ-polarized) has been achieved for σ-polarized emission at 1562-nm wavelength. The modulator has been operated with a DC-bias voltage to give maximum optical extinction and an AC-switching voltage (square wave) of amplitude V_π and about 5% duty cycle in the frequency range 1–10 kHz. No evidence of prelasing could be identified. The Q-switched pulses have been attenuated by 50 dB in a cascade of fiberoptic splitters and attenuators to ensure linearity of the detector, a biased PIN-photodiode of 1.5-GHz bandwidth. The detector signal has been measured using a scope of 1.5-GHz bandwidth. In Fig. 18, experimental results (a) and theoretical results

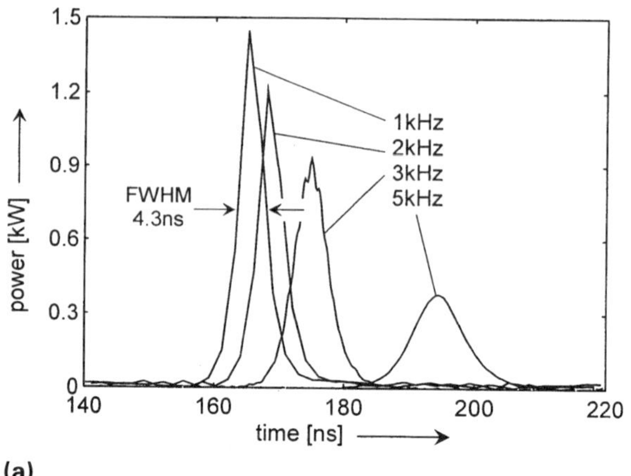

(a)

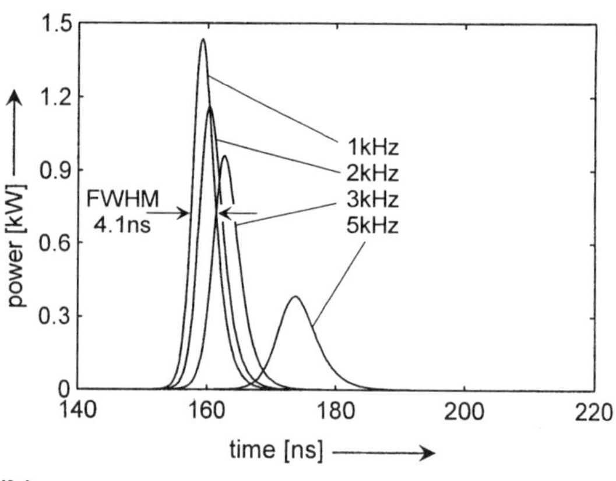

(b)

Figure 18 Output power of a Q-switched Ti:Er:$LiNbO_3$ waveguide laser versus time. (a) Measured through the pump/output coupler in the backward direction. The zero of the abscissa scale coincides with the leading edge of the electrical switching pulse; the parameter of the set of graphs is the pulse repetition frequency. (b) Numerically simulated results for comparison. (From Ref. 43, © 1998 IEEE.)

(b) of Q-switched operation of the Ti:Er:$LiNbO_3$ waveguide laser are presented for 145-mW incident pump power. The zero point of the abscissa of the diagrams correspond to the leading edge of the electrical switching pulse. At 1-kHz repetition rate, up to 1.44-kW of peak power has been measured. The build-up time is about 165 ns, and the pulse width is 4.3 ns (FWHM). With increasing repetition rate, the peak power degrades and the pulse width and the build-up time increase, respectively. Good agreement between measured and calculated peak power levels and pulse widths has been achieved.

For π-polarized pumping, π-polarized emission at 1575 nm has been observed with significantly lower peak powers (up to $\simeq$500 W), compared to the σ-polarized emission at 1562-nm wavelength. Diode-pumped integrated Q-switched lasers are efficient, miniaturized sources of short optical pulses with a variety of possible applications. They can be used as pump sources for parametric nonlinear frequency conversion, as sources for optical time-domain reflectometry (OTDR), and for light detection and ranging (LIDAR).

5.5 Narrow-Linewidth Distributed Bragg Reflector Lasers

By replacing one of the dielectric endface mirrors of a Fabry–Perot type of laser by a (first-order) Bragg grating etched into the surface of Ti:Er:$LiNbO_3$ channel guides, distributed Bragg reflector (DBR) lasers have been fabricated. Their emission wavelength is determined by the periodicity of the grating. Even single-frequency operation can be expected in the case of a grating response narrower than the frequency spacing of the longitudinal laser modes.

With Bragg gratings of 352-nm and 346-nm periodicity, DBR lasers for λ_s = 1561 nm and λ_s = 1531 nm emission wavelengths have been developed in $\hat{Z}$-cut $LiNbO_3$ [44]. It was even possible to fabricate two lasers of both periods on a common substrate. The gratings have been holographically defined in a resist layer and subsequently transferred into the waveguide surface by sophisticated masking and dry-etching techniques [45]. The depth of grooves was limited by redeposition effects during etching to about 420 nm. The inset of Fig. 19 shows an SEM top view of a grating of 352-nm periodicity etched 300 nm deep into the surface of $LiNbO_3$.

Figure 19 also shows the power characteristics (left) and the highly resolved spectrum (right) of one of the 1561-nm lasers. The laser has been fiber-pigtailed, packaged, and temperature-stabilized. Diode pumping at $\lambda_p \simeq$ 1480 nm in TE polarization has been used. At about 45-mW pump power, laser oscillation sets in; with 110 mW, about 1.4 mW of output power has been measured. This poor efficiency is a consequence of the not-yet-optimized laser cavity, with excess losses for pump and signal radiation induced by the grating. Moreover, only single-pass pumping is possible with the present design.

The linewidth of the laser emission is very narrow. Using a scanning Fabry–Perot resonator as spectrum analyzer, single-frequency operation of the laser could be verified; however, the true linewidth could not be resolved in this way. Therefore, delayed self-heterodyne detection with a 26-km-long fiber as the delay line in one arm of a fiber-optical Mach–Zehnder interferometer has been used for high-resolution spectral analysis. An integrated acousto-optical filter was in-

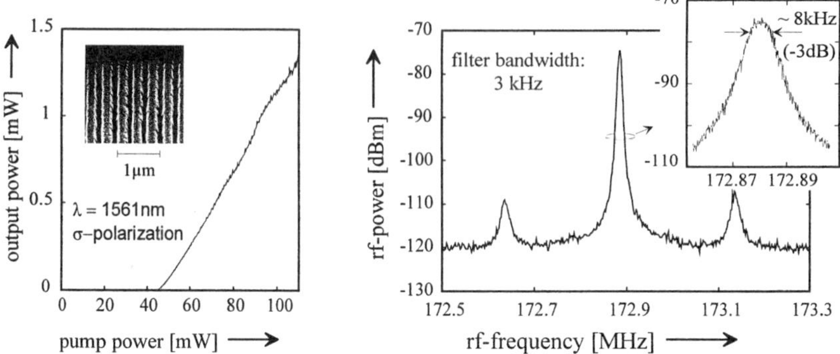

Figure 19 Power characteristics (left) and self-heterodyne beat spectrum (right) of the emission of a diode-pumped, pigtailed and packaged single-frequency DBR Ti:Er: $LiNbO_3$ waveguide laser. Left inset: SEM image of the dry-etched first-order surface-relief grating used as Bragg reflector. Right inset: Peak of the beat spectrum with higher resolution. (From Ref. 51, © 1996 IEEE.)

serted into the other arm of the interferometer as frequency shifter. The interferometer output was measured, and the resulting electronic beat spectrum is shown on the right of Fig. 19. From the 3-dB bandwidth of the spectral power density, the laser linewidth can be determined; it is narrower than 8 kHz. Feedback-controlled pumping and a good optical isolation of the laser output were absolutely necessary to achieve this result.

Distributed Bragg reflector lasers with etched surface gratings suffer from several drawbacks: The fabrication technology is complicated [45]. Grating inhomogeneities induce extra losses of the lasing mode. The overlap of the grating and the lasing mode is very small, requiring a long interaction length. The pump mode is partially coupled to substrate modes, resulting in high extra losses; therefore, pumping through the Bragg grating is not possible.

Photorefractive gratings, as successfully used in fiberoptic DBR and DFB lasers [46], are a very promising alternative, avoiding all the drawbacks just mentioned. We describe in the following the first DBR waveguide laser (λ = 1531 nm) in Er-diffusion-doped $LiNbO_3$ with a fixed photorefractive grating in an Fe-doped Ti-diffused strip waveguide; the device is pumped by a laser diode ($\lambda_p \simeq$ 1480nm).

A schematic diagram of the laser is presented in Fig. 20. It is fabricated in a 70-mm-long X-cut $LiNbO_3$ substrate, which has been Er doped over 43 mm by an in-diffusion of a 15-nm-thick vacuum-deposited Er layer at 1120°C during 120 h. Subsequently, the remaining surface has been iron-diffusion doped (33 nm, 1060°C, 72 h) to increase the photorefractive sensitivity for grating fabrica-

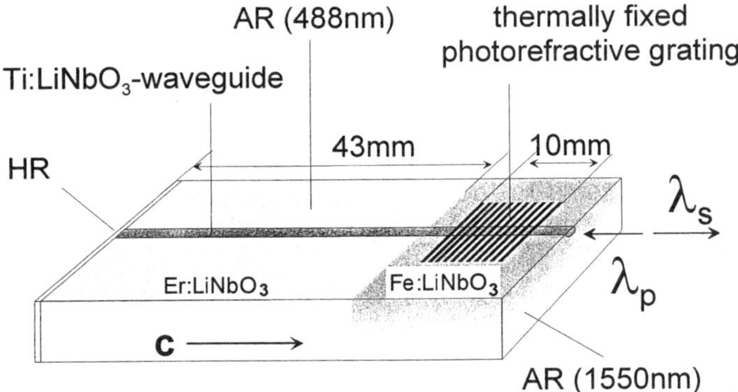

Figure 20 Schematic structure of the Ti:Er:$LiNbO_3$ DBR waveguide laser with photorefractive grating in the Fe-doped section. HR: highly reflecting dielectric mirror; AR: antireflection coating. (From Ref. 52, © 1998 OSA.)

tion. Finally, an 8-mm-wide, 97-nm-thick, photolithographically defined Ti stripe parallel to the *c*-axis has been in-diffused forming the optical channel guide. The sample has been annealed at 500°C for 3 h in flowing Ar (0.5 L/min) to enhance the Fe^{2+}/Fe^{3+} ratio, which determines the photorefractive susceptibility.

The laser resonator consists of a broadband dielectric high reflector on the polished waveguide endface of the Er-doped section and of the narrow-band grating reflector in the Fe-doped section. The endface on the right side has been antireflection (AR) coated for fiber butt coupling. Finally, the upper and lower sample surfaces have been antireflection coated as well to avoid interference effects during the grating fabrication.

The grating has been written using a holographic setup with an Ar laser (λ = 488 nm). The periodic illumination leads to a corresponding excitation of electrons from Fe^{2+} states; they are redistributed by drift, diffusion, and the photovoltaic effect in $LiNbO_3$. The last is the dominant transport mechanism along the optical *c*-axis. Finally, the electrons are trapped by acceptor states (Fe^{3+} ions) in areas of low optical intensity. This redistribution generates a periodic space charge field that modulates the refractive index via the electro-optical effect and generates in this way a narrow-band Bragg reflector grating.

A grating fabricated at room temperature is not stable. Therefore, it has been written by a 2-h exposure at 170°C. At this temperature, protons in the crystal become mobile and compensate for the periodic electronic space charge [47,48]. After cooling to room temperature, these ions are frozen at their high-temperature positions. Homogeneous illumination with the collimated beam of a 100-W halide lamp then leads to a nearly homogeneous redistribution of the electronic charge, developing in this way a stable ionic grating as a replica of inverse polarity of the initial electronic space charge distribution.

The spectral characteristics of the holographically written grating have been determined by slightly pumping the device and measuring the back-scattered amplified spontaneous emission from the Er-doped section transmitted through the grating. This yielded a minimum transmission of about 40%, corresponding to a reflectivity of about 60%. Due to an electronic compensation of the ionic grating, the reflectivity drops slowly as a function of time. However, this compensation can easily be reversed by another homogeneous illumination. Figure 21 shows the specific result measured for the optimum output coupling of the DBR laser. The half-width of the grating is about 0.11 nm.

To operate the DBR-laser, a pigtailed diode laser ($\lambda_p \approx$ 1480nm) has been used for pumping. A fiber-optic wavelength-division multiplexer (WDM) launched up to 110 mW of pump power into the DBR laser and simultaneously extracted the laser emission in the backward direction. Laser emission could be achieved in both TE and TM polarization. TE polarization for both pump and emission has yielded the maximum output power (see Fig. 22), because the smaller TE modes result in a better overlap with the erbium concentration profile. To suppress TM emission, a stripe of silver paste operating as a TE-pass polarizer has been deposited across the waveguide close to the high reflector. The bandwidth of the grating of $\approx$0.11 nm has led to the simultaneous emission of three longitudinal modes (see inset of Fig. 22), with a central wavelength of 1531.7 nm. The maximum TE-polarized output power of 5 mW has been measured for a grating reflectivity of about 45%. The saturation of the output power at high pump power levels is

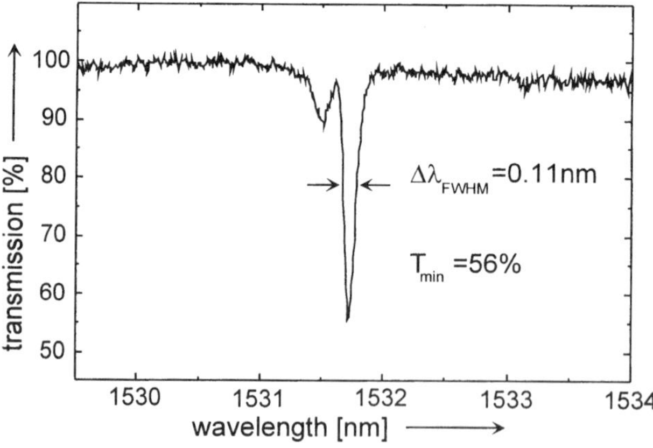

Figure 21 Transmission of a fixed photorefractive grating versus wavelength. (From Ref. 52, © 1998 OSA.)

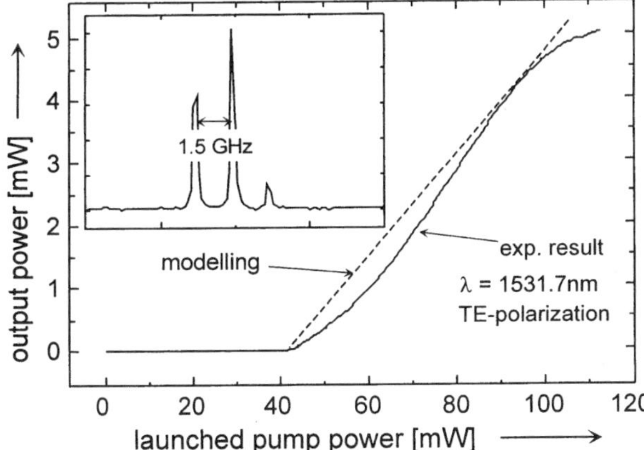

Figure 22 Power characteristics of the DBR laser. Inset: axial mode spectrum (TE polarized). (From Ref. 52, © 1998 OSA.)

due to a change of the pump spectrum (broadening and shifting to longer wavelengths) caused by back-reflections into the nonisolated diode laser. As a result the pump absorption efficiency and hence the slope efficiency of the DBR laser is reduced.

Modeling results predict a significant potential for improvements. Assuming waveguide scattering losses of 0.1 dB/cm, an optimized Er diffusion, and output coupling through the grating, up to 32% slope efficiency and 15 mW of output power at 80 mW of launched pump power seem feasible at the same emission wavelength. However, for practical applications the slow electronic compensation of the ionic space charge grating has to be suppressed to stabilize the Bragg response.

6 INTEGRATED OPTICAL CIRCUITS

Distributed Bragg reflector lasers, as presented in Sec. 5.5, are key components to develop optically powered, monolithic, integrated optics in $LiNbO_3$. Circuits of higher functionality and complexity can be designed by combining lasers and further active and passive devices on the same substrate. This new field is of growing interest and open for many ideas to be realized.

As a first example, an integrated transmitter unit has recently been demonstrated consisting of a narrow-linewidth DBR laser and a Mach–Zehnder type of encoding modulator on the same Er-doped substrate [49]. This is a very attractive combination for long-haul fiber-optic communication because the extracavity electro-optical waveguide modulator includes only a negligible wavelength chirp of the laser emission.

The structure of the integrated laser/modulator unit is sketched schematically on the left of Fig. 23; it has been fabricated by 3-inch full-wafer technology. The $\hat{Z}$-cut $LiNbO_3$ substrate has been Er doped over the full transmitter length of 66 mm by in-diffusion of a sputtered Er layer during 100 h at 1100°C. The laser cavity is comprised of an 8-mm-long first-order ($\lambda_B = 1561$ nm) Bragg grating of 352.5-nm period dry-etched 300 nm deep into the waveguide surface and a dichroic dielectric mirror at the pump input side. This mirror has a transmittivity of 60% at the pump wavelength ($\lambda_p \simeq 1480$ nm) and a high reflectivity at the laser emission wavelength ($\lambda_s = 1561$ nm). For the grating mirror, a transmission drop at the Bragg wavelength of more than 3.5 dB has been measured for TE-polarized light, corresponding to a peak reflectivity of over 50%. The gain section between the mirrors is 46 mm long. Pump power not absorbed in this section is utilized to provide loss compensation of the DBR laser signal in the subsequent extracavity Mach–Zehnder type of encoding modulator, which is also Er doped.

On the right side of Fig. 23, the measured laser performance is summarized. The laser has a threshold of 54.8 mW of incident pump power. At a pump power level of 145 mW, up to 0.63 mW of output power has been measured at the modulator output. The laser emission wavelength is determined by the Bragg wavelength of the grating reflector, as shown in the upper left inset of the right-hand diagram. Stable single-longitudinal-mode operation has been confirmed us-

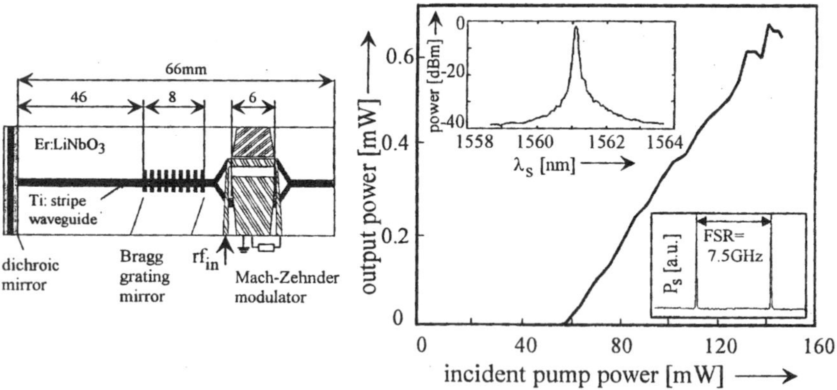

Figure 23 Structure (schematic) and properties [(cw power characteristics and spectrum (insets)] of an integrated Ti:Er:$LiNbO_3$ DBR laser modulator unit (also see text). (From Ref. 51, © 1996 IEEE.)

ing a scanning Fabry–Perot interferometer of 7.5-GHz free spectral range and 37.5-MHz resolution (see lower right inset).

The encoding modulator has traveling-wave Au electrodes of 6-mm length and 10-Ω serial resistance. An insulating SiO_2 buffer layer was deposited prior to the electrode fabrication to minimize excess losses in the modulator. Due to the TE-polarized laser emission, only the smaller r_{31} electro-optical cooefficient of $LiNbO_3$ could be utilized, resulting in a fairly high half-wave voltage $V_\pi >$ 20 V. As a result of this high drive voltage, the modulator could be operated only in a lumped electrode configuration. The frequency response of the integrated laser/modulator unit under electro-optical modulation has been measured using an electro-optical spectrum analyzer. A 3-dB optical bandwidth of ~3 GHz has been achieved.

A significant improvement of the integrated optical transmitter unit will be possible by using $\hat{X}$-cut substrates with Bragg gratings of high reflectivity for TE modes. This configuration would allow one to utilize the largest electro-optical coefficient r_{33} for the modulation and thus permit the application of traveling-wave configurations of very high bandwidth. Moreover, the use of 4-inch-diameter $LiNbO_3$ wafers would allow the fabrication of longer modulators of a significantly reduced V_π.

7 CONCLUSIONS

Erbium diffusion doping of $LiNbO_3$ allows one to develop a variety of attractive integrated optical devices. Besides optically pumped waveguide amplifiers and relatively simple Fabry–Perot type of waveguide lasers, more advanced devices can be designed by combining optical amplification and/or lasing with electro- or acousto-optically controlled functions in the same waveguide (laser) structure. Examples are amplifying electro-optical phase and intensity modulators, high-finesse Fabry–Perot type of optical spectrum analyzers, and tunable, acousto-optical, polarization-dependent and -independent wavelength filters. Furthermore, mode-locked, Q-switched, and tunable lasers have been developed in this way. Additional functions can be expected by combining optical amplification with different nonlinear effects. All-optical switching bistable behavior and parametric frequency conversion might be achieved in well-designed erbium-doped structures.

Moreover, Er-doped DBR lasers can easily be combined with additional active and passive devices on the same substrate to form optical circuits of higher functionality. Laser–modulator combinations are the first examples. More sophisticated circuits will include heterodyne interferometers for optical metrology and vibration analysis with a DBR laser and up to 11 additional devices [50]. It is a challenge for the future to design and to develop a complex, optically powered, monolithic integrated optics in (selectively) Er-doped $LiNbO_3$ with new application-specific optical circuits.

REFERENCES

1. E. Lallier, J. P. Pocholle, M. Papuchon, C. Grezes-Besset, E. Pelletier, M. De Micheli, M. J. Li, Q. He, and D. B. Ostrowsky. Laser oscillation of single mode channel waveguide in Nd:MgO:$LiNbO_3$. Electron. Lett. 25:1491–1492, 1989.
2. R. Brinkmann, W. Sohler, H. Suche, and Ch. Wersig. Single mode Ti-diffused optical strip guides and lasers in Nd:MgO:$LiNbO_3$. Tech. Dig. Integrated Photonics Res. 5:116–117, 1990.
3. E. Lallier, J. P. Pocholle, M. Papuchon, Q. He, M. De Micheli, D. B. Ostrowsky, C. Grezes-Besset, and E. Pelletier. Integrated Nd:MgO:$LiNbO_3$ FM mode-locked waveguide laser. Electron. Lett. 27:936–937, 1991.
4. E. Lallier, D. Papillon, J. P. Pocholle, M. Papuchon, M. De Micheli, and D. B. Ostrowsky. Short pulse, high power Q-switched Nd:MgO:$LiNbO_3$ waveguide laser. Electron. Lett. 29:175–176, 1993.
5. J. Amin, M. Hempsted, J. E. Román, and J. S. Wilkinson. Tunable coupled-cavity waveguide laser at room temperature in Nd-diffused Ti:$LiNbO_3$. Opt. Lett. 19: 1541–1543, 1994.
6. J. P. de Sandro, J. K. Jones, D. P. Shepherd, J. Webjörn, M. Hempstead, J. Wang, and A. C. Tropper. Tm^{3+} indiffused $LiNbO_3$ waveguide lasers. Proceedings 7th European Conference on Integrated Optics (ECIO '95), Delft, The Netherlands, 1995, post-deadline papers, pp. 17–20.
7. J. K. Jones, J. P. De Sandro, M. Hempstead, D. P. Shepherd, A. C. Large, A. C. Tropper, and J. S. Wilkinson. Channel waveguide laser at 1 μm in Yb-indiffused $LiNbO_3$. Opt. Lett. 20:1477–1479, 1995.
8. W. Sohler. Erbium-doped waveguide amplifiers and lasers in $LiNbO_3$. Tech. Dig. Integrat. Phot. Res. (OSA Tech. Dig. Series 7):212–214, 1995, and references cited therein.
9. H. Suche, D. Hiller, I. Baumann, and W. Sohler. Integrated optical spectrum analyzer with internal gain. IEEE Photon. Technol. Lett. 7:505–507, 1995.
10. R. Brinkmann, M. Dinand, I. Baumann, Ch. Harizi, W. Sohler, and H. Suche. Acoustically tunable wavelength filter with gain. Phot. Techn. Lett. 6:519–521, 1994.
11. I. Baumann, R. Brinkmann, M. Dinand, W. Sohler, and S. Westenhöfer. Ti:Er: $LiNbO_3$ waveguide laser of optimized efficiency. IEEE J. Quantum Electron. 32: 1695–1706, 1996.
12. H. Suche, I. Baumann, D. Hiller, and W. Sohler. Modelocked Er:Ti:$LiNbO_3$-waveguide laser. Electron. Lett. 29:1111–1112, 1993.
13. H. Suche, R. Wessel, S. Westenhöfer, W. Sohler, S. Bosso, C. Carmannini, and R. Corsini. Harmonically modelocked Ti:Er:$LiNbO_3$-waveguide laser. Opt. Lett. 20: 596–598, 1995.
14. I. Baumann, D. Johlen, W. Sohler, H. Suche, and F. Tian. Acoustically tunable Ti: Er:$LiNbO_3$-waveguide laser. Proceeding of 20th European Conference on Optical

Communication (ECOC '94), Florence, Italy, 1994, 4: Post Deadline Papers, pp. 99–102.
15. W. Sohler. Integrated optical circuits with Er:$LiNbO_3$ amplifiers and lasers. Optical Fiber Communication Conference, 2, 1996 OSA Technical Digest Series, p. 251.
16. L. F. Johnson and A. A. Ballman. Coherent emission from rare earth ions in electro-optic crystals. J. Appl. Phys. 40:297–302, 1969.
17. N. F. Evlanova and L. N. Rashkovich. Domain structure of lithium metaniobate crystals. Sov. Phys.—Solid State 13:223–224, 1971.
18. R. Brinkmann, Ch. Buchal, St. Mohr, W. Sohler, and H. Suche. Annealed erbium-implanted single-mode $LiNbO_3$ waveguides. Technical Digest on Integrated Photonics Research, 1990, 5, post-deadline paper PD1.
19. W. Sohler and H. Suche. Rare-earth-doped lithium niobate waveguide structures. European Patent No. 0569353 and U.S. Pat. Appln. Serial No. 08/094,1991.
20. I. Baumann, R. Brinkmann, Ch. Buchal, M. Dinand, M. Fleuster, H. Holzbrecher, W. Sohler, and H. Suche. Er-diffused waveguides in $LiNbO_3$. Proceedings of European Conf. on Integrated Optics (ECIO '93), paper: 3–14, Neuchâtel, Switzerland, 1993.
21. I. Baumann, L. Beckers, Ch. Buchal, R. Brinkmann, M. Dinand, Th. Gog, H. Holzbrecher, M. Fleuster, M. Materlik, K. H. Müller, H. Paulus, W. Sohler, H. Stolz, W. von der Osten, and O. Witte. Erbium incorporation in $LiNbO_3$ by diffusion-doping. Appl. Phys. A64:33–44, 1997.
22. T. Gog, M. Griebenow, and G. Materlik. X-ray standing wave determination of the lattice location of Er diffused into $LiNbO_3$. Phys. Lett. A 181:417–420, 1993.
23. O. Witte, H. Stolz, and W. von der Osten. Upconversion and site-selective spectroscopy in erbium-doped $LiNbO_3$. J. Phys. D: Appl. Phys. 29:561–568, 1996.
24. P. F. Bordui, R. G. Norwood, C. D. Bird, and G. D. Calvert. Compositional uniformity in growth and poling of large-diameter lithium niobate crystals. J. Crystal Growth 113:61–68, 1991.
25. V. T. Gabrielyan, A. A. Kaminskii, and L. Li. Absorption and Luminescence spectra and energy levels of Nd^{3+} and Er^{3+} ions in $LiNbO_3$ crystals. Phys. Stat. Sol. (a) 3: K37–K42, 1970.
26. R. Brinkmann, I. Baumann, M. Dinand, W. Sohler, and H. Suche. Erbium-doped single- and double-pass Ti:$LiNbO_3$ waveguide amplifiers. J. Quantum Electron. 30: 2356–2360, 1994.
27. M. Dinand and W. Sohler. Theoretical modeling of optical amplification in Er-doped Ti:$LiNbO_3$ waveguides. J. Quantum Electron. 30:1267–1276, 1994.
28. H. Okamura and K. Iwatsuki. A finesse-enhanced Er-doped-fiber ring resonator. IEEE J. Quantum Electron. QE-9:1554–1560, 1991.
29. K. Hattori. Planar silica waveguide amplifiers and related circuits. Techn. Digest of 5th Optoelectronics Conference, OEC '94, Chiba, Japan, 1994, pp. 162–163.
30. K. W. Cheung, S. C. Liew, C. N. Lo, D. A. Smith, J. E. Baran, and J. J. Johnson. Simultaneous five-wavelength filtering at 2.2-nm separation using acousto-optic tunable filter with subcarrier detection. Electron. Lett. 25:636–637, 1989.
31. H. Herrmann, P. Müller-Reich, R. Reimann, R. Ricken, H. Seibert, and W. Sohler. Integrated Optical, TE- and TM-Pass Acoustically Tunable, Double-Stage Wavelength Filters in $LiNbO_3$. Electron. Lett. 28:642–643, 1992.

32. R. Brinkmann, W. Sohler, and H. Suche. Continuous-Wave Erbium-Diffused $LiNbO_3$ Waveguide-Laser. Electron. Lett. 27:415–416, 1991.
33. P. Becker, R. Brinkmann, M. Dinand, W. Sohler, and H. Suche. Er-diffused Ti: $LiNbO_3$ waveguide laser of 1563 and 1576 nm emission wavelengths. Appl. Phys. Lett. 61:1257–1259, 1992.
34. K. Schäfer, I. Baumann, W. Sohler, H. Suche, and S. Westenhöfer. Diode-Pumped and Packaged Acoustooptically Tunable Laser of Wide Tuning Range. IEEE J. Quantum Electron. 33:1636–1641, 1997.
35. D. J. Kuizenga and A. E. Siegman. FM and AM mode locking of the homogeneous laser—part I: theory. IEEE J. Quantum Electron. 6:694–708, 1970.
36. H. Suche, A. Greiner, W. Qiu, R. Wessel, and W. Sohler. Integrated optical Ti:Er: $LiNbO_3$ soliton source. IEEE J. Quantum Electron. 33:1642–1646, 1997.
37. M. Dinand, Ch. Schütte, R. Brinkmann, W. Sohler, and H. Suche. Relaxation Oscillations in Er-diffused Ti:$LiNbO_3$ Waveguide Lasers. Proceedings of Europ. Conf. on Integrated Optics (EOS, IEEE-LEOS, OSA), ECIO, 93, Neuchâtel, Switzerland, 1993, paper 3–20.
38. M. Becker, D. J. Kuizenga, and A. E. Siegman. Harmonic mode locking of the Nd: YAG laser. IEEE J. Quantum Electron. 8:687–693, 1972.
39. R. Wessel, K. Rochhausen, H. Suche, and W. Sohler. Single supermode harmonically modelocked Ti:Er:$LiNbO_3$ waveguide laser. Proceedings of Conference on Lasers and Electro-optics, CLEO—Europe, Glasgow, Scotland, 1998, postdeadline papers, paper CPD 1.5.
40. J. S. Wey, J. Goldhar, and G. L. Burdge. Active harmonic modelocking of an Erbium fiber laser with intracavity Fabry–Perot filters. IEEE J. Lightwave Technol. 15:1171–1180, 1997.
41. F. Matera, M. Romagnoli, and B. Daino. Alternate polarization soliton transmission in standard dispersion fibre links with no in-line control. Electron. Lett. 31:1172–1174, 1995.
42. D. L. Veasey, J. M. Gary, J. Amin, and J. A. Aust. Time-dependent modeling of erbium-doped waveguide lasers in lithium niobate pumped at 980 and 1480 nm. IEEE J. Quantum Electron. 33:1647–1662, 1997.
43. H. Suche, T. Oesselke, J. Pandavenes, R. Ricken, K. Rochhausen, W. Sohler, S. Balsamo, I. Montrosset, and K. K. Wong. Efficient Q-switched Ti:Er:$LiNbO_3$ waveguide laser. Electron. Lett. 34:1228–1229, 1998.
44. J. Söchtig, R. Gross, I. Baumann, W. Sohler, H. Schütz, and R. Widmer. DBR waveguide laser in erbium-diffusion-doped $LiNbO_3$. Electron. Lett. 31:551–552, 1995.
45. J. Söchtig, H. Schütz, R. Widmer, R. Lehmann, and R. Gross. Grating reflectors for erbium-doped lithium niobate waveguide lasers. Proceedings of SPIE Conf. ''Nanofabrication and Device Integration. 1994, pp. 98–107.
46. J. Hübner, P. Varming, and M. Kristensen. Five wavelength DFB fiber laser source for WDM systems. Electron. Lett. 33:139–140, 1997.
47. J. J. Amodei and D. L. Staebler. Holographic pattern fixing in electro-optic crystals. Appl. Phys. Lett. 18:540–542, 1971.
48. K. Buse, S. Breer, K. Peithmann, S. Kapphan, M. Gao, and E. Krätzig. Origin of thermal fixing in photorefractive lithium niobate crystals. Phys. Rev. (B) 56:1225–1235, 1997.

49. J. Söchtig, H. Schütz, R. Widmer, R. Corsini, D. Hiller, C. Carmannini, G. Consonni, S. Bosso, and L. Gobbi. Monolithically integrated DBR waveguide laser and intensity modulator in erbium-doped $LiNbO_3$. Electron. Lett. 32:899–900, 1996.
50. F. Tian, R. Ricken, St. Schmid, and W. Sohler. Integrated acousto-optical heterodyne interferometers in $LiNbO_3$. Laser in der Technik, Proc. Congress Laser '93, Munich, June 1993, W. Waidlich (ed.), Springer Verlag, 1993, pp. 725–728.
51. I. Baumann, S. Bosso, R. Brinkmann, R. Corsini, M. Dinand, A. Greiner, K. Schäfer, J. Söchtig, W. Sohler, H. Suche, and R. Wessel. Er-doped integrated optical devices in $LiNbO_3$. IEEE J. Selected Topics Quantum Electron. 2:355–366, 1996.
52. Ch. Becker, A. Greiner, Th. Oesselke, A. Pape, W. Sohler, and H. Suche. Integrated optical $Ti:Er:LiNbO_3$ distributed Bragg reflector laser with a fixed photorefractive grating. Optics Letters 23:1194–1196, 1998.

7

Indium Phosphide–Based Photonic Circuits and Components

Anat Sneh and Christopher R. Doerr

Lucent Technologies
Holmdel, New Jersey

1 INTRODUCTION

Semiconductor optoelectronic integration is an important branch of integrated optics. The large versatility in device design and functionality, manifested in the ability to provide in a compact size optical gain, absorption, and substantial high-speed refractive index change, has always been a compelling driver for semiconductor device research and development. Many important advances have been consequently achieved in material growth, device fabrication and design, and packaging techniques. In particular, the constant need for increased bandwidth in optical communication systems has accelerated these advances in the past decade, especially in the InP-based material system. In this chapter we narrow our review of integrated semiconductor devices to InP-based devices, due to the authors' greater familiarity with this field, and since recent demonstrations with enhanced device functionality for broadband wavelength-division multiplexed (WDM) communication systems are primarily in this material system. This refers to structures containing InGaAsP layers with varying compositions grown epitaxially on InP wafers. We first introduce some basic concepts of this technology briefly, and then give a selection of advanced device demonstrations, many WDM related, that are intended to indicate the status of progress achieved so far. We also limit our discussion to fully (monolithically) integrated photonic circuits, although an important aspect of InP-based device technology is hybrid integration, since the latter is discussed in detail in Chapter 9.

So far, significant maturity has been achieved in several applications leading to commercial availability by numerous manufacturers. However, this progress relates mainly to the two ends of a transmission system, namely, to sources and receivers. As networks evolve from point-to-point WDM systems to dense-WDM systems with optical networking functionalities in the optical layer, new applications are required to reach a high degree of maturity. This poses new challenges in all aspects of device development, such as growth, processing, and design. For example, as device size and functionality grow, so do its on-chip loss. Hence, an important aspect of the InP material system in such applications is the ability to provide gain for internal loss compensation. This usually involves design trade-offs between the active and nonactive sections of the device structure. Such device trade-offs and the resulting performance limitations are discussed in this chapter. Challenges in achieving successful integration lie in areas such as improving the yield of integrating structures with varying compositions, particularly those containing multiple quantum wells (MQWs), reducing chip-to-fiber coupling loss, and, in some applications, achieving polarization independence. New concepts—such as selective area growth (SAG) of MQWs, which allows for the simultaneous growth of MQWs with varying bandgap wavelengths on the same base wafer, and integrated beam expansion sections, which substantially improve external coupling efficiencies and tolerances—are examples of possible solutions for some of these difficulties.

2 WAVEGUIDE DESIGN AND OPTICAL PROPERTIES

2.1 Refractive Index Determination

The refractive index variation between the epitaxially grown layers, required for the formation of optical waveguides, is achieved primarily by controlling their alloy composition. In the $In_{1-x}Ga_xAs_yP_{1-y}$ material system, lattice-matched to InP by setting $x = 0.46y$, the bandgap energy of lattice-matched compositions is given by the following empirical expression [1]:

$$E_g = 1.35 - 0.72y + 0.12y^2 \quad (1)$$

where E_g [eV] $= 1.2395/\lambda_g$ [μm] is the bandgap energy, with λ_g being the bandgap wavelength. A theoretical model calculating refractive index values as a function of the incident wavelength and the composition parameter y was suggested by Adachi [2,3]. Another method, which determines the refractive index in the transparent region using photoluminescence excitation, was developed by Henry et al. [4]. A two-oscillator model was used to fit the data to the following closed-form formula:

$$n = \left[1 + \frac{A_1}{1 - \left(\dfrac{1/\lambda}{1/\lambda_p + 2.0208} \right)^2} + \frac{A_2}{1 - \left(\dfrac{1/\lambda}{1/\lambda_p + 0.13215} \right)^2} \right]^{1/2} \tag{2}$$

where n is the refractive index of the material, λ [μm] is the free-space propagating wavelength, λ_p [μm] is the photoluminescence wavelength of the investigated material composition, and the fitting parameters A_1 and A_2 are

$$A_1 = 13.3510 - \frac{6.7620}{\lambda_p} + \frac{1.8946}{\lambda_p^2} \tag{3}$$

$$A_2 = 0.7140 - \frac{0.4470}{\lambda_p} \tag{4}$$

The refractive index is also affected by the free-carrier concentration in the semiconductor layer, via plasma and band filling/shrinkage effects. The main contribution to carrier-induced index change comes from electrons rather than holes, introduced into the layer by current injection or through n-type doping. As discussed in more detail in Sec. 3.1, the overall index change due to the various free-carrier effects has a negative sign for photon energies both well below and close to the bandgap of $In_{1-x}Ga_xAs_yP_{1-y}$ alloys. Away from the bandgap, the rate of index change Δn with electron density N is only weakly wavelength dependent and is typically estimated at $\Delta n/N = -1 \times 10^{-20}$ cm^{-3} for carrier densities up to about 1×10^{18} cm^{-3} [5–9]. At larger concentrations, the rate of change is larger; however, the absorption increase due to free-carrier intraband and inter–valence band transitions results in loss levels that are impractical for passive devices operating in the transparent region. For operation close to the bandgap, band filling and shrinkage effects are responsible for a larger change in the refractive index, amounting to values that may be twice as large as the change obtained in the transparent region with similar carrier densities [6].

2.2 Optical Loss

The main mechanisms determining the propagation loss in straight semiconductor waveguides are absorption and scattering loss. In the transparent region, where the propagating photon energy is sufficiently smaller than the layer bandgap energy E_g, absorption loss is due primarily to free-carriers effects. As the photon energy comes closer to E_g, below-bandgap absorption increases rapidly due to band-tail (Urbach) effects [10] that effectively lower the nominal bandgap energy. In the following, the dominant loss mechanisms in the transparent region are discussed, i.e., carrier-induced absorption and scattering loss.

2.2.1 Carrier-Induced Absorption

As discussed in the previous section, free carriers affect both the refractive index and the absorption properties of a semiconductor layer. There are various carrier-induced absorption mechanisms, involving intraband and interband transitions, both in the conduction and in the valence band. Unfortunately, there is very little carrier-induced below-bandgap absorption data for the InGaAsP system, particularly for n-type material. The dominant loss mechanism below bandgap for InGaAsP alloys is intervalence-band absorption due to electron transitions from the split-off band to holes in the heavy-hole band [11]. The carrier-induced absorption coefficient α is, in general, proportional to the carrier density. It was found in Ref. 12 that the intervalence-band absorption coefficient is independent of InGaAsP composition, and by fitting the below-bandgap experimental absorption spectra in [12] we can get

$$\alpha_{IVB}\,[\mathrm{cm}^{-1}] = 4.25 \times 10^{-16} \exp\left(\frac{-4.533}{\lambda}\right) P \tag{5}$$

where λ [μm] is again the photon wavelength and P is the hole density, in cm^{-3}. As an example, for a p-doped material with hole concentration of $10^{18}\ \mathrm{cm}^{-3}$, this gives an absorption coefficient of 13 cm^{-1} at $\lambda = 1.3$ μm, and 25 cm^{-1} at $\lambda = 1.6$ μm, yielding a layer propagation loss of 56 and 109 dB/cm at 1.3 and 1.6 μm, respectively. Since the epitaxial layer scheme of most semiconductor photonic devices consists of a p-i-n structure, it contains a p-doped cladding layer. Hence, low-loss propagation can be obtained only if the optical mode tail has negligible penetration into the p-cladding. This is often achieved by introducing a nonintentionally doped ($P \sim 10^{16}\ \mathrm{cm}^{-3}$) buffer layer between the core and the p-cladding layer.

The loss coefficient due to n-doping or electron injection is substantially lower, though still high at large carrier concentration levels. Figure 1 shows a qualitative representation of the absorption coefficient, as influenced by electron transitions in the conduction band. There are two types of transitions involved, which are indicated schematically in the Fig. 1 insert: inter–conduction-band transitions from the conduction-band minimum to higher-lying minima and free-carrier intraband transitions. The latter is affected by three different types of phonon scattering assisting the indirect transition, and those are plotted separately in the figure. As shown, the inter–conduction-band absorption affects the intermediate-wavelength region closer to the bandgap, while the long-wavelength region is bound by free-carrier absorption (also known as the *plasma effect*). The latter has a λ^p dependence, with p varying between 1.5 and 3.5 for III–V semiconductors [14,15], which deviates from the classical electron plasma λ^2 dependence. For InGaAsP compositions with bandgap wavelengths varying between $\lambda_g = 1$

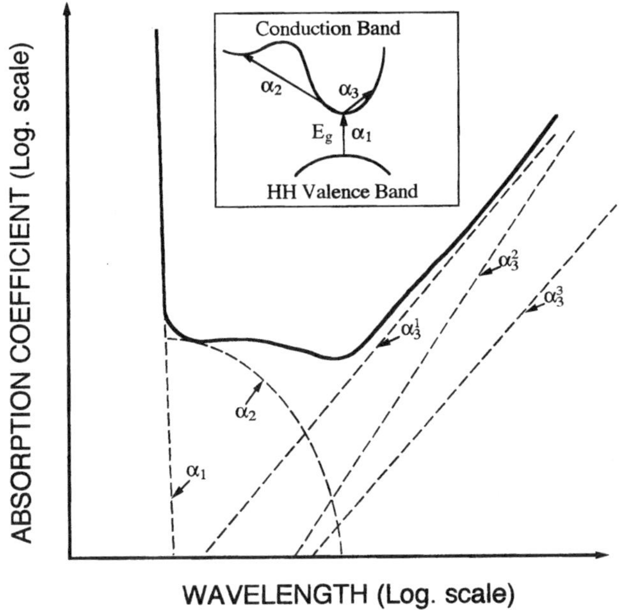

Figure 1 Schematic representation of the absorption coefficient α as is influenced by electronic transitions in the conduction band. Broken line: asymptotic wavelength dependence due to individual transitions. Solid line: combined wavelength dependence. Insert: illustration of individual transitions; α_1 direct valence-to-conduction-band transition; α_2 indirect inter–conduction-band transition; α_3 indirect free-carrier intraband transition assisted by optical phonons (α_3^1), ionized impurities (α_3^2), and acoustical phonons (α_3^3). (After Ref. 13.)

and $\lambda g = 1.3$ μm, we can estimate from Ref. 13 that the rate of absorption coefficient change with free-electron concentration N would be in the range $\alpha/N = 1 - 2 \times 10^{-18}$ [cm^2] at $\lambda = 1.55$ μm. For an n-doped layer with electron concentration of 1×10^{18} cm^{-3}, this gives a propagation loss of about 4–8 dB/cm.

2.2.2 Scattering Loss

Rough interfaces in etched waveguide surfaces is a source for scattering loss. Both the waveguide sidewalls and the etched surfaces beside them can give rise to scattering. Since layer scattering grows with increasing refractive index contrast at the rough interface and decreasing core layer thickness [16,17], semiconductor waveguides are particularly susceptible to high scattering loss. The lowest-propagation-loss waveguides in the InP/InGaAsP system were implemented with undoped epitaxial layers containing a thick (1–3-μm) core layer, a low index change between the core and the cladding layers ($\sim 10^{-2}$), and relatively wide ribs (4–8 μm). Propagation losses of less than 0.25 dB/cm were reported with these geometries [18–20], using both bulk and multiple quantum well (MQW) waveguides. The use of MQWs in the guiding core can help reduce losses by ''diluting'' the core-to-cladding index contrast. This is achieved by choosing thin wells with substantially thicker barriers of InP, hence yielding an average core index that can be tailored closer to the InP cladding index.

Relatively smooth waveguide boundaries can be achieved with wet chemical etching. However, the high material and crystalline orientation selectivity can be a limiting factor in some applications, in which case dry-etching would be a preferred method. Propagation losses as low as 1 dB/cm were obtained for deeply dry-etched rib waveguides, using an alternating etching/polymer-descumming reactive ion etching (RIE) process for optimal control of etching-produced roughness [21]. In addition to roughness control, RIE processing usually requires postetching treatment, such as removal of any formed damage layer by nonselective (and possibly combined with selective) wet chemical etch [22], and reactivation of acceptors by annealing for doped p-i-n waveguide structures [23].

2.3 Waveguide Structures

The design of optical waveguides in semiconductor materials requires the consideration of various aspects associated with the fabrication techniques and the desired functionality of the device. Photonic integrated circuits (PICs) with varying degrees of integration complexity are generally comprised of both passive and active waveguides. In the following, waveguides are considered passive if their core material has a bandgap energy that is larger than the photon energy of the propagating light, while active waveguides contain core material with bandgap energy similar or smaller than the photon energy.

Since active waveguides usually have quite different requirements than those of passive waveguides, the typical structures employed in each case can differ greatly. For example, gain media structures in lasers and amplifiers, where gain is achieved by carrier population inversion via external current injection, are comprised of highly doped layers in a p-i-n junction configuration. The doping profile and the composition of the various layers are restricted to only particular sequences that can provide an effective gain. In contrast, passive waveguides can have a more flexible layer design, yet they require low doping levels both in their core and their cladding, to avoid substantial carrier-induced absorption. In addition, the type of guiding structure most suitable for lasers and amplifiers, the buried heterostructure, is strongly guiding in the lateral dimension and is often fabricated wide enough to support more than a single fundamental mode. While gain media naturally filter out the higher-order modes because the large variation in mode confinement within the active core translates to substantial variation in effective modal gain, this is not the case for passive waveguides. Moreover, the strong lateral confinement is highly undesirable for adiabatic mode-evolving or interferometric-type devices, such as Y-branch switches and directional couplers, since it necessitates tight-spacing waveguide placement and imposes difficult-to-reach fabrication tolerances.

Some examples of typical waveguide cross-sectional schematics are given in Fig. 2. The tightly guiding rib waveguide in (a) is used in applications requiring compact structures to minimize wafer real estate [24–26]. It is usually fabricated via dry-etching techniques; hence care should be taken to ensure low scattering losses from both the rib side walls and the etched surface [21]. The strip-loading ridge waveguide (b) is typically used in passive couplers and switches, which require accurate dimensional control and a more weakly guided mode. The degree of mode confinement in the core is controlled primarily by adjusting the thickness of the slab section directly above the core layer. As the thickness of the slab

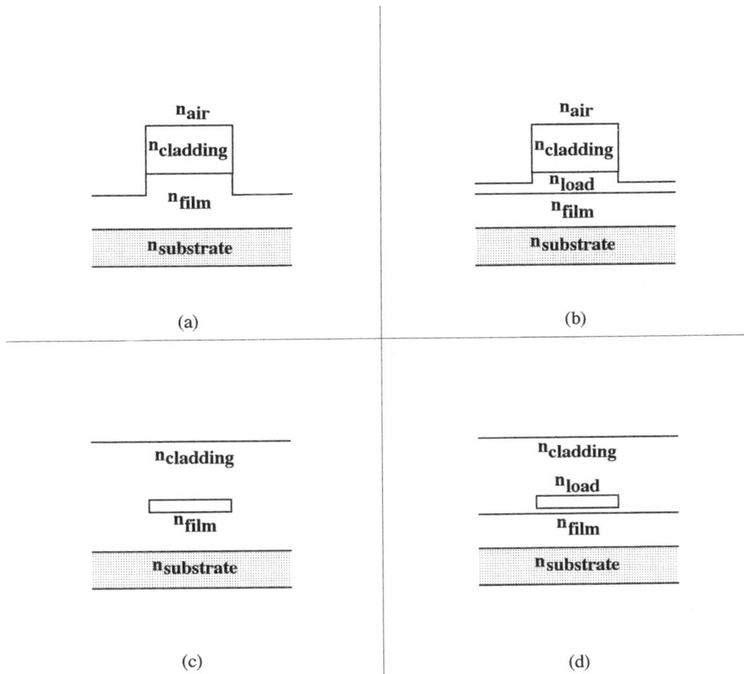

Figure 2 Examples of schematic waveguide cross sections: (a) rib guide, (b) strip-loading ridge guide, (c) buried heterostructure guide, (d) buried rib guide. In all cases, n_{film}, $n_{load} > n_{substrate}$, $n_{cladding}$.

section grows to reduce mode confinement, the requirement on the fabrication tolerance of its thickness can become as low as several hundreds of angstroms. This level of accuracy can easily be achieved by incorporation of thin etch-stop layers or by contrasting the material composition of the slab section to that of the top strip section during the growth of the epitaxial layers. In that case, these waveguide structures are formed by an initial dry-etching step followed by a material-selective wet chemical etch that stops at the compositional change in the slab section, thereby achieving accurate dimensional control in both lateral and vertical directions (see, for example, Refs. 27 and 28).

The buried heterostructure (c) is commonly used as an active laser or amplifier waveguide, while the buried rib (d) is often utilized as a ''backbone'' interconnecting waveguide in monolithic integration of active laser/amplifier sections with other various passive sections. A processing sequence for such monolithically integrated devices [29,30] includes first the longitudinal definition of the passive and active sections. Then a lateral definition of the waveguide structures is obtained by etching a shallow mesa in the backbone waveguide sections and a deeper mesa in the active sections. Following the waveguide etching process, the etched mesa is buried in two epitaxial growth steps. First, a current blocking

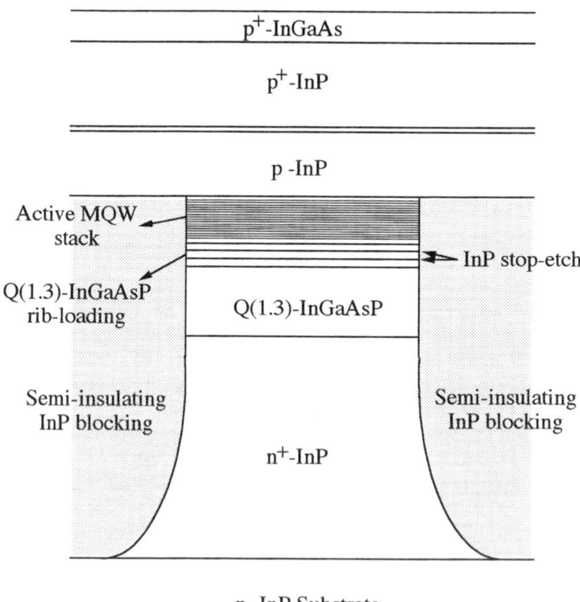

Figure 3 Schematic cross section of a channel mesa buried heterostructure (CMBH) laser containing active MQW layers and passive waveguide ($\lambda_p = 1.3\ \mu m$) layers.

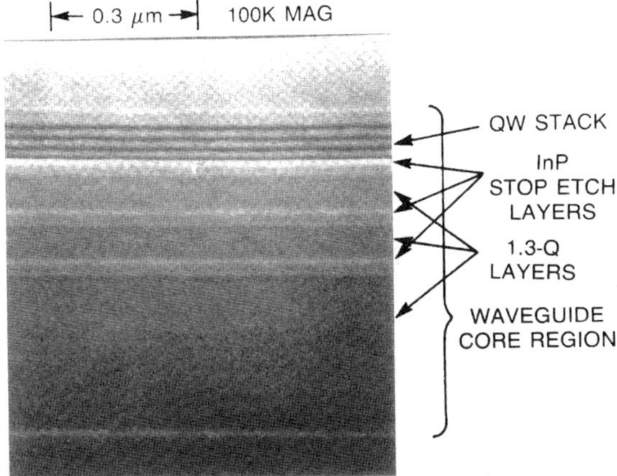

Figure 4 Scanning electron micrograph showing epitaxial layers of a CMBH waveguide structure with an MQW active guiding stack and a passive $\lambda_p = 1.3$-μm guiding layer, containing InP etch-stop layers for possible additional processing in the passive waveguide section. (Obtained with permission, © 1991, IEEE.)

layer is overgrown on the passive waveguides, while it covers only the mesa side walls of the active waveguides (due to a growth-inhibiting oxide mask on top of the active waveguides). Then a second growth of p^+-doped cladding buries the active mesa and covers the already buried passive sections. Figure 3 shows a schematic cross section of a typical implementation of a channel mesa buried heterostructure (CMBH) laser containing passive waveguide ($\lambda_p = 1.3$ μm) layers [30]. Figure 4 shows a scanning electron microscope (SEM) micrograph of a similar waveguide structure, containing additional InP etch-stop layers to allow for possible further processing in the passive waveguide section [30].

2.4 Active/Passive Waveguide Transitions

The transition between active and passive waveguide sections has a critical effect on the performance of many PICs. High coupling efficiency and low reflection at the active/passive interface are crucial, particularly in laser-integrated structures. Various coupling methods have been investigated by many research groups, including butt-joint coupling [31–33], active-layer removal [29,34], selective area growth (SAG) of quantum wells [35–38], and quantum wells intermixing [39,40]. The more frequently used coupling methods are the first two, which are discussed in the following.

The most flexible method in terms of device design is butt-joint coupling. In this coupling scheme the active waveguide epilayers are grown in the first-base-wafer growth, and they are subsequently removed by wet chemical etching in the passive regions. The passive waveguide layers are then regrown in alignment with the active sections in a second epitaxial growth step. Since the passive waveguide layers are grown in a step separate from the active layers, each section can be designed independently for optimal layer composition, carrier concentration, and modal properties. This also provides a theoretical coupling efficiency of 100% if the active and passive waveguide modes are perfectly matched. Coupling efficiencies higher than 90% are frequently obtained [34,41]. Figure 5 shows an example of a butt-joint distributed-feedback (DFB) laser and an electroabsorption modulator [31]. In this device example, the active 1.55-μm DFB laser section is coupled to an electroabsorption section with an absorption layer of $\lambda_p = 1.40$-μm composition. Figure 6 shows the smooth transition obtained between the regrown butt-joint layers in a SEM micrograph of the longitudinal device cross section. Recently, RIE was used in the etching process of the active layer removal [42], for precise and reproducible control of the etched facet shape and the active layer undercut, both of which affect the subsequent regrowth quality at the active/passive interface. Average coupling efficiencies of about 91% with standard deviation of 1.5% was obtained for more than 100 samples of integrated butt-coupled amplifiers and passive waveguides.

A second common method is the active-layer removal scheme, illustrated schematically in Fig. 7 [30,34]. In this method, the active layers are grown on top of the passive ones during the base-wafer epitaxial growth sequence, and they are subsequently etched away in the intended passive section. If the active

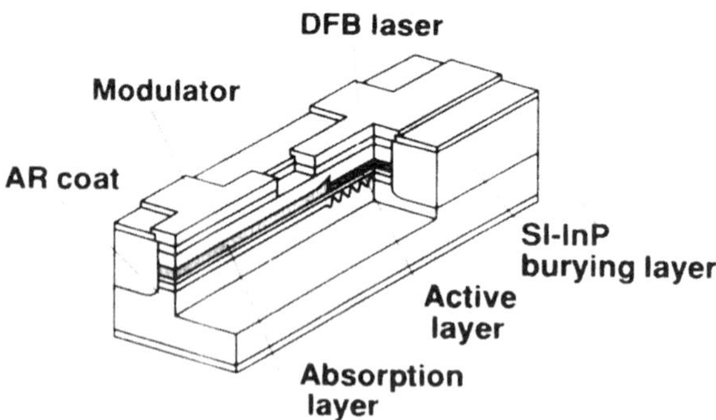

Figure 5 Device structure of a monolithically integrated butt-joint DFB laser/electroabsorption modulator. (From Ref. 31.)

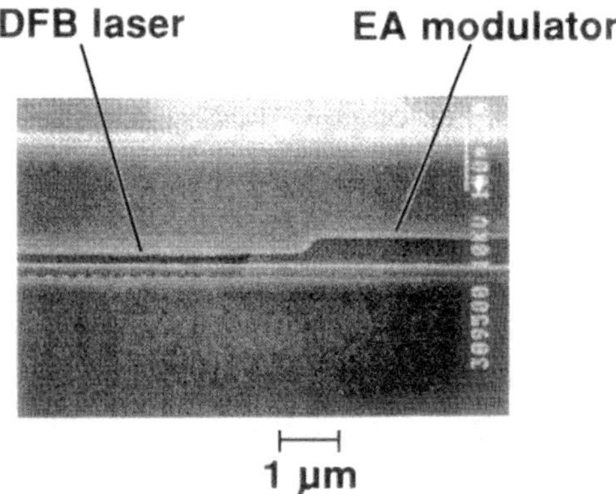

Figure 6 SEM longitudinal cross section of the butt-joint DFB/electroabsorption modulator device. (From Ref. 31.)

core layer is thin relative to the passive core, a smooth transition with coupling efficiencies of more than 92% can be achieved [30,34]. This approach has the advantage of simpler processing, for one less regrowth step is required. However, since the active sections contain both passive and active core layers, some compromise in device design may be required.

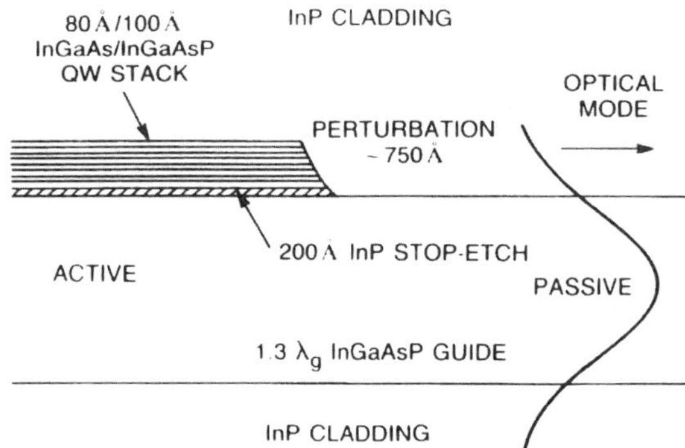

Figure 7 Longitudinal cross section of an active-layer removal device example. The active MQW stack is selectively removed from the passive section using an InP etch-stop layer.

3 MODULATORS AND SWITCHES

3.1 Electro-Optical Index Change

There are several effects that can be used for refractive index modulation in III–V semiconductors: the linear electro-optical (or Pockels) effect, the Franz–Keldysh effect in bulk semiconductors, and carrier-induced effects due to free-carrier injection or depletion.

3.1.1 Linear Electro-Optical Effect

The InGaAsP material system has a zinc-blende crystal structure ($\bar{4}3m$ point group symmetry), which is cubic ($n_x = n_y = n_z = n_o$), and possesses a single electro-optical tensor element, r_{41}. This Pockels coefficient was found to have a weak composition dependence and an even weaker wavelength dependence in the transparent region [43,44]. Representative reported values are in the range of $r_{41} = -1.3$ to -1.6×10^{-12} m/V.

In a common driving scheme, shown in Fig. 8, a (100)-oriented crystal is used with cleavage planes (011) and (01$\bar{1}$). For an electric field E applied along <100>, the principal axes and the refractive indices along them become [45]:

$$
\begin{aligned}
n_x' &= n_o && \langle 100 \rangle \\
n_y' &= n_o + \frac{1}{2} n_o^3 r_{41} E && \langle 01\bar{1} \rangle \\
n_z' &= n_o - \frac{1}{2} n_o^3 r_{41} E && \langle 011 \rangle
\end{aligned}
\tag{6}
$$

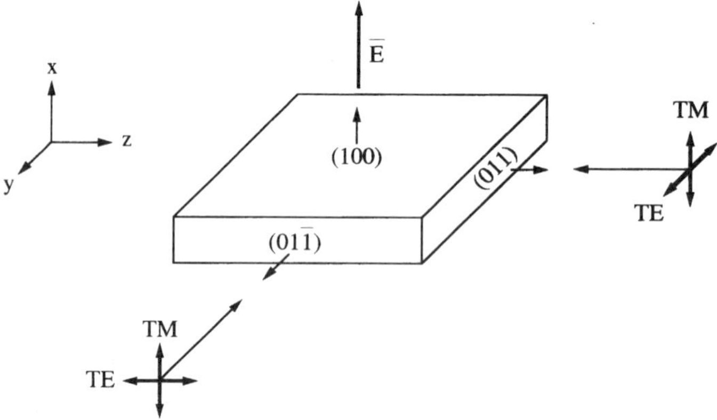

Figure 8 Typical linear electro-optical driving scheme for a (100)-oriented crystal.

Since r_{41} is negative, a negative index change is obtained for a TE-polarized <011>-propagating optical field (see Fig. 8), while for a TE field with $<01\bar{1}>$ propagation the index change is positive. In the case of TM polarization, no index change is obtained in either direction.

3.1.2 Franz–Keldysh and Quantum-Confined Stark Effect

The Franz–Keldysh effect is an electric field–induced extension of the bulk semiconductor absorption tail deeper into the bandgap [46]. It can be explained as resulting from the tilting of energy bands under an electric field, which allows the electron wavefunction to penetrate evanescently into the bandgap, thereby inducing an effective red shift in the bandgap energy. This changes both the absorption constant of the semiconductor (electroabsorption) and its refractive index (electrorefraction). The electrorefraction effect has an approximately quadratic field dependence for photon energies well below the bandgap and for moderate electric fields. The amount of index change is strongly dependent on the energy difference between the bandgap and the photon energy, particularly at optical energies close to the bandgap [44,47–51]. The calculated energy-difference dependence of the Franz–Keldysh electrorefraction in InGaAsP is shown in Fig. 9(a), for several applied electric field strengths [50]. This figure shows that the index change well below bandgap is positive, changing its sign as the energy detuning from the bandgap becomes very small.

As the photon energy approaches the band edge, the index change becomes stronger, however it is accompanied by increased electroabsorption. A useful figure of merit is thus the ratio $\Delta n/\Delta k$, where Δn and Δk are the real and imaginary parts of the complex refractive index change, respectively, and the extinction coefficient k is related to the absorption coefficient α by $\alpha = 4\pi k/\lambda$. Figure 9(b) shows the calculated dependence of this ratio on energy difference at the same electric field values used in Fig. 9(a) [50]. For efficient phase modulation with low loss, a large $\Delta n/\Delta k$ ratio should be used, which is obtained with larger-energy detuning from the bandgap. In contrast, an electroabsorption intensity modulator with low chirp should be designed to operate much closer to bandgap, with $|\Delta n/\Delta k| \sim 1$

Significantly enhanced electrorefraction and electroabsorption can be obtained in multiple quantum well (MQW) structures [52,53]. Moreover, this so-called quantum-confined Stark effect (QCSE) involves an energy red shift of a relatively narrow excitonic absorption resonance, with much smaller associated exciton broadening, as compared to the bulk Franz–Keldysh effect. Hence, it is possible to obtain large $\Delta n/\Delta k$ ratios closer to the bandgap. This was demonstrated in InGaAsP/InP MQW heterostructure waveguides [54], along with enhanced electrorefraction coefficients both at 1.3- and 1.5-μm wavelength ranges. The QCSE is polarization sensitive—i.e., TE exhibits a larger response than TM—unlike the Franz–Keldysh effect, which may be only mildly sensitive to polarization.

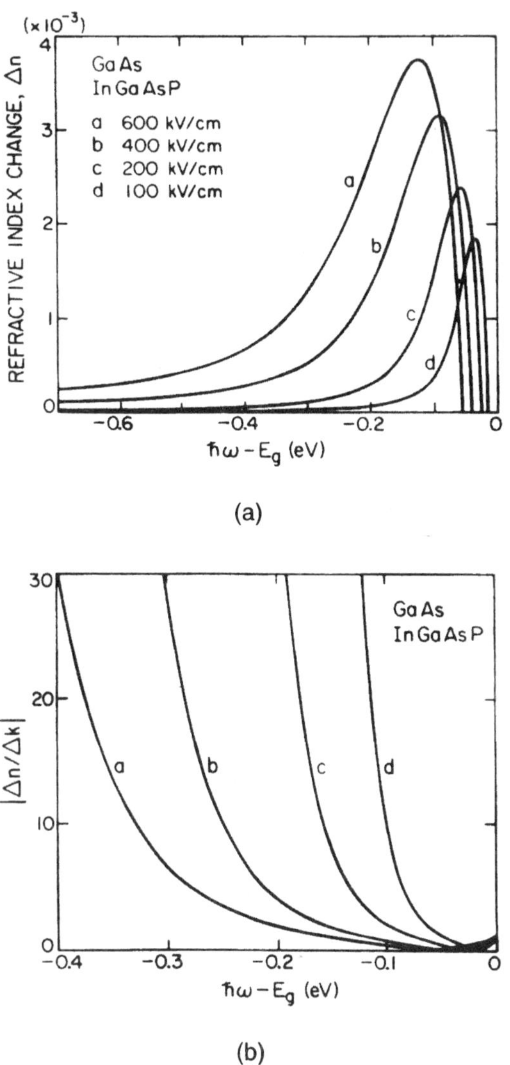

Figure 9 (a) Calculated Franz–Keldysh electrorefractive index change in bulk InGaAsP (and GaAs) as a function of photon energy detuning from the bandgap. (b) Calculated $\Delta n/\Delta k$ ratio dependence.

3.1.3 Carrier-Induced Index Change

The main contributions to the carrier-induced index change are due to free-carrier plasma and bandfilling/shrinkage effects. These effects are polarization insensitive in bulk semiconductors, and their speed is determined by the time it takes to eliminate free carriers from the semiconductor layer (limited to approximately nanoseconds for current injection).

The first contribution is due to indirect intraband transitions in the conduction band (see Sec. 2.2). The index change resulting from this plasma effect is usually estimated approximately by the classical Drude model for electron gas, which can be expressed as

$$\Delta n = -b\lambda^2\left(\frac{N}{m_e} + \frac{P}{m_h}\right) \qquad (7)$$

where N and P are the free electron and hole densities, respectively, m_e and m_h are their respective effective masses, and b is a proportionality constant. Since the hole effective mass is much larger than that of the electron, the hole contribution to the index change is negligible.

Band filling in the conduction band and valence band due to carrier injection causes an absorption shift to higher photon energies, known as the *Burnstein shift* or the *Burnstein–Moss effect* [55]. This effect is characterized by absorption decrease with increasing carrier concentration for photon energies slightly above the nominal bandgap. Through the Kramer–Kronig relation, this absorption shift causes a substantial index change even for energies below bandgap. Since the density of the states in the conduction band is much smaller than that in the valence band, electrons in the conduction band fill it to a much deeper level. Hence, this index change is also dominated by electron contributions, rather than holes. For large electron concentrations ($N > 10^{18}$ cm^{-3}), band filling is compensated by bandgap shrinkage due to electron interactions [8].

The combined effects result in an overall index change in InGaAsP alloys that is negative for photon energies in the transparent region as well as close to bandgap. Figure 10 shows the calculated index change at $\lambda = 1.56$ μm due to several carrier concentrations, as a function of InGaAsP bandgap wavelength [5]. As shown, the index change well below bandgap is weakly wavelength dependent, while close to bandgap band filling/shrinkage effects dominate, substantially increasing the amount of index change. Typical calculated values as well as relatively scarce experimental values for the rate of index change in the transparent region [5–9] are approximately $\Delta n/N = -1 \times 10^{-20}$ cm^{-3} for carrier densities up to about 1×10^{18} cm^{-3}. As shown in Fig. 10, the rate of change is higher for larger concentrations; however, as discussed in Sec. 2.2, it is also accompanied by a substantial absorption increase.

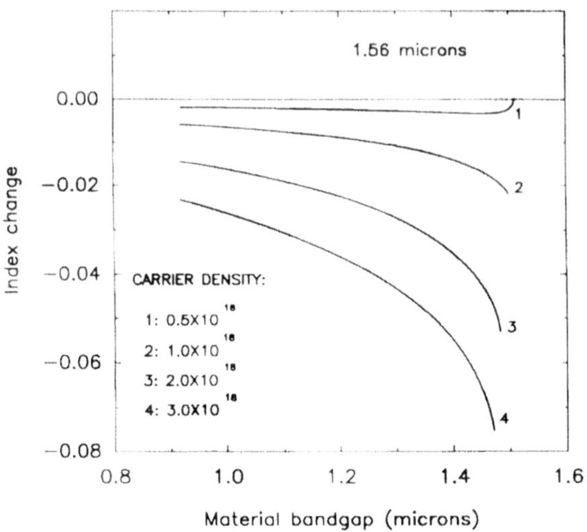

Figure 10 Calculated carrier-induced index change at $\lambda = 1.56\ \mu m$ as a function of InGaAsP composition for four different electron densities. (Obtained with permission, © 1994, IEEE.)

3.2 Electroabsorption Modulators

Electroabsorption (EA) due to the Franz–Keldysh effect in bulk, or to the QCSE in multiple quantum wells (MQWs), both described in the previous section, can be used effectively for external laser modulation. Electroabsorption modulators have proven to be efficient external modulators with low drive voltages, typically lower than 4 V for bulk-type devices, and only 1–2 V for MQW modulators. Their modulation speed is essentially limited by the lumped-*RC* time constant of the device, and since very short modulator lengths are required to obtain a high extinction ratio (100–300 μm), modulation bandwidths of 30 GHz and higher were obtained by several groups [56–59]. The modulators' small footprint is likewise beneficial for cost-effective laser-modulator integration.

Electroabsorption modulator design involves several trade-offs. In order to increase modulator response to applied driving voltage, either the EA material bandgap can be designed closer to the operating wavelength or a longer modulator can be used. This will increase the insertion loss of the device, and in the latter case would also reduce the modulation bandwidth due to increased device capacitance. Ultrahigh-speed modulation at bit rates of more than 10 Gb/s requires a device length smaller than 150 μm. These trade-offs can be alleviated by the use of MQW-EA modulators, since the QCSE provides enhanced field response. In addition, MQW structures allow for greater flexibility in modulator design: by adjusting simultaneously the well width and the built-in layers' strain, the transmission performance can be optimized [60,61].

Intensity modulation with EA is accompanied by phase modulation due to the associated electrorefraction, resulting in frequency chirping. The chirp parameter for EA modulators can be defined as the ratio of phase modulation to amplitude modulation [62], yielding

$$\alpha_m = \frac{d\phi/dt}{(1/E)dE/dt} = \frac{dn}{dk} \tag{8}$$

where ϕ and E are the phase and amplitude of the electric field at the output of the modulator, respectively. This small-signal chirp parameter cannot fully predict the transmission performance of the modulator, since it relies on the assumption of bias-independent chirp. However, the actual chirp parameter and its specific sign depend not only on the proximity of the operating wavelength to the modulator bandgap, but also on the bias voltage and the modulation depth [63,64]. This is illustrated in Fig. 11 [64], which shows the measured chirp parameter of a InGaAsP/InGaAsP MQW-EA modulator as a function of applied bias at several wavelengths. For an ideal bias-independent chirp parameter, values in

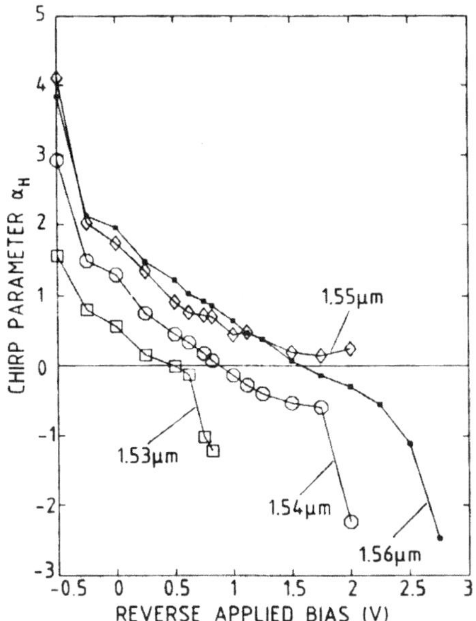

Figure 11 Measured chirp parameter of an MQW electroabsorption modulator as a function of reverse bias at several wavelengths. (Obtained with permission, © 1994, IEEE.)

the range 0 and -1 provide optimal transmission performance at the 1.55-μm operating wavelength range. Average effective chirp parameters for bias-dependent modulators, however, are typically measured to be positive. General solutions to this involve the reduction of the small-signal chirp parameter of the modulator at the *off* (transparent) state, such that it becomes negative for a larger range of applied voltages. This can be obtained by operating closer to the band edge or by ingenious design of the quantum-well strain and well width [61].

As stand-alone modules, EA modulators suffer from a high coupling loss, and their polarization sensitivity requires polarization control or, alternatively, a polarization-independent design, which can sacrifice device performance. To alleviate the high insertion loss, integration of modulators with short, tapered, passive waveguide sections can be used, as shown in Fig. 12 [58]. However, the most cost-effective and low-loss approach is to integrate monolithically the laser and the modulator on a single chip. Several integration techniques have been employed [31,65], including butt joint [58] and selective area growth (SAG) [66–69]. The main requirements for integrated laser/modulator devices is to maintain high coupling efficiency with low back-reflection between the laser and the modulator sections, as well as sufficiently high electrical isolation between the two sections (>10 kΩ). The SAG technique is particularly suitable for satisfying these requirements. This approach allows for the simultaneous epitaxial growth of MQW layers with varying bandgap wavelengths, without necessitating additional regrowth steps. The bandgap control is achieved longitudinally across the wafer by controlling the quantum-wells thickness with longitudinal variation of a growth-inhibiting dielectric mask opening. A schematic layout of a MQW DFB

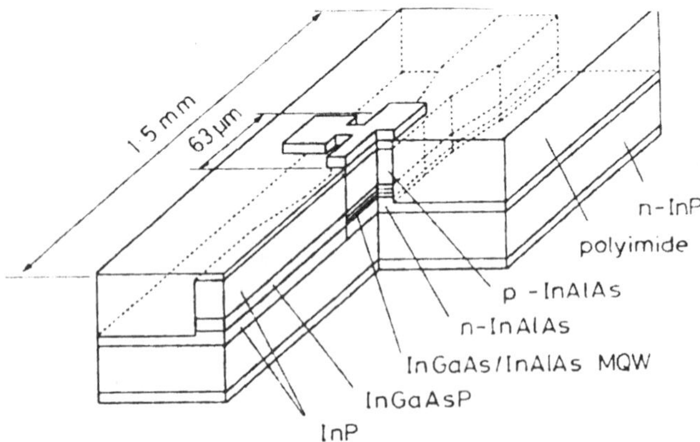

Figure 12 Schematic structure of an MQW electroabsorption modulator with integrated short, tapered, passive waveguide sections. (From Ref. 58.)

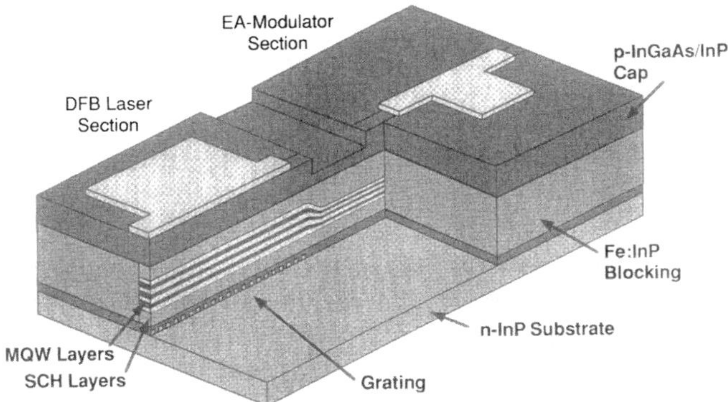

Figure 13 Schematic layout of a selective-area-growth MQW structure for an integrated DFB laser/electroabsorption modulator. (From Ref. 66.)

laser/EA modulator integration using SAG is shown in Fig. 13 [66]. Laser/EA modulator modules monolithically integrated with various integration techniques are commercially available today from several manufacturers.

3.3 Mach–Zehnder Modulators

Mach–Zehnder (MZ) modulators are attractive external modulators due to their ability to provide a widely adjustable chirp parameter. This capability is obtained in a push–pull driving configuration, where both arms of the MZ interferometer are biased around a common value, and each one is driven to provide a phase change with an opposing sign to the other. Using the same chirp parameter definition given in the first of Eqs. (8), the chirp parameter of a MZ modulator can be written approximately in terms of the phase change in each arm as

$$\alpha_m = \frac{\Delta\phi_1 + \Delta\phi_2}{\Delta\phi_1 - \Delta\phi_2}$$

By fixing the phase difference in the denominator, a constant extinction ratio is obtained, while independently adjusting the actual value of the phase change in each arm provides a widely tunable chirp, from negative to positive values [70]. In particular, a symmetrical driving configuration, i.e., when $\Delta\phi_1 = -\Delta\phi_2$, yields chirp-free modulation. External modulation with $LiNbO_3$-based MZ modulators has been studied extensively, and it was found that for transmission in the nonzero fiber dispersion wavelength range of 1.55 μm, optimal transmission performance is obtained with a small negative chirp parameter ($-1 < \alpha_m < 0$) due to the mild pulse compression it provides [71].

Indium phosphide–based MZ modulators possess some attractive characteristics, such as small size, low driving voltage, and compatibility to laser/modulator integration [72,73]. Typical implementation is shown in Fig. 14 [74], where deeply etched rib waveguides are used and the phase modulation in the electrode sections is provided by electrorefraction. For enhanced index change, multiple quantum well waveguides are used [75]. Typical phase modulation section lengths are in the range of 400–600 μm, while the overall device length is about 1.5 mm. Unlike $LiNbO_3$ modulators, which rely on the linear electro-optical effect, the electro-optical phase change provided by electrorefraction has a nonlinear relationship to the applied voltage. As a result, the phase change obtained in each MZ arm is strongly bias dependent, particularly for multiple quantum well modulators that utilize the QCSE. This means that in the case of the symmetrical push–pull driving configuration, the MZ output phase is unbalanced, with the stronger phase change obtained in the arm that is modulated between the higher voltage values. Unfortunately, this imbalance produces a small positive chirp, as illustrated in Fig. 15(a). A remedy to this situation can be to shift the operating point of the modulator by introducing a fixed π-phase shift between the arms. This is obtained by elongating one MZ arm with respect to the other, rather than changing its voltage bias point [76]. Figure 15(b) demonstrates that a negative chirp parameter is obtained in this case.

In addition to bias-dependent phase change, the electroabsorption associated with the electrorefractive index change is also bias dependent. This provides an imbalance of power in the two arms, which reduces the extinction ratio obtained. By changing the splitting ratio at the Y-junctions of the MZ, this effect can be somewhat compensated for; however, it concurrently affects the chirp para-

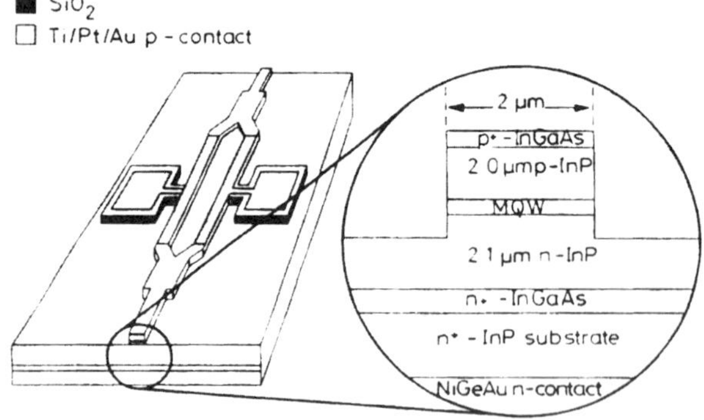

Figure 14 Schematic layout and waveguide cross section of an MQW Mach-Zehnder modulator.

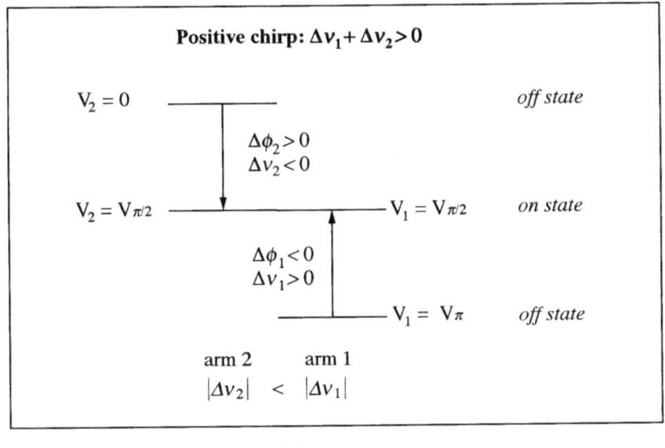

(a)

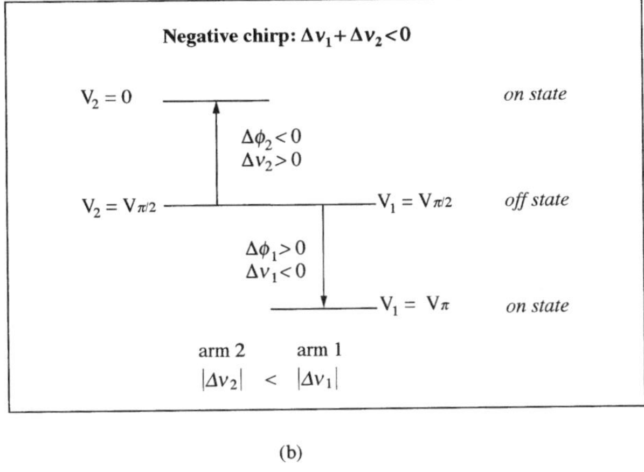

(b)

Figure 15 Frequency chirp of a semiconductor Mach–Zehnder modulator. (a) Symmetrical push–pull configuration: positive chirp, since the phase change in arm 1 is larger than that in arm 2. (b) π-Shifted configuration results in negative chirp. (After Ref. 76.)

meter. Hence, a modulator design should be based on considering the impact of the variation in both extinction ratio and chirp on system performance. An example is shown in Fig. 16 [77] for a dual-drive MZ modulator with symmetrical arm lengths, where the splitting ratio is varied to maximize receiver sensitivity. The optimal splitting ratio is clearly dependent not only on the amount of electroabsorption in the arms, but also on the chirp level variation with splitting ratio.

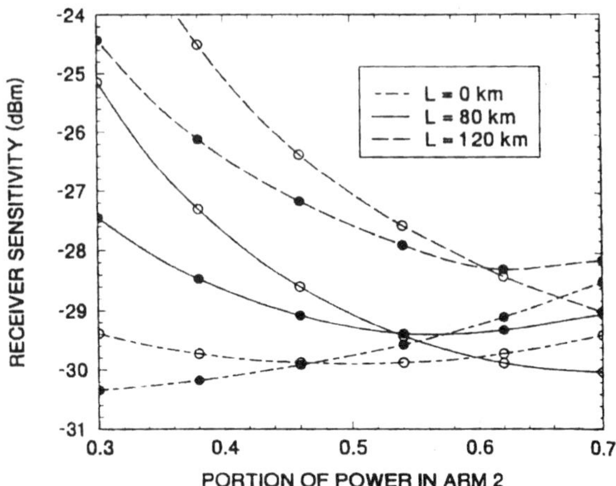

Figure 16 Receiver sensitivity versus power splitting ratio of a symmetrical dual-drive MZ modulator for three different transmission fiber lengths. Solid circles: modulator with nonzero attenuation due to electroabsorption; open circles: modulator with zero absorption change. (Obtained with permission, © 1994, IEEE.)

System experiments with bit rates up to 10 Gb/s have been reported by several groups using InP-based MZ modulators [74,76,78,79]. Recently, wavelength-independent transmission at 10 Gb/s was demonstrated for the 1530–1560-nm wavelength range with a InP/InGaAsP phase-shifted MQW modulator [80].

3.4 Reduced-Size Components

In implementing photonic integrated circuits in the InP material system, one must take into account the relatively high propagation loss and one's manufacturing limitations in obtaining large-area device uniformity (typical InP wafer diameter is 2–3-in.). Hence, reducing the components' size has great importance. An example of photonic circuits that require particularly large chip area is switch arrays and wavelength-(de)multiplexing devices utilizing waveguide grating routers (WGRs, discussed in greater detail in Sec. 5). A large part of these circuits' chip area is devoted to passive waveguide interconnections that include bends or curves. To reduce the size of the interconnection real estate, several methods have been suggested. Sharper bends and curves can generally be achieved by increasing the mode confinement in the waveguide core, which is typically obtained by increasing the etch depth of rib waveguides [25,81]. Even sharper bends can be obtained by utilizing corner mirrors [82–85]. Another method is to use

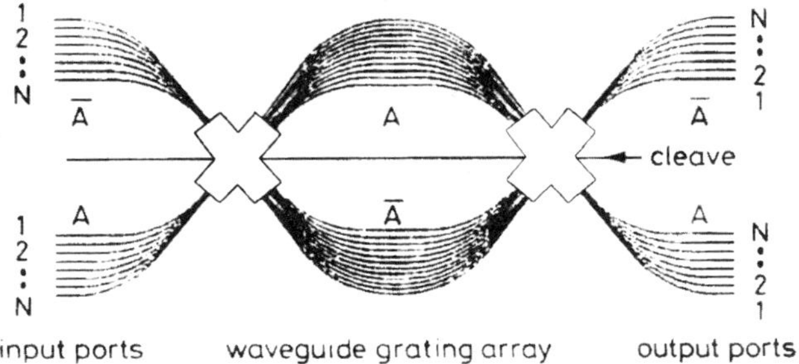

Figure 17 Two compact routers obtained by cleaving the two interlaced routers A and $\bar{A}$ through the center of their free-space region.

the circuit symmetry to save chip real estate. This is demonstrated in Fig. 17, where WGR size was reduced to half by using total internal reflection at the cleavage plane, which halves the two interlaced routers (A and $\bar{A}$) into compact size routers $A\bar{A}A$ and $\bar{A}A\bar{A}$ [86].

For components based on modal interference, reduction of the modal interaction length can lead to increased sensitivity to fabrication tolerances and hence can sacrifice device robustness. For example, shortening the interaction length of directional couplers by reducing the gap between waveguides results in a severe penalty of reduced fabrication tolerances, which are already low due to the relatively tight guiding that is typical of semiconductor waveguides. Utilizing zero-gap multimode interference (MMI) couplers presents an elegant solution to this problem. The MMI coupler, shown schematically in Fig. 18, relies on the

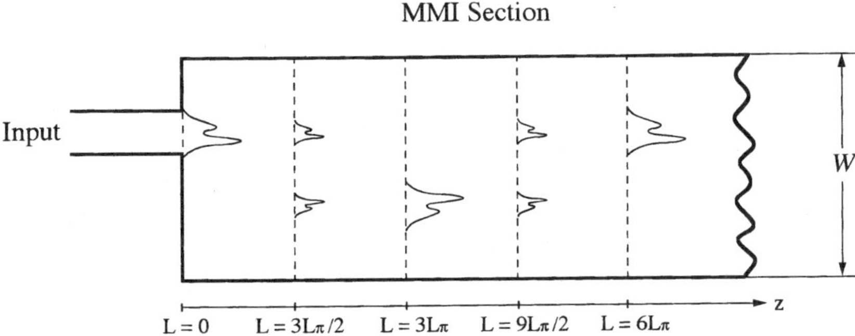

Figure 18 Self-imaging in an MMI coupler structure as a function of propagation distance along the multimode section.

self-imaging property of multimode waveguide propagation [87–90]. An incident field at the input of the multimode section excites the eigen modes supported by this region, which then propagate and interfere to form various power distribution patterns at the output plane. The output pattern is determined by the accumulated phase difference between the modes, given approximately by

$$(\beta_m - \beta_0)L = -\frac{\pi m\,(m + 2)L}{3L_\pi} \tag{9}$$

where β_m is the propagation constant of the mode m ($m = 0, 1, 2, \ldots$) and $L_\pi = 4nw^2/3\lambda$ is the beat length between the first two modes, $m = 0, 1$. The MMI section parameters are its width w, length L, and guiding region effective index n. As shown in Fig. 18, all the modes interfere constructively at a distance of $L = 6L_\pi$, forming a self-image of the input field. At half this distance, $L = 3L_\pi$, the odd and even modes are out of phase and hence form a mirror image of the input, while an N-fold image is formed at distances $L = 3L_\pi/N$.

This self-image formation is independent of the input field excitation position, since all the modes that can be excited experience the same phase change. It was shown that by restricting the input field excitation to specific locations, such that only certain mode sets are excited, shorter imaging distances can be obtained. This is illustrated in Fig. 19, which shows three different types of MMI couplers [91]. The $N \times N$ coupler [Fig. 19(a)] is obtained with the *general self-imaging* configuration discussed earlier, where the input field position is arbitrary and the imaging distance is $L = 3L_\pi/N$. The $1 \times N$ coupler uses a *symmetrical power splitting* configuration [Fig. 19(b)], where the input waveguide is positioned at the center of the MMI section and self-imaging forms at $L = 3L_\pi/4N$ [90,92]. *Restricted self-imaging* $2 \times N$ couplers have input waveguides at the $w/3$ and $2w/3$ locations [see Fig. 19(c)], such that self-imaging occurs for $L = L_\pi/N$ [89].

Unlike conventional directional couplers, MMI couplers use zero-gap strongly multimoded interference structures, characterized by larger spacings between the modal propagation constants. Consequently, this allows for very compact devices. Typical InP-based 3-dB couplers have 200–500-μm-long MMI sections. Low-loss MMI coupler length as short as 90 μm was demonstrated with an insertion loss of 0.7 dB, implemented with deep etching for increased guiding strength in the MMI region [25]. Multimode interference couplers possess important properties, such as low polarization and temperature sensitivity, broadband (~100 nm for 3-dB couplers [91]), and relatively robust splitting ratio characteristics (as compared to conventional directional couplers). They are fabrication tolerant to etch depth, input/output waveguide width, and coupler length, while their largest sensitivity is to the MMI region width w (e.g., ±0.5 μm for up to 1-dB excess loss and unbalance in restricted-type 3-dB couplers) [91]. These robust and compact features have led to the incorporation of MMI couplers in a number

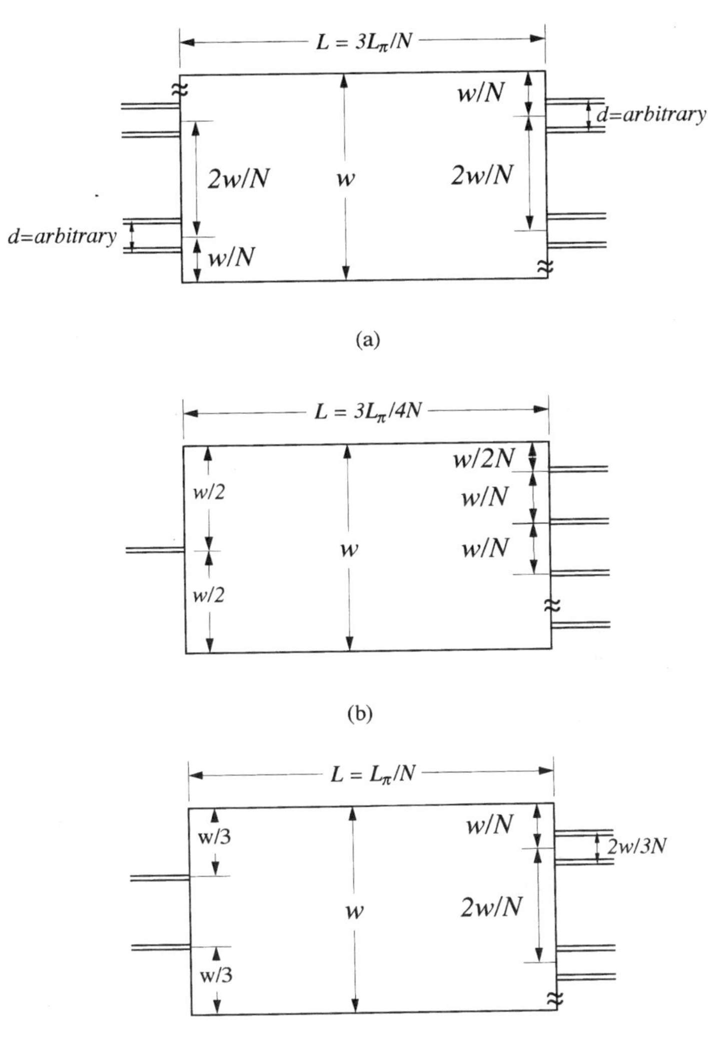

Figure 19 MMI coupler configurations. (a) $N \times N$ general self-imaging with arbitrary spacing between input waveguides. (b) $1 \times N$ symmetrical power splitting with input at the center of the coupler. (3) $2 \times N$ restricted self-imaging with inputs at $w/3$ and $2w/3$.

of applications, such as power splitters and combiners and polarization-independent switches [93–96].

3.5 Optical Switches

Optical switches are needed in a growing number of near-term applications, particularly in signal routing and time-division signal processing. In these applications, the optical transparency of photonic switching modules offers simplified and higher-capacity system operation, by allowing networking functions to be lowered into the optical layer.

Important parameters of optical switches are their optical crosstalk, insertion loss, polarization sensitivity, and optical bandwidth. A large variety of switch types has been implemented in the InP material system. These include total internal reflection (TIR) switches [97,98], digital optical switches (DOS) [27,99–101], Mach–Zehnder interferometer switches [93–96,102], directional coupler switches [103–108], amplifier-based gate switches [109–115], and electroabsorption-based gate switches [116]. Some illustrative examples of advanced switch demonstrations are described in this section.

Switch elements with desired polarization-independent loss and extinction ratio characteristics were demonstrated in a number of techniques. Since the switching curves of Y-shaped ''digital'' optical switches have relatively low polarization and wavelength dependence, owing to their reliance on modal evolution rather than modal interference, their performance is generally more fabrication tolerant. Y-switches based on carrier injection have been demonstrated, achieving polarization-independent crosstalk levels below −12 dB for a low injection current of 6 mA [99] and better than −20 dB crosstalk with 50–100-mA obtained in a larger switch angle and a different PIN junction design [27]. A polarization-independent Y-switch was also demonstrated in a InGaAsP/InGaAsP MQW structure [28], where the polarization sensitivity of the QCSE was compensated for by balancing the TE and TM switching curves, exploiting the relaxed polarization sensitivity of the Y-switch. Adequate crosstalk levels in Y-switches can be achieved with sufficiently adiabatic modal evolution, which requires rather long devices. Reducing the voltage–length product of the switch by shaping the Y-junction profile was suggested and implemented in lithium niobate [117,118]. Similar techniques were implemented in InP to achieve switch lengths below 2 mm [28,100,101].

Polarization independence in interferometric-type switches requires equalizing the contribution of each polarization state to the refractive index change. This was demonstrated in MQW switches by applying tensile strain to enhance the (otherwise weaker) electrorefractive response in TM [93,108] or by using a wide-well structure [102]. In bulk material, current injection can be used to provide polarization independence [103], as long as the optical mode overlap with the injected-current profile is the same for both polarizations. Reverse-bias operation, on the other hand, is polarization dependent, due to the substantial contribution of the linear electro-optical effect, which is strongly polarization sensitive (see Sec. 3.1). This problem was circumvented in a Mach–Zehnder switch by choosing a propagation direction that equalizes the contributions of the linear electro-optical effect and the slightly polarization-dependent Franz–Keldysh effect [95].

Another important parameter in determining switch performance is its extinction ratio, or crosstalk. This should be distinguished from the system crosstalk at the output of a switch fabric, which accounts for all possible coherent and incoherent contributions of unwanted leakage signals in the switch matrix. Recent simulations have shown that WDM networks with narrow laser linewidths and multigigabit rates place severe restrictions on the allowed switch fabric crosstalk, indicating that unwanted leakage power levels at each switch fabric output should be greater than 40 dB below the signal power [119,120]. The required crosstalk from an individual switch element depends to a large extent on the switch matrix architecture [121,122], varying from −20 dB to as low as −40 dB for gate switches. While Y-branch digital optical switches feature robust polarization-independent operation, they render greater difficulty to obtain low crosstalk levels, particularly below −20 dB. A crosstalk-reduced digital optical switch was suggested, which utilizes a mode-converting section to mitigate crosstalk-producing second-order mode excitation at abrupt junctions [123]. Crosstalk improvement for typical switch parameters can be as low as −20 dB, though at the expense of added complexity and possibly enhanced polarization sensitivity. Dilated architectures that integrate additional switch stages can be employed in a similar manner to their utilization in lithium niobate. However, due to the higher on-chip loss of semiconductor devices, the switch size should be made compact.

Another approach to crosstalk reduction that manifests the versatility of the InP material system is to use absorption switching to suppress crosstalk terms. This can be implemented in conjunction with electro-optical switching, using current injection to induce gain/absorption and index change [105,124] or using reverse biasing for electroabsorption combined with electrorefraction [125]. A simpler implementation is to use gate switches in a passive space-routing fabric [109–115]. Amplifier gate switches can provide excellent extinction ratio (>40 dB), broad bandwidth (>50 nm), and large gain to compensate for the passive fabric loss. Their gain can be made polarization insensitive by equalizing the mode confinement in the active core of bulk amplifiers or by equalizing the gain coefficients of MQW amplifiers with alternating tensile and compressive strain. Some demonstrations of amplifier-gate switch arrays are discussed in the next section. Polarization independence can be achieved more readily with electroabsorption gate switches, since their *on* state is below bandgap and hence unaffected by gain or absorption parameters [116]. However, these switches display narrower optical bandwidths (~15 nm) and increased insertion loss, and, most importantly, no gain compensation is provided to mitigate the large passive fabric loss.

3.6 Switch Array Demonstrations

One of the near-term applications most likely to accelerate the deployment of photonic switching systems is optical cross-connecting that offers transparent protection switching and network reconfiguration. Compact size and the ability to integrate active gain compensation are important advantages of semiconductor switch arrays for such network applications. Indium phosphide–based $N \times N$ and $1 \times N$ waveguide switch arrays have been demonstrated in a number of various techniques. The first demonstrations of polarization-independent 4×4 switch arrays have used electro-optical switching based on current injection [126,127]. Low-current directional coupler switches with 15-dB crosstalk were used in a rearrangeble nonblocking architecture with six such switch elements [126]. A strictly nonblocking tree structure architecture that provides better crosstalk suppression was implemented with 24 polarization-independent Y-branch digital optical switches [126]. The array size was about 40 mm $\times$ 2.25 mm, which was later reduced to half this size by incorporating 45° corner mirrors [82], yielding a fiber-to-fiber array loss of 15 dB. Triple-core waveguides were later used to improve fiber-to-chip coupling loss and alignment tolerances [128]. This switch array was used in a 4×4 optical cross-connect experiment operating with four wavelengths at 10 Gbit/s [129].

Several attempts to compensate for array loss by integrating active gain sections were reported. Lossless 1×4 arrays based on twin-guide amplifier switches, which function as active gain-incorporated directional coupler switches [105], have been demonstrated [130]. A 1×16 amplifier-gate switch was demonstrated [111], incorporating a passive star coupler with 16 gate amplifiers at its output and one common booster amplifier at the input. Gain compensation was achieved for the on-chip loss, estimated at 21.5 dB. A lossless and low-crosstalk 4×4 switch array was reported by Kirihara et al. [124]. The schematic structure of the 12-mm $\times$ 2-mm crossbar switch containing 16 gain-incorporated switch units is shown in Fig. 20. Each 2×2 switch element consists of two current-injection total internal reflection (TIR) Y-switches and an optical amplifier, placed in the common branch of the two Y-structures [see Fig. 20(b)]. The amplifier provides gain compensation in the *on* state and crosstalk suppression via absorption in the *off* state. This configuration yielded lossless switching for seven of the 16 possible paths and crosstalk suppression in the range of 40–54 dB.

The first monolithically integrated 4×4 amplifier-gate switch matrix was demonstrated by Gustavson et al. [110], achieving zero fiber-to-fiber insertion loss and typical crosstalk levels of -40 dB in a 7-mm $\times$ 3-mm compact chip size. The strictly nonblocking tree structure array, shown in Fig. 21, consists of

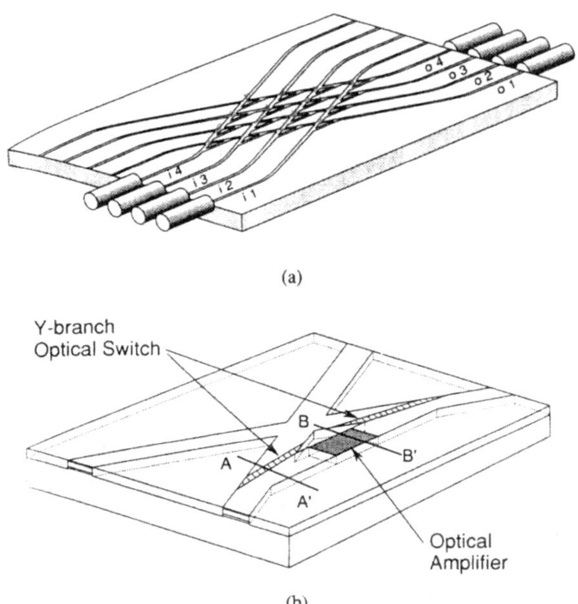

Figure 20 Schematic structure of a 4 × 4 crossbar switch. (a) Array layout; (b) single switch element. (From Ref. 124.) (Obtained with permission, © 1994, IEEE.)

passive waveguide Y-branch splitters and combiners, integrated with 16 gate-switch amplifiers at the center of the array, and 8 booster amplifiers at both input and output sections. Several experimental evaluations of these switches were performed, including 10-Gb/s transmission experiment [131] and crosstalk investigation of a fully loaded switch [132]. A polarization-independent version was demonstrated more recently, utilizing bulk amplifiers with square-shaped core cross sections [133]. Good extinction ratio, nearly lossless switching, and polar-

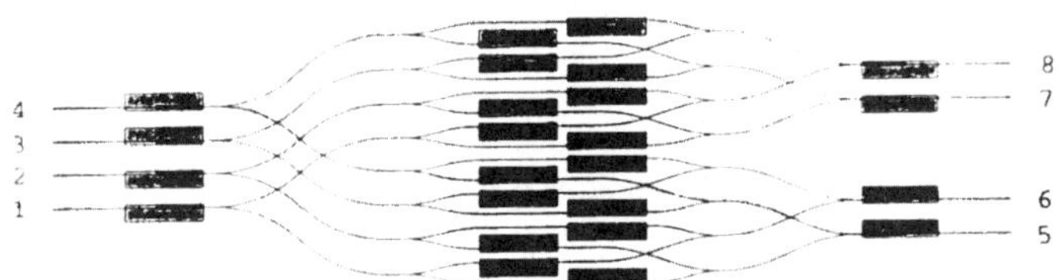

Figure 21 Strictly nonblocking 4 × 4 amplifier-gate switch array. Amplifiers are indicated by dark rectangles.

ization sensitivity as low as 1 dB were achieved, but the array exhibited low output saturation power (about −18.5 dB) due to the high modal confinement in the active core and, possibly, to less efficient current injection.

Incorporating passive splitting and combining arrays with amplifier-gate switches is a very appealing configuration due to the simplicity of the switch fabric, compact array size, and built-in gain compensation. These properties have led to a large number of associated performance analysis studies, which indicated several fundamental and fabrication-oriented performance issues that limit the size and cascadability of these arrays. Limited dynamic range, bound on the low side by amplified spontaneous emission (ASE) noise and on the other side by amplifier saturation, is a major limiting factor. In particular, the amplifiers' fast gain recovery time can lead to bit-pattern-dependent signal gain when operating near saturation, which causes signal distortion at high-speed data rates. Low noise and high output saturation power are thus required characteristics for large-array scalability, which may be obtained with strained MQW amplifiers. Another issue is crosstalk suppression. The passive array structure gives rise to a large number of crosstalk components, interfering coherently and incoherently at the array output. Recent crosstalk characterization of a fully loaded 4 × 4 amplifier-gate switch has shown a mean switch fabric crosstalk of about −30 dB, with several crosstalk components as high as and beyond −20 dB [132]. The crosstalk mapping for this switch is shown in Fig. 22. This crosstalk accumulation problem can be greatly alleviated by integrating more gate-switch amplifiers at splitting and combining stages inside the fabric but, more significantly, by reducing the leakage at waveguide crossings to levels below the required system crosstalk.

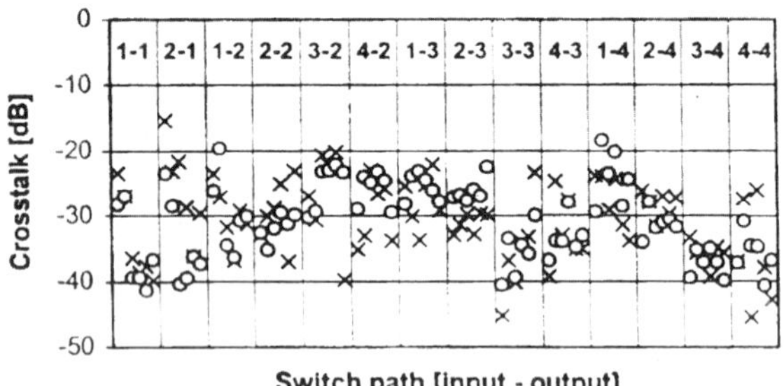

Figure 22 Crosstalk components of a fully loaded 4 × 4 amplifier-gate switch array. Shown are 14 input–output connections (out of 16 possible), which are indicated at the top. For each connection of main signal there are six different combinations for connecting the other three inputs, each combination generating copropagating crosstalk signals that interfere to a crosstalk level indicated in the diagram. Crosses: input powers of −10 dB; circles: input powers of −20 dB. (Obtained with permission, © 1997, IEEE.)

4 Fully Integrated Wavelength-Division Multiplexed Sources

For WDM networks, single-frequency (i.e., single-mode) lasers with accurate wavelengths are indispensible. Integrated InP WDM lasers offer the possibility of one or more wavelength channels per compact, reliable, large-scale manufacturable chip. In this section we discuss the main categories of present-day integrated WDM lasers in InP and their applications. Because we confine our discussion to integrated sources, we regrettably leave out discussion of such sources as fiber grating semiconductor lasers [134] and external-cavity tunable lasers.

4.1 Categories

We will consider four main categories of integrated WDM lasers: (1) solitary distributed-feedback (DFB) lasers; (2) arrays of DFB lasers coupled together into a single output; (3) tunable intracavity filter (TIF) lasers; and (4) shared angular dispersive element (SADE) lasers. Some WDM lasers that do not fit neatly into one of these categories, such as arrays of tunable lasers [135–137], are not discussed here.

4.1.1 Distributed-Feedback Lasers

Distributed-feedback lasers consist only of a gain section through which a waveguide passes with a quarter-wave-shifted Bragg grating inside [see Fig. 23(a)].

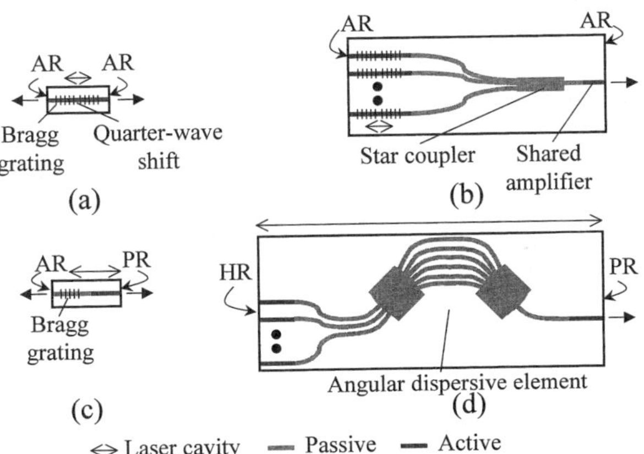

Figure 23 Block diagrams of integrated WDM laser categories: (a) solitary DFB laser; (b) DFB laser array coupled into a single output; (c) tunable intracavity filter laser; and (d) shared angular dispersive element laser. AR = antireflection coating, PR = partial reflection coating, and HR = high-reflection coating.

The Bragg grating is a corrugated index of refraction along the waveguide,* and a quarter-wave-shifted Bragg grating consists of two Bragg gratings of the same period placed in series, with a gap between them that has a width of one-half of the period. As a lightwave passes through a Bragg grating, side modes, in terms of propagation constant (the number of radians per distance) spaced by the spatial periodicity of the grating, are generated. When a side mode equals the negative propagation constant, i.e., when the light wavelength equals twice the grating spatial period, a strong backward-propagating wave is generated. Thus a laser cavity is made around the quarter-wave "gap," the lasing frequency constrained to be at the Bragg wavelength. The quarter-wave shift can be made using techniques such as a phase mask. Another way to make an effective quarter-wave shift is to cleave a Bragg grating at a transition and have a mirror there. Since the required accuracy of such a cleave is nearly impossible, manufacturers sometimes cleave at random and discard about half the lasers.

DFB lasers are currently the most-used lasers in commercial WDM systems. They reliably oscillate in a single longitudinal mode, since the cavity modes and intracavity filter (the Bragg grating) are intrinsically locked together and the intracavity filter is quite narrow compared to the cavity mode spacing—provided there are no extraneous reflections, such as from the chip ends. Distributed-feedback laser cavities are short, though, and so as the real part of the refractive index of the gain medium changes with drive current, age, etc., the oscillation wavelength changes significantly. It is recommended that DFB lasers be stabilized to an external filter for narrow channel spacing ($\leq$100 GHz) systems. Telecom-grade DFB lasers have been reported to output over 100 mW [138], and they tune in wavelength $\sim$+0.11 nm/°C and are thus limited to a tuning range via temperature of $\sim\pm$1 nm.

4.1.2 Distributed-Feedback Laser Arrays Coupled into a Single Output

Distributed-feedback laser arrays coupled into a single output are a set of *N* DFB lasers with Bragg gratings of different-period lengths whose outputs are coupled together on-chip using an optical power combiner [139,140] [see Fig. 23(b)]. Coupling together via a wavelength multiplexer has not been demonstrated, because it is difficult to wavelength-align both the Bragg gratings and the multiplexer. Also, unless the number of lasers is larger than $\sim$5, a power combiner can have less loss than the multiplexer, and a power combiner is much smaller than a multiplexer. In some designs, there is an optical amplifier in the output of the power combiner to boost the output power. The usual choice for the power combiner is the star coupler [141] (see Appendix 1) because of its fabrication

* Actually, some DFB lasers have the corrugation in the gain material, in which case the laser is said to be complex coupled. In such a case, the quarter-wave shift may not be necessary to obtain single-mode operation.

robustness (e.g., symmetrical undercutting does not change the coupling ratios) and small size, although a cascade of Y-branches has also been used. The advantage of the DFB laser array is the selection of single-mode wavelengths it gives the user. The disadvantages are the difficulty of achieving accurate channel spacings and the inherent $1/N$ loss in the power combiner. Such arrays typically contain six lasers, but have contained up to 18 lasers [139], and they output up to 0 dBm when only one channel is operated and there is a shared amplifier [142].

4.1.3 Tunable Intracavity Filter Lasers

Examples of tunable intracavity filter (TIF) lasers are distributed Bragg reflector (DBR) lasers [143], vertical filter lasers [144], and sampled DBR lasers [145]. They consist of a gain section, an electrically tunable filter (which is not in the gain section), and an optional phase-tuning section, all inside a Fabry–Perot cavity [see Fig. 23(c)]. For the DBR laser, the filter is a Bragg grating whose index is changed by forward current injection. The index can be changed up to ~0.7% of the index, and thus at 1550 nm the wavelength tuning range is ~10 nm (~1.2 THz). Sampled DBR lasers achieve a wider tuning range by having two Bragg gratings at both ends of the cavity that are periodically modulated (such as by blanking out parts of the grating) with different periods. The laser thus has two intracavity filters consisting of multiple peaks, and it is tuned by controlling which peaks overlap. Vertical filter lasers also achieve an increased tuning range, by using a grating-assisted phase matching. One can make a TIF laser that has extremely fast tuning (<400 ps) by including a reverse-voltage-operated electro-refractive device inside the cavity [146].

The price paid by TIF lasers for having more tuning range than the DFB laser is complexity in controlling the spectrum. By uncoupling the filter from the gain section, the user must now worry about the filter-to-cavity mode alignment, which can drift with age. Although TIF lasers have longer cavities than DFB lasers, they still have cavity mode spacings on the order of 100 GHz; and thus, for long-term operation with narrow channel spacings, not only must the filter-to-cavity mode alignment be monitored, but the cavity length must be monitored (via the oscillation wavelength), as well. Like DFB lasers, TIF lasers tune about +0.12 nm (15 GHz)/°C.

4.1.4 Shared Angular Dispersive Element Lasers

Shared angular dispersive element (SADE) lasers (called geometric λ-selection lasers in Ref. 147) consist of an array of N amplifiers, an angular dispersive element, and an optional shared amplifier, all between a set of mirrors [see Fig. 23(d)]. There are N separate laser cavities, all sharing the dispersive element. Turning on a particular array amplifier induces laser oscillation at a particular wavelength (or wavelengths spaced the free-spectral range of the SADE, as is discussed shortly) through the dispersive element in one of the N cavities. The earliest SADE laser in InP used a reflective corrugated grating as the SADE [148]. However, fabricating an on-chip mirror with vertical walls and accurate position is difficult. Later, the waveguide grating router (WGR; see Appendix 2) was proposed as the SADE and subsequently demonstrated [149,150]. The WGR laser (also called the *multifrequency laser*) consists of a WGR with the ports terminated in semiconductor optical amplifiers and mirrors [see Fig. 23(d)]. Like DFB and TIF lasers, SADE lasers tune about +0.12 nm (15 GHz)/°C.

The filter response from any input port to any output port of a conventional WGR consists of a periodic sequence of Gaussian-shaped passbands. To make the WGR have only one passband per input–output combination (in order to control the oscillation wavelength) without increasing its size, loss, or passband width, one can chirp the WGR [151,152,153] (see Appendix 3).

The most influential feature on the performance of SADE lasers is their long cavity. SADE laser cavities typically have cavity mode spacings of 3 GHz, more than 30 times longer than typical DFB and DBR lasers. The round-trip time in SADE cavities approaches the spontaneous emission time of the carriers in the gain medium, ~1 ns. Subsequently, SADE lasers behave substantially differently than DFB and conventional TIF lasers, in three main ways. First, while the optical power/drive current bandwidth (direct modulation speed) is determined largely by photon lifetime in short-cavity lasers [154], it is determined largely by the cavity round-trip time in SADE lasers [155]. The direct modulation bandwidth is typically one-third of the cavity-mode spacing in SADE lasers.

Figure 24 shows the theoretical small-signal optical power/drive current vs. modulation frequency for (a) a short-cavity laser (cavity-mode spacing of 90

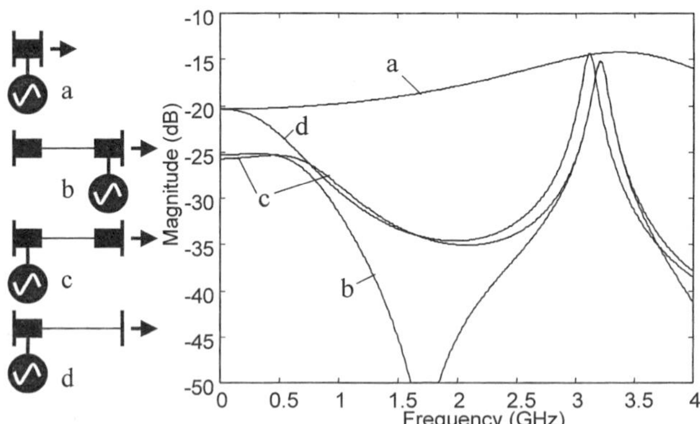

Figure 24 Calculated small-signal responses of output power over drive current vs. modulation frequency for a semiconductor laser. In (a) the cavity-mode spacing is 90 GHz, and in (b)–(d) the cavity-mode spacing is 3 GHz.

GHz), (b)–(c) a long-cavity laser (cavity-mode spacing of 3 GHz) with a gain section at each end of the cavity (but only one modulated), and (d) the same long-cavity laser but with only one gain section. Part (b) is for the output on the modulated amplifier side, and (c) and (d) are for the output on the opposite side. For (a), there is no noticeable difference between the two outputs. Note that for the short-cavity laser there is a resonance peak at a lower frequency than the round-trip frequency, while for the long-cavity laser the resonance peak exists at a higher frequency than the round-trip frequency. This can be used as the definition of a long-cavity laser [155]. Also, note that for the long-cavity laser, the highest modulation bandwidth is given by the output furthest from the modulated amplifier.*

Second, the linear concept that the laser oscillation frequency(ies) are in the cavity mode(s) nearest the intracavity filter peak does not apply to SADE lasers. Instead, nonlinearities result in a strong hysteresis (memory) of the oscillation frequency. It is quite possible for a SADE laser to oscillate in a single longitudinal mode that is multiple cavity modes away from the filter peak. With proper design, the hysteresis can guarantee stable, single-longitudinal-mode behavior in SADE lasers (see Appendix 4) [156].

Third, the narrow cavity-mode spacing implies that the oscillation wavelength shifts very little in wavelength with amplifier changes, and this may eliminate the need for external filter stabilization.

4.2 Applications

4.2.1 Direct Modulation of a Solitary Wavelength

Direct modulation is modulation of the drive current to one or more gain sections inside the laser cavity with the desired transmission signal. Distributed-feedback lasers are the fastest, with demonstrations up to 25-GHz bandwidth at 1.55-μm wavelength [157]. They exhibit significant wavelength excursions because of their short cavities, though. This chirp limits the typical transmission distance to ~160 km at 2.5 Gb/s in conventional fiber (dispersion of 17 ps/nm/km) [158]. The chirp has been reduced by various techniques, including detuning the band-gap wavelength to reduce the linewidth enhancement parameter [157], placing a narrow filter after the laser [159], or externally modulating the output with a periodic signal [160].

SADE lasers have lower chirp (determined almost solely by the linewidth

* Here is the explanation for why the output response depends on its location with respect to the modulated amplifier: The output closest to the modulated amplifier sees the power halfway through the change in power, while the furthest output sees the power after it has passed through the modulated amplifier twice in rapid succession. When the modulation frequency is half the round-trip frequency, the halfway point of the transition is a constant; thus, the modulation response is zero in the closest output, while it is finite in the furthest.

enhancement parameter of the gain sections [155]) but also have much lower modulation speeds. Tunable intracavity filter lasers fall somewhere in between DFB lasers and SADE lasers. Ideally, one should choose the laser cavity length adequate for the desired bit rate and no shorter.

It is important to note that there are two main ways to modulate lasers directly: falling below and staying above oscillation threshold during the ''0''s. Going below threshold gives a high extinction ratio but is limited to low speeds, such as 155 Mb/s. In nearly all the reported high-speed results, the laser never falls below threshold.

4.2.2 Direct Modulation of Many Wavelengths Simultaneously

Out of the four laser categories covered here, this application allows only DFB laser arrays and SADE lasers. For both categories, in order to avoid gain compression-induced crosstalk [161,162], one generally should not include a shared amplifier. One can use a feed-forward scheme of driving the shared amplifier with the sum of the currents of the drives to the lasers [163], but it requires extra electronics and careful adjustments. Also, when running many channels simultaneously, the power per channel may actually be reduced by having a shared amplifier.

With DFB laser arrays, the modulation speeds can be very high because of the short laser cavities, but it can be difficult to obtain an accurate channel spacing because of mutual heating, etc., and high enough output power when running many channels simultaneously because of the inherent power-combining loss. The optical spectrum from a direct modulation experiment of an array of six DFB lasers, each laser modulated at 622 Mb/s, is shown in Fig. 25(a) [164].

With SADE lasers the modulation speeds are typically limited to $\leq$1 Gb/s because of the great cavity lengths, but the channel spacings are accurate, and output powers can be reasonable for a large number of channels. The optical spectrum from a direct modulation experiment of a 16-channel WGR laser, each channel modulated at 622 Mb/s, is shown in Fig. 25(b) [164]. Note that in this experiment, the SADE passband was very wide (110 GHz), and so some of the channels were oscillating in multiple cavity modes (cavity mode spacing = 2.5 GHz), and the data eyes are less open than in the DFB case because of the limited modulation speed of the long-cavity SADE laser.

For both the DFB laser array and SADE lasers, controlling all the channel powers is not straightforward, because of mutual heating. One elegant monitoring method is to measure the channel powers using waveform correlation [165].

4.2.3 External Modulation of a Solitary Wavelength

This application implies a continuous-wave (cw) laser followed by an intensity modulator. Currently, the top choice for this application is the solitary DFB laser.

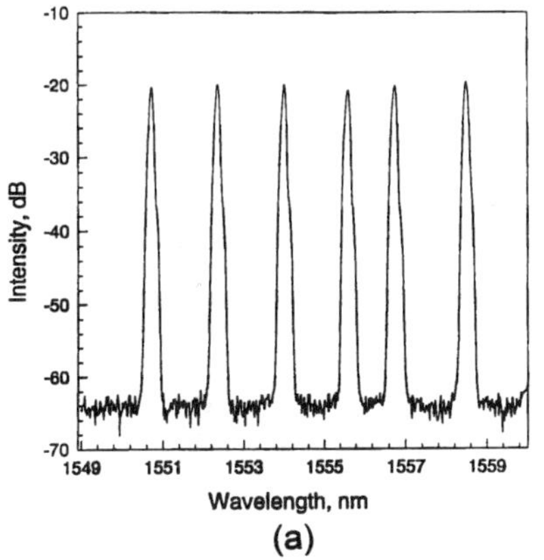

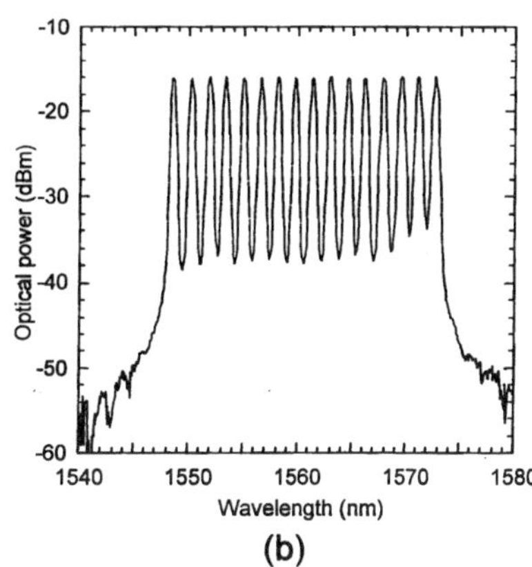

Figure 25 Measured optical spectra from simultaneous direct modulation experiments of (a) a DFB laser array and (b) a SADE (WGR) laser. Modulation speed is 622 Mb/s for both cases. (From Ref. 164.)

It is the smallest chip and can have high power. However, as the number of wavelengths grows, the network operator must keep many DFB lasers of different wavelengths to provide backup to any lasers that need to be replaced or have their wavelength changed. A solution to this is the wavelength-selectable laser. This universal laser is envisioned to be able to provide backup (or even main service) for any desired wavelength, with the wavelength selectable by a computer, without human intervention.

The DFB laser array can accomplish this with a relatively small chip size, with the DFB wavelengths spaced so that any wavelength in the span can be reached by substrate temperature changing. The drawback is that to cover a large wavelength span, the array must contain many lasers, and thus the splitting loss of the integrated power combiner will be high. There can be an integrated shared amplifier in the output to boost the power, but any residual reflection from the output facet can cause multimode oscillation if the shared amplifier gain is too high. An example of a DFB laser with an integrated electroabsorption modulator used as a wavelength-selectable laser is given in Ref. 166.

Tunable intracavity filter lasers are also wavelength-selectable lasers. Advantages include a small chip size and the possibility of extremely fast tuning [146]. The drawbacks, though, are the limited tuning range of simple TIF lasers, such as the DBR laser ($\sim$10 nm), and the risk of a mode hop to another channel during aging. Tunable intracavity filter lasers with wider tuning range have an even greater risk of a mode hop to another channel because of the larger number of controls.

Shared angular dispersive element lasers, also wavelength-selectable lasers, can easily cover the desired wavelength range, and they have the additional advantage of being able to provide fast tuning (~2 ns) [167] because of their inherently accurate channel spacing. For this application, SADE lasers can either contain a shared amplifier to boost the output power or not have a shared amplifier and have an extremely high signal-to-noise ratio because of the filtering of sponanteous emission by the SADE (see upcoming Fig. 27). Also, it is not possible for a SADE laser to mode hop to the wrong channel. The drawbacks are the lower power of SADE lasers (typically 0 dBm [153]) and the lack of extensive field-trial experience concerning the nonlinear means of obtaining single-longitudinal-mode oscillation.

There are three main ways to connect the laser to the modulator. The first way is to couple the laser and the modulator via a fiber. The second way is to employ hybrid integration, in which the laser and the modulator are on separate chips but in the same package. The third way is fully to integrate the laser and the modulator on the same chip. The main issue with using a modulator is prevention of optical feedback. Modulator reflections that go back into the laser can cause multimode oscillation and can change the oscillation wavelength by affecting the optical power in the cavity (static chirp*). The sure solution is to place an isolator between the laser and the modulator, which is easily done in the fiber-coupled case and has been demonstrated in the hybrid integration case for SADE lasers and DFB laser arrays. To make up for coupling losses in hybrid integration, one can integrate an optical amplifier with the modulator [168]. An example of a SADE (WGR) laser hybridly integrated with a modulator is given in Ref. 169. See Sec. 3.3 and Ref. 170 for examples of fully integrated lasers and modulators. Great care must be taken to minimize modulator back-reflections in fully integrated laser-modulators, via antireflection (AR) coatings, windows, and/or angled facets combined with beam expanders.

4.2.4 External Modulation of Many Wavelengths Simultaneously

This application implies a source providing many cw wavelengths simultaneously in one output, which is followed by a demultiplexer, an array of modulators, and a multiplexer. Like the simultaneous direct modulation application, this application allows only DFB laser arrays and SADE lasers. Also like the simultaneous direct

* *Static chirp* is the oscillation wavelength difference between when the modulator is at a "1" or a "0". *Dynamic chirp*, on the other hand, is the wavelength change during the transitions between the modulator levels. Modulator back-reflections actually do cause both static chirp and dynamic chirp.

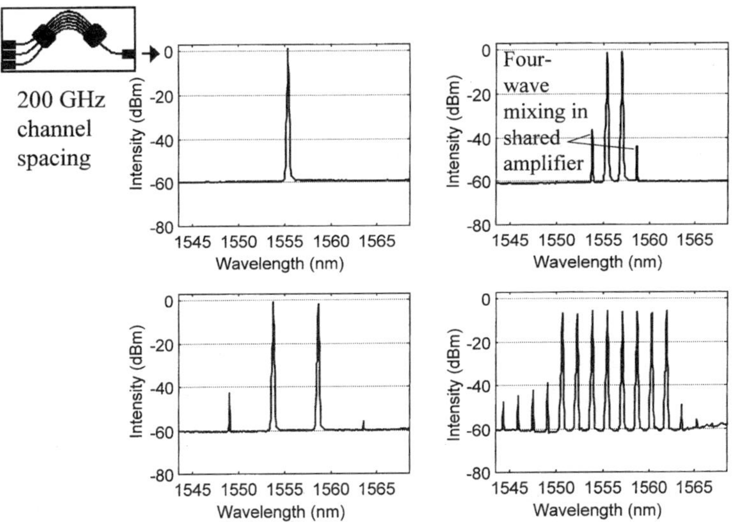

Figure 26 Measured optical spectra from a SADE (WGR) laser with a shared amplifier with various channel configurations lasing. One can see the mixing products that are generated in the shared amplifier. (From Ref. 171.)

modulation application, one does not want to have a shared amplifier in either case. This is because strong mixing products are created in the shared amplifier (see Fig. 26), and a shared amplifier does not provide much extra power per wavelength when many wavelengths are sent through the amplifier. In fact, with SADE lasers, one does not even want a shared intracavity passive waveguide. This is because wave mixing in the passive shared waveguide can cause laser instabilities when the mixing products overlap with the channels or their neighboring cavity modes (see Fig. 27) [171].

Again, as with the simultaneous direct modulation application, DFB laser arrays have a difficult time providing a precisely spaced wavelength comb together with equal channel powers. Shared angular dispersive element lasers can accomplish this, but either the channels must be slightly nonevenly spaced or the shared waveguide must be extremely short to avoid the wave-mixing-induced instabilities. Figure 28 shows the setup and results of a successful 16-wavelength long-haul transmission experiment using two WGR SADE lasers [153,172]. Although there were not enough modulators to modulate each channel individually, this configuration was simulated by having a single modulator before the demultiplexer–multiplexer pair.

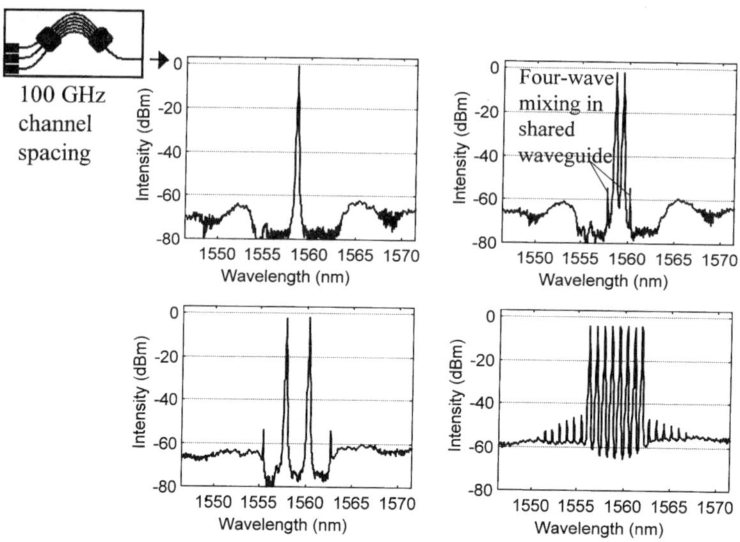

Figure 27 Measured optical spectra from a SADE (WGR) laser without a shared amplifier but still with a 3-mm-long shared passive waveguide. One can see the mixing products that are generated in the shared waveguide. (From Ref. 171.)

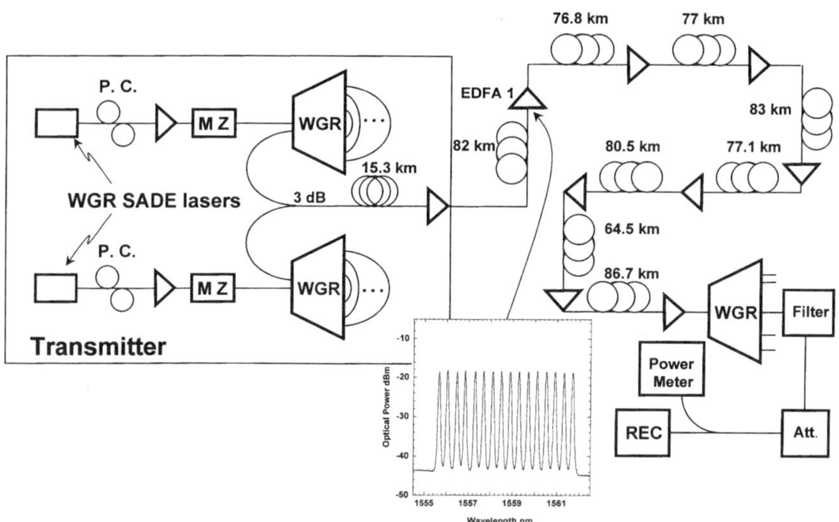

Figure 28 Experimental setup and measured optical spectrum of a long-haul experiment using two interleaved 8-channel SADE (WGR) lasers externally modulated at 2.5 Gb/s. Note the demultiplexer/multiplexer pair to simulate a real transmitter in which the channels would be demultiplexed, with each channel individually modulated, and then remultiplexed. The 16 channels were successfully transmitted over 627 km of conventional fiber. (From Ref. 172.)

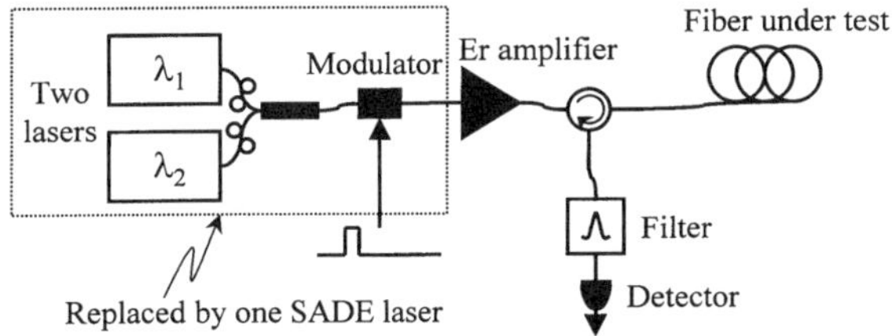

Figure 29 Apparatus for measuring fiber chromatic dispersion maps.

4.2.5 Miscellaneous

Another application for integrated WDM lasers is a fiber dispersion-map measurer. Mollenauer et al. devised a method for measuring chromatic dispersion maps in optical fiber based on phase mismatch in four-wave mixing (see Fig. 29) [173]. The technique requires a ~1-μs pulse at a rate of a few kilohertz of two copolarized wavelengths whose frequency difference is highly controlled. Shared angular dispersive element lasers are ideally suited for providing such wavelengths. However, if one attempts to obtain the pulse by directly modulating the intracavity semiconductor optical amplifiers, then in order to attain the extinction ratio required for the pulse to obtain a high peak power after passing through an erbium-doped fiber amplifier, the oscillation must be extinguished between pulses. Shared angular dispersive element lasers are not guaranteed to resume oscillation in the same cavity modes, and so one will observe significant amplitude and frequency fluctuations from pulse to pulse. Thus the only way to achieve the desired pulses is to run the lasers continuous-wave and employ an extracavity amplitude modulator, like the application of the previous section. Because a high extinction ratio and only modest speeds are required, the modulator can be an amplifier. For a WGR SADE laser, the modulator can be integrated by attaching it to the output star coupler of the WGR, one Brillouin zone width away from the shared waveguide connection, as in Ref. 174. For such an application, one can use a shared amplifier (both intra- and extracavity), because the mixing products amplify the measurements results [175]. Figure 30 shows an example of such a SADE laser design [176].

4.3 Conclusions for WDM Sources

Wavelength-division multiplexed lasers and high-speed modulators are ideal devices for the InP material system. Currently, only single-channel DFB lasers are commercially deployed. It is expected that multichannel InP lasers with high complexity will be deployed in the next several years.

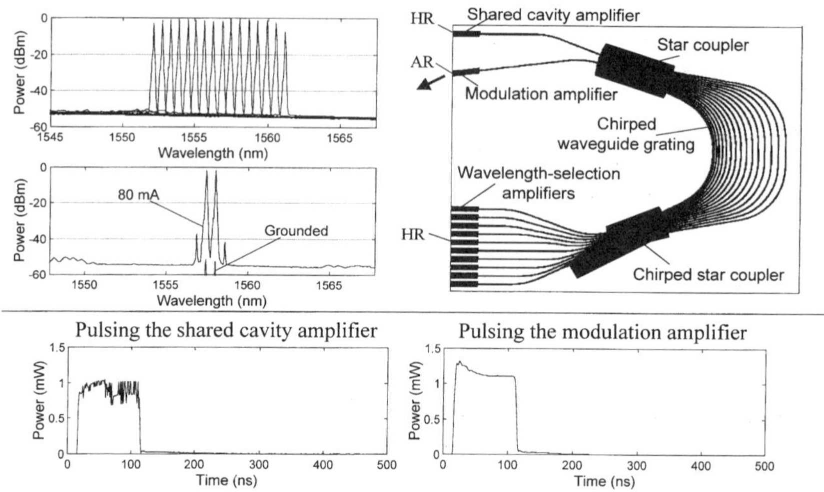

Figure 30 SADE laser and measured results for a source to using the apparatus of Fig. 29. Upper right shows the idea of the laser design (although there are actually 16 channels); upper left shows the measured optical spectra, with top showing all 16 channels and bottom showing two channels on, with the modulation amplifier at 80 mA and grounded; and bottom shows the waveform when the shared cavity amplifier is pulsed, showing pulse-to-pulse fluctuations, and when the modulation amplifier is pulsed, showing a clean signal. (From Ref. 176.)

5 DYNAMIC WAVELENGTH-DIVISION MULTIPLEXED CONTROL AND ROUTING DEVICES

This section concentrates on WDM devices for optical networking. Unlike the lasers of the previous section, WDM optical networking devices usually have the additional requirements of being low loss and polarization insensitive. This section thus first covers the issue of achieving polarization-independent WDM multi/demultiplexers, and then discusses two of the many categories of dynamic WDM routing devices: WDM channel power controllers and multichannel WDM add–drop and cross-connects. Other categories of WDM routing devices include single-channel WDM switching devices using tunable Bragg gratings [177].

5.1 Polarization-Insensitive Multi/Demultiplexers

One generally requires the multi/demultiplexer response to be identical for the TE and TM polarizations. It is generally not difficult to make the transmissivity for the TE and TM polarizations the same, but birefringence in the waveguides shifts the TE response with respect to the TM response, typically by ~3 nm.

There are three main approaches to fix this shift: employ a nonbirefringent waveguide structure, which has the same TE and TM indices at all points in the filter; employ two different waveguide structures such that the entire filter response vs. wavelength is the same; or employ a periodic response filter and have the filter period match the TE–TM wavelength shift [178].

The simplest way to make a nonbirefringent waveguide is to make a square core in a uniform cladding [179,180]. However, the lateral waveguide dimensions are usually small (unless one uses a very small index difference between the core and the cladding [181], which is difficult to grow precisely), and, as discussed in Sec. 2, it is difficult to make the vertical side walls, which require dry-etching, and still achieve good regrowth. Another way is to make a ridge waveguide with an air or glass upper cladding. It has been shown that such a waveguide can be nonbirefringent. However, the dimensions are critical, are different for straights and bends, and require high side walls. Examples of InP WGRs made with such a structure is given in Ref. 182, and a 64-channel version is found in Ref. 183.

To make a polarization-insensitive WGR using the second approach of two types of waveguides, one can use a patch on top of the waveguide grating [184]. In such a case the effective path-length difference between adjacent waveguide grating arms in terms of ΔL_1 and ΔL_2, for waveguide types 1 and 2, respectively, is designed to be

$$\beta_{TE_1}\,\Delta L_1 + \beta_{TE_2}\,\Delta L_2 = \beta_{TM_1}\,\Delta L_1 + \beta_{TM_2}\,\Delta L_2$$

where β is the propagation constant. Thus the filter responses for TE and TM are identical and are of the same grating order, but the total absolute path length through the grating is different for TE and TM. The second waveguide type is usually differentiated from the first by etching away the upper cladding, leaving only a narrow layer above the core. This tends to increase the index difference between TE and TM, and thus ΔL_2 has the opposite sign to ΔL_1. This results in an inverted triangle-like patch of etched cladding material on top of the waveguide grating. Although this compensation scheme is attractive, since it simply involves only a few extra fabrication steps, in practice it is difficult to cancel the polarization wavelength shift to the degree required in many WDM systems (typically <10 GHz).

5.2 Dynamic Wavelength-Division Multiplexed Filters

As discussed in Sec. 3, a semiconductor optical amplifier can be used as a gate switch. One potentially useful device is a demultiplexer and a multiplexer connected by an array of amplifiers [185] [see Fig. 31(a)]. The multiplexed signals enter the device through a single waveguide and are demultiplexed (by a WGR in this case). Each channel can then be absorbed (actually, it can be detected by using the amplifier as a detector) or sent through with a controlled amount of

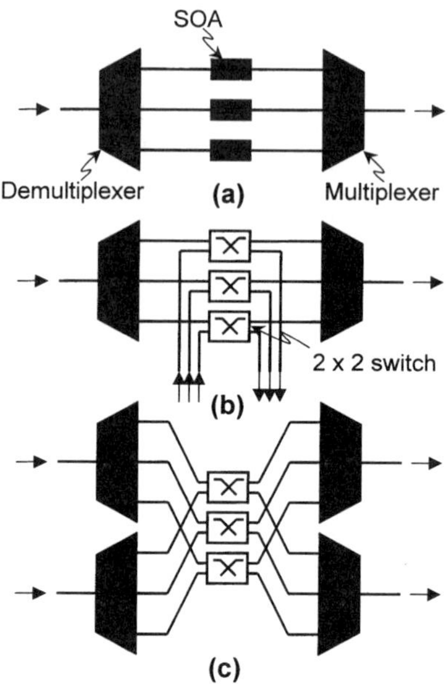

Figure 31 Block diagrams of (a) a channel controller, (b) a programmable optical add–drop, and (c) a WDM cross-connect. SOA = semiconductor optical amplifier.

gain. The channels are then remultiplexed and exit the device through a single waveguide. Although the reported device of Ref. 185 had high losses, it could theoretically provide gain for all the channels and be used to control the channel powers. The reported device was polarization sensitive.

5.3 Dynamic Wavelength-Division Multiplexed Add–Drops and Cross-Connects

A phase shifter is a device that can change its refractive index with an external signal by means that were discussed in Sec. 3.1. The phase shifters can be operated by forward injection current, resulting in a negative index change, or reverse voltage, resulting in a positive index change (for example, see Ref 186). Forward injection usually gives a larger index change, but it is slow (response time of ~1 ns) and dissipates power; the heating from the dissipated power, which has a positive index change, eventually limits the total possible phase shift. Reverse voltage is much faster and dissipates no power in the steady state, but the applied voltage is limited to the breakdown voltage (typically 25 V). Typical phase shifters in InP shift −0.4 rad/mA with forward operation [187] and +0.2 rad/mm/

V with reverse operation [186,187], although these numbers depend on many parameters (see Sec. 3.1).

Phase shifters can be used to make optical routing switches. A useful type of 2×2 optical switch is the Mach–Zehnder interferometer. It consists of two 2×2, equal-power-dividing couplers, connected by two waveguides containing phase shifters. A lightwave entering a port on the left side can be sent to either or both ports on the right by controlling the phase shift between the two central waveguides.

Using two multiplexers, such as WGRs, connected by an array of 2×2 switches, one can make a programmable channel-dropping-and-adding filter, often called a *programmable add–drop*, as shown in Fig. 31(b). Using four multiplexers and the switches, one can make a WDM cross-connect [188],* as shown in Fig. 31(c). Such devices have been demonstrated in InP. Figure 32(a) shows a 4-channel, 400-GHz channel spacing programmable add/drop [189], and Fig. 32(b) shows a 4-channel WDM cross-connect [190]. In the add–drop, the same WGR is used for the demultiplexing and the multiplexing; in the cross-connect, the two multi/demultiplexers for the two lines are placed together and are also used for both the multiplexing and the demultiplexing. In both, the couplers in the Mach–Zehnder switches are multimode interference (MMI; see Sec. 3.4) couplers. The add–drop of Ref. 189 is polarization sensitive, but the cross-connect of Ref. 190 was made polarization insensitive by using the patch scheme in the WGR [184], described in Sec. 5.1, and placing the switches at a specific angle to the InP crystal axis (see Sec. 3.5).

One difficulty with Mach–Zehnder switches is that the switching extinction ratio degrades as the couplers deviate from having perfectly equal power division.* One way to relax the coupler tolerance is to employ a dilated switch (replacing each 2×2 switch with two pairs of 2×2 switches in series), as is often done in silica and lithium niobate circuits. Unfortunately, in InP, unlike in silica, one cannot easily make evanescent 2×2 couplers with precise splitting ratios because of the tight required gap tolerance. The more robust choices (multimode interference couplers or star couplers) tend to be lossy (~0.5–1-dB excess loss), and so the loss of a dilated switch on InP can be on the order of 3 dB or more.

One solution is to use a three-arm Mach–Zehnder. As long as the sum of the powers in the two arms with lower power is greater than that of the third, the phase shifters can always be adjusted to achieve a perfect extinction ratio. This is explained in Fig. 33. Figure 33(a) shows the case of a perfect two-arm Mach–

---

* The term *cross-connect* applies to many devices. For the case here, we define a WDM cross-connect as a device that can swap WDM channels between lines.

* If both couplers deviate exactly the same from equal power division, the bar state can still have a good extinction ratio. One can understand this immediately by considering the case of no coupling at all.

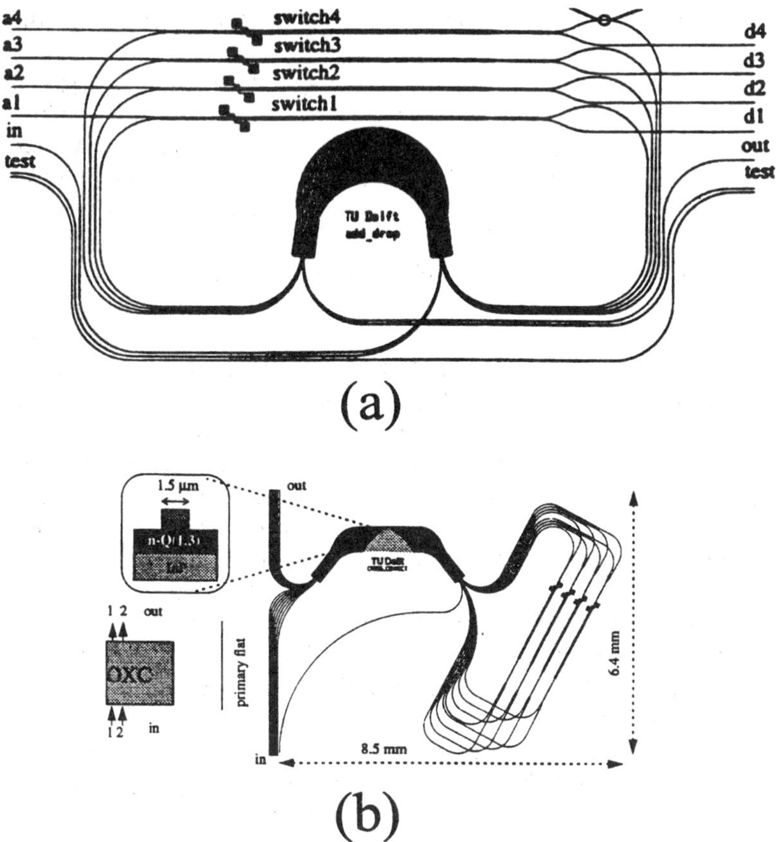

Figure 32 Waveguide and metal layouts for (a) a programmable optical add–drop in InP, and (b) a WDM cross-connect in InP. (From Refs. 189 and 190.)

Zehnder, in which the power-exiting single output port can be made exactly zero; Fig. 33(b) shows how the extinction ratio is reduced when a loss is introduced into one arm. Figure 33(c) shows the regular case of a three-arm Mach–Zehnder; Fig. 33(d) shows how when a loss is introduced into one arm, the power can still be made exactly zero in one output port (a triangle can still be made) by readjusting the phase shifters. The drawback over a dilated switch is increased sensitivity to any long-term drifts in the phase shifters. A star coupler (see Appendix 1) is an appropriate coupler for this type of switch. Since there are already star couplers in the waveguide grating router, one can combine the three-arm Mach–Zehnder with the router to create a compact WDM cross-connect or add–drop

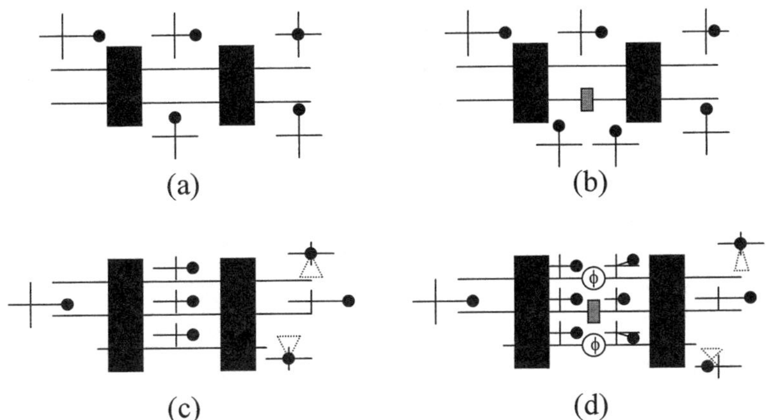

Figure 33 Mach–Zehnder switches showing robustness to a loss (represented by a small grey rectangle) in an arm. Parts (a) and (b) show a two-arm Mach–Zehnder; (c) and (d) show a three-arm Mach–Zehnder.

with high switching extinction ratio. This is accomplished by employing an "interleave chirp" in the WGR [191].

Such a WDM cross connect is shown in Fig. 34. The interleave chirp in the WGR consists of giving every other grating arm an additional path length of $\lambda/4$, which results in the WGR's creating two images for each wavelength in each Brillouin zone Ω_i of the star couplers (see Appendix 1). The images are collected by the equal-length waveguides, each of which contains a phase shifter, connecting the WGRs. The three centermost images are connected, since they contain almost all the power. There are input and output waveguides spaced by the width of the image spacing on the other sides of both WGRs. A channel entering one of the input ports can be switched to either output port by controlling the relative phases between the three connecting waveguides for that channel. For instance, in Fig. 34, channel 1 is controlled by phase shifters 3, 7, and 11 (start counting from the top waveguide). Figure 35 shows measured results for TE-polarized light (the device was not made polarization insensitive) from a six-channel interleave-chirp cross-connect that was made in InP [187].

Even though InP-based routine devices that use only phase shifting for switching have much lower nonlinearities than amplifier-gate switch-based InP devices, the nonlinearity of passive InP waveguides still must be considered, as found in the shared waveguide of the WGR SADE laser [171]. The nonlinear index of typical InP passive waveguides is $n_2 = +1 \times 10^{-12}$ cm^2/W [171] (this is for a bandgap of 1.3 μm and reportedly flips sign as the bandgap approaches the propagation wavelength [192]), and the typical effective cross-sectional area is $1 \times$

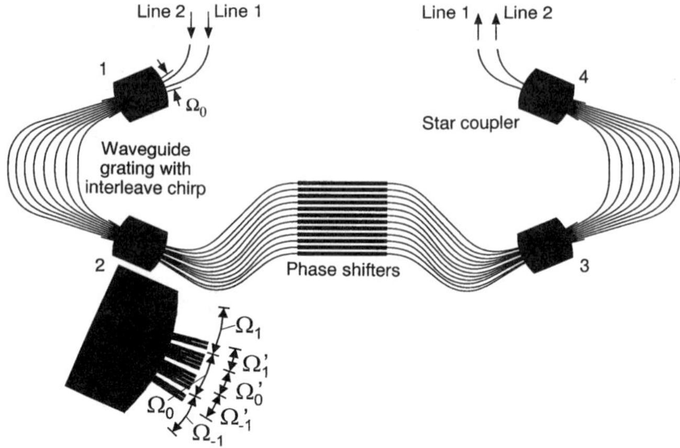

Figure 34 Waveguide and metal layout (simplified) for a WDM cross connect using interleave-chirped WGRs. (From Ref. 191.)

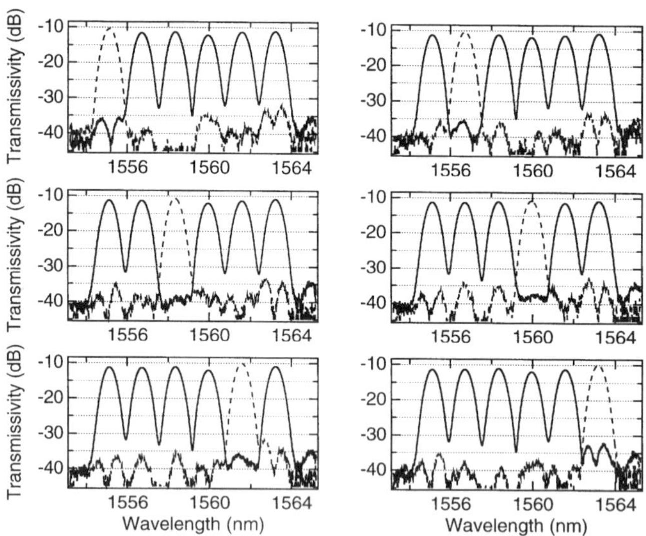

Figure 35 Measured on-chip transmissivity (TE polarization) of a six-channel WDM cross-connect using interleave-chirped WGRs in InP for various settings of the phase shifters. Solid and dashed lines are for line 1 to line 1 and line 1 to line 2, respectively. (From Ref. 187.)

10^{-8} cm^2. The nonlinear index of conventional fiber is $n_2 = +2.6 \times 10^{-16}$ cm^2/W, and the effective cross-sectional area is 8×10^{-7} cm^2 [193]. Thus the total nonlinear phase shift encountered in 1 cm of typical InP passive waveguide is equivalent to that encountered in 3 km of conventional fiber! For long-haul networks, one may have either to limit the input power or to ensure that all the waveguides in which many channels pass together are as short as possible.

5.4 Conclusions for Dynamic WDM Devices

The InP WDM control and routing devices discussed in the previous section face tough competition from silica and lithium niobate planar lightwave devices and bulk optic devices. The main advantages InP devices have over silica and bulk are a compact size, much faster switches, and lower switching powers. However, the difficulties that InP technology must overcome are that the fiber coupling to InP is lossy, the InP waveguide losses are not negligible, polarization insensitivity is difficult to achieve, InP has strong nonlinearities, and many applications do not require very fast switching. The next, and final, section discusses InP devices that detect optical signals, which is something that, like optical amplification, silica, lithium niobate, and bulk devices cannot readily do.

6 INTEGRATED WAVELENGTH-DIVISION MULTIPLEXED RECEIVERS/MONITORS

A WDM receiver/monitor is a device that converts the wavelength channels on a transmission line to an array of electrical signals. A monitor is distinguished from a receiver in that it operates at low speed and measures channel characteristics, such as power and signal-to-noise ratio, rather than the signal itself. The possible components that can be integrated together are an optical preamplifier, a demultiplexer, a detector array, and amplifying electronics. We discuss demonstrated devices that have used some combination of these components.

6.1 Demultiplexer and Detectors

One of the first WDM receivers was a WGR integrated with an array of detectors [194]. This device had a TE–TM wavelength shift of 4.8 nm, though. Since then, several receivers and monitors integrating a WGR and an array of detectors, achieving polarization insensitivity via making the WGR and the detectors polarization insensitive, have been demonstrated. In some cases the patch on the WGR was employed [195], in some, square buried waveguides [196], and in others, a special air-clad structure [197]. Reference 197 reports an eight-channel, 200-GHz channel-spacing device using a special air-clad structure with a responsivity of 0.23 A/W. However, in that case, as in many reported integrated InP WDM receivers, the detector speeds were low, so it can be used only as a monitor. Also, a few WDM receivers using demultiplexers other than the WGR have been demonstrated [198].

To reduce the required wavelength alignment of the source to the receiver and to reduce the required minimization of the TE–TM wavelength shift, one can employ rectangular passbands in the WGR. The significant advantage of a receiver in this case is that the rectangular passbands can be achieved by using multimoded waveguides that lead from the WGR to the photodetectors [199]. If the photodetector is large enough, it will simply add the photocurrents from all the waveguide modes.

One often uses an optical preamplifier, such as an Er-doped fiber amplifier, before the integrated receiver. The noise in an optically preamplified receiver comes from signal shot noise, signal–spontaneous emission beat noise, spontaneous emission shot noise, spontaneous–spontaneous emission beat noise, and receiver electrical noise [200,201]. The demultiplexer will filter out most of the spontaneous emission of the preamplifier. However, there will still be a sizable contribution to the spontaneous emission shot noise and spontaneous–spontaneous emission beat noise from the other free-spectral range modes of the demultiplexer (if the demultiplexer is a WGR, for instance). This can be eliminated by placing a band-pass filter before the device. A solution that adds no extra components or size for the case when the demultiplexer is a WGR is to chirp the WGR [202]. Chirping will reduce the spontaneous–spontaneous emission beat noise but will leave the spontaneous emission shot noise unchanged. This reduction becomes helpful when the spontaneous emission power is equal to or greater than the received signal power. This in turn occurs when the detected bit rates are low (<1 Gb/s), and thus the required received power is low.

6.2 Optical Preamplifier, Demultiplexer, and Detectors

One of the primary advantages of integrated WDM receivers in InP is the ability to have an integrated optical preamplifier to boost the receiver sensitivity. Polarization-sensitive versions have been demonstrated with reflection gratings [203] and WGRs [204]. A very-wide-channel-spacing, polarization-insensitive preamplified WDM receiver, using a polarization-insensitive optical amplifier and a reflection grating as the demultiplexer, has been demonstrated [205].

All of the previously discussed polarization-insensitive multi/demultiplexers in this chapter have channel spacings ≤ 200 GHz. For narrower channel spacings, it becomes highly difficult to employ successfully the various methods of making the multi/demultiplexer polarization insensitive. A means of attaining polarization insensitivity that does not require polarization-insensitive components is polarization diversity [206]—the splitting of the incoming signal into two orthogonal polarizations, detecting each polarization independently, and adding the photocurrents. One convenient means of accomplishing this is to use the design

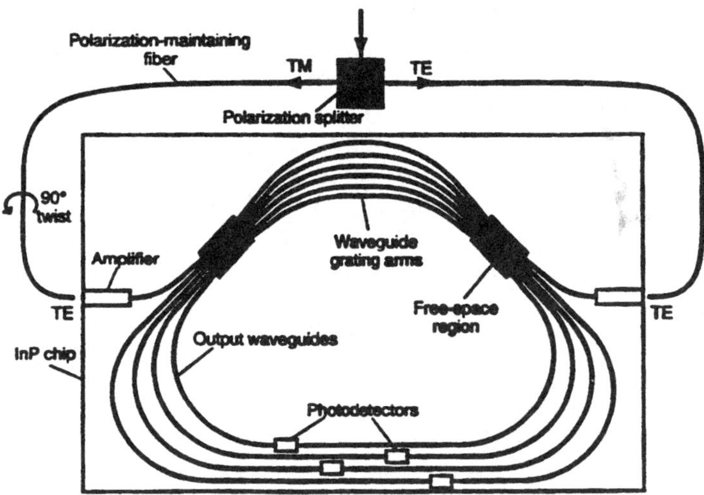

Figure 36 Diagram of a WDM receiver with an integrated optical preamplifier that employs polarization diversity to achieve polarization insensitivity. (From Ref. 202.)

of Fig. 36 [202]. In this design, the incoming signal TE and TM components are injected into the chip from opposite sides of the WGR, with the TM rotated to be TE so that only TE polarization is on the chip. Thus all the components operate only on TE-polarized light. The two components reach the photodetectors from opposite sides, so their photocurrents are automatically added. Because the components go through the same grating, there is no wavelength shift between them. The main difficulties with the design are that if the AR coatings are not good enough, laser oscillation will occur between the two input facets, the two input fibers must have the same length, and the input powers must be low enough to avoid cross gain compression in the amplifiers.

6.3 Demultiplexer, Detectors, and Amplifying Electronics

One can integrate transistors with photodetectors on InP to amplify the received currents [207]. An important advantage of this is the elimination of the inductance of bond wires, improving receiver speed and noise. However, transistor circuits usually require an insulating substrate. To achieve this in InP, one usually uses an Fe-doped InP substrate. Fe-doped InP increases the waveguide losses because of free-carrier absorption, though. A monolithically integrated WGR, receiver array, and transistor array (each electronic preamplifier contained five transistors) has been reported, with a responsivity of ~0.08 A/W, and is described in Ref. 208.

6.4 Conclusions

Although this section discussed fully integrated WDM receivers, unless one includes the integrated preamplifier, a strong competitor for WDM receivers is a hybrid version, consisting of a silica or bulk demultiplexer and an InP detector array [209,210]. The main disadvantages of fully integrated InP-based WDM receivers are the higher losses of InP, the difficulty of achieving polarization insensitivity, and the compromises that must be made for full integration. For WDM monitors, fully integrated versions in InP are highly attractive because of their compact size and potential ease of mass fabrication.

APPENDIX 1: THE STAR COUPLER

The star coupler began with microwave theory and was applied to planar lightwave circuits [141,211]. It was demonstrated first in silica [212] and later in InP [213]. It consists of two arrays of waveguides terminating on arcs of radii R, with the center of each arc falling on the arc of the other. It couples light from a waveguide on one side of the coupler to all the waveguides on the other side. The star coupler is appropriate for the small dimensions of InP because the coupling ratios are very robust. The coupling ratios of most couplers, such as the multimode interference coupler and the evanescent coupler, change with symmetrical etching errors, while in the star coupler the coupling ratios remain approximately the same.

Here we calculate the transmissivities through the star coupler. We will use scalar wave theory, and thus all feature sizes must be larger than a wavelength, and we must measure diffracted fields only at distances further than several wavelengths from features. An amplitude distribution $u(x)$ (x is the distance along the arc) launched into free space (the slab in the middle of the star coupler) is equivalent to an infinite sum of plane waves. The amplitude of the plane wave traveling in the direction ϕ (referenced perpendicular to the arc) after a distance R is

$$\tilde{u}(\phi) = \frac{e^{jkR}}{2\pi k} \int_{-\infty}^{\infty} u(x) \exp\,(jkx \sin\phi)\, dx \tag{10}$$

where k is the propagation constant of the free space.

The transmissivity between a left and right mode at angles θ_1 and θ_2 and with amplitude distributions u_1 and u_2, respectively, is the integral of the products of the codirectional normalized plane wave amplitudes from the two guides:

$$t_{\text{star}}(\theta_1,\theta_2) = \frac{\displaystyle\int_{-\pi}^{\pi} \tilde{u}_1(\theta_1 - \phi)\tilde{u}_2(\theta_2 + \phi) \exp\,[jkR(\theta_1\theta_2 + \cos\phi)]\, d\phi}{\sqrt{\displaystyle\int_{-\pi}^{\pi} |\tilde{u}_1(\phi)|^2\, d\phi \int_{-\pi}^{\pi} |\tilde{u}_2(\phi)|^2\, d\phi}} \tag{11}$$

We have assumed that $\theta_1, \theta_2 \ll 1$. Assuming $kR \gg 1$, we can use the principle of stationary phase to integrate Eq. (11). Because

$$\begin{aligned}\int_{-\pi}^{\pi} f(\phi) \exp(jA \cos\phi)\, d\phi &\approx f(0) \int_{-\varepsilon}^{\varepsilon} \exp[jA(1 - \phi^2/2)]\, d\phi \\ &\approx \sqrt{\frac{2\pi}{jA}} f(0) \exp(jA)\end{aligned} \tag{12}$$

for $A \gg 1$, then Eq. (11) becomes, dropping constant phase shifts,

$$t_{\text{star}}(\theta_1,\theta_2) = e^{-jkR\theta_1\theta_2} \frac{\int_{-\infty}^{\infty} u_1(x)e^{jkx\theta_2}\, dx \int_{-\infty}^{\infty} u_2(x)e^{jkx\theta_1}\, dx}{\sqrt{\frac{2\pi R}{k} \int_{-\infty}^{\infty} |u_1(x)|^2\, dx \int_{-\infty}^{\infty} |u_2(x)|^2\, dx}} \tag{13}$$

Now we must find $u_1(x)$ and $u_2(x)$ so that t_{star} represents the transmissivity from a waveguide on the left to one on the right. If the waveguides on both sides are very widely spaced, u_1 and u_2 are simply the waveguide modes. When the waveguides are close enough that the waveguide modes significantly overlap, then one choice is to use numerical beam propagation to find the spatial mode at the port inlet (i.e., start the waveguide mode where the waveguides are effectively uncoupled and numerically propagate until reaching the free-space region). If the waveguide mode overlap is only slight, then another approach is to let u_1 and u_2 be weighted sums of the solitary waveguide modes and their nearest neighbors, with the weights calculated by requiring mode orthogonality [214]. For example, let

$$u(x) = \varepsilon_1 u_w(x - a) + u_w(x) + \varepsilon_2 u_w(x + a) \tag{14}$$

where a is the spacing between waveguides and u_w is the solitary waveguide mode. ε_1 and ε_2 can be found by enforcing orthogonality:

$$\int_{-\infty}^{\infty} u_1(x)u_2^*(x \pm a)\, dx = 0 \tag{15}$$

This is very similar to the Gram–Schmidt procedure used to obtain a set of orthonormal functions from a set of linearly independent square-integrable functions [215]. It is important to realize that this mode orthogonalization is a trick that tends to work only because it obeys power conservation. It results in a π phase shift between the center of the mode and the components in the adjacent waveguides, which is incorrect, since coupling between waveguides necessarily results in a $\pi/2$ phase shift [216]. But when the mutual coupling is small, the transmissivities it gives do happen to be close to that given by a more accurate calculation.

Equation (13) shows that the transmissivity through the star coupler is proportional to the Fourier transform of the field distributions along each arc. Thus if one side of the star coupler has an array of waveguides spaced by a, the transmissivity to the other side has an angular periodicity (under an envelope given by the Fourier transform of one coupled waveguide element) of $2\pi/(ka)$. One arc-width of $2\pi/(ka)$ is called a Brillouin zone. For a conventional star coupler, the highest transmissivities occur between the central Brillouin zones of both sides, and power that is scattered outside of the central Brillouin zone is lost.

Finally, consider the case of a star coupler in which the waveguide connections on the right and left sides of the free space are shifted radially by distances d_1 and d_2, respectively from R. In such a case, Eq. (13) is modified to read

$$t_{\text{star}}(\theta_1,\theta_2,d_1,d_2) \approx \exp\left\{jk\left[d_1 + d_2 - R\theta_1\theta_2 - R\frac{d_1\theta_2^2 + d_2\theta_1^2}{2(R + d_1 + d_2)}\right]\right\} \frac{\int_{-\infty}^{\infty} u_1(x)\exp\left[jkx\left(\frac{R\theta_2 + d_2\theta_1 + d_2\theta_2}{R + d_1 + d_2}\right)\right]dx \times \int_{-\infty}^{\infty} u_2(x)\exp\left[jkx\left(\frac{R\theta_1 + d_1\theta_1 + d_1\theta_2}{R + d_1 + d_2}\right)\right]dx}{\sqrt{\frac{2\pi(R + d_1 + d_2)}{k}\int_{-\infty}^{\infty}|u_1(x)|^2\,dx\int_{-\infty}^{\infty}|u_2(x)|^2\,dx}} \quad (16)$$

This is needed to calculate the transmissivities through some types of chirped WGRs (see Appendix 3).

APPENDIX 2: THE WAVEGUIDE GRATING ROUTER

The waveguide grating router (WGR) consists of two star couplers connected by an array of waveguides of increasing path length. The first waveguide gratings were published by Smit [217] and also by Takahashi et al. [218] Dragone then combined the waveguide grating with two star couplers to make the $N_1 \times N_2$ WGR [219]. The WGR routes wavelengths between its left and right ports. The WGR is relatively fabrication robust in terms of crosstalk and wavelength because it consists of star couplers with robust coupling ratios and long paths between the star couplers over which small fabrication errors can average out. Essentially, the WGR trades device size (it is large) for fabrication tolerance (it is fabrication tolerant).

If we let the transmissivity through the mth waveguide with length $L(m)$, star-coupler connection angle $\alpha(m)$, and propagation constant β be $\exp[j\beta L(m)]$, then the total transmissivity through the device from a port at angle θ_1 on the left side to a port at angle θ_2 on the right is

$$t_{WGR}(\theta_1,\theta_2) = \sum_{m=1}^{M} t_{star}[\theta_1,\alpha_1(m)] \exp [j\beta L(m)] t_{star} [\alpha_2(m),\theta_2] \quad (17)$$

The conventional router has linear distributions of grating waveguide angle α in the star couplers and length L:

$$\alpha(m) = [m - (M + 1)/2] \Delta\alpha \qquad \text{and} \qquad L(m) = m \Delta L + \text{constant}$$

$\Delta\alpha$ is usually chosen so that the powers in the outermost waveguides are roughly 10 times lower than the powers in the central guides of the grating, $\Delta L$is chosen to give a desired free spectral range, and M is chosen to give a desired passband width. The passband shape is approximately the Fourier transform of the power distribution in the arms, which is in turn approximately the magnitude squared of the Fourier transform of a port waveguide mode. Since a typical waveguide mode can be closely approximated by a Gaussian, the transmissivity passband shape is approximately Gaussian. Between ports with θ_1-θ_2, the passband repeats in frequency with a free spectral range of $c/(n_{group} \Delta L)$, where c is the speed of light, n is the refractive index, and $n_{group} \equiv n - \lambda\, dn/d\lambda$, in which λ is the free-space wavelength. For typical InP waveguides, $n = 3.25$ and $n_{group} = 3.61$ at $\lambda = 1.55$ μm. The free spectral range changes slightly for other port combinations.

To design an actual WGR, one must first decide on the waveguide widths. A wide waveguide has more than one real β; i.e., it is multimode. As the waveguide width is decreased, the guided mode with the lowest β, the fundamental mode, decreases in transverse width. Beyond a certain waveguide width, the mode width increases again. As the waveguide width decreases even further, all the modes except the fundamental are no longer guided; i.e., the waveguide becomes single transverse mode. The narrower the waveguide, the more the fundamental mode touches the sides, the higher the loss because of side-wall roughness, and the more sensitive β is to waveguide width variation. Thus, for nearly the entire structure, the straight waveguide widths are typically wide enough to support the fundamental and second-order modes. Occasionally, the waveguides are adiabatically tapered down to the single-mode waveguide width to remove any higher-order modes generated by scattering. For the bends, it was shown by Dragone [220] that the smallest low-loss bend radius is achieved by using wide waveguides in the bends. To transition between straights and bends, one can use either abrupt offsets or adiabatic tapers, although the latter are more complex to program, use more space, and are highly robust against lithography imperfections. For the waveguide grating connections to the star couplers, to maximize the efficiency for all the channels, one wishes to maximize the coupling between waveguides at the star coupler. This implies minimizing both the spacings between waveguides and the waveguide widths. However, because of the small dimensions in InP, it is difficult to control the waveguide spacings accurately. When the waveguide coupling is strong, small imperfections in waveguide spacings result in large distortions in the star-coupler field distributions. Thus, applications that are willing to sacrifice passband uniformity and crosstalk for lower loss should use narrow waveguides at the star-grating junctions, while applications that are willing to sacrifice low loss for high-quality filter response should use wide waveguides (accomplished by tapering outwards). In both cases, one wishes to make the waveguide gaps as small as the lithography will allow [221].

APPENDIX 3: CHIRPING OF THE WAVEGUIDE GRATING ROUTER FOR PASSBAND SELECTION

If a conventional WGR is employed in a SADE laser, then in which passband oscillation occurs among the infinite number of passbands spaced by the WGR free spectral range is determined mainly by the amplifier gain spectrum. Because the gain spectrum can be quite broad, oscillation in multiple or undesired passbands is often observed in such WGR lasers. One could argue that the difficulty can be solved by making the WGR free spectral range extremely wide. However, this increases the device size, since to keep a specified passband width, widening the free spectral range entails adding more grating waveguides. Another approach, one that has no penalty in size or loss, is to chirp the WGR [151,152].

A representation of the waveguide array and one of the star couplers of the WGR is shown in Fig. 37 (the arc on which the waveguides terminate has been equivalently replaced by a thin lens). Suppose that the phases of all the lightwaves in the waveguides are equal at the entrances of the waveguides. If all the waveguides have the same length, then the light focuses at a single point, regardless of wavelength (we are considering only the central Brillouin zone) [Fig. 37(a)].

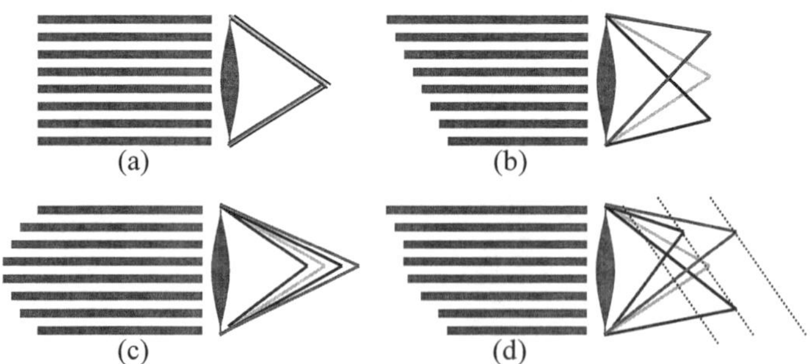

Figure 37 Representations of the travel of a lightwave through a waveguide array and a thin lens. Path-length distribution is (a) a constant, (b) linear, (c) purely quadratic, and (d) linear plus quadratic.

Figure 38 Ways to distort the star coupler to achieve linear performance in a desired free spectral range. (a) Focal length adjustment and (b) angle chirping.

If the path length increases linearly from waveguide to waveguide, as in the conventional WGR, then the focal spot moves along an arc, with the wavelength, the far arc of the star coupler, starting over again every time a phase difference of 2π between adjacent arms (free spectral range) is reached [Fig. 37(b)]. Thus, multiple wavelengths still focus to the same point, which is the ambiguity we wish to correct. If the waveguide lengths have a purely quadratic distribution $\{L(m) = [m - (M + 1)/2]^2 \, \Delta L\}$, then the focal spot moves radially with wavelength [Fig. 37(c)]. This is the same as having a thin lens with chromatic aberration at the output of the waveguides. In most cases, the free spectral range of the focus change is extremely large. If the waveguide lengths have the sum of a linear and a quadratic distribution, i.e., a parabola, the focal spot will move both laterally and radially with wavelength [Fig. 37(d)]. The focal spot will follow tilted curves and jump to another tilted curve every time the free spectral range of the linear part of the distribution is reached. This is termed a *linearly chirped waveguide grating*.*

We want all the channels (right–left port combinations) in the laser to have only one passband with high peak transmissivity, with all the other passbands having low peak transmissivity. We have two options. The first is to place the port connections to the star coupler along one of the focal curves. Such a design is shown in Fig. 38(a). In this way, only one wavelength is in focus for each port waveguide, with all the other waveguides being out of focus. We term this method *focal length adjustment* [152]. The other option is to arrange the grating waveguide connection points in the star coupler with a parabolic distribution in angle such that if the waveguides were stretched straight back, their ends would fall on a straight line. This method, which we call *angular chirping* [153], is shown in Fig. 38(b). One can implement one or the other or a combination of both.

For the case of linear chirp in the WGR, focal length adjustment and angle chirping have exactly the same effect of making the WGR appear nonchirped for one grating order, with out-of-focus passbands in all the other grating orders.

* This type of chirping was first called *M-grating chirp* and then later *parabolic chirp. Linear chirp* is actually the most correct label.

(Angle chirping is actually more powerful than focal length adjustment in that it can compensate for any type of chirp, although rapid changes in $L(m)$ with m will generally result in reduced transmissivities through grating–mode mismatches. For an example, see Ref. 222). However, both chirp compensation techniques require geometric distortions: focal length adjustment requires variations in port waveguide width, and angle chirping requires variations in grating waveguide width. These distortions can lead to slightly increased WGR loss and physical size. To minimize the distortions, one can use half focal length compensation and half angle chirping compensation. Figure 38 shows a linearly chirped WGR with angle chirping and focal length adjustment. Figure 39 shows calculated transmissivities from the central port on the left side to the 9 ports on the right side of a 9-channel WGR, with $M = 68$, a channel spacing of 150 GHz, and a free spectral range of 9×150 GHz. Waveguide index step $\Delta n/n = 0.0086$, $n = 3.25$, nominal grating arm widths = 3.1 μm (with spacings between 0–0.8 μm), $\Delta\alpha = 0.19/M$, and nominal port waveguide widths $w_0 = 2.0$ μm. Figure 39(a) shows the case of $\gamma = 0$; Fig. 39(b) shows the case of $\gamma = 6.6$ for the grating, with $\gamma = 3.3$ for focal length adjustment and $\gamma = 3.3$ for angle chirping.

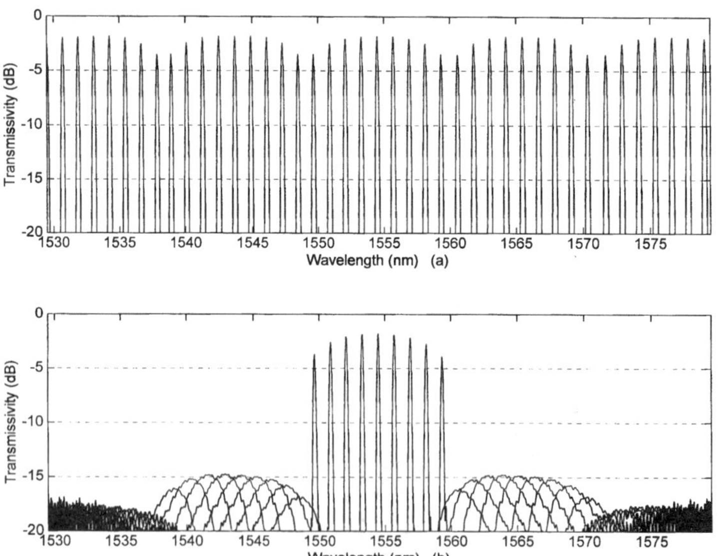

Figure 39 Calculated transmissivity through (a) a conventional WGR and (b) a linearly chirped WGR with focal length adjustment and angle chirping.

The rest of this appendix explains how the chirp is calculated. The distribution of the grating waveguide lengths $L(m)$ for an arbitrarily chirped WGR that has channels with high peak transmissivities centered around β_c is given by

$$L(m) = \text{round}\left\{ [m + g(m)] \frac{\Delta L \beta_c}{2\pi} \right\} \frac{2\pi}{\beta_c} + \text{constant} \tag{18}$$

where $g(m)$ is an arbitrary function and "round" is a function that rounds its argument to the nearest integer. For linear chirping (a parabolic arm-length distribution),

$$g(m) = \gamma \left[\left(\frac{m - \dfrac{M+1}{2}}{1 - \dfrac{M+1}{2}} \right)^2 - \frac{1}{3} \right] \tag{19}$$

where γ is a constant that determines the amount of chirp. The "1/3" is irrelevant to the arm-length chirping, but it is used later in the angle chirping.

We now analyze the case of focal length adjustment. Neglecting overall phase shift, the phase shift upon passing through a one-dimensional thin lens of focal length f is $-kx^2/(2f)$, where x is the dimension along the lens. The phase shift encountered in the grating arms is $\beta L(m)$. Thus, expanding β as $\beta_c + \Delta\beta$, we see that the chirp of Eq. (19) is equivalent to placing a lens of focal length

$$f_c = \frac{-k\left[R\,\Delta\alpha \left(1 - \dfrac{M+1}{2} \right) \right]^2}{2\gamma\,\Delta\beta\,\Delta L} \tag{20}$$

against the grating arms. The overall focal length f of two thin lenses of focal lengths f_1 and f_2 placed next to each other is given by $(1/f) = (1/f_1) + (1/f_2)$. Thus the overall focal length of the grating arms is $(1/f) = (1/R) + (1/f_c)$. Defining $f \equiv R + d$, where d is the radial offset of the focus from R, then

$$\begin{aligned} d &= R \frac{2\gamma\,\Delta\beta\,\Delta L}{kR\left[\Delta\alpha \left(1 - \dfrac{M+1}{2} \right) \right]^2 - 2\gamma\,\Delta\beta\,\Delta L} \\ &= R \frac{2\gamma\theta}{\Delta\alpha \left(1 - \dfrac{M+1}{2} \right)^2 - 2\gamma\theta} \end{aligned} \tag{21}$$

where the second "equals" sign holds true for the $\Delta L\ \beta_c/(2\pi)$ grating order. $\theta = \Delta\beta\ \Delta L/(kR\ \Delta\alpha)$.

Thus, as explained before, the port entrances can be placed along the tilted focal line given by Eq. (21). As the port waveguides are moved radially, the size of the focal spot varies, and thus the port waveguide width must be varied. A reasonable approach is to change the waveguide width w as $w_0(R + d)/R$, provided that w_0 is somewhat larger than the waveguide width that gives the minimum mode width.

To calculate the transmissivity for a WGR with a focal length adjustment, one can use Eq. (17), with Eq. (16) for the second star coupler with $d_1 = 0$ and $d_2 = d$.

We next analyze the case of angle chirping. The phase accumulation in the WGR from the central port ($\theta_1 = 0$) on the left side to a port at θ_2 on the right side through arm m is given by

$$\beta L(m) - kR\alpha_2(m)\theta_2 \tag{22}$$

[see Eq. (17)]. Let $\beta = \beta_0 + \Delta\beta$ and $k = k_0 + \Delta k$. If we constrain $\Delta k\ R\alpha_2\theta_2 \ll 2\pi$, then, neglecting integer multiples of 2π, the phase becomes

$$L(m)\ \Delta\beta - \alpha_2(m)k_0R\theta_2 \tag{23}$$

From Eq. (23) we see that if we choose $\alpha(m + 1) - \alpha(m)$ to be proportional to $L(m + 1) - L(m)\ \forall\ m$, then θ is proportional to $\Delta\beta$, as in a conventional WGR, even if $L(m)$ is nonlinear in m, i.e., even if the WGR is chirped. For example, if $L(m)$ is given by Eq. (18) and $\alpha_2(m)$ is given by

$$\frac{\alpha_2(m)}{\Delta\alpha_2} = \mathrm{round}\left\{[m + g(m)]\frac{\Delta L\ \beta_c}{2\pi}\right\}\frac{2\pi}{\beta_c} - \frac{M + 1}{2} \tag{23}$$

then the WGR will act as a conventional WGR for the channel passbands in the grating order centered around β_c.

Linear chirp [Eq. (19)] corresponds to a linear increase in increment. So to maximize the transmissivity, the grating waveguide connections to the second star coupler get wider (or narrower if $\gamma < 0$) in a linear fashion going from the shortest to the longest arm. The $-1/3$ in the expression for $g(m)$ in Eq. (19) was found empirically and maximizes the overlap of the effective modes in the grating arms from the two star couplers.

For both focal length compensation and angle chirping, the value of γ that gives the flattest passband suppression outside the desired grating order is $\sim M/8$, which occurs approximately when the focal line tilt becomes steeper than the angular spread of the port mode in the free-space region of the star coupler.

APPENDIX 4: SINGLE-LONGITUDINAL-MODE OSCILLATION IN SHARED ANGULAR DISPERSIVE ELEMENT LASERS

All gain media are limited in the total power they can provide. Thus as the input power to the amplifier increases, the gain decreases. This encourages stability and suppresses power fluctuations in lasers [223]. There are three main conditions a laser can be in. The first is when it is oscillating in only one cavity mode. In such a case the power is a constant. The second is when it is oscillating in multiple cavity modes simultaneously with the total power a constant (i.e., pure phase modulation). The third is when it is oscillating in multiple cavity modes simultaneously with the total power fluctuating. If the laser contains an intracavity filter, then the gain compression and the filtering will always drive the second two cases toward the first.

In reality, the situation is complicated by the fact that the real part of the index also changes as the gain changes. The ratio of the real index change to the imaginary index change in the gain medium is called the *linewidth enhancement parameter*, α [224]. Although it is defined to be positive in semiconductor amplifiers (typically 3–10), the index decreases as the gain increases. Reference 156 calculates the stability of single-mode oscillation for a laser consisting of an amplifier and a filter using a Jacobian matrix. The results show that there is a finite region of stability centered on the slightly low-frequency side of the filter. Single-mode oscillation in this frequency range will stay single mode in the face of perturbations. If the oscillating mode exits the region, such as by moving the filter frequency with respect to the cavity modes, single-mode oscillation becomes unstable, and the laser mode hops, eventually to a frequency that is somewhere back in the stability region. The narrower the filter and the shorter the cavity, the deeper the stability region. It is found that the depth of the stability region is approximately maximized when the ratio (cavity mode spacing)^3/(filter bandwidth) is maximized.

Another way to understand the self-stability is to realize that in order to mode-hop, the intensity in the cavity must go to exactly zero at some instant. This can be proven by imagining a single-frequency oscillating mode with frequency ω as a rotating phasor in the complex plane, $e^{j\omega t}$. This phasor rotates about the origin $\omega t/(2\pi)$ times each unit of time t. If the cavity length or index were changed, the oscillating frequency would follow it. This is similar to the idea that the cosmic background radiation temperature is constantly downshifting as the universe expands. However, if the oscillation must shift to a different cavity mode, the number of rotations by the phasor about the origin per cavity round trip must change. In order to accomplish this, the phasor must pass through the origin. Thus the laser power must drop to exactly zero for an instant. This is traumatic for the laser and is highly resisted, since the gain will be very high in such a low-intensity state. If the intracavity filter is very broad compared to the gain response time, the power fluctuation can happen so quickly that the gain barely realizes it. This is why the filter must be kept narrow.

The preceding discussion ignored two main anti-single-mode forces that exist in real semiconductor lasers: spatial hole burning and extra reflections. Spatial hole burning occurs in Fabry–Perot lasers, which all semiconductor lasers, except for rings, are. When the oscillation is single mode, there is significantly high gain at the standing-wave nodes, where the optical intensity is always zero. This encourages different oscillating frequencies, which have different standing wave periods, to oscillate as well, resulting in multimode oscillation. This effect is much stronger when the amplifier is in the center of the cavity than at the ends. As long as the gain medium distance from an end mirror is less than the equivalent intracavity filter length (i.e., filter path-length difference), spatial hole burning cannot cause multimode oscillation [225]. This is true for typical SADE lasers, since the amplifiers are adjacent to the mirrors. When a laser is multimode because of spatial hole burning, its output power is usually a constant, and so the multimode spectrum consists of pure phase modulation.

Reflections from other than the two facets can cause multimoding by creating a stiuation in which higher powers receive higher gain. This can happen when a lightwave reflected off an unwanted mirror interferes with the main oscillating lightwave, and one of the lightwaves experiences a power dependent-phase shift in an amplifier due to the amplifier's finite linewidth enhancement parameter. When a laser is multimode due to extra reflections, the power is usually fluctuating (at the round-trip frequency and its harmonics), unlike with spatial hole burning.

REFERENCES

1. R. E. Nahory, A. M. Pollack, W. D. Johnson Jr., and R. L. Barns. Band gap versus composition and demonstration of Vegard's law for InGaAsP lattice-matched to InP. Appl. Phys. Lett. 33:659–661, 1978.
2. S. Adachi. Optical functions of InGaAsP. In: Properties of Indium Phosphide, EMIS Datareviews Series No. 6. New York: INSPEC, 1991, pp. 408–428.
3. S. Adachi. Refractive Indices of III-V compounds: Key properties of InGaAsP relevant to device design. J. Appl. Phys. 53:5863–5869, 1982.
4. C. H. Henry, L. F. Johnson, R. A. Logan, and D. P. Clarke. Determination of the refractive index of InGaAsP epitaxial layers by mode line spectroscopy. J. Quantum Electron. QE-21:1887–1892, 1985.
5. J. P. Weber. Optimization of the carrier-induced effective index change in InGaAsP waveguides—Application to tunable Bragg filters. IEEE J. Quantum Electron. 30: 1801–1816, 1994.
6. J. Manning, R. Olshansky, and C. B. Su. The carrier-induced index change in

AlGaAs and 1.3 μm InGaAsP Diode Lasers, IEEE J. Quantum Electron. QE-19: 1525–1530, 1983.

7. G. Schraud, G. Muller, L. Stoll, and U. Wolff. Simple measurement of carrier induced refractive index change in InGaAsP pin ridge waveguide structures. Electron. Lett. 27:297–298, 1991.
8. D. Botteldooren and R. Baets. Influence of bandgap shrinkage on the carrier induced refractive index change in InGaAsP. Appl. Phys. Lett. 54:1989–1991, 1989.
9. B. R. Bennett, R. A. Soref, and J. A. D. Alamo. Carrier-induced change in refractive index of InP, GaAs, and InGaAsP. IEEE J. Quantum Electron. 26:113–122, 1990.
10. J. D. Dow and D. Redfield. Toward a unified theory of Urbach's rule and exponential absorption edges. Phys. Rev. B 5:594–610, 1972.
11. M. Asada, A. R. Adams, K. E. Stubkjaer, Y. Suematsu, Y. Itaya, and S. Arai. The temperature dependence of the threshold current of InGaAsP/InP DH lasers. IEEE J. Quantum Electron. QE-17:611–618, 1981.
12. C. H. Henry, R. A. Logan, F. R. Merritt, and J. P. Luongo. The effect of intervalence band absorption on the thermal behavior of InGaAsP lasers. IEEE J. Quantum Electron. QE-19:947–952, 1983.
13. F. Fiedler and A. Schlachetzki. Optical parameters of InP-based waveguides. Solid-State Electron. 30:73–83, 1987.
14. H. Y. Fan. Effect of free carriers on optical properties. In: R. K. Willardson and A. C. Beer, eds. Semiconductors and Semimetals. Vol. 3. New York: Academic Press, 1967, Ch. 9.
15. W. Szuszkiewicz. Absorption due to free carriers in InP. In: Properties of Indium Phosphide, EMIS Datareviews Series No. 6, New York: INSPEC, 1991, pp. 144–148.
16. R. J. Deri, R. J. Hawkins, and E. Kapon. Rib profile effects on scattering in semiconductor optical waveguides. Appl. Phys. Lett. 53:1483–1485, 1988.
17. P. K. Tien. Light waves in thin films and integrated optics. Appl. Opt. 10:2395–2419, 1971.
18. R. J. Deri, E. Kapon, R. Bhat, M. Seto, and K. Kash. Low-loss multiple quantum well GaInAs/InP optical waveguides. Appl. Phys. Lett. 54:1737–1739, 1989.
19. R. J. Deri and E. Kapon. Low-loss III-V semiconductor optical waveguides. IEEE J. Quantum Electron. 27:626–640, 1991.
20. H. Angenent, M. Erman, J. M. Auger, R. Gamonal, and P. J. A. Thijs. Extremely low loss InP/GaInAsP rib waveguides. Electron. Lett. 25:628–629, 1989.
21. Y. S. Oei, L. H. Spiekman, F. H. Groen, I. Moerman, E. G. Metaal, and J. W. Pedersen. Novel RIE-process for high quality InP-based waveguide structures. In: Proc. Eur. Conf. on Integrated Optics (ECIO '95), 1995, pp. 205–208.
22. M. Itoh, Y. Kondo, K. Kishi, M. Yamamoto, and Y. Itaya. High-quality 1.3 μm GaInAsP-BH-lasers fabricated by MOVPE and dry-etching technique. IEEE Photon. Technol. Lett. 8:989–991, 1996.
23. T. R. Hayes, W. C. Dautremont-Smith, H. S. Luftman, and J. W. Lee. Passivation of acceptors in InP resulting from CH4/H2 reactive ion etching. Appl. Phys. Lett. 55:56–58, 1989.

24. C. Rolland, G. Mak, K. E. Fox, D. M. Adams, and A. J. Springthorpe. Analysis of strongly guiding rib waveguide s-bends: Theory and experiment. Electron. Lett. 25:1256–1257, 1989.
25. L. H. Spiekman, Y. S. Oei, E. G. Metall, F. H. Groen, I. Moerman, and M. K. Smit. Extremely small multimode interference couplers and ultrashort bends on InP by deep etching. IEEE Photon. Technol. Lett. 6:1008–1010, 1994.
26. C. van Dam, L. H. Spiekman, F. P. G. M. van Ham, F. H. Groen, J. J. G. M. v. d. Tol, I. Moerman, W. W. Pascher, M. Hamacher, H. Heidrich, C. M. Weinert, and M. K. Smit. Novel compact InP-based polarisation converters using ultrashort bends. In: Tech. Dig. of Integrated Optics Research (IPR '96), Vol. 6, 1996, pp. 468–471.
27. W. H. Nelson, A. N. M. M. Choudhury, M. Abdalla, R. Bryant, E. Meland, and W. Niland. Wavelength- and polarization-independent large angle InP/IngaAsP digital optical switches with extinction ratios exceeding 20 dB. IEEE Photon. Technol. Lett. 6:1332–1334, 1994.
28. A. Sneh, J. E. Zucker, B. I. Miller, and L. W. Stulz. Polarization-insensitive InGaAsP/InGaAsP MQW digital optical switch. IEEE Photon. Technol. Lett. 9: 1589–1561, 1997.
29. U. Koren, T. L. Koch, B. I. Miller, and A. Shahar. Processes for large scale photonic integrated circuits. In: Tech. Dig. of Integrated and Guided Wave Optical Conf., paper MDD2, 1989.
30. T. L. Koch and U. Koren. Semiconductor photonic integrated circuits. IEEEE J. Quantum Electron. 27:641–653, 1991.
31. H. Soda, M. Furutsu, K. Sato, N. Okazaki, Y. Yamazaki, H. Nishimoto, and H. Ishikawa. High-Power and high-speed semi-insulating BH structure monolithic electroabsorption modulator/DFB laser light source. Electron. Lett. 26:9–10, 1990.
32. M. Suzuki, Y. Noda, H. Tanaka, S. Akiba, Y. Kushiro, and H. Isshiki. Monolithic integration of InGaAsP/InP distributed feedback laser and electroabsorption modulator by vapor phase epitaxy. J. Lightwave Technol. LT-5:1277–1285, 1987.
33. Y. Tohmori, F. Kano, M. Oishi, Y. Kondo, M. Nakao, and K. Oe. Narrow lindwidth and low chirping characteristics in high power operating butt joint DBR lasers grown by MOVPE. Electron. Lett. 24:1481–1483, 1988.
34. T. L. Koch and U. Koren. Photonic integrated circuits. In: M. Dagenais, R. F. Leheny, and J. Crow, eds. Integrated Optoelectronics. San Diego: Academic Press, 1995, Ch. 15.
35. T. Kato, T. Sasaki, N. Kida, K. Komatsu, and I. Mito. Novel MQW DFB laser diode/modulator integrated light source using bandgap energy control epitaxial growth technique. In: Proc. ECOC/IOOC '91, 1991, pp. 429–432.
36. M. Aoki, M. Suzuki, M. Takahashi, H. Sano, T. Ido, T. Kawano, and A. Takai. High speed (10 Gb/s) and low-drive-voltage (1 V peak to peak) InGaAs/InGaAsP MQW electroabsorption-modulator integrated DFB laser with sem-insulating buried heterostructure. Electron. Lett. 28:1157–1158, 1992.
37. Y. D. Galeuchet and P. Roentgen. Selective area MOVPE of GaInAs/InP heterostructures on masked and nonplanar (100) and (111) substrates. J. Cryst. Growth 107:147–150, 1991.

38. C. H. Joyner. Semiconductor laser growth and fabrication technology. In: I. P. Kaminow and T. L. Koch, eds. Optical Fiber Telecommunications IIIB. San Diego: Academic Press, 1997, pp. 179–186.
39. T. Miyazawa, H. Iwamura, and N. Naganuma. Integrated external-cavity InGaAs/InP lasers using cap-annealing disordering. IEEE Photon. Technol. Lett. 3:421–423, 1991.
40. J. E. Zucker, B. Tell, K. L. Jones, M. D. Divino, K. F. Brown-Goebeler, C. H. Joyner, B. I. Miller, and M. G. Young. Large blueshifting of InGaAs/InP quantum-well band gaps by ion implantation. Appl. Phys. Lett. 60:3036–3038, 1992.
41. H. Takeuchi, K. Kasaya, and K. Oe. Experimental evaluation of the coupling efficiency between monolithically integrated DFB lasers and waveguides. In: Tech. Dig. of 7th Int. Conf. on Int. Opt. and Optical Fiber Communications, paper 20B1-8, 1989.
42. J.-H. Ahn, K. R. Oh, J. S. Kim, S. W. Lee, H. K. Kim, K. E. Pyun, and H. M. Park. Uniform and high coupling efficiency between InGaAsP/InP buried heterostructure optical amplifier and monolithically butt-cuopled waveguide using RIE. IEEE Photon. Technol. Lett. 8:200–202, 1996.
43. S. Adachi. Linear electro-optic properties of InGaAsP. In: Properties of Indium Phosphide, EMIS Datareviews Series No. 6. New-York: INSPEC, 1991, pp. 429–431.
44. J. E. Zucker. Linear and quadratic electrooptic coefficients in InGaAsP. In: P. Bhattacharya, ed. Indium Gallium Arsenide. EMIS Datareviews Series No. 8. Exeter, Eng.: INSPEC, 1993, pp. 213–218.
45. A. Yariv. Optical Electronics, 3rd ed. New York: Holt, Reinhart and Winston, 1985, pp. 539–542.
46. D. E. Aspenes and N. Bottka. Electric-field effects on the dielectric function of semiconductors and isolators. In: R. K. Willardson and A. C. Beer, eds. Semiconductors and semimetals, Vol. 9. New York: Academic Press, 1972, pp. 457–543.
47. T. E. Van Eck, L. M. Walpita, W. S. C. Chang, and H. H. Wieder. Franz–Keldysh electrorefraction and electroabsorption in bulk InP and GaAs. Appl. Phys. Lett. 48: 451–453, 1986.
48. S. Adachi and K. Oe. Quadratic electro-optic (Kerr) effects in zincblende-type semiconductors: key properties of InGaAsP relevant to device design. J. Appl. Phys. 56:1499–1504, 1984.
49. D. J. Robbins. Franz–Keldysh effect in InGaAsP. In: Properties of Indium Phosphide, EMIS Datareviews Series No. 6. New-York: INSPEC, 1991, pp. 435–437.
50. A. Alping and L. A. Coldren. Electrorefraction in GaAs and InGaAsP and its application to phase modulators. J. Appl. Phys. 61:2430–2433, 1987.
51. B. R. Bennet and R. A. Soref. Analysis of Franz–Keldysh electrooptic modulation in InP, GaAs, GaSb, InAs, and InSb. Proc. SPIE 836:158, 1988.
52. D. A. B. Miller, D. S. Chemla, T. C. Damen, A. C. Gossard, W. Wiegmann, T. H. Wood, and C. A. Burrus. Bandedge electroabsorption in quantum well structures: the quantum-confined Stark effect. Phys. Rev. Lett. 53:2173–2176, 1984.

53. D. A. B. Miller, D. S. Chemla, and S. Schmitt-Rink. Relation between electroabsorption in bulk semiconductors and in quantum wells: the quantum-confined Franz–Keldysh effect. Phys. Rev. B 33:6976–6982, 1986.
54. J. E. Zucker, I. Bar-Joseph, B. I. Miller, U. Koren, and D. S. Chemla. Quaternary quantum wells for electrooptic intensity and phase modulation at 1.3 and 1.55 μm. Appl. Phys. Lett. 54:10–12, 1989.
55. T. S. Moss, G. J. Burrell, and B. Ellis. Semiconductor Opto-Electronics. New York: Wiley, 1973, pp. 48–94.
56. O. Mitomi, I. Kotaka, K. Wakita, S. Nojima, K. Kawano, Y. Kawamura, and H. Asia. 40-Ghz bandwidth InGaAs/InAlAs multiple quantum well optical intensity modulator. Appl. Opt. 31:2030–2035, 1992.
57. F. Devaux, P. Bordes, A. Ougazzaden, M. Carre, and F. Huet. Experimental optimization of MQW electroabsorption modulators with up to 40 GHz bandwidths. Electron. Lett. 30:1347–1348, 1994.
58. T. Ido, S. Tanaka, M. Suzuki, and H. Inoue. MQW electroabsorption modulator for 40 Gb/s modulation. Electron. Lett. 31:2124–2125, 1995.
59. H. Takeuchi, K. Tsuzuki, K. Sato, M. Yamamoto, Y. Itaya, A. Sano, M. Yoneyama, and T. Otsuji. Very high speed light-source module up to 40 Gb/s containing an MQW electroabsorption modulator integrated with a DFB laser. IEEE. J. Sel. Topics Quantum Electron. 3:336–343, 1997.
60. T. Ido, H. Sano, D. J. Moss, S. Tanaka, and A. Takai. Strained InGaAs/InAlAs MQW electroabsorption modulators with large bandwidth and low driving voltage. IEEE Photon. Technol. Lett. 6:???, 1994.
61. T. Yamanaka, K. Wakita, and K. Yokoyama. Potential chirp-free characteristics (negative chirp parameter) in electroabsorption modulation using a wide tensile-strained quantum well structure. Appl. Phys. Lett. 68:3114–3116, 1996.
62. F. Koyama and K. Iga. Frequency chirping in external modulators. IEEE J. Lightwave Technol. 6:87–92, 1987.
63. F. Devaux, Y. Sorel, and J.-F. Kerdiles. Simple measurements of fiber dispersion and chirp parameters of intensity modulated light emitter. J. Lightwave Technol. 11:1937–1940, 1993.
64. F. Dorgeuille and F. Devaux. On the transmission performances and the chirp parameter of a multiple-quantum-well electroabsorption modulator, IEEE J. Quantum Electron. 30:2565–2572, 1994.
65. K. Wakita, K. Sato, I. Kotaka, M. Yamamoto, and T. Kataoka. 20 Gb/s 1.55 μm strained-InGaAsP MQW modulator integrated DFB laser module. Electron. Lett. 30:302–303, 1994.
66. J. E. Johnson, P. A. Morton, T. Nguyen, O. Mizuhara, S. N. G. Chu, G. Nykolak, T. Tanbun-Ek, W. T. Tsang, T. R. Fullowan, D. F. Sciortino, A. M. Sergent, K. W. Wecht, and R. D. Yadvish. 10-Gb/s transmission using an integrated electroabsorption modulator/DFB laser grown by selective area epitaxy. In: Tech. Dig. Optical Fiber Communication (OFC '95), Vol. 8, 1995, pp. 21–23.
67. K. Komatsu, T. Kato, M. Yamaguchi, T. Sasaki, S. Takano, H. Shimizu, N. Watanabe, and M. Kitamura. DFB-LD/modulator integrated light sources fabricated by bandgap-energy-controlled selective MOVPE with stable fiber transmission charac-

teristics. In:Tech. Dig. Optical Fiber Communication (OFC '94), Vol. 4, 1994, pp. 8–9.
68. M. Aoki and H. Sano. High-performance modulator/integrated light sources grown by in-plane bandgap energy-control technique. In: Tech. Dig. Optical Fiber Communication (OFC '95), Vol. 8, 1995, pp. 25–26.
69. H. Haisch, W. Baumert, C. Hache, E. Kühn, M. Klenk, K. Satzke, M. Schilling, J. Weber, and E. Zielinski. 10 Gb/s standard fiber TDM transmission at 1.55 μm with low chirp monolithically integrated MQW electroabsorption modulator/DFB laser realized by selective area MOVPE. In: Proc. Eur. Conf. Optical Communication (ECOC '94), 1994, pp. 801–804.
70. S. K. Korotky, J. J. Veselka, C. T. Kemmerer, W. J. Minford, D. T. Moser, J. E. Watson, C. A. Mattoe, and P. L. Stoddard. High-speed low-power optical modulator with adjustable chirp parameter. Integrated Photo. Res., Tech. Dig. Series, Vol. 8, pp. 53, 1991.
71. A. H. Gnauck, S. K. Korotky, J. J. Veselka, J. Nagel, C. T. Kemmerer, W. J. Minford, and D. T. Moser. Dispersion penalty reduction using an optical modulator with adjustable chirp. IEEE Photon. Technol. Lett. 3:916–918, 1991.
72. J. E. Zucker, K. L. Jones, M. A. Newkirk, R. P. Gnall, B. I. Miller, M. G. Young, U. Koren, C. A. Burrus, and B. Tell. Quantum well interferometric modulator monolithically integrated with 1.55 μm tunable distributed Bragg reflector laser. Electron. Lett. 28:1888–1889, 1992.
73. D. M. Adams, C. Rolland, and S. Bradshaw. Mach–Zehnder modulator integrated with a gain-coupled DFB laser for 10 Gb/s, 100 km NDSF transmission at 1.55 μm. Electron. Lett. 32:485–486, 1996.
74. C. Rolland, R. S. Moore, F. Shepherd, and G. Hillier. 10 Gb/s, 1.56 μm multiple quantum well InP/InGaAsP Mach-Zehnder optical modulator. Electron. Lett. 29: 471–472, 1993.
75. J. E. Zucker. High-speed quantum-well interferometric modulators for InP-based photonic integrated circuits. Microwave Opt. Technol. Lett. 6:6–14, 1993.
76. J. Yu, C. Rolland, D. Yevick, A. Somani, and S. Bradshaw. Phase-engineered III-V MQW Mach–Zehnder modulators. IEEE Photon. Technol. Lett. 8:1018–1020, 1996.
77. J. C. Cartledge, C. Rolland, S. Lemerle, and A. Solheim. Theoretical Performance of 10 Gb/s lightwave systems using a III-V semiconductor Mach–Zehnder modulator. IEEE Photon. Technol. Lett. 6:282–284, 1994.
78. P. Delansay, D. Penninckx, S. Artigaud, J.-G. Provost, J.-P. Hebert, E. Boucherez, J.-Y. Emery, C. Fortin, and O. L. Gouezigou. 10 Gb/s transmission over 90–127 km in the wavelength range 1530–1560 nm using an InP-based Mach–Zehnder modulator. Electron. Lett. 32:1820–1821, 1996.
79. H. Sano, H. Inoue, S. Tsuji, and K. Ishida. InGaAs/InAlAs MQW Mach–Zehnder optical modulator for 10 Gb/s long-haul transmission systems. Tech. Dig. Optical Fiber Communications, OFC '92, Paper ThG4, 1992.
80. D. Penninckx, P. Delansay, E. Boucherez, C. Fortin, and O. L. Gouezigou. InP/InGaAsP phase-shifted Mach–Zehnder modulator for wavelength independent (1530–1560 nm) propagation performance at 10 Gb/s over standard dispersive fiber. Electron. Lett. 33:697–698, 1997.

81. E. C. M. Pennings, R. J. Deri, and R. J. Hawkins. Simple method for estimating usable bend radii of deeply etched optical rib waveguides. Electron. Lett. 27:1532–1533, 1991.
82. J.-F. Vinchant, A. Goutelle, B. Martin, F. Gaborit, P. Pagnod-Rossiaux, J. Peyre, J. L. Bris, and M. Renaud. New compact polarization insensitive 4 × 4 switch matrix on InP with digital optical switches and integrated mirrors. In: Proc. Eur. Conf. Opt. Comm. (ECOC '93), postdeadline paper ThC 12.4, 1993.
83. A. L. Burness, P. H. Loosemore, S. N. Judge, I. D. Henning, S. E. Hicks, G. F. Doughty, M. Asghari, and I. White. Low loss mirrors for InP/InGaAsP waveguides. Electron. Lett. 29:520–521, 1993.
84. E. Gini, G. Guekos, and H. Melchior, Low loss corner mirrors with 45° deflection angle for integrated optics. Electron. Lett. 28:499–501, 1992.
85. R. van Roijen, G. L. A. van der Hofstad, M. Groten, J. M. M. van der Heyden, P. J. A. Thijs, and B. H. Verbeek. Fabrication of low loss integrated optical corner mirrors. Appl. Opt. 32:3246–3248, 1993.
86. T. Brenner, C. H. Joyner, and M. Zirngibl. Compact design waveguide grating routers. Electron. Lett. 32:1660–1661, 1996.
87. R. Urlich and G. Ankele. Self-imaging in homogenous planar optical waveguides. Appl. Phys. Lett. 27:337–339, 1975.
88. R. Urlich and T. Kamiya. Resolution of self-images in planar optical waveguides. J. Opt. Soc. Am. 68:583–592, 1978.
89. L. B. Soldano, F. B. Veerman, M. K. Smit, B. H. Verbeek, A. H. Dubost, and E. C. M. Pennings. Planar monomode optical couplers based on multimode interference effects. J. Lightwave Technol. 10:1843–1850, 1992.
90. L. B. Soldano and E. C. M. Pennings. Optical multi-mode interference devices based on self-imaging: Principles nad applications. J. Lightwave Technol. 13:615–627, 1995.
91. P. A. Besse, M. Bachmann, H. Melchior, L. B. Soldano, and M. K. Smit. Optical bandwidth and fabrication tolerances of multimode interference couplers. J. Lightwave Technol. 12:1004–1009, 1994.
92. L. B. Soldano, M. Bouda, M. K. Smit, and B. H. Verbeek. New small-size single-mode optical power splitter based on multi-mode interference. In: Proc. Eur. Conf. on Opt. Comm. (ECOC '92), 1992, pp. 465–468.
93. J. E. Zucker, K. L. Jones, T. H. Chiu, B. Tell, and K. Brown-Goebler. Polarization-independent electro-optic waveguide switch using strained InGaAs/InP quantum wells. In: Integrated Photonic Research (IPR '92), postdeadline papers, 1992, pp. 21–24.
94. M. Bachmann, M. K. Smit, P. A. Besse, E. Gini, H. Melchior, and L. B. Soldano. Polarization-insensitive low-voltage optical waveguide switch using InGaAsP/InP four-port Mach–Zehnder interferometer. In: OFC/IOOC '93 Tech. Dig., Vol. 4, 1993, pp. 32–33.
95. R. Krähenbühl, R. Kyburz, W. Vogt, M. Bachmann, T. Brenner, E. Gini, and H. Melchior. Low-loss polarization-insensitive InP-InGaAsP optical space switches for fiber optical communication. IEEE Photon. Technol. Lett. 8:632–634, 1996.
96. T. Uitterdijk, D. H. P. Maat, F. H. Groen, and H. v. Brug. Dilated, polarization

insensitive InP-based space switch. In: Proc. Eur. Conf. on Int. Opt. (ECIO '97), 1997, pp. 551–554.

97. H. Inoue, H. Nakamura, K. Morosawa, Y. Sasaki, T. Katsuyama, and N. Chinone. An 8 mm length nonblocking 4 × 4 optical switch array. IEEE J. Select. Areas Commun. 6:1262–1265, 1988.

98. H. Inoue, T. Kirihara, Y. Sasaki, and K. Ishida. Carrier-injection type optical S3 switch with travelling-wave amplifier. IEEE Photon. technol. Lett. 2:214–215, 1990.

99. J.-F. Vinchant, P. P. Rossiaux, J. I. Bris, A. Goutelle, H. Bissessur, and M. Renaud. InP digital optical switch: Key element for guided-wave photonic switching. IEEE Proc. 140:301–307, 1993.

100. M. N. Khan, J. E. Zucker, T. Y. Chang, N. J. Sauer, M. D. Divino, T. L. Koch, C. A. Burrus, and H. M. Presby. Design and demonstration of weighted-coupling digital Y-branch optical switches in InGaAs/InGaAlAs Electron transfer waveguides. J. Lightwave Technol. 12:2032–2038, 1994.

101. A. Sneh, J. E. Zucker, and B. I. Miller. Compact, low crosstalk and low propagation loss quantum-well Y-branch switches. IEEE Photon. Technol. Lett. 8:1644–1646, 1996.

102. N. Yoshimoto, Y. Shibata, S. Oku, S. Kondo, Y. Noguchi, K. Wakita, K. Kawano, and M. Naganuma. Fully polarization insensitive Mach–Zehnder optical switches using a wide-well InGaAlAs/InAlAs MQW structure. In: Tech. Dig. of Photonics in Switching (PS '96), Vol. 1, 1996, pp. 78–79.

103. L. Stoll, G. Müller, U. Wolff, B. Sauer, S. Eichinger, and S. Sürgec. Compact and polarization independent optical switch on InP/InGaAsP. In: Proc. Eur. Conf. Opt. Commun. (ECOC '92), 1992, pp. 337–340.

104. P. J. Duthie, N. Show, M. J. Wale, and I. Bennion. Guided wave switch array using electrooptic and carrier depletion effects in InP. Electron. Lett. 27:1747–1748, 1991.

105. D. A. H. Mace, M. J. Adams, J. Singh, M. A. Fisher, I. D. Henning, and W. J. Duncan. Twin-ridge laser amplifier crosspoint switch. Electron. Lett. 25:987–988, 1989.

106. G. Glastre, D. Rondi, A. Enard, E. Lallier, R. Blondeau, and M. Papuchon. Monolithic integration of 2 × 2 switch and optical amplifier with 0 dB fiber to fiber insertion loss grown by LP-MOCVD. Electron. Lett. 29:124–126, 1993.

107. K. Kawano, S. Sekine, H. Takeuchi, M. Wada, M. Kohtoku, N. Yoshimoto, T. Ito, M. Yanagibashi, S. Kondo, and Y. Noguchi. 4 × 4 InGaAlAs/InAlAs MQW directional coupler waveguide switch modules integrated with spot-size converters and their 10 Gb/s operation. Electron. Lett. 31:96–97, 1995.

108. T. Aizawa, Y. Nagasawa, and T. Watanabe. Polarization-independent switching operation in directional coupler using tensile-strained multi-quantum wells. IEEE Photon. Technol. Lett. 7:47–49, 1995.

109. S. Oku, Y. Yoshino, M. Ikeda, M. Okamoto, and T. Kawani. Design and performance of monolithic LD optical matrix switches. In: Photonic Switching. 1990, pp. 58–61.

110. M. Gustavsson, B. Lagerström, L. Thylén, M. Janson, L. Lundgern, A.-C. Mörner,

R. Rask, and B. Stoltz. Monolithically integrated 4 × 4 InGaAsP/InP laser amplifier gate switch arrays. Electron. Lett. 28:2223–2225, 1992.

111. U. Koren, M. G. Young, B. I. Miller, M. A. Newkirk, M. Chien, M. Zirngible, C. Dragone, B. Glance, T. L. Koch, B. Tell, K. Brown-Goebler, and G. Raybon. 1 × 16 photonic switch operating at 1.55 μm wavelength based on optical amplifiers and a passive optical splitter. Appl. Phys. Lett. 61:1613–1615, 1992.
112. G. Sherlock, J. D. Burton, P. J. Fiddyment, P. C. Sully, A. E. Kelly, and M. J. Robertson. Integrated 2 × 2 optical switch with gain. Electron. Lett. 30:137–138, 1994.
113. K. Hamamoto and K. Komatsu. Insertion-loss-free 2 × 2 InGaAsP/InP optical switch fabricated using bandgap energy controlled selective MOVPE. Electron. Lett. 31:1779–1781, 1995.
114. F. Dorgeuille, B. Mersali, M. Feuillade, S. Sainson, S. Slempkes, and M. Foucher. Novel approach for simple fabrication of high performance InP-switch matrix based on laser-amplifier gates. IEEE Photon. Technol. Lett. 8:1178–1180, 1996.
115. I. H. White, J. J. S. Watts, and J. E. Carroll. InGaAsP 400 × 200 μm acrive cross-point switch operating at 1.5 μm using novel reflective Y-coupler components. Electron. Lett. 26:617–618, 1990.
116. T. Ido, M. Koizumi, S. Tanaka, M. Suzuki, and H. Inoue. Polarization and wavelength insensitive MQW electro-absorption optical gate switches for WDM applications. In: Tech. Dig. of Photonics in Switching (PS '96), Vol. 1, 1996, pp. 80–81.
117. W. K. Burns. Shaping the digital switch. IEEE Photon. Technol. Lett. 4:861–863, 1992.
118. H. Okayama and K. Kawahara. Reduction of voltage-length product for Y-branch digital optical switch. J. Lightwave Technol. 11:379–387, 1993.
119. E. L. Goldstein and L. Eskildsen. Scaling limitations in transparent optical networks due to low-level crosstalk. IEEE Photon. Technol. Lett. 7:93–94, 1995.
120. C. P. Larsen, L. Gillner, and M. Gustavsson. Scaling limitations in optical multiwavelength meshed networks and ring networks due to their crosstalk. In: Tech. Dig. Photon. Switching '96, paper PMB4, 1996.
121. R. A. Spanke. Architectures for large nonblocking optical space switches. IEEE J. Quantum Electron. QE-22:964–967, 1986.
122. D. J. Blumenthal, P. Granestrand, and L. Thylén. BER floors due to heterodyne coherent crosstalk in space photonic switches for WDM networks. IEEE Photon. Technol. Lett. 8:284–286, 1996.
123. H.-P. Nolting and M. Gravert. Architecture of crosstalk-reduced digital optical. IEEE Photon. Technol. Lett. 11:1294–1296, 1995.
124. T. Kirihara, M. Ogawa, H. Inoue, H. Kodera, and K. Ishida. Lossless and low crosstalk characteristics in an InP-based 4 × 4 optical switch with integrated single-stage optical amplifiers. IEEE Photon. Technol. Lett. 6:218–220, 1994.
125. D. Hoffmann, C. Bornholdt, F. Kappe, L. Mörl, and F.-W. Reier. Novel digital optical switches with crosstalk below −40 dB based on absorptive switching. In: Proc. 21st Eur. Conf. on Opt. Comm. (ECOC '95), 1995, pp. 107–110.
126. J.-F. Vinchant, M. Renaud, A. Goutelle, J. Peyre, P. Jarry, M. Erman, P. Svennson,

and L. Thylén. First polarization insensitive 4 × 4 switch matrix on InP with digital optical switches. In: Proc. Eur. Conf. on Opt. Comm. (ECOC '92), 1992, pp. 341–344.

127. M. Cada, G. Müller, A. Greil, L. Stoll, and U. Wolff. Dynamic switching characteristics of a 4 × 4 InP/InGaAsP matrix switch. Electron. Lett. 28:2149–2150, 1992.
128. J.-F. Vinchant, P. Pagnod-Rossiaux, J. L. Bris, A. Goutelle, H. Bissessur, and M. Renaud. Low-loss fiber-chip coupling by InGaAsP/InP thick waveguides for guided wave photonic integrated circuits. IEEE Photon. Technol. Lett. 6:1347–1349, 1994.
129. A. Jourdan, G. Soulage, G. DaLoura, B. Clesca, P. Doussiere, C. Duchet, D. Leclerc, J.-F. Vinchant, and M. Sotom. Experimental assessment of a 4 × 4 four-wavelength all-optical crossconnect at 10 Gb/s line rate. In: Tech. Dig. Optical Fiber Communication Conference (OFC '95), 1995, pp. 277–278.
130. D. A. O. Davies, P. S. Mudhar, M. A. Fisher, D. A. H. Mace, and M. J. Adams. Integrated lossless InP/InGaAsP 1 to 4 optical switch. Electron. Lett. 28:1521–1522, 1992.
131. M. Gustavsson, M. Janson, and L. Lundgern. Digital transmission experiment with monolithic 4 × 4 InGaAsP/InP laser amplifier gate switch array. Electron. Lett. 29:1083–1085, 1993.
132. C. P. Larsen and M. Gustavsson. Linear crosstalk in 4 × 4 semiconductor optical amplifier gate switch matrix. J. Lightwave Technol. 15:1865–1870, 1997.
133. W. van Berlo, M. Janson, L. Lundgern, A.-C. Mörner, J. Terlecki, M. Gustavsson, P. Granestrand, and P. Svensson. Polarization-insensitive, monolithic 4 × 4 InGaAsP-InP laser amplifier gate switch matrix. IEEE Photon. Technol. Lett. 7: 1291–1293, 1995.
134. D. M. Bird, J. R. Armitage, R. Kashyap, and R. M. A. Fatah. Narrow line semiconductor laser using fibre grating. Electron. Lett. 27:1115–1116, 1991.
135. U. Koren, T. L. Koch, B. I. Miller, G. Eisenstein, and R. H. Bosworth. Wavelength division multiplexing light source with integrated quantum well tunable lasers and optical amplifiers. Appl. Phys. Lett. 54:2056–2508, 1989.
136. M. G. Young, U. Koren, B. I. Miller, M. A. Newkirk, M. Chien, M. Zirngibl, C. Dragone, B. Tell, H. M. Presby, and G. Raybon. A 16 × 1 wavelength division multiplexer with integrated distributed Bragg reflector lasers and electroabsorption modulators. IEEE Photon. Technol. Lett. 5:908–910, 1993.
137. J.-M. Verdiell, T. L. Koch, D. M. Tennant, K. Feder, R. P. Gnall, M. G. Young, B. I. Miller, U. Koren, M. A. Newkirk, and B. Tell. 8-wavelength DBR laser array fabricated with a single-step Bragg grating printing technique. IEEE Photon. Technol. Lett. 5:619–621, 1993.
138. T. R. Chen, J. Ungar, J. Lannelli, et. al. High Power Operation of InGaAsP/InP Multiquantum Well DFB Lasers at 1.55-μm Wavelength. Electron. Lett. 32:898–898, 1996.
139. C. E. Zah, F. J. Favire, B. Pathak, R. Bhat, C. Caneau, P. S. D. Lin, A. S. Gozdz, N. C. Andreakdakis, M. A. Koza, and T. P. Lee. Monolithic integration of multiwavelength compressive-strained multiquantum-well distributed-feedback laser array with star coupler and optical amplifiers. Electron. Lett. 28:2361–2362, 1992.

140. M. Zirngibl, C. Dragone, C. H. Joyner, M. Kuznetsov, and U. Koren. Efficient 1 × 16 optical power splitter based on InP. Electron. Lett. 28:1212–1213, 1992.
141. C. Dragone. Efficient $N \times N$ star couplers using Fourier optics. J. Lightwave Technol. 7:479–489, 1989.
142. M. R. Amersfoort, C. E. Zah, B. Pathak, F. Favire, P. S. D. Lin, A. Rajhel, N. C. Andreadakis, R. Bhat, C. Caneau, and M. A. Koza. Wavelength accuracy and output power of multiwavelength DFB laser arrays with integrated star couplers and optical amplifiers. Integrated Photonics Research Conference, Optical Society of America, 1996, pp. 478–481.
143. T. L. Koch and U. Koren. Semiconductor lasers for coherent optical fiber communications. J. Lightwave Technol. 8:274–293, 1990.
144. R. C. Alferness, U. Koren, L. L. Buhl, B. I. Miller, M. G. Young, T. L. Koch, G. Raybon, and C. A. Burrus. Broadly tunable InGaAsP/InP laser based on a vertical coupler filter with 57 nm tuning range. Appl. Phys. Lett. 60:3209–3211, 1992.
145. V. Jayaraman, A. Mathur, L. A. Coldren, and P. D. Dapkus. Extended tuning range in sample grating DBR lasers. IEEE Photon. Technol. Lett. 5:489–491, 1993.
146. T. Tanbunek, L. E. Adams, and G. Nykolak. Broad Band Tunable Electroabsorption Modulated Laser for WDM Application. IEEE S. T. Qu. 3:960–967, 1997.
147. T. L. Koch. Laser sources for amplified and WDM lightwave systems. In: Optical Fiber Telecommunications IIIB. San Diego: Academic Press, 1997, pp.
148. J. B. D. Soole, K. Poguntke. A. Scherer, H. P. LeBlanc, C. Chang-Hasnain, J. R. Hayes, C. Caneau, R. Bhat, and M. A. Koza. Multi-strip array grating integrated cavity (MAGIC) laser. A new semiconductor laser for WDM applications. Electron. Lett. 28:1805–1807, 1992.
149. M. Zirngibl, C. H. Joyner, L. W. Stulz, U. Koren, M.-D. Chien, M. G. Young, and B. I. Miller. Digitally tunable laser based on the integration of a waveguide grating multiplexer and an optical amplifier. IEEE Photon. Technol. Lett. 6:516–518, 1994.
150. M. R. Amerfoort, J. B. D. Soole, C. Caneau, H. P. LeBlanc, A. Rajhel, C. Youtsey, and I. Adesida. Compact arrayed waveguide grating multifrequency laser using bulk active material. Electron. Lett. 33:2124–2126, 1997.
151. C. R. Doerr and C. H. Joyner. Double-chirping of the waveguide grating router. IEEE Photon. Technol. Lett 9:776–778, 1997.
152. C. R. Doerr, M. Shirasaki, and C. H. Joyner. Chromatic focal plane displacement in the parabolic chirped waveguide grating router. IEEE Photon. Technol. Lett. 9: 625–627, 1997.
153. C. R. Doerr, C. H. Joyner, L. W. Stulz, and J. C. Centanni. Wavelength selectable laser with inherent wavelength and single-mode stability. IEEE Photon. Technol. Lett. 9:1430–1432, 1997.
154. I. P. Kaminow and R. S. Tucker. Mode-controlled semiconductor lasers. In: Guided-Wave Optoelectronics. Berlin: Sprinter-Verlag, 1990, pp.
155. C. R. Doerr. Direct modulation of long-cavity semiconductor lasers. J. Lightwave Technol. 14:2052–2061, 1996
156. C. R. Doerr. Theoretical stability analysis of single-mode operation in uncontrolled mode-selection semiconductor lasers. IEEE Photon. Technol. Lett. 9:1457–1459, 1997.

157. P. A. Morton, T. Tanbun-Ek, R. A. Logan, N. Chand, K. W. Wecht, A. M. Sergent, and P. F. Sciortino. Packaged 1.55 μm DFB laser with 25 GHz modulation bandwidth. Electron. Lett. 30:2044–2046, 1994.
158. C. Y. Kuo, M. L. Kao, J. S. French, R. E. Tench, and T. W. Cline. 1.55 μm, 2.5 Gb/s direct detection repeaterless transmission of 160 km nondispersion shifted fiber. IEEE Photon. Technol. Lett. 2:911–913, 1990.
159. P. A. Morton, G. E. Shtengel, L. D. Tzeng, R. D. Yadvish, T. Tanbun-Ek, and R. A. Logan. 38.5 km error free transmission at 10 Gbit/s in standard fibre using a low chirp, spectrally filtered, directly modulated 1.55 μm DFB laser. Electron. Lett. 33:310–311, 1997.
160. C. R. Doerr and R. Monnard. Method for improving transmission distance of directly modulated lasers for WDM systems. 24th European Conference on Optical Communication, 1998.
161. G. Grosskopf, R. Ludwig, and H. G. Weber. Crosstalk in optical amplifiers for two-channel transmission. Electron. Lett. 22:900–901, 1986.
162. R. M. Jopson, K. L. Hall, G. Eisenstein, G. Raybon, and M. S. Whalen. Observation of two-color gain saturation in an optical amplifier. Electron. Lett. 23:510–512, 1987.
163. C. R. Doerr, C. H. Joyner, M. Zirngibl, L. W. Stulz, and H. M. Presby. Elimination of signal distortion and crosstalk from carrier density changes in the shared semiconductor amplifier of multifrequency signal sources. IEEE Photon. Technol. Lett. 7:1131–1133, 1995.
164. G. Raybon, K. Dreyer, U. Koren, B. I. Miller, M. Chien, D. Tennant, and K. Feder, personal communication; R. Monnard, C. R. Doerr, C. H. Joyner, M. Zirngibl, and L. W. Stulz. Direct modulation of a multifrequency laser up to 16 $\times$ 2.5 Gb/s. IEEE Photon. Technol. Lett. 9:815–817, 1997.
165. C. R. Giles and M. Zirngibl. Multichannel stabilization of an integrated WDM laser transmitter through correlation feedback. IEEE Photon. Technol. Lett. 10:150–152, 1998.
166. G. Raybon, U. Koren, B. I. Miller, et. al. Reconfigurable Optoelectronic Wavelength Translation Based on an Integrated Electroabsorption Modulated Laser Array. IEEE Photon. Technol. Lett 10:215–217, 1998.
167. R. Monnard, M. Zirngibl, C. R. Doerr, C. H. Joyner, and L. W. Stulz. Demonstration of an eight-wavelength fast packet switching transmitter of 2.5-Gb/s data stream. IEEE Photon. Technol. Lett. 10:430–432, 1998.
168. U. Koren, B. I. Miller, M. G. Young, M. Chien, G. Raybon, T. Brenner, R. Ben-Michael, K. Dreyer, and R. J. Capik. Polarization insensitive semiconductor optical amplifier with integrated electroabsorption modulators. Electron. Lett. 32:111–112, 1996.
169. C. R. Doerr, L. W. Stulz, C. H. Joyner, and U. Koren. Characterization of the hybrid integration of a shared dispersive element laser and a modulator. J. Lightwave Technol. 16:2401–2406, 1998.
170. C. H. Joyner, M. Zirngibl, and J. C. Centanni. An 8-channel digitally tunable transmitter with an electroabsorption modulated output by selective area epitaxy. IEEE Photon. Technol. Lett 7:1013–1015, 1995.
171. C. R. Doerr, R. Monnard, C. H. Joyner, and L. W. Stulz. Simultaneous cw operation

of shared angular dispersive element WDM lasers. IEEE Photon. Technol. Lett. 10:501–503, 1998.

172. R. Monnard, A. K. Srivastava, C. R. Doerr, C. H. Joyner, L. W. Stulz, M. Zirngibl, Y. Sun, J. W. Sulhoff, J. L. Zyskind, and C. Wolf. 16-channel 50 GHz spacing long-haul transmitter for DWDM systems. Electron. Lett. 34:765–766, 1998.
173. L. F. Mollenauer, P. V. Mamyshev, and M. J. Neubelt. Method for facile and accurate measurement of optical fiber dispersion maps. Opt. Lett. 21:1724–1726, 1996.
174. M. Zirngibl. Wavelength grating router with output coupler. U.S. Patent 5,600,742, 1997.
175. J. Gripp and L. F. Mollenauer. Enhanced range for OTDR-like dispersion-map measurements. Submitted to Opt. Lett.
176. C. R. Doerr, C. H. Joyner, L. W. Stulz, and J. Gripp. Multifrequency laser having integrated amplified output coupler for high-extinction-ratio modulation with single-mode behavior. To appear in IEEE Photon. Technol. Lett.
177. G. Raybon, U. Koren, B. I. Miller, et. al. A wavelength tunable semiconductor amplifier/filter for add/drop multiplexing in WDM networks. IEEE Photon. Technol. Lett. 9:40–42, 1997.
178. M. Zirngibl, C. H. Joyner, L. W. Stulz, Th. Gaiffe, and C. Dragone. Polarization-independent 8×8 waveguide grating multiplexer on InP. Electron. Lett. 29:201–202, 1993.
179. J. B. D. Soole, M. R. Amersfoort, H. P. LeBlanc, N. C. Andeadakis, A. Rajhel, C. Caneau, M. A. Koza, R. Bhat, C. Youtsey, and I. Adesida. Polarization-independent InP arrayed waveguide filter using square cross-section waveguides. Electron. Lett. 32:323–324, 1996.
180. E. Gini, W. Hunziker, and H. Melchior. Polarization independent InP WDM multiplexer/demultiplexer module. J. Lightwave Technol. 16:625–630, 1997.
181. E. Gini, W. Hunziker, and H. Melchior. Polarization independent InP WDM multiplexer/demultiplexer module. J. Lightwave Technol. 16:625–630, 1998.
182. B. H. Verbeek, A. A. M. Staring, E. J. Jansen, R. Van Roijen, J. J. M. Binsma, T. Van Dongen, M. R. Amersfoort, C. Van Dam, and M. K. Smit. Large-bandwidth, polarization-independent, and compact 8-channel PHASAR demultiplexer/filter. OFC '94, Paper PD13-1, 1994.
183. M. Kohtoku, H. Sanjoh, S. Oku, Y. Kadota, Y. Yoshikuni, and Y. Shibata. InP-based 64-channel arrayed waveguide grating with 50 GHz channel spacing and up to −20 dB crosstalk. Electron. Lett. 33:1786–1787, 1997.
184. M. Zirngibl, C. H. Joyner, and P. C. Chou. Polarization-compensated waveguide grating router on InP. Electron. Lett. 31:1662–1664, 1995.
185. M. Zirngibl, C. H. Joyner, and B. Glance. Digitally tunable channel dropping filter/equalizer based on waveguide router and optical amplifier integration. IEEE Photon. Technol. Lett. 6:513–515, 1994.
186. J.-F. Vinchant, J. A. Cavailles, M. Erman, P. Jarry, and M. Renaud. InP/GaInAsP guided-wave phase modulators based on carrier-induced effects: theory and experiment. J. Lightwave Technol. 10:63–69, 1992.
187. C. R. Doerr, C. H. Joyner, L. W. Stulz, and R. Monnard. Wavelength-division multiplexing cross connect in InP. IEEE Photon. Technol. Lett. 10:117–119, 1998.

188. K. Vreeburg. InP-based Photonic Integrated Circuits for Wavelength Routing and Switching. Ph.D. dissertation, Delft University of Technology, 1997.
189. C. G. M. Vreeburg, T. Uitterdijk, Y. S. Oei, M. K. Smit, F. H. Groen, E. G. Metaal, P. Demeester, and H. J. Frankena. First InP-based reconfigurable integrated add-drop multiplexer. IEEE Photon. Technol. Lett. 9:188–190, 1997.
190. C. G. P. Herben, C. G. M. Vreeburg, D. H. P. Maat, X. J. M. Leijtens, Y. S. Oei, F. H. Groen, J. W. Pedersen, P. Demeester, and M. K. Smit. A compact integrated InP-based single-phasar optical crossconnect. IEEE Photon. Technol. Lett. 10:678–680, 1998.
191. C. R. Doerr. Proposed WDM cross-connect using a planar arrangement of waveguide grating routers and phase shifters. IEEE Photon. Technol. Lett. 10:528–530, 1998.
192. M. J. LaGasse, K. K. Anderson, C. A. Wang, H. A. Haus, and J. G. Fujimoto. Femtosecond measurements of the nonresonant nonlinear index in AlGaAs. Appl. Phys. Lett. 56:417–419, 1990.
193. F. Forghieri, R. W. Tkach, and A. R. Chraplyvy. Fiber nonlinearities and their impact on transmission systems. In: Optical Fiber Telecommunications IIIA. San Diego: Academic Press, 1997.
194. M. R. Amersfoort, C. R. de Boer, B. H. Verbeek, P. Demeester, A. Looyen, J. J. G. M. van der Tol. Low-loss phased-array based 4-channel wavelength demultiplexer integrated with photodetectors. IEEE Photon. Technol. Lett. 6:62–64, 1994.
195. C. A. M. Steenbergen, C. van Dam, A. Looijen, C. G. P. Herben, M. de Kok, M. K. Smit, J. W. Pedersen, I. Moerman, R. G. F. Baets, and B. H. Verbeek. Compact low loss 8 × 10 GHz polarization independent WDM receiver. 22nd European Conference on Optical Communication, 1996, pp. 1.129–1.131.
196. J. B. D. Soole, M. R. Amersfoort, H. P. LeBlanc, N. C. Andreakis, A. Rajhel, and C. Caneau. Polarization-independent monolithic eight-channel 2 nm spacing WDM detector based on compact arrayed waveguide demultiplexer. Electron. Lett. 31: 1289–1290, 1995.
197. M. Kohtoku, H. Sanjoh, S. Oku, Y. Kadota, and Y. Yoshikuni. Low-loss polarization-insensitive packaged WDM monitor. Integrated Photonics Research Conference, PD-2, 1998.
198. J. B. D. Soole, H. P. LeBlanc, N. C. Andreadakis, C. Caneau, R. Bhat, and M. A. Koza. High speed monolithic WDM detector for 1.5 μm fiber band. Electron. Lett. 31:1276–1277, 1995.
199. M. R. Amersfoort. Phased-Array Wavelength Demultiplexers and Their Integration with Photodetectors. Ph.D. dissertation, Delft University of Technology, 1994.
200. H. A. Haus and J. A. Mullen. Quantum noise in linear amplifiers. Phys. Rev. 128: 2407–2413, 1962.
201. Y. Yamamoto. Noise and error rate performance of semiconductor laser amplifiers in PCM-IM optical transmission systems. IEEE J. Quantum Electron. QE-16:1073–1081, 1980.
202. C. R. Doerr, M. Zirngibl, C. H. Joyner, L. W. Stulz, and H. M. Presby. Polarization diversity waveguide grating receiver with integrated optical preamplifiers. IEEE Photon. Technol. Lett. 9:85–87, 1997.
203. J. M. Verdiell, T. L. Koch, B. I. Miller, M. G. Young, U. Koren, F. Storz, and

K. F. Brown-Goebeler. A WDM receiver photonic integrated circuit with net on-chip gain. IEEE Photon. Technol. Lett. 6:960–962, 1994.
204. M. Zirngibl, C. H. Joyner, and L. W. Stulz. WDM receiver by monolithic integration of an optical preamplifier, waveguide grating router and photodiode array. Electron. Lett. 31:581–582, 1995.
205. R. Schimpe, C. Cremer, L. Hoffman, D. Romer, M. Shier, G. Baumeister, G. Ebbinghaus, G. Kristen, and M. Morvan. InP-based 2.5 Gbit/s optical preamplified WDM receiver. Electron. Lett. 32:1141–1142, 1996.
206. B. Glance. Polarization independent coherent optical receiver. J. Lightwave Technol. LT-5:274–276, 1987.
207. S. Chandrasekhar and M. A. Pollack. Optoelectronic and photonic integrated circuits. In: Perspectives in Optoelectronics. River Edge, NJ: World Scientific, 1995.
208. S. Chandrasekhar, M. Zirngibl, A. G. Dentai, C. H. Joyner, F. Storz, C. A. Burrus, and L. M. Lunardi. Monolithic eight-wavelength demultiplexed receiver for dense WDM applications. IEEE Photon. Technol. Lett. 7:1342–1344, 1995.
209. J. G. Bauer, G. Baumeister, P. C. Clemens, G. Heise, L. Hoffmann, M. Klein, W. Kunkel, R. Marz, H. Michel, A. Reichelt, M. Schier, and H. W. Schneider. 8-channel DWDM-receiver frontend based on a SiO_2/Si spectrograph and an InGaAs/InP PIN-photodiode-array. In: Proc. 20th European Conf. Opt. Commun., Firenze, Italy, 1994, pp. 751–754.
210. Y. Kanabar and N. Baker. Demonstration of novel optical multichannel grating demultiplexer receiver (MGD) for HDWDM systems. Electron. Lett. 25:817–819, 1989.
211. C. Dragone. Optimum design of a planar array of tapered waveguides. J. Opt. Soc. Am. A 7:2081–2092, 1990.
212. C. Dragone, C. H. Henry, I. P. Kaminow, and R. C. Kistler. Efficient multichannel integrated optics star coupler on silicon. IEEE Photon. Technol. Lett. 1:241–243, 1989.
213. M. Zirngibl, C. Dragone, C. H. Joyner, M. Kuznetsov, and U. Koren. Efficient 1×16 optical power splitter based on InP. Electron. Lett. 28:1212–1213, 1992.
214. C. Dragone, personal communication.
215. B. Friedman. Principles and Techniques of Applied Mathematics. New York: Wiley, 1956, p. 16.
216. H. A. Haus. Waves and Fields in Optoelectronics. Prentice Hall NJ: Englewood Cliffs, 1984.
217. M. K. Smit. New focusing and dispersive planar component based on an optical phased array. Electron Lett. 24:385–386, 1988.
218. H. Takahashi, S. Suzuki, K. Kato, and I. Nishi. Arrayed-waveguide grating for wavelength division multi/demultiplexer with nanometer resolution. Electron. Lett. 26:87–88, 1990.
219. C. Dragone. An $N \times N$ optical multiplexer using a planar arrangement of two star couplers. IEEE Photon. Technol. Lett. 3:812–815, 1991.
220. C. Dragone. Optimum planar bends. Electron. Lett. 29:1121–1122, 1993.
221. J. C. Chen and C. Dragone. A study of fiber to fiber losses in waveguide grating routers. J. Lightwave Technol. 15:1895–1899, 1997.

222. M. Zirngibl, C. R. Doerr and C. H. Joyner. Demonstration of a splitter/router based on a chirped waveguide grating router. IEEE Photon. Technol. Lett. 10:87–89, 1998.
223. R. F. Kazarinov, C. H. Henry, and R. A. Logan. Longitudinal mode self-stabilization in semiconductor lasers. J. Appl. Phys. Lett. 53:4631–4644, 1982.
224. C. H. Henry. Theory of the linewidth of semiconductor lasers. IEEE J. Quant. Electron. QE-18:259–264, 1982.
225. F. Kartner, personal communication.

8

Polymeric Thermo-Optic Digital Optical Switches

Mart Diemeer
Akzo Nobel Central Research, Arnhem, The Netherlands
Peter De Dobbelaere
Akzo Nobel Electronic Products Inc., Sunnyvale, California
Rien Flipse
Akzo Nobel Photonics, Arnhem, The Netherlands

1. INTRODUCTION

Photonic switches for routing are key components in optical fiber communication systems. From reliability considerations, solid-state planar waveguide components are preferred for that application, and the thermo-optic (t.o.) effect is preeminently suited as their operational principle because it is polarization independent and sufficiently strong in a lot of waveguide materials, while the speed of t.o. waveguide components is adequate for all routing applications. The digital optical switch (DOS) [1] has, unlike interferometric switches, a steplike switching characteristic and is wavelength and polarization independent. In optical fiber communication systems stable, dual transmission window- and polarization-independent operation of the components are important requirements. Therefore the DOS is very well suited for applications in these systems. The only class of materials that allows the realization of a t.o. DOS with specs that meet all the requirements of optical fiber communication systems is that of polymeric materials.

In this chapter, we provide, in Sec. 2, an overview of the progress in solid-state t.o. switches and a basic understanding of how a polymeric t.o. DOS operates. It will be followed by an overview of the reliability and environmental stability aspects of polymeric a t.o. DOS in Sec. 3. The chapter will conclude, in Sec. 4, with a discussion on the applications and requirements for optical switches in the optical layer of telecommunication systems.

2 PHYSICS OF POLYMERIC THERMO-OPTIC DIGITAL OPTICAL SWITCHES*

2.1 Thermo-Optic Waveguide Switches

In 1981 the t.o. effect was used for the first time, by Haruna and Koyama, in a $LiNbO_3$ multimode channel cutoff modulator [2] and in 1982 in a multimode modulator and a 1 × 2 switch, both based on glass [3], followed by a single-mode interferometric modulator a year later [4]. The silica-on-silicon technology provides low-propagation-loss and low-fiber-coupling-loss waveguides. It has been used for the realization of t.o. 2 × 2 switches by NTT researchers in 1988 [5]. These switches, based on Mach–Zehnder interferometers (MZIs) equipped with two directional couplers, had rise and fall times of 0.8 ms and required 0.8 W for switching. This building block was used for the fabrication of a low-loss (av. 7 dB), high isolation (−55 dB, by using double MZI switching units, each consuming ~1 W) t.o. 16 × 16 matrix switch [6]; 8 × 8 matrix switches are commercially available [7].

In 1989, Diemeer et al. demonstrated for the first time the strong t.o. effect in polymer waveguides in a single-mode planar total-reflection switch [8]. They realized the first polymeric 1 × 2 t.o. DOS in 1991 [9] and 2 × 2 versions in 1993 [10]. These channel waveguide devices were based on nonlinear optical polymers where the channel waveguides were defined by photobleaching [11]. The channel waveguides were not matched to single-mode optical fibers. NTT researchers fabricated a polymeric version of their silica-on-silicon 2 × 2 interferometric switch and demonstrated fiber compatibility of the waveguides and a hundredfold reduction in switch power (5 mW) [12].

The strong t.o. effect in polymers, combined with their low thermal conductivity, allows the fabrication of an efficient DOS. A pigtailed, fiber-compatible 1 × 2 DOS based on photo-bleachable polymers was demonstrated by researchers from Akzo Nobel and Bosch in 1994 [13,14]. A power consumption of 30 mW for an isolation of −20 dB and a switching time of about 1 ms has been reported.

The first polymeric t.o. 4 × 4 switch matrix was presented by researchers from the Heinrich Hertz Institut [15] in 1994. The component was based on five directional coupler switches with a response time below 1 ms and a power consumption of 70 mW. The channel waveguides were defined by photopatterning a photoinitiator-doped polymethylmethacrylate (PMMA) core layer. In 1995 Akzo Nobel presented a pigtailed and packaged 1 × 8 DOS that could be used in both the 1300-nm window and the erbium-doped fiber-optic amplifier (EDFA) window [16]. The channel waveguides were defined by reactive ion etching (RIE) to allow

---

* Sections 2.2–2.8 are reprinted from: Mart B. J. Diemeer. Polymeric thermo-optic space switches for optical communications. Optic. Mater. 9:192–200, 1998, with permission from Elsevier Science.

the application of polymers without photosensitive groups. These polymers are photochemically far more stable than the photobleachable polymers.

A fiber-compatible waveguide concept, consisting of a shallow, inverted polymer-rib waveguide core with high refractive index contrast cladding layers, allowed channel waveguide patterning without structuring the polymer directly [17]. A rib was etched into a silicon substrate that was oxidized thereafter to provide the bottom cladding. The good planarization property of polymers was utilized to create the inverted rib waveguide by spin-coating onto the etched substrates. The waveguides were single moded, despite the large index steps between core and cladding. The applied polymers were highly temperature-stable benzocyclobutenes, originally developed by Dow Chemical Co. under the trade name Cyclotene for thin-film dielectric applications. The DOS realized in this technology showed good performance with respect to insertion loss and isolation. A drawback of the concept is the relatively large power of more than 200 mW that was required for switching, caused by the proximity of the waveguide core to the silicon substrate. An advantage is the small switching time of less than 0.25 ms, which can be attributed to the proximity of the heater to the core. This has been demonstrated in a 4×4 switching matrix [18].

There are reports on the DOS in various different polymers in the literature. NTT researchers have explored the use of deuterated fluormethacrylates in channels made by RIE for low propagation losses around 1300 nm (0.1 dB/cm) [19]. The insensitivity of the isolation in a DOS for polymer moisture (de/)sorption has been demonstrated by them in Ref. 20. They have also realized DOS's based on deuterated silicone resin channels that were defined by RIE [21]. This material has an estimated intrinsic absorption loss of less than 0.25 dB/cm in the spectral window from 1530 to 1650 nm, which makes it suitable for applications in gain-shifted EDFA systems (1570–1600 nm). The DOS's exhibit an insertion loss of less than 2.5 dB with an isolation of -30 dB at 1550 nm. Fluorinated polyimides are very stable materials due to their high glass-transition temperatures ($>350°C$), thermal degradation temperature ($>300°C$), and mechanical strength, as well as their low water absorption (0.2%). Hitachi has realized a 1×8 DOS using this material with an average insertion loss of 6.5 dB and a worst-case isolation of -18 dB [22]. The relatively high polarization-dependent loss (PDL) in the *on* channel of maximal 0.5 dB might be attributed to the high birefringence in polyimide waveguides.

The most advanced polymeric DOS switch matrix is the 8×8 matrix realized by Akzo Nobel in 1996 [23]. It had a strictly nonblocking architecture, based on the recursive tree structure, comprising 112 DOS's integrated on a 4-in. wafer. The component had an average fiber-to-fiber insertion loss of 10.7 dB with an average isolation of -30 dB. Akzo Nobel presented the first 1×3 polymeric t.o. DOS in 1997 [24]. This basic switch structure can yield bigger matrices for a given wafer size. Its design was based on the concept of *effective temperature*, which is an optimization of the position of the heaters with respect to the induced effective refractive index differences in the output branches. Isolation values of −20 dB for all the switching states had been achieved at switch powers of 75 mW for switching to the outer branches and 370 mW for switching to the central branch.

2.2 Thermo-Optic Waveguide Materials

Thermo-optic control of optical waveguide devices is attractive from the viewpoint of simplicity and flexibility. The t.o. effect is present in all practically used waveguide materials [25]. This allows materials with good waveguide characteristics (low loss and cost) to be selected for t.o. component realization. The most important examples of such materials are polymers and fused silica.

Thermo-optic space switches in these materials are commercially available [26]. The polymer devices are marketed under the trade name BeamBox™ by Akzo Nobel Photonics. The fused silica devices are derived from the well-established silica-on-silicon technology for passive waveguide components [27], whereas the polymer components are developed primarily for t.o. space-switching applications utilizing the high t.o. coefficient of polymers.

In both cases the refractive index of the waveguide materials can be selected freely from a range that is wide enough to make optical fiber–compatible channel waveguides. These waveguides have refractive index differences between waveguide core and cladding of 0.25–0.75% at core sizes of 8 × 8 to 6 × 6 μm, respectively, resulting in coupling losses with a standard single mode telecom optical fiber that are smaller than 0.05 dB to 0.5 dB, respectively. Channel waveguides in both the polymer and the silica-on-silicon technology are made by first depositing a bottom cladding layer, followed by the deposition of a core layer. Reactive ion etching is used to etch the core ridge out of the core layer. This is followed by an overcoating with the cladding layer, resulting in an embedded channel waveguide.

Propagation losses in silica-on-silicon channel waveguides can be less than 0.02 dB/cm in the low-loss windows of the optical fiber, whereas for polymer channel waveguides losses can be less than 0.1 dB/cm at 1300 nm [28] and 0.5 dB/cm at 1500 nm [29,30]. The polymers in the last two cases have been optimized with respect to absorption losses by reducing the C—H bond content. This is done by replacing hydrogen for heavier atoms, which shifts the absorption peaks due to C—H bond overtones in the optical fiber windows to longer wavelengths. The film waveguide losses in BeamBox™ materials are below 0.1 dB/cm for both windows, indicating that the channel waveguide losses can still be improved by reducing scattering from wall roughness through an optimization of the RIE process.

In the commercially available t.o. components, silicon wafers are used as sub-

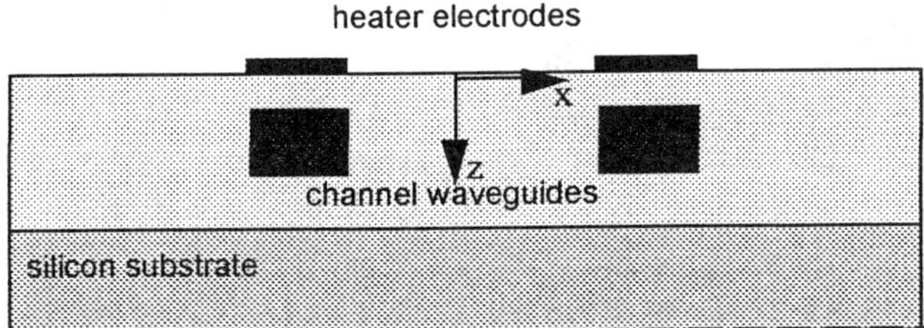

Figure 1 Cross section of a thermo-optic waveguide device.

strates because of their compatibility with standard IC process equipment, their good surface quality, and their excellent heat-conducting property. The last property is important for t.o. components become it provides them with a good heat spreader. Figure 1 shows the cross section of a t.o. waveguide device.

Switching between the channels can be induced by creating a difference in their effective refractive indices. This can be done by activating one of the resistive stripe heater electrodes that are deposited above the waveguides, on top of the cladding layer. If the silicon substrate acts as a perfect heat sink, the rise in temperature of the heater electrode is directly proportional to the dissipated electrical power density and inversely proportional to the thermal conductivity, λ, of the waveguide material. The thermal conductivity of glassy polymers is typically $\lambda = 0.2$ W/m·K, while for fused silica $\lambda = 1.4$ W/m·K [31]. Therefore, to induce the same difference in the temperatures of the channel waveguides, it takes for a silica-on-silicon waveguide seven times more power than for a polymer waveguide on silicon. With the substrate acting as a perfect heat sink, the temperature distribution in the waveguide layer for a given heater stripe temperature is independent of its thermal conductivity. Since the t.o. coefficient for silica is an order of magnitude lower than for polymers (see Sec. 4), it can be concluded that takes about two orders of magnitude more power to induce thermo-optically in silica-on-silicon waveguides the same difference in (effective) refractive index as in polymer waveguides.

2.3 Thermo-Optic Switch Principles

There are two different waveguide switch principles that can be used in practical t.o. devices:

The Mach–Zehnder interferometer (MZI) switch principle
The mode-evolution, or digital optical switch (DOS), principle.

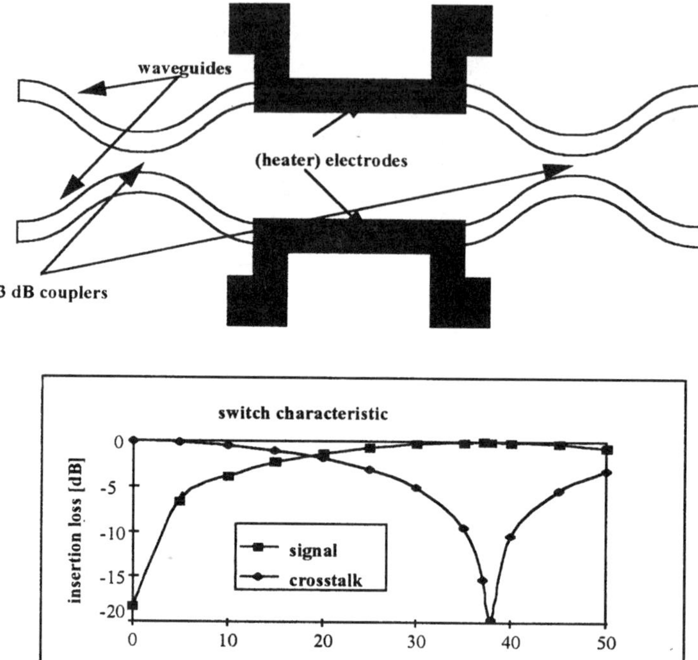

Figure 2 2 × 2 MZI space switch with its switch characteristic.

In Fig. 2 a 2 × 2 MZI space switch is plotted, together with its typical switch characteristic. The interferometric principle of the device results in the recursive character of the switch characteristic, with high extinction peaks only at well-defined heater powers. For the same reason, the switch is wavelength sensitive. The main advantage of the MZI switch over a DOS is the lower power required for switching. For a fiber-compatible polymeric MZI, a switch power as low as 5 mW has been demonstrated [32]. The silica-on-silicon devices typically require 0.5 W, which is two orders of magnitude more power.

In Fig. 3 a 1 × 2 DOS is plotted with a typical switch characteristic. The operation of this switch is based on the adiabatic evolution of the (system) mode of the two waveguides in the branching section to the (fundamental) mode of the output channel. The output channel will be the channel with the highest effective refractive index. For a polymer DOS this is the unheated branch, because the t.o. coefficient in polymers is negative.

The typical "digital" behavior of the switch characteristic can clearly be seen in Fig. 3. When the switch is powered further than required, the light remains

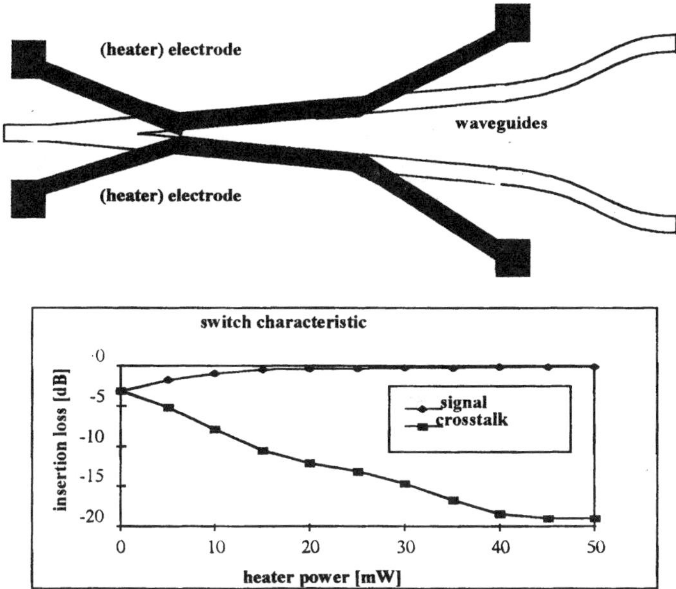

Figure 3 1 × 2 DOS space switch with its switch characteristic.

in the intended output channel with high extinction. The advantage of the DOS over the MZI switch is its insensitivity for drive power fluctuations and, associated with that, for temporal refractive index changes caused, for example, by water desorption or material relaxation phenomena. In addition to that, the DOS principle results in wavelength and polarization insensitivity. These features make the DOS highly compatible with optical fiber systems, where operation in both windows is often required and the polarization is random. A disadvantage of the DOS as compared with the MZI is the higher power required for switching. This is typically an order of magnitude higher. The switch power for polymeric DOS (Fig. 3) is 50–100 mW. A silica-on-silicon DOS would require 5 W, which is extremely high and excludes the practical use of the DOS principle in t.o. silica-on-silicon space switches.

It can be concluded that the DOS principle has very attractive features for switch applications in optical communication systems; however, it requires strong effects, such as the t.o. effect in polymers.

2.4 Thermo-Optic Effect

The change in refractive index, n, of a material with temperature, T, is due to the change in density, ρ, and due to the temperature change itself. Therefore the t.o. coefficient dn/dT can be written as

$$\frac{dn}{dT} = \left(\frac{\delta n}{\delta \rho}\right)_T \left(\frac{\delta \rho}{\delta T}\right) + \left(\frac{\delta n}{\delta T}\right)_\rho \tag{1}$$

or

$$\frac{dn}{dT} = -\left(\frac{\rho \delta n}{\delta \rho}\right)_T \gamma + \left(\frac{\delta n}{\delta T}\right)_\rho \tag{2}$$

where γ is the coefficient of volume expansion of the material.

From the Lorentz–Lorenz (LL) equation, the following expression for $(\rho\delta n/\delta\rho)_T$ can be derived:

$$\left(\frac{\rho \delta n}{\delta \rho}\right)_T = (1 - \Lambda_0)\frac{(n^2 + 2)(n^2 - 1)}{6n} \tag{3}$$

where Λ_0 is the strain polarizability constant that has been introduced by Mueller [33] to take into account the effect of density changes on the atomic polarizability of the material. For polymers, Λ_0 is small compared to unity as a consequence of the weak interaction between the molecular units. Values of $\Lambda_0 = 0.18$ for polycarbonate and $\Lambda_0 = 0.15$ for polymethylmethacrylate have been determined [34]. As the refractive index of most polymers is approximately 1.5, it can be seen from Eq. (3) that

$$\left(\frac{\rho \delta n}{\delta \rho}\right)_T \sim 0.5 \quad \text{for glassy polymers} \tag{4}$$

Polymers have a relatively large coefficient of expansion. For most polymers in the glassy state, $\gamma \sim 2 \times 10^{-4}/°C$ [31]. Therefore the first term in Eq. (2) becomes:

$$-\left(\frac{\rho \delta n}{\delta \rho}\right)_T \gamma \sim -10^{-4}/°C \quad \text{for glassy polymers} \tag{5}$$

The thermal change of the refractive index at constant density, the second term in Eq. (2), is small. For polycarbonate $(\delta n/\delta T)_\rho \sim 9 \times 10^{-6}/°C$, and for polymethylmethacrylate $(\delta n/\delta T)_\rho \sim -4 \times 10^{-6}/°C$ [34]. Therefore it can be concluded from Eq. (2) that the t.o. coefficient in polymers has a large negative value, because it is determined predominantly by density changes caused by the strong thermal expansion in these materials. The value of

$$\frac{dn}{dT} \sim -\left(\frac{\rho\delta n}{\delta\rho}\right)_T \gamma \sim -10^{-4}/^\circ\mathrm{C} \quad \text{for glassy polymers} \tag{6}$$

For fused silica it was found that $\Lambda_0 = 0.4$ [35]. The refractive index for fused silica is approximately 1.5. From Eq. (3) it follows that

$$\left(\frac{\rho\delta n}{\delta\rho}\right)_T \sim 0.3 \quad \text{for fused silica} \tag{7}$$

The thermal expansion coefficient in this material is low: $\gamma \sim 10^{-6}/^\circ\mathrm{C}$. Therefore the first term in Eq. (2) becomes:

$$-\left(\frac{\rho\delta n}{\delta\rho}\right)_T \gamma \sim -0.3 \times 10^{-6}/^\circ\mathrm{C} \quad \text{for fused silica} \tag{8}$$

In fused silica, the t.o. effect is due mainly to the second term in Eq. (2), which originates from the thermal changes in the polarizability. Its value is

$$\frac{dn}{dT} \sim -\left(\frac{\delta n}{\delta T}\right)_\rho \sim 10^{-5}/^\circ\mathrm{C} \quad \text{for fused silica} \tag{9}$$

If this result is compared to that for glassy polymers (Eq. 6), it can be concluded that the t.o. coefficient in fused silica is, in absolute value, an order of magnitude smaller than in polymers. Its sign is positive, opposite to that of the coefficient in polymers.

2.5 Thermo-Optic Strain Effects

Because of the fact that the t.o. effect in polymers is due to thermal expansion alone, constrained thermal expansion will reduce the effect. If a planar polymer waveguide (thin film) on a silicon substrate is heated, the substrate will prevent the in-plane expansion of the polymer film due to its much larger stiffness and lower thermal expansion. The expansion will only be out of plane, leading to a reduction of the (apparent) t.o. coefficient. In a t.o. channel waveguide switch, the t.o. effect will be induced locally in a channel waveguide. In that case, mainly the nonheated polymer environment will restrict the lateral expansion of the heated region. To determine the magnitude of these effects, the stress/strain distributions have to be calculated.

According to Timoshenko [36], Hooke's law in the presence of thermal strains is as follows:

$$\begin{aligned} \varepsilon_x &= \frac{1}{E}\{\sigma_x - \nu(\sigma_y + \sigma_z)\} + \frac{\gamma}{3}\Delta T \\ \varepsilon_y &= \frac{1}{E}\{\sigma_y - \nu(\sigma_x + \sigma_z)\} + \frac{\gamma}{3}\Delta T \\ \varepsilon_z &= \frac{1}{E}\{\sigma_z - \nu(\sigma_y + \sigma_x)\} + \frac{\gamma}{3}\Delta T \end{aligned} \tag{10}$$

where the ε's are the strains, the σ's are the stresses, E is Young's modulus, ν is the Poisson ratio of the material, and ΔT is the temperature difference.

In a planar (xy) geometry (film), in-plane expansion is not possible while the out-of-plane (z) stress is zero due to the free expansion in the out-of-plane direction:

$$\begin{aligned} \varepsilon_x &= \varepsilon_y = 0 \\ \sigma_z &= 0 \end{aligned} \tag{11}$$

Combining Eqs. (10) and (11), we obtain relations for stress/strain for a planar configuration (film):

$$\begin{aligned} \varepsilon_z &= \left(\frac{1+\nu}{1-\nu}\right)\frac{\gamma}{3}\Delta \mathrm{T} \\ \varepsilon_x &= \varepsilon_y = 0 \end{aligned} \tag{12}$$

and

$$\begin{aligned} \sigma_z &= 0 \\ \sigma_x &= \sigma_y = -\frac{\gamma}{3}\Delta \mathrm{T}\,\frac{E}{1-\nu} \end{aligned} \tag{13}$$

For small strains, the following relation between the volume change and the strain can be derived:

$$\frac{\Delta V}{V} = \varepsilon_x + \varepsilon_y + \varepsilon_z \tag{14}$$

From Eqs. (12) and (14) it follows that the apparent thermal expansion coefficient for a planar configuration γ^* is:

$$\gamma^* = \frac{\Delta V}{V\Delta T} = \frac{1}{3}\left(\frac{1+\nu}{1-\nu}\right)\gamma \tag{15}$$

Table 1 Material Parameters of Waveguide and Substrate

Material	Polymer	Silicon
Thickness	t = 26.5 μm: waveguide film	t = 525 μm: substrate
Young's modulus	E = 1.7 GPa	E = 131 GPa
Poisson ratio	ν = 0.43	ν = 0.28
Cubic expansion coefficient	$\gamma = 2.04 \times 10^{-4}/°C$	$\gamma = 8.1 \times 10^{-6}/°C$
Thermal conductivity	λ = 0.2 W/m·K	λ = 84 W/m·K

Due to the relatively high Poisson ratio of glassy polymers ν ~ 0.4 (for fused silica ν = 0.16), the reduction factor will be only 0.8.

Note that in elastomeric films (ν ~ 0.5) this reduction effect is not present. In addition, the t.o. effect for polymers in the rubbery state is three to five times larger than for polymers in the glassy state because the thermal expansion is increased by this factor.

The refractive index changes in the channel waveguides of t.o. waveguide switches is due to local heating of the polymer film by means of heater stripes that are deposited on top of the channel waveguides. In that case there are thermal gradients in both the lateral and the transverse directions, and the magnitude of thermal stresses and strains can be obtained only by computer calculations using, for instance, finite element (FE) analysis. The constraining will be less than in the case of uniform heated film on a nonexpanding substrate, because there is some in-plane lateral expansion of the heated polymer stripe possible if the width of the stripe is small compared to the film thickness. Finite element calculation of the temperature and the stress/strain distribution in a typical t.o. fiber–compatible polymer channel waveguide configuration have been performed. Table 1 lists the data for this configuration. Figure 1 shows the cross section. The heater is 6 μm wide, infinitely long in the y-direction, and positioned directly above the channel waveguide. The channel waveguide is centered in the middle of the polymer layer stack.

Table 2 Thermal Stresses in Various Configurations with Polymeric Materials

Stress per degree temp. increase	Bulk	Plane	Stripe
σ_x/T	0	0.2 MPa/°C	0.08 MPa/°C
σ_y/T	0	0.2 MPa/°C	0.15 MPa/°C
σ_z/T	0	0	0.01 MPa/°C

Table 3 Thermal Strains in Various Configurations with Polymeric Materials

Strain per degree temp. increase	Bulk linear/cub. exp. coeff.	Plane	Stripe
ε_x/T	0.68×10^{-4}/°C	0	0.56×10^{-4}/°C
ε_y/T	0.68×10^{-4}/°C	0	0
ε_z/T	0.68×10^{-4}/°C	1.70×10^{-4}/°C	1.30×10^{-4}/°C
γ^*/T	2.04×10^{-4}/°C	1.70×10^{-4}/°C	1.86×10^{-4}/°C
Thermal expansion reduction factor			
γ^*/γ	1	0.83	0.91

The result of the calculation for the center of the channel waveguide is listed in Tables 2 and 3 together with the results for an unconstrained (bulk) sample and the planar configuration using Eqs. (12), (13), and (15). As expected, it can be seen that for the stripe geometry the reduction factor for the cubic thermal expansion and t.o. coefficient (0.91) lies in between the value for the unconstrained expansion (1) and the planar configuration (0.83).

The FE calculation predicts a raise of the polymer surface under the heater

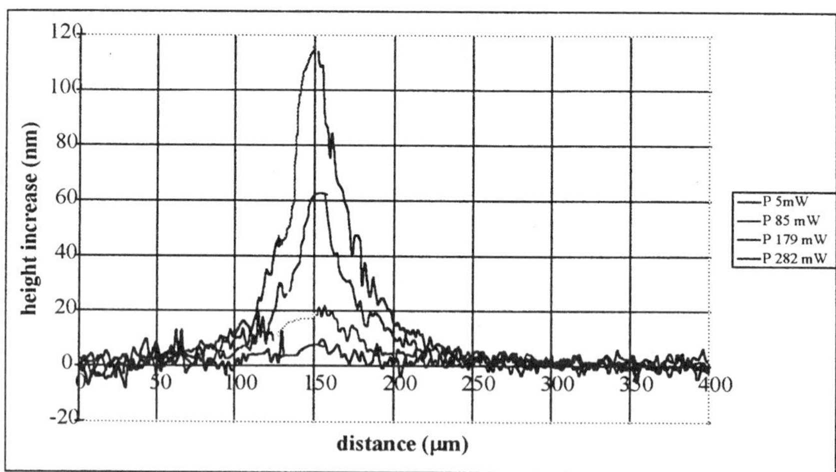

Figure 4 Height profile of the polymer surface around a heater stripe at various dissipated powers P. Heater stripe position is at 150 µm; its width is 6 µm, its area is 0.078 mm^2.

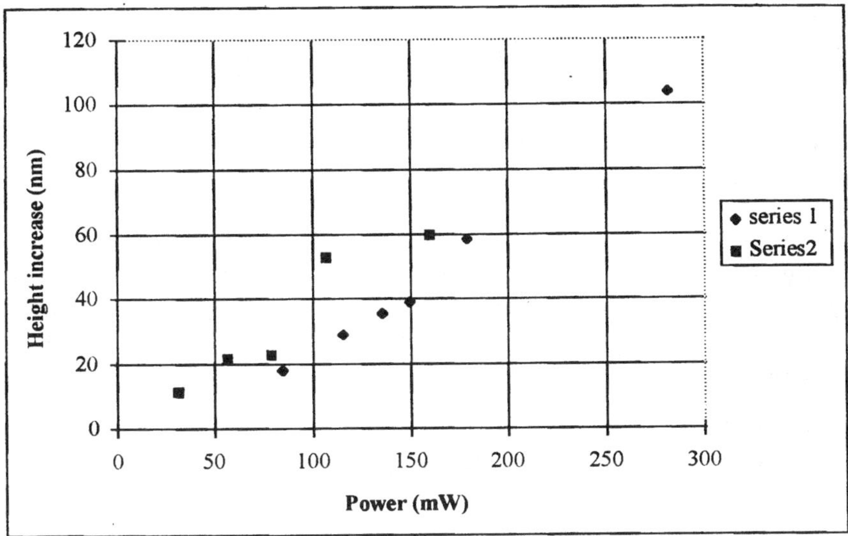

Figure 5 Height increase of the surface of a polymer layer under a heater stripe as a function of dissipated power. Heater stripe width is 6 μm; its area is 0.078 mm^2.

stripe by 0.035 μm per W/mm^2 dissipated power of the heater stripe. This has been verified experimentally in a practical sample realized with the data of Table 1, by observing the raise through an interference microscope. The height increase of the polymer surface around the heater stripe at various power dissipations can be seen in Fig. 4. In Fig. 5 the expansion directly under the heater stripe as a function of the heater power is plotted. From this plot a height increase of 0.33 μm/W can be determined. With a heater area of 0.078 mm^2, this results in a measured increase of 0.026 μm per W/mm^2, which is in good agreement with the calculated value.

2.6 Thermo-Optic and Thermal Stress Effects

From Table 2 it can be seen that the (compressive) t.o. induced stresses are of the order of 0.1 MPa/°C. A typical value for the refractive index change required for switching in a DOS is 0.001. From Eq. (6) it can be seen that this will be induced by a (local) temperature increase of 10°C. Consequently, the thermal stresses during t.o. switching are of the order of 1 MPa.

The result for the planar configuration in Table 2 shows that temperature excursions will induce thermal stresses of the order of 0.2 MPa/°C in the polymer layer of a waveguide chip. In some telecommunication system applications, temperature excursions ranging from −40°C to +85°C can be expected.

Assuming stress-free conditions at room temperature, the maximum temperature swing of 65°C will induce a stress of 13 MPa in the polymer layer. Because all glassy polymers have comparable t.o. coefficients, the choice of polymer for the t.o. waveguide components can be based on its tensile strength too, in order to guarantee the reliability of the components. The typical tensile strengths at yield for engineering polymers is 50 MPa [31]. This is well above the induced stress of 13 MPa. As a breakpoint test, BeamBox™ t.o. chips have been repeatedly (up to 40 times) removed from a hotplate at 130°C and submerged in liquid nitrogen (−196°C). This will induce tensile stresses of more than 40 MPa in the polymer layer. No cracks could be found in the polymer layer after this extreme test. This clearly demonstrates that the mechanical robustness of the polymers for t.o. switches can be excellent.

2.7 Repetitive Switching

Solid-state optical switches are switches that don't contain moving parts that suffer from wear and tear. Polymeric t.o. waveguide switches are ranged under this class of switches. From the foregoing it can be concluded that, strictly speaking, this is not true, since the polymer under a powered heater stripe expands. To see if this leads to wear effects, t.o. 1 × 2 switches have been switched up to 10 million times at a frequency of 5 Hz. In Fig. 6 the switch insertion loss has been plotted after different numbers of switch cycles. It can be seen that there is no change, thus demonstrating the absence of wear effects.

2.8 Thermo-Optic Stress-Birefringence

The occurrence of a nonuniform stress distribution in a polymer channel waveguide during t.o. switching will induce birefringence, because polymers have a nonzero-stress optical coefficient C. This coefficient relates the difference in refractive index parallel ($n_{\parallel}$) and perpendicular ($n_{\perp}$) to the stress.

$$n_{\parallel} - n_{\perp} = C\sigma \tag{16}$$

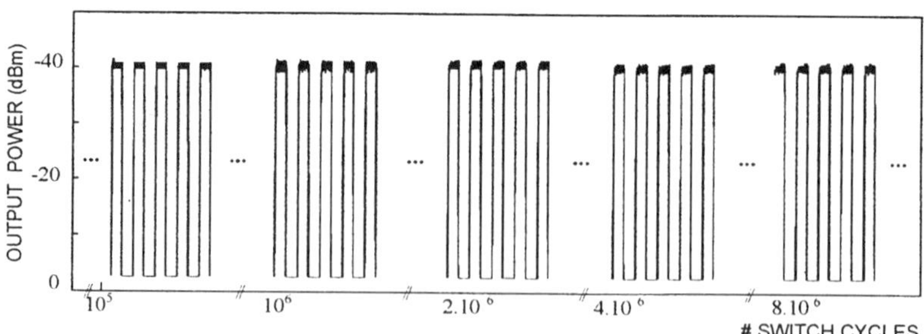

Figure 6 Optical output of 1 × 2 switch switched at 5Hz vs. the number of switch cycles.

The coefficient C for several polymers can be found in Ref. 31. For glassy polymers that don't contain benzene rings in the main chain, this coefficient is low. As an example, for polymethylmethacrylate $C \sim +4 \times 10^{-12}$ Pa^{-1}. For polymers with benzene rings in their main chain, the stress optical coefficient is much higher. For instance, for polycarbonate: $C \sim -111 \times 10^{-12}$ Pa^{-1}.

The normalized polarization dependence, PD, defined as the ratio of the stress birefringence temperature coefficient to the t.o. coefficient, is a measure of the birefringence induced by the t.o. effect in a constrained expansion configuration:

$$\mathrm{PD} = \frac{d(n_{\parallel} - n_{\perp})/dT}{dn/dT} = \frac{Cd\sigma/dT}{dn/dT} \tag{17}$$

For the typical t.o. polymer channel waveguide of above, the result is

$$\mathrm{PD} = \frac{C(\sigma_z - \sigma_x)/T}{dn/dT} = \frac{-0.07 \times 10^6 C}{-10^{-4}} = 0.7 \times 10^9 \mathrm{C} \quad (\mathrm{C\ in\ Pa^{-1}}) \tag{18}$$

This results in:

PD $\sim +0.3\%$ for a polymethylmethacrylate t.o. channelwaveguide

PD $\sim -8\%$ for a polycarbonate t.o. channel waveguide

In calculating these values, the results of the FE stress calculation of Table 2 have been used for both materials. Although there are some differences in the material parameters, the order of magnitude of the PDs is correct.

Figure 7 shows the best-case and worst-case insertion loss with rotated polarization of the *off* and *on* channels of a BeamBox™ (digital optical) 1 × 2 switch. Its (absolute value of) PD is about 10%.

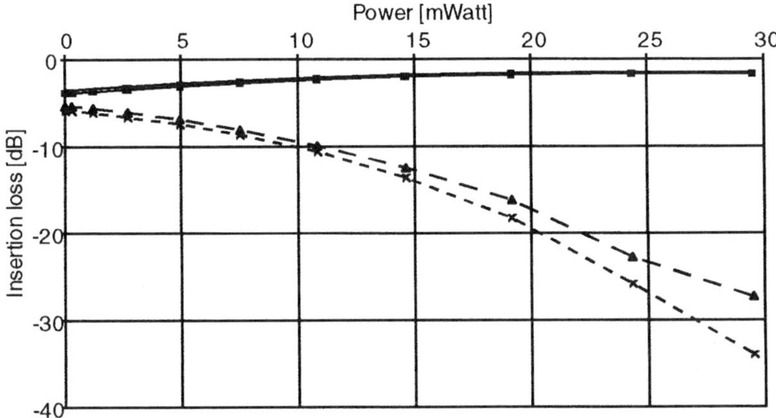

Figure 7 Best-case and worst-case insertion loss of a BeamBox™ (digital optical) 1 × 2 switch.

Inoue et al. [37] have calculated and measured the PD for a fused-silica fiber-compatible t.o. channel waveguide. They found:

PD ~ 3% for a fused silica t.o. channel waveguide

The PD of a t.o. switch gives the shift, as fraction of the heater power of the out-of-plane polarization (TM) with respect to in-plane polarization (TE), of the switching characteristic (heater power vs. extinction) of the switch. In the fused-silica MZI switches this results in a reduction of the switch isolation when randomly polarized light is used. In a (polymeric) DOS, the effect is clearly visible only when the switch is operated below the power required to reach its digital regime, as is demonstrated in Fig. 7.

2.9 Switching Time

The switching time that can be expected in t.o. switches can be estimated from the temperature rise in time, $\Delta T(t)$, of a waveguide layer stack (thickness L) due to heating with a constant flux, F_0, by a plane heater that is deposited onto the waveguide stack. It is assumed that there is no heat flow to the medium above the heater, which is a good assumption in the case when this medium is air at rest, due to its low thermal conductivity. In addition, the substrate is considered to be a perfect heat sink. This a good assumption for a silicon wafer, due to its high thermal conductivity. In this case the analytical expression for the temperature rise is given in Ref. 38, p. 113, Eq. 5:

$$\frac{\Delta T(t)}{\Delta T_0(\infty)} = \frac{2\sqrt{\kappa t}}{L} \sum_{n=0}^{\infty} (-1)^n \left\{ \mathrm{ierf}\left[\frac{(2n+1)\cdot L + z - L}{2\sqrt{\kappa t}}\right] - \mathrm{ierf}\left[\frac{(2n+1)\cdot L - z + L}{2\sqrt{\kappa t}}\right]\right\} \tag{19}$$

where

$\mathrm{ierf}(x) = -x + x \cdot \mathrm{erf}(x) + 1/\sqrt{\pi}\exp(-x^2)$ and $\mathrm{erf}(x)$ is the error function

$z = 0$ is the interface between the waveguide stack and the heater (surface)

$z = L$ is the interface between the layer stack and the substrate

$\kappa = \lambda/\rho c_p$ is the thermal diffusivity of the waveguide stack material and c_p its specific heat capacity

The temperature rise is normalized to the steady-state temperature rise of the surface, ΔT_0 (∞):

$$\Delta T_0(\infty) = \frac{F_0 L}{\lambda} \tag{20}$$

The analytical expression in the case where the layer stack under the heater extends to infinity (semi-infinite layer) is given in Ref. 38, p. 75, Eq. 7:

$$\frac{\Delta T(t)}{\Delta T_0(\infty)} = \frac{2}{L}\left\{\sqrt{\frac{\kappa t}{\pi}}\exp\left(-\frac{z^2}{4\kappa t}\right) - \frac{z^2}{2}\left[1 - \operatorname{erf}\left(\frac{z}{2\sqrt{\kappa t}}\right)\right]\right\} \tag{21}$$

In order to compare the results of Eqs. (19) and (21), Eq. (21) is just like Eq. (19) normalized to the steady-state temperature of the heat-sinked layer (Eq. 20).

Figure 8 shows the temperature increments as a function of the dimensionless time parameter, $\kappa t/L^2$, as determined with Eqs. (19) and (21). The results are calculated for $z = 0.5L$, which is typically the position of the waveguide core buried in a layer stack on a substrate. The initial rise for the (semi-) infinite layer (dotted line) and the layer stack on the heat sink (solid line) are identical. The temperature of the (semi-) infinite layer increases continuously. For the configuration with the heat sink, it levels off to half the temperature increase of the

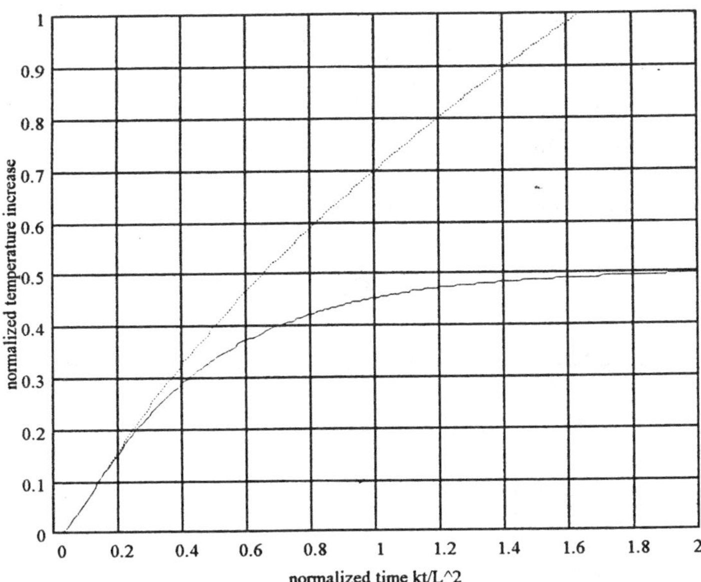

Figure 8 Temperature rise due to a plane heater on a heat-sinked layer of thickness L(solid) and a semi-infinite layer (dashed). The temperatures are calculated at a depth of 0.5 L from the heater (surface) and are normalized to the steady-state temperature of the heat-sinked layer.

surface, as could be expected from the linear steady-state temperature profile. This clearly demonstrates the stabilizing effect of a heat-sinking substrate (silicon).

From Fig. 8 it can be seen further that in the middle of the layer stack the rise time, τ, defined as the time to reach $(1 - 1/e)$ of the steady-state temperature, can be determined from:

$$\frac{\kappa\tau}{L^2} = 0.47 \tag{22}$$

With a typical value for the thickness of a fiber-compatible layer stack of $L = 26.5\ \mu m$ and with $\kappa = 13.5 \times 10^{-8}\ m^2/s$, which is a typical value for polymers [31, p. 400], a rise time $\tau = 2.4$ ms can be calculated. For silica with $\kappa = 75.4 \times 10^{-8}\ m^2/s$ [39], the rise time would be six times smaller: $\tau = 0.4$ ms.

From Fig. 8 it can be concluded that for $t < \tau$, the temperature rise in the middle of heat-sinked layer stack due to a plane heater can be approximated reasonably by that at the depth of a semi-infinite layer. In practical components, the temperature increase is induced by stripe heaters of finite width. This width is typically 1–4 times the waveguide channel width. Consequently, for fiber-compatible channels, heater widths of 6–24 μm are typical. It can be expected that the behavior of the initial temperature increase induced by these heaters is comparable to that of plane heaters; i.e. the initial temperature increase in the middle of a layer stack is independent of the presence of a heat-sinking substrate, and the rise time for the heat-sinked layer stack can be derived from the solution of the heat equation for a semi-infinite layer. This temperature rise (by a stripe heater on a semi-infinite layer) can be expressed as the integration over the heater width, w, of the solution for line sources [38, p. 261]:

$$\Delta T(t) = \frac{-f_0}{2\pi\lambda w} \cdot \int_{-w/2}^{+w/2} \mathrm{Ei}\left[-\frac{z^2 + (x - \phi)^2}{4\kappa t}\right] d\phi \tag{23}$$

where f_0 is the dissipated power per unit length and $\mathrm{Ei}(x)$ is the exponential integral function.

The solution for the single line source is given by:

$$\Delta T(t) = \frac{-f_0}{2\pi\lambda} \cdot \mathrm{Ei}\left(-\frac{z^2 + x^2}{4\kappa t}\right) \tag{24}$$

For small values of the argument in the Ei function, the following expression is a good approximation for Eq. (24).

$$\Delta T(t) = \frac{-f_0}{2\pi\lambda}\left[\ln\left(\frac{z^2 + x^2}{4\kappa t}\right) + 0.5772 - \left(\frac{z^2 + x^2}{4\kappa t}\right)\right] \tag{25}$$

Figure 9 shows the temperature increments (arbitrary units), as calculated with of Eqs. (23)–(25) for the same heater dissipation per unit length, plotted as a function of $\kappa t/L^2$ at the position $z = 0.5L$, $x = 0$; this is perpendicular under the heater. The width of the heater used in Eq. (23) was $0.266L$, corresponding with 6 μm when $L = 26.5$ μm. The results for the stripe heater (Eq. 23) and the line heater (Eq. 24) coincide completely (solid line). The approximate expression for the line source, Eq. (25) (dotted line), is good for $\kappa t/L^2 > 0.1$. Even for a stripe heater with $w = L$, the difference from the line source results is no more than 20%.

In conclusion, the approximate line source expression for a semi-infinite layer (Eq. 25) can be used for the determination of the thermal rise time for a waveguide channel in the middle of a heat-sinked layer stack of a t.o. switch.

The operation of a t.o. 1 × 2 DOS is based on the effective index difference and, consequently, the temperature difference between the two output branches in the *y*-junction. From the onset of the junction, the separation between the channels increases very gradually from zero. The branch angle is typically 0.1°. In fiber-compatible components, optical power transfer to the *on* channel occurs

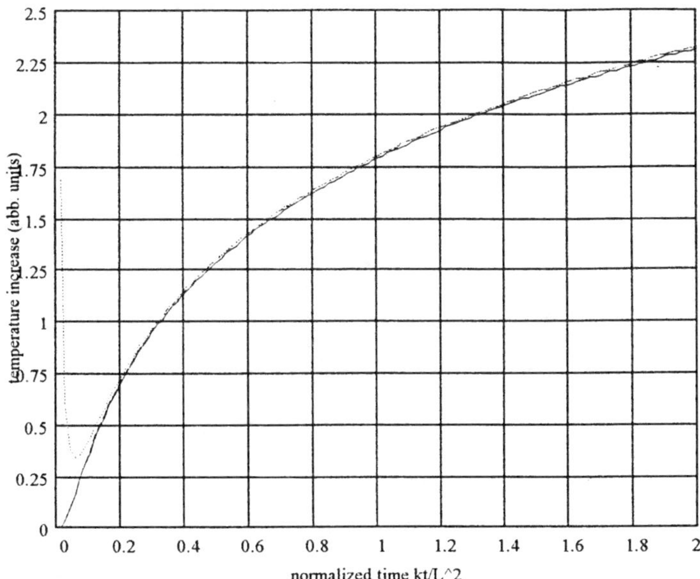

Figure 9 Temperature rise due to a stripe heater of width $0.266L$ (solid line) and a line heater (solid line), both on a semi-infinite layer and at a depth of $0.5L$ and both dissipating the same power per unit length. The dotted line is an approximation for the line source.

as long as the separation is about 15 μm. In Fig. 10 the temperature increments due to the heater with $w = 0.266\ L$, as calculated with Eq. (26) (stripe heater) at the position $z = 0.5L$, $x1 = 0$ (channel 1, dashed) and at a lateral shifted position $z = 0.5L$, $x2 = 0.19L$ (5 μm, for $L = 26.5$ μm) (channel 2, dash-dot) can be seen, together with their difference $\Delta T1 - \Delta T2$ (solid).

Because the line source expressions would give identical results, the approximate line source expression (Eq. 25) can be used to derive a limiting value for $t \to \infty$ for the temperature difference:

$$\Delta T1(\infty) - \Delta T2(\infty) = \frac{-f_0}{2\pi\lambda} \cdot \ln\left(\frac{z^2 + x1^2}{z^2 + x2^2}\right) \tag{26}$$

The results in Fig. 10 are normalized to this limiting value using equal heater power dissipation/unit length for the line source (Eq. 26) and the stripe source (Eq. 23).

From Fig. 10 it can be concluded that the temperature difference reaches a steady-state value in spite of the continuous temperature increase in the two chan-

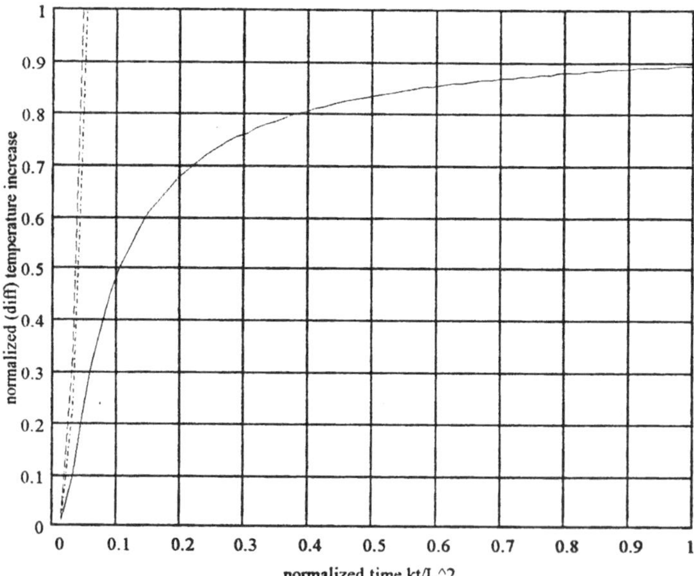

Figure 10 Temperature rise due to a heater with $w = 0.266L$ on a semi-infinite layer, as calculated with Eq. (26) (stripe heater) at a depth of $0.5L$, directly under the heater (channel 1, dashed) and at a $0.19L$, lateral shifted position (channel 2, dash-dot). The solid line is the difference. The results are normalized to the steady state of the difference.

nels. Since the distance of the shifted waveguide to the heater is smaller than the distance of the heater to the heat sink (the layer stack thickness), this stabilizing effect in the temperature difference between the channels will take place before the stabilizing effect of the heat sink is noticeable. This is an additional justification to apply the formulas for the semi-infinite layer for the calculations of initial temperature differences between waveguide channels in a heat-sinked layer stack of a t.o. DOS.

The rise time, τ, for the temperature difference in Fig. 10 can be determined from:

$$\frac{\kappa\tau}{L^2} = 0.16 \tag{27}$$

Using the approximate line source expressions Eqs. (25) and (26), an analytical expression for the rise time, τ, can be derived:

$$\frac{\Delta T1(t) - \Delta T2(t)}{\Delta T1(\infty) - \Delta T2(\infty)} \cong \frac{\ln\left(\dfrac{z^2 + x1^2}{z^2 + x2^2}\right) - \dfrac{x1^2 - x2^2}{4\kappa\tau}}{\ln\left(\dfrac{z^2 + x1^2}{z^2 + x2^2}\right)} = 1 - \frac{1}{e} \tag{28}$$

Placing channel 1 at the origin; $x1 = 0$, yields:

$$\tau \cong \frac{e.x2^2}{4\kappa \ln\left(1 + \dfrac{x2^2}{z^2}\right)} \tag{29}$$

Using in Eq. (32) the same parameters as for the results in Fig. 10; $x2 = 0.19L$; $z = 0.5L$, gives:

$$\frac{\kappa\tau}{L^2} = 0.18 \tag{30}$$

which is in good correspondence with Eq. (30).

A DOS is switched mostly from a (steady) *on* state of one of the outputs to the *on* state of the other output (switching over). In this case the switching time depends on the rise time of the heated channel and the fall time of the channel that is cooling down. The temperature change in the layer stack after switching off the heater at time $t0$, $\Delta T\text{off}(t)$, can be determined by applying Duhamel's theorem [38, p. 30]:

$$\begin{aligned} &\Delta T\text{off}(t) = \Delta T(t) - \Delta T(t - t0) \\ &t > t0 \end{aligned} \tag{31}$$

The temperature difference between channel 1 and channel 2, after switching at $t = t0$ from a heater on channel 2 to the heater on channel 1 (switching over), can then be written as:

$$\Delta T1(t) - \Delta T2(t) = \Delta T2(t) - \Delta T1(t) + 2[\Delta T1(t) - t0) - \Delta T2(t - t0)] \quad t > t0 \tag{32}$$

This formula includes the differential heating and cooling effects of the heaters on channels 1 and 2.

In Fig. 11 this difference is plotted (solid line) using the same parameters as for the results in Fig. 10. The switching action starts with power dissipation in heater 2 for a normalized time of 1.5, which will result in a steady-state temperature difference between the channels. For comparison, the behavior for switching from the passive state (switching on), as was already shown in Fig. 10, is included (dashed line). The dotted line is the result of calculating Eq. (32) with the approximate Eq. (25).

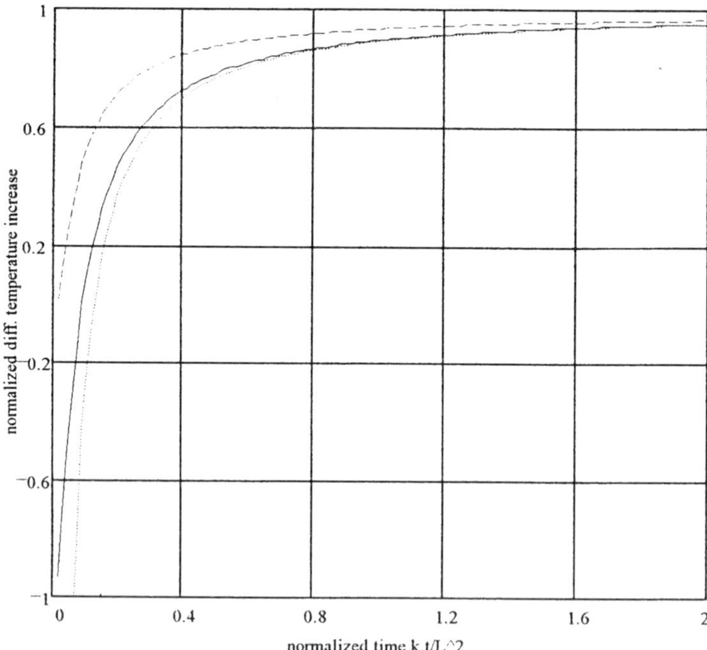

Figure 11 Normalized temperature difference for switching on (dashed) and switching over (solid) with parameters as in Fig. 10. The dotted line is the approximate result for switching over.

Figure 11 demonstrates that switching on is faster than switching over, in spite of the steeper gradient in time of the latter switching action. The rise time, τ, for switching over is at

$$\frac{\kappa\tau}{L^2} = 0.31 \tag{33}$$

which is nearly a factor of 2 lower than for switching on. The approximation yields a comparable result:

$$\frac{\kappa\tau}{L^2} = 0.35 \tag{34}$$

From the approximations Eqs. (27) and (28), the analytical expression for the rise time, τ, can be derived from:

$$\frac{\Delta T1(t) - \Delta T2(t)}{\Delta T1(\infty) - \Delta T2(\infty)} \cong \frac{\ln\left(\dfrac{z^2 + x2^2}{z^2 + x1^2}\right) + \dfrac{x2^2 - x1^2}{4\kappa}\left(\dfrac{1}{\tau} - \dfrac{2}{\tau - t0}\right)}{\ln\left(\dfrac{z^2 + x1^2}{z^2 + x2^2}\right)} = 1 - \frac{1}{e} \tag{35}$$

This yields:

$$\frac{1}{\tau} - \frac{2}{\tau - t0} = \frac{4\kappa}{e\left(x2^2 - x1^2\right)} \ln\left(\frac{z^2 + x1^2}{z^2 + x2^2}\right) \tag{36}$$

Since we are interested in $\tau - t0$, the rise time after the switch-over action, a redefinition is made: $\tau \rightarrow \tau - t0$. In most practical situations, the redefined τ is much smaller than t. Implementing these modifications and taking $x1 = 0$ results in:

$$\tau \cong 2 \frac{e.x2^2}{4\kappa \ln\left(1 + \dfrac{x2^2}{z^2}\right)} \tag{37}$$

This is twice as large as the approximate switching time for switching on (Eq. 29). Equation (37) can be simplified further, because for the greater part of the active y-junction in the switch $x2^2 \ll z^2$, leading to the following approximate expression for the rise time for switching over:

$$\tau \cong \frac{e.z^2}{2\kappa} \tag{38}$$

Due to the digital switching characteristic (output channel isolation vs. dissipated heater power) of a t.o. DOS, the component is switched once it reaches a certain specified isolation value. This is at a certain minimal dissipated power *P*dig, and the component will maintain or improve this isolation value when the heater power is increased further until the isolation drops under the specified level at very high dissipations. The region in which the switch is within specifications is called the *digital regime*. This is illustrated in Fig. 12 for a 1 × 8 polymeric t.o. DOS.

In practice a DOS is operated in the middle of this regime, at *P*op, to utilize its bandwidth to absorb changing switch parameters (like drive voltage, refractive index changes). Using this definition of switching time, Eq. (41) has to be modified to:

$$\tau \cong \frac{Pop \cdot z^2}{2\kappa(Pop - Pdig)} \tag{39}$$

Equation (39) is the approximate expression of the time for switching over in a t.o. DOS. Using typical values for a polymeric DOS: $\kappa = 13.5 \times 10^{-8}$ m²/s,

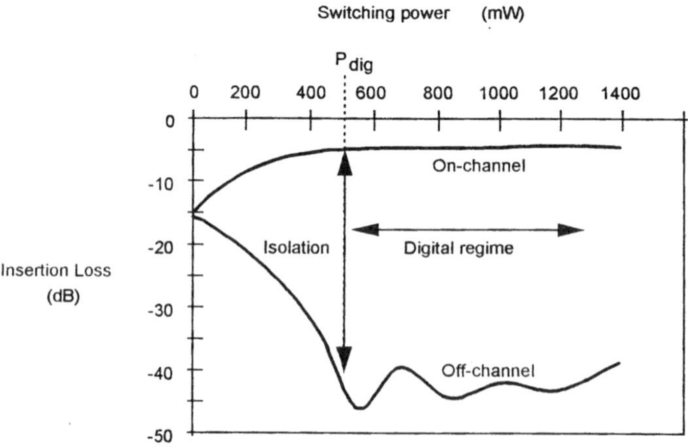

Figure 12 Switching characteristic of a 1 × 8 t.o. DOS showing the digital regime.

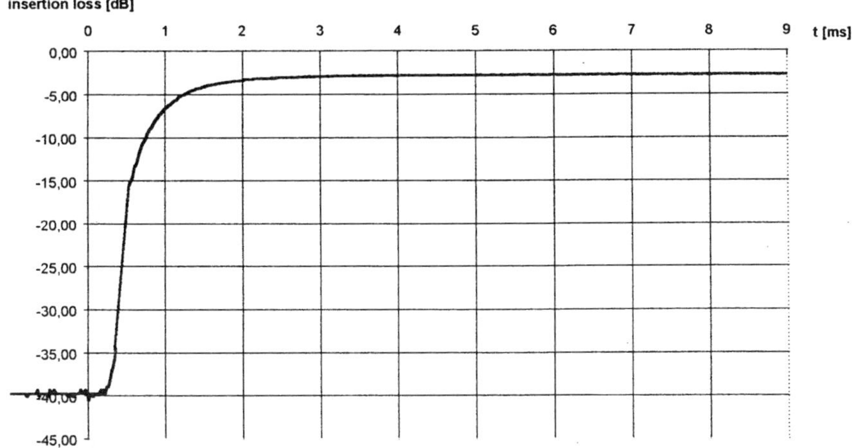

Figure 13 Insertion loss versus time for a channel in a fiber-compatible polymeric t.o. DOS that is switched at $t = 0$ from *off* to *on*.

$z = 13.25$ μm, and $P\text{op} = 1.5 \cdot P\text{dig}$, yields $\tau = 2$ ms. Figure 13 shows the insertion loss of one of the channels in such a polymeric DOS as a function of time. At $t = 0$, the channel is switched over from the *off* state to the *on* state. The switching time that can be determined from this figure is in good correspondence with the theoretical value.

Equation (39) yields a very simple expression for the switching time in a t.o. DOS. It shows that, apart from the material diffusivity, it depends exclusively on the distance between the heater and the core waveguide and the parameters *P*dig and *P*op of the switching characteristic. *P*dig should be minimized with respect to *P*op with an optimized switch design to improve the switching time. However, more effective is a reduction of the distance, z, of the waveguide core to the heater, due to the quadratic dependence. This can be achieved by applying a thinner top buffer layer with a lower refractive index. However, this will be at the expense of fiber-chip coupling efficiency if it causes a mismatch of the mode in the fiber and the chip.

The relatively small thermal diffusivity of polymers is not preferred for short switching times. However, it is a consequence of the small thermal conductivity in polymers, because the diffusivity is proportional to it. The advantage of the small thermal conductivity (low switch power) greatly compensates this disadvantage, because from Eq. (39) it can be concluded that a fiber-compatible polymeric DOS will have a switching time in the range 0.5–2 ms, which is acceptable for all routing applications in optical communication systems.

2.10 Switching Power

Equation (26) yields an approximation for the steady-state temperature difference between two close-by points in a polymer layer stack and can be used to estimate the power required for switching. Assuming that the heaters are positioned directly above the waveguide channels, the following expression for $f0$, the power dissipation per unit length that is required to get a certain temperature difference, can be derived from Eq. (26):

$$f0 = \frac{2\pi\lambda[\Delta T1(\infty) - \Delta T2(\infty)]}{\ln\left(\frac{1 + x2^2}{z^2}\right)} \tag{40}$$

From beam propagation method (BPM) analysis of the mode evolution behavior of the y-junction in fiber-compatible t.o. DOS's, it can be derived that the refractive index change in the core that is required to reach the digital regime is $\sim 10^{-3}$ [40]. Using a typical polymer t.o. coefficient of -10^{-4}/°C, it can be seen that this corresponds to a temperature difference in the cores of ~10°C. A more detailed analysis based on the effective temperature method [24], yields a temperature difference of ~5°C. From BPM analysis it can also be derived that most of the optical power transfer to the *on* channel takes place at a separation $x2$, in the range 10–15 μm. Taking 12.5 μm as the average value for the channel separation with 5°C as the average channel core temperature difference and combining these values in Eq. (40) with the typical values for a polymeric DOS—$\lambda = 0.2$ W/m · K, $z = 13.25$ μm—yields $f0 \sim 100$ mW/cm. Finite difference computer calculations of a real situation, i.e., the temperature difference in the cores of channels, separated by 12.5 μm and in the middle of a 26.5-μm-thick heat-sinked polymer layer stack, due to power dissipation of 100 mW/cm in a 6-μm-wide stripe heater, yields 5°C, which confirms the correctness of the approach that has been followed. Typical t.o. DOS heater electrode lengths range from 0.5 to 1.5 cm, corresponding to switching powers of 50—150 mW. These are indeed values that can be found in practical components.

Equation (40) shows that switching power reduction can be achieved in the same way as switching time reduction, i.e., by reducing the distance between the core and the heater.

2.11 Heater Temperature

From reliability considerations, it is important to know the maximum temperature to which the polymer (and the heater) in a t.o. switch is exposed. This is the temperature of the polymer surface that is in contact with the heater stripe. An expression for the temperature rise, ΔT, of the heater stripe on a heat-sinked layer can be found in Ref. 41:

$$\Delta T = \frac{f0}{\lambda\left(\dfrac{w}{L} + 0.88\right)} \tag{41}$$

This expression is valid for $w/L > 0.4$. For $L = 26.5\ \mu m$, the equation is valid for heater widths over 10 μm. A calculation with Eq. (45) using $f0 = 100$ mW/cm and a heater width of 6 μm yields a temperature increase of 45°C. A finite difference calculation gives 54°C. This is a considerable temperature rise. According to Eq. (45), it can be reduced most effectively by widening the heater stripe to, say, 20 μm. This reduces the temperature increase to 31°C, while it hardly affects the temperature distribution in the middle of the layer stack, as was shown in Sec. 2.9.

2.12 Conclusions

In the previous sections, the fundamentals of polymeric digital optical switches have been presented, and expressions for the t.o. effect, thermal stress effects, and t.o. stress and strain effects, switching time, switching power, and heater temperature have been derived. It can be conlcluded that polymers possess a unique combination of properties, i.e., a high t.o. coefficient and a low thermal conductivity. These are both required for the successful realization of t.o. digital optical switches. The expressions show how these switches can be optimized and what their fundamental limitations are. It has been shown that they can meet all functional requirements for routing switches in optical communication systems. In the following sections, it will be shown that they can also meet the reliability requirements for these applications.

3 RELIABILITY AND ENVIRONMENTAL STABILITY OF POLYMERIC THERMO-OPTIC DIGITAL OPTICAL SWITCHES

3.1 Introduction

It will be shown in Sec. 4 that the major applications for photonic space switches are in the field of telecommunications. Because of the availability requirements for telecommunication systems, all the components used in these systems must be very reliable [42,43]. Specific applications for photonic switches, such as optical cross-connects, make them the heart of telecommunications systems. Therefore the impact of switch reliability on system availability is considerable. Another application is protection switching, which improves photonic network availability by building redundancy into the network [44]. For these applications, reliable operation of the optical switches is vital for properly directing the optical data stream from one fiber to another in the case of a line failure.

3.2 Reliability Basics

Before we discuss the reliability of t.o. solid-state optical switches in more detail, we will define some important basic reliability concepts, such as *failure, failure mechanism, failure mode*, and *failure rate and median life* [45,46].

A *failure* of a device can be a sudden ending of its operation, or it can be a gradual degradation, where a critical parameter exceeds a certain limit value. Which critical parameters are taken into account, the respective range of the operation interval in which the parameters may vary, and under which environmental conditions the device should operate are prescribed in relevant documents (for optical switches, e.g., Ref. 47) or must be agreed upon between vendor and user.

A *failure mode* is the effect by which a failure is observed. In the case of an optical switch, a failure mode may be, e.g., an increase of insertion loss or an increase of back-reflection above the maximum allowed value.

The physical processes behind failures are called *failure mechanisms*. Examples of such processes are corrosion, electromigration, diffusion, and fatigue. In the case of electronic devices and housings, many of these processes are extensively described in the literature [48–50].

Component failure phenomena are stochastic processes; therefore they have to be described by their respective statistical probability distributions. The *cumulative failure distribution function* $F(t)$ is defined as the probability that a failure occurred in the interval $[0,t]$. The *failure probability density function* $f(t)$ is defined as the first time derivative of the cumulative distribution function. In reliability studies, exponential, Weibull, and log-normal distribution functions are often used to describe the statistical behavior of failures. An overview of these distributions and their relevant parameters is given in Table 4.

The *reliability* $R(t)$ of a device is defined as the probability that the device has not yet failed at moment $\tau = t$ when it started operation at $\tau = 0$. The reliability function is given by: $R(t) = 1 - F(t)$.

The *failure rate* is the number of failures per unit time. The instantaneous failure rate $\lambda(t)$ is defined so that $\lambda(t).\ dt$ is the conditional probability that a component fails in the time interval $[t, t + dt]$ under the condition that it did not yet fail in the interval $[0, t]$. The failure rate is often expressed in FIT units (failure in time), 1 FIT = 1 failure in 10^9 operating hours. The failure rate is given by the equation

Table 4 Frequently Used Distribution Functions in Reliability Studies, and Their Relevant Parameters

Distribution	$f(t)$	$F(t)$	Mean time to failure	Median time to failure
Exponential	$\lambda e^{-\lambda t}$	$1 - e^{-\lambda t}$	$\frac{1}{\lambda}$	$\frac{\ln 2}{\lambda}$
Log-normal	$\frac{1}{\sigma t\sqrt{2\pi}} e^{-1/2[\ln(t/\mu)/\sigma]^2}$	$N\frac{\ln(t/\mu)}{\sigma}$	$\mu e^{\sigma^2/2}$	μ
Weibull	$\frac{\beta}{t}\left[\frac{t}{\alpha}\right]^{\beta} e^{-[t/\alpha]^{\beta}}$	$1 - e^{-[t/\alpha]^{\beta}}$	$\alpha\Gamma\left(1 + \frac{1}{\beta}\right)$	$\alpha(\ln 2)^{1/\beta}$

$$\lambda(t) = \frac{f(t)}{1 - F(t)} \tag{42}$$

In the case of exponential distributions, the failure rate is a constant number equal to the reciprocal value of the mean time to failure (MTTF).

The *median life* or *median time to failure* of a component is defined as the length of the time interval until 50% of the components from a population have failed. Figure 14 shows the functional behavior of the basic statistical functions in the case of log-normal statistics.

In reality the failure-rate curve of a device often looks different, since in many cases different failure mechanisms, with their respective statistical distributions, are active, and each contributes to the total failure distribution of a device. A

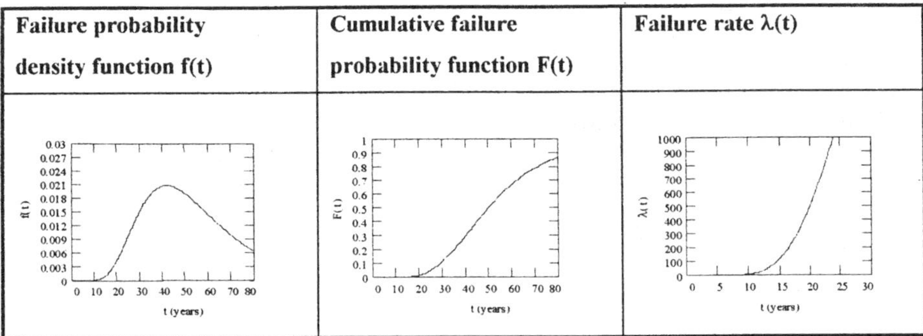

Figure 14 Basic reliability functions as a function of time.

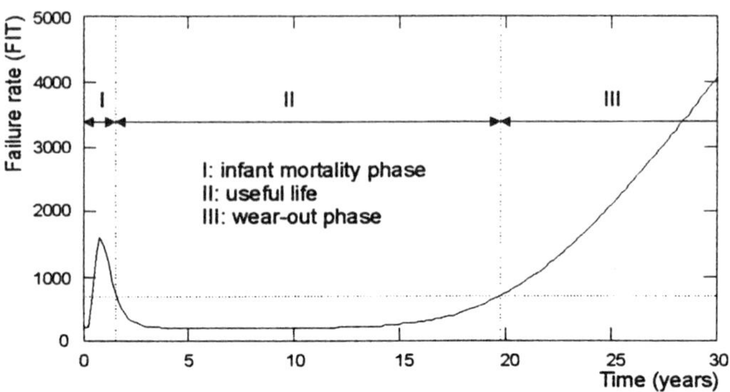

Figure 15 Realistic failure-rate curve, as a function of time, showing the different phases in the life of a device.

typical curve is shown in Fig. 15. It often shows three phases: the early life with potential infant mortality, the useful life period, and, finally, the wear-out period.

Infant mortality in a component population is often due to weak components that have slipped through quality testing. Current production practices eliminate such flawed components by a burn-in procedure. *Burn-in* consists of a brief stressing of the components (e.g., operation at elevated temperature, thermal cycling) followed by a screening procedure, when failed components are removed. This reduces infant mortality considerably.

During the useful life, the failure rate is low and almost constant in time. In this phase the failure statistics can be adequately described by an exponential distribution:

$$\lambda = \frac{1}{\text{MTTF}} = \frac{1}{\text{MTBF}} \tag{43}$$

where λ is the constant failure rate, MTTF is the median time to failure, and MTBF is the mean time between failures. For example, when a component has an MTBF of 10^6 hours, its failure rate is 1000 FIT. The constant failure rate of a component is also often called the *random failure rate*.

Wear-out failure mechanisms start to show their influence at the end of a component's life, resulting in a rapidly increasing failure rate.

Experimental verification of the reliability of components can turn out to be a costly and time-consuming task, especially when high reliability must be verified. In order to obtain statistically meaningful data, many components must be tested over a long period of time. Therefore, life testing of higly reliable components can become prohibitively expensive.

One method to reduce test time is to perform accelerated tests in which elevated temperatures are used to increase the rate of failure mechanisms (failure acceleration). The relation between the accelerated rates and those under operational conditions is made using the Arrhenius equation. The Arrhenius equation is an empirical result, but in some cases it can be theoretically derived [51]:

$$\text{MTTF} = Ae^{E_a/kT} \qquad \text{or} \qquad \lambda = Be^{-E_a/kT} \tag{44}$$

where A is a constant, E_a (in joules) is the activation energy, k is Boltzmann's constant (1.38×10^{-23} J/K), and T is the absolute temperature (in K).

The graph in Fig. 16 represents typical Arrhenius plots (log MTTF as a function of reciprocal temperature). The slope of the curves yields the activation energy. One can easily derive the acceleration relation:

$$\text{MTTF}(T_2) = \text{MTTF}(T_1)e^{E_a/k[T_2 - T_1]} \quad \text{or} \quad \lambda(T_2) = \lambda(T_1)e^{E_a/k[T_1 - T_2]} \tag{45}$$

It is important to mention that one has to be careful when using these equations. First, more than one failure mechanism may have an influence on the failure behavior of a component. In that case one has to trace each failure back to its corresponding failure mechanism. For each independent failure mechanism, the Arrhenius equation can be used, and the combined statistics of the different failure mechanisms yield the failure rate. Second, the Arrhenius equation is not always valid. This is, for instance, the case for processes taking place in polymers.

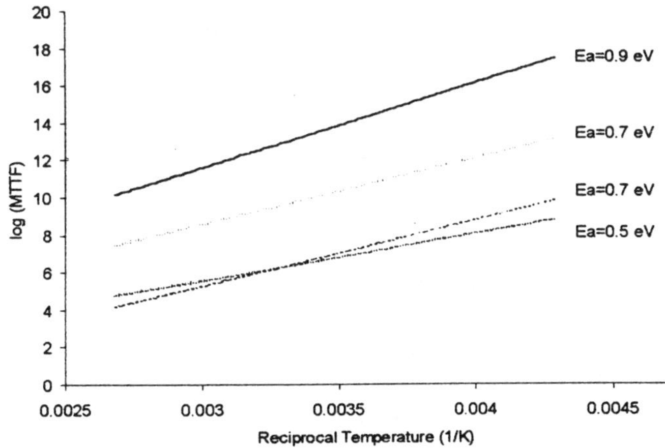

Figure 16 Examples of Arrhenius plots, showing the influence of the activation energy E_a. The two parallel lines corresponding to the same activation energy have a different constant factor A.

When one attempts to extrapolate results from above the glass-transition temperature of a polymer toward lower temperatures, the Arrhenius equation cannot be used, since processes in polymers may change drastically at this transition temperature. In these cases the temperature behavior is better described by the Williams, Landed, Ferry (WLF) model [52].

Direct estimation of the *failure rate of a system* is often difficult, because of the complexity and cost of complete systems. In that case, a bottom-up approach is used to determine system reliability. In this approach, the system is split into its elements, and the reliability of each element is determined independently. The failure rate of the system is calculated as a combination of the failure rates of the elements. In that case, a reliability diagram of the system must be used showing how the reliability of an individual component affects the reliability of the system. In many cases, this can be obtained by a series/parallel combination. Failure-rate data for individual electronic components can be found in handbooks such as Refs. 53 and 54. Illustrative calculation examples can be found in Ref. 55. Unfortunately, in the case of fiber-optic and integrated optic components, reliability data is still relatively scarce. Often, only estimates or preliminary reliability studies are available [56]. But as the implementation of integrated optic components in the field increases, more accurate data are expected to become available.

The *availability of a system* $A(t)$ is the probability that the system performs its function at a given instant of time t. It depends on a number of parameters, including: the failure rate, the repair rate, and the failure detection rate. In Ref. 57 an unavailability $[U(t) = 1 - A(t)]$ requirement of $U < 5.7 \times 10^{-6}$ is proposed; this corresponds with a mean down time of the cross-connect of <3 min/year.

Methods for calculating telecommunication system availability are described in Ref. 55. In Ref. 56 some examples are given for the availability of optical cross-connects for different system architectures. These results show that the required availability is challenging for the used-component reliability parameters. This can be overcome by using system architectures that allow a sufficient degree of redundancy.

3.3 Environmental Requirements for Photonic Components

The environment under which a component or a system is operated is of major importance for its performance and reliability. In Refs. 58 and 59, different environmental conditions that can be encountered in telecommunications systems are described and classified. Most of the applications for integrated optic circuits are in a controlled environment. In Table 5 a classification of different environments is given.

One of the most important environmental parameters is temperature. As already discussed, temperature has an important effect on the rate of failure mechanisms and therefore on the reliability of components. It may also have an influ-

Table 5 Classification of Environments

Environment	Long term	Short term	Humidity
Controlled	+4°C/+38°C	+2°C/+49°C	20–50% RH
Uncontrolled	−40°C/+70°C	−40°C/+85°C	5–95% RH
Uncontrolled (ventilated)	−40°C/+50°C	−40°C/+65°C	5–95% RH

Source: Ref. 60.

ence on the performance parameters of a component. For example, parameters such as insertion loss and return loss may vary during temperature excursions. These variations should remain between well-defined boundaries within a certain temperature range. During their life, components may be exposed to extreme temperature during transport or storage (often within the range −40°C–+85°C. These exposures should not result in any damage to the components.

Thermo-optic switches are by definition heat-dissipating devices. Low-complexity devices, such as 1 × 2 switches, dissipate about 100 mW; more complicated devices, such as 1 × 8 switches dissipate about 1000 mW. Here the chip temperature (which is important for reliability estimation and for defining the operational temperature window) depends not only on the ambient temperature, but also on the dissipated power. The temperature can be determined from an electrical equivalent of the thermal problem, as shown in Fig. 17. R_{cp} is the thermal resistance between the chip and the package and is determined mainly by the design of the package and submount. R_{pa} is the thermal resistance between the package and the environment. R_{pa} depends on the package design and the ways heat can be carried away from the component to the ambient environment (e.g., air convection, heat conduction through the pins to the printed circuit board).

Besides temperature, the humidity level is also an important environmental factor, in particular for nonhermetically packaged components. Humidity may have a major influence on failure mechanisms such as delamination, corrosion, and electromigration. This is one of the reasons we chose a fully hermetic package for BeamBox™ solid-state optical switches. Hermeticity immediately excludes a large family of failure mechanisms and should yield more reliable components. The combined influence of temperature and humidity has been described by various models [61]. An example is:

$$\text{MTF} = A \times (\%\text{RH})^n e^{E_a/kT} \tag{46}$$

where n ranges between −2 and −5. Additional environmental factors that have to be taken into account include electrostatic charges, which might cause electrostatic damage (ESD), dust, airborne contaminants, vibration, mechanical stresses, and rough handling. Test procedures to verify whether components are durable

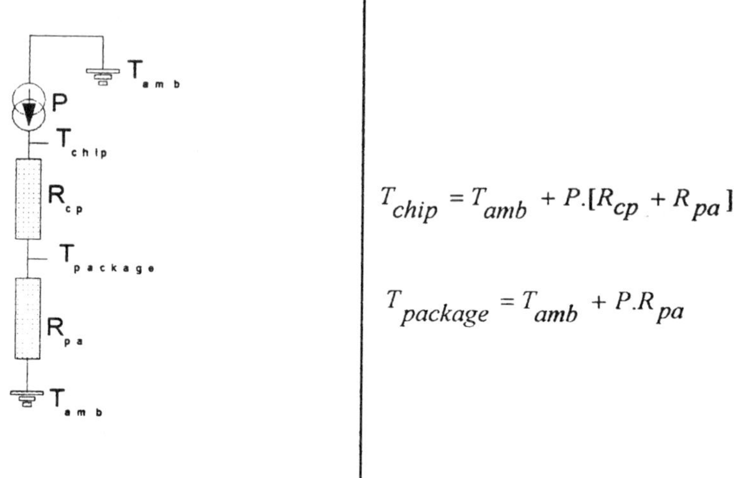

Figure 17 Simplified electrical equivalent of the thermal model of a BeamBox™ solid-state optical switch. P, R_{cp}, and R_{pa} depend on the type of switch. For a 1 × 8 switch, typical values for R_{cp} and R_{pa} are about 10 K/W (free convection and conduction through pins to printed circuit board).

enough to withstand these environmental factors are described in the following relevant Refs.: 47 and 62–64.

3.4 Design for Optimum Reliability and Stability

In an early stage of the development phase of BeamBox™ Solid State Optical Switches, we performed a detailed analysis of the component and generated a list of the potential failure modes, failure mechanisms, and the impact these mechanisms may have on component reliability and stability. Such an analysis is often called the failure mode, effects and criticality analysis (FMECA). A partial list of these potential failure mechanisms and their related failure modes is given in Table 6. Figure 18 shows the basic elements and parts that have been reviewed.

In order to get insight into how important the influence of a certain failure mechanism was, experimental data were gathered by performing accelerated tests on different parts and components. We tested, among others, the heater elements, the polymer waveguide stability, and the pigtail adhesive bond. Where possible, corrective actions for reliability problems were immediately implemented in the prototypes.

3.4.1 Optochip: Waveguide Material

3.4.1.1 Photochemical Stability. Certain polymers may be sensitive to irradiation with light of certain wavelengths. For a number of nonlinear polymers used for electro-optic integrated optic devices it is known that exposure to light with

Table 6 Potential Failure Mechanisms and Failure Modes of BeamBox™ Solid-State Optical Switches

Component part	Subelement	Potential failure mechanisms	Related potential failure modes	Measures taken to avoid failure
Waveguide optochip	Polymer waveguides	Modifications waveguide polymer (thermal, optical)	Increase insertion loss Decrease isolation Increase back-reflection	Use highly stable polymer for waveguide fabrication (thermally, optically, mechanically) Package components in hermetic housing
	Heater elements	Corrosion Electromigration	Decrease isolation Increase insertion loss	Package components in hermetic housing Use heater elements made out of material with high electromigration robustness
Fiber array unit	Pigtail adhesive bond Si parts and fibers	Delamination Mechanical movement due to creep adhesive	Increase back-reflection Change insertion loss	Package components in hermetic housing Use reliable pigtail adhesive combined with suitable surface treatment of parts (reliability) Use adhesive with high glass transition temperature and high elasticity modulus (stability)
Electrical connections	Lead frame and conductive adhesive bonds	Wet electromigration	Opens, shorts	Place components in hermetic package Use suitable conductive adhesive
Hermetic solder seal	Lid and solder seal Fiber feedthroughs	Mechanical breaks	Increase back-reflection Change insertion loss	Careful mechanical design of solder seal and fiber feedthroughs
Submount and other parts	All parts	Mechanical stresses or strain exceeding the material maximum levels Temperature rise	Breaks, cracks Increase of failure rate	Careful mechanical design of parts and joints by matching thermomechanical parameters of materials used. Optimize thermal management in the component.

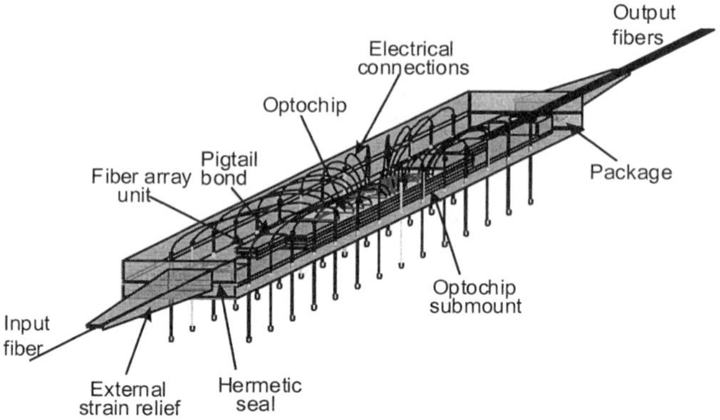

Figure 18 A BeamBox™ solid state optical switch showing the basic elements and parts.

a wavelength around 1300 nm in combination with the presence of oxygen may result in chemical changes due to the formation of singlet oxygen [65]. Advances toward more photostable nonlinear polymers have been reported [66].

Since for the fabrication of t.o. switches no complex side-chain polymers with light-sensitive chromophores are required, we chose for BeamBox™ components a polymer system without these chromophores that is optimized for high stability and low loss and allows straightforward processing. In order to verify the photochemical stability of the waveguide polymer, we performed a number of verification tests on slab and channel waveguides at high optical powers in the wavelength bands of interest for applications (1300-nm and 1550-nm bands). A single-mode component was exposed to +20 dBm optical input at 1319 nm and another to +17 dBm at 1550 nm. No change in the performance of the switches was observed, as illustrated in the graph of Fig. 19.

The only possible effect of high optical power density on the waveguide polymer is the heating of the material due to absorption of light. We calculated that in our waveguide structure, the resulting temperature rise due to the injection of +27 dBm in the single-mode waveguide results in a temperature rise of only a few degrees celsius. This is due to the low intrinsic absorption of the polymer at the wavelengths of interest and the heat-sinking effect of the Si substrate. We also verified experimentally that switching high optical powers presents no problem.

3.4.1.2 Thermochemical Stability. When exposed to elevated temperatures, polymers may show changes in chemical composition. Since we chose a highly stable polymer with a high glass-transition temperature, the probability of failure due to this mechanism is largely reduced. In order to investigate the thermochemical stability of the waveguide polymer, we performed extensive accelerated tests.

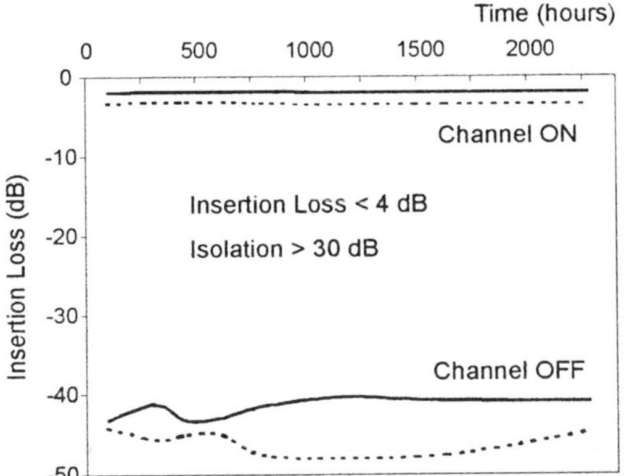

Figure 19 Behavior of a BeamBox™ solid-state optical switch under high-optical-flux conditions. The switch is exposed to +17 = dBm input power of 1550 nm. The performance is measured at 1300 nm (solid line) and 1550 nm (dashed line).

Samples of the waveguide polymer were exposed to elevated temperatures, and the chemical composition was monitored at regular intervals by Fourier transform IR spectroscopy. This experiment allowed us to determine the kinetics of the chemical reactions that take place at these elevated temperatures. The graph in Fig. 20 shows the main results. Clearly, even at the highest operational temperatures, the time constants are so large that no failures due to thermochemical degradation of the polymer are expected. This is because of the high glass-transition temperature of the waveguide polymer. Below the glass-transition temperature the mobility in the polymer is much lower, so chemical reactions are more than linearly decelerated in the Arrhenius plot [52].

3.4.2 Optochip: Heater Elements

The heater elements in BeamBox™ solid-state optical switches are basically thin-film metallic resistors. As for all electrical conductors, they are prone to a failure mechanism called *electromigration*. In microelectronics, electromigration is one of the most important failure mechanisms of Al-based conductor paths [68,69]. It is caused by the exchange of momentum between electrons (electron wind) and metal ions. Practically, electromigration leads to the formation of voids near the negative contact and eventually hillocks near the positive contact. Obviously, voids can lead to failure modes such as increase of electrical resistance or even open circuits. Hillocks can lead to shorts.

The kinetics of the electromigration mechanism can be described by an Arrhenius-like equation [69]:

$$\mathrm{MTF} = AJ^{-2}e^{E_a/kT} \qquad (47)$$

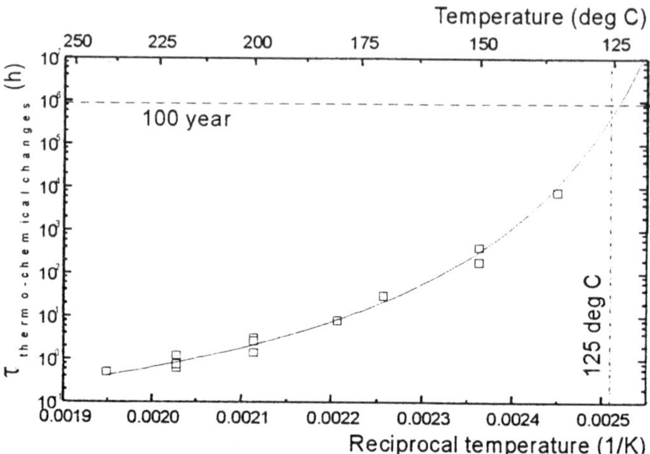

Figure 20 Kinetics of thermochemical changes in the waveguide polymer used in BeamBox™ solid-state optical switches

where MTF stands for the median time to failure (time elapsed before 50% of the heater electrodes under test have failed), A is a constant, E_a (in joules) is the activation energy of the electromigration process, J (in A/m^2) is the current density, k is Boltzmann's constant (1.38×10^{-23} J/K), T (in K) is the temperature of the conductor (not the ambient temperature).

Black's equation allows prediction of the intrinsic median lifetime of the electrodes under specified conditions of temperature and current density. This can be done starting from experimental lifetime data obtained by accelerated testing (higher current densities and temperatures than under normal operation).

In an early stage we found that the old design for the heater elements that were based on Au conductors seemed unsuitable for long-term reliable operation. In Fig. 21 we show an example of an accelerated-life test performed on Au elements. In this case, the MTTF was about 265 hours. Consequently, a new type of heater element based on Ni has been developed. These elements turned out to be extremely robust, as proven by a number of verification tests, including an accelerated-life test where four groups of elements were tested under different conditions in which the heater temperature ranged between 300 and 650°C (see Fig. 22). The current density J ranged between 0.85 and 1.1 MA/cm^2. Because of the severity, these tests were performed on SiO_x-coated substrates instead of polymer-coated wafers. Extrapolation of these results toward even the most extreme practical operational conditions yielded a negligible contribution to the total failure rate of the component.

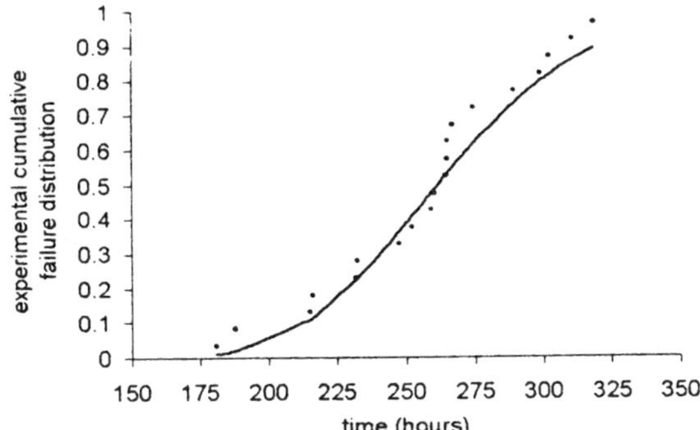

Figure 21 Experimental cumulative failure distribution of Au heater elements when exposed to elevated temperatures (121°C) and current densities (0.7 MA/cm²).

3.4.3 Attachment of Optical Fibers: Pigtail Adhesive Bond

An important part of each integrated optic component is its connection with optical fibers. The quality, stability, and reliability of this bond affects considerably the performance of the integrated optic device.

Two parameters most affected by this bond are the insertion loss and the return loss of the device. The variations of these parameters must be kept between narrow limits (often an insertion loss stability better than 0.5 dB (peak-to-peak) is required). Therefore, the adhesive bond must maintain the critical alignment between the waveguides and the fibers over the whole life of the component and under all allowed environment conditions.

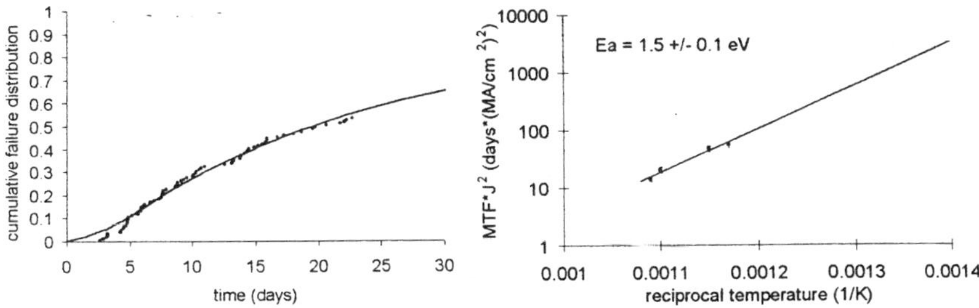

Figure 22 (a) Experimental cumulative failure distribution of Ni-based heater elements during an extreme accelerated test (heater element temperature of 650°C and current density of 0.85 MA/cm²). (b) Arrhenius plot of the electromigration process in the Ni-based heater elements. The slope of the curve yields an activation energy of 1.5 eV.

In the BeamBox™ production process, the fibers and the waveguides are aligned using an active alignment technique, where the position of the two parts is optimized by optimizing the optical throughput of one or more channels. Since most of the BeamBox™ components have more than one optical input or output per endface, the necessary fibers are arranged in a fiber array unit, which consists of a Si V-groove part and two reinforcement parts. The V-grooves are used to ensure high accuracy in the relative position of the fiber cores. The fibers are fixed in the grooves by a suitable adhesive. Further, since the fibers are all held together in the fiber array unit, the manipulation of the fibers during assembly is also considerably simplified. In order to obtain a strong bond and to obtain high-quality endfaces, the surface of the bond is enlarged by adding top and bottom plates (see Fig. 23).

The choice of the adhesive used for the pigtail bond is very important. The adhesive must fulfill a number of requirements, including: high thermal stability, stable mechanical properties in the operational temperature interval, UV curability, low shrinkage during curing, low thermal expansion coefficient, good adhesion on glass, silicon, and polymers.

In Ref. 63 some guidelines are given for the choice and quality control of adhesives for pigtail applications. It is stated that the glass-transition temperature should be larger than 95°C; this ensures that the mechanical properties of the adhesive remain constant in the operational temperature ranges. Above the glass-transition temperature, the thermal expansion coefficient may increase considerably while the elasticity modulus may drop over several decades, which is to be avoided when highly stable adhesive bonds must be obtained. The glass-transition temperature can be determined by methods such as differential scanning calorimetry (DSC) and dynamic mechanical thermal analysis (DMTA) [70]. In the litera-

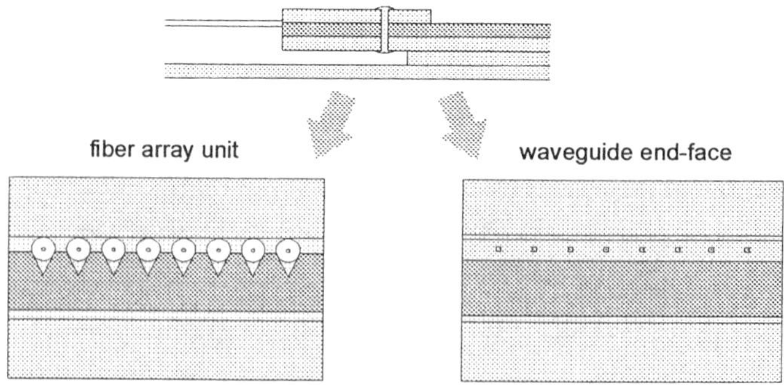

Figure 23 Schematic drawing of the pigtail adhesive bond. The upper drawing shows a side view of the construction. The two drawings below show the front view of, respectively, the fiber array unit and the optochip endfaces.

ture, a number of suitable adhesives are described that may fulfill these requirements [71,72]. Table 7 shows some results obtained during thermal stability tests on BeamBox™ pigtailed devices. Fig. 24 plots the stability of such a device. The results of the stability tests indicate that BeamBox™ solid-state optical switches easily fulfill the 0.5-dB peak-to-peak stability requirement in their extreme operational temperature window (−10°C/+70 °C).

Besides optimizing the stability of the pigtail bonds, we also optimized the reliability of the pigtail joint. Potential failure mechanisms for a pigtail joint are delamination and creep of the adhesive. Delamination of the adhesive joint is often caused by the diffusion of moisture. The moisture may react with the UV adhesive and finally lead to detachment of the bond [73,74]. This failure mechanism is obviously humidity and temperature dependent. Creep of the adhesive is caused by residual stresses in the adhesive bond and poor mechanical properties of the adhesive (especially at higher temperatures). It results in a slowly degrading alignment between the fiber and the waveguide [73].

In order to quantify the reliability of the pigtail adhesive bond we performed accelerated tests on the pigtail bond. The insertion loss of a number of components was monitored at elevated temperatures (110°C, 120°C, and 140°C) under dry conditions. As failure criterion, we chose a peak-to-peak variation in insertion loss of larger than 0.5 dB. In Fig. 25 is plotted the failure rate of the pigtail adhesive bond as a function of reciprocal temperature. From this graph, the failure rate at 40°C could be extrapolated. An upper limit lower than 10 FIT per pigtail was determined with a 60% confidence level. These results have been obtained

Table 7 Some Experimental Results Obtained During Thermal Stability Tests on Pigtailed Devices

			IL change during test (dB)		IL change before/after	
# samples	Condition	Duration (hours)	Typical	Max.	Typical	Max.
Dry Heat Test						
9	85°C	88	0.15	0.4	0.0	0.2
6	100°C	100	0.15	0.4	0.0	0.0
6	100°C	140	0.15	0.3	0.0	0.3
6	110°C	100	0.15	0.7	0.0	0.0
6	120°C	100	0.15	0.7	0.0	0.0
11	130°C	100	0.5	1.0	0.2	0.5
Temperature Cycling						
3	−45–+75°C	800	0.3	0.45	0.0	0.1
5	−55–+100°C Info only	80	0.3	0.6	0.0	0.0
5	−60–+125°C Info only	80	0.3	0.75	0.0	0.0

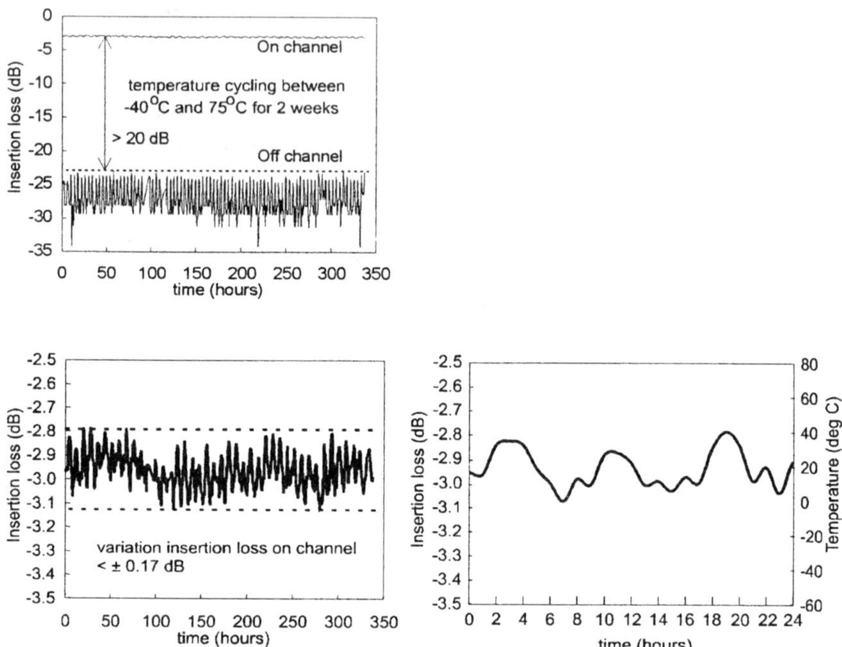

Figure 24 Experimental data showing the stability of a t.o. 1 × 2 switch during a thermal cycling experiment. Graph (a) shows the insertion losses of both the *on* and the *off* channels. Graph (b) shows a more detailed view of the insertion loss of the *on* channel. Graph (c) shows in detail how the insertion loss varies during the different phases of thermal cycling.

by a proper choice of adhesives and surface preparation. The influence of humidity on the pigtail bond is virtually eliminated by the use of a hermetic housing for the components. Some results obtained by other groups on nonhermetically sealed pigtails are given in Refs. 73 and 75–77.

In order to obtain reliable operation under high optical fluxes through the component, one has to ensure that no absorbing particles are present in the optical path. Therefore it is necessary that the endfaces of both the fiber array units and the optochip be thoroughly cleaned and that the adhesive be properly filtered.

3.4.4 Hermetic Package

3.4.4.1 Hermeticity. As just mentioned, BeamBox™ solid-state optical switches are packaged in a hermetic housing. It is expected that this has a major beneficial impact on the component reliability, since many failure mechanisms

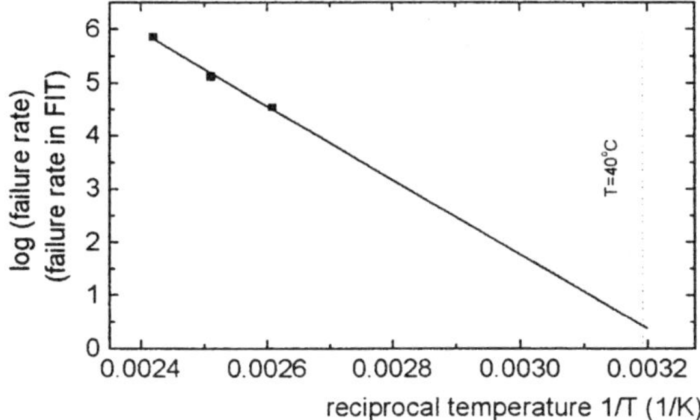

Figure 25 Failure rate of pigtail adhesive bond as a function of reciprocal temperature. Extrapolation toward 40°C operation yields a failure rate of less than 10 FIT per fiber bond (based on the data currently available).

are excluded. The package is made of standard alumina, and the lid is made of kovar. In our technology a proprietary solder process is used to obtain a hermetic feedthrough for the optical fibers. Seals that do not consist of a metallic or glass barrier are not considered truly hermetic, since other materials and especially organic materials are permeable for gases, including moisture [78]. Therefore these materials cannot be used as a long-term barrier against humidity. In order to exclude the possibility of moisture being enclosed in the package during sealing, the components are vacuum baked and sealed in a dry N_2 atmosphere with a dew point of less than −40°C. This sealing method results in reliable and reproducible hermetic packages fulfilling the requirements of MIL-STD-883D [79] for the leak rate ($<1.10^{-6}$ cc(air)/s/atm, method 1014) and for moisture content ($<$5000 ppm, method 1018/1).

3.4.4.2 Mechanical Stresses in the Package and Component. A careful analysis has been performed of the stresses that may occur in the components during thermal excursions. It was theoretically and experimentally verified that the stresses (or strains) did not exceed their allowed values. The theoretical analysis first consisted of an analytical approach in which approximate formulas were used to evaluate potential problems with excessive stresses or strains. A good reference for such an analysis is Ref. 47, where many package reliability issues are addressed, including thermomechanical stress analysis and solder bond fatigue. We also performed a finite element (FE) stress analysis of the package. This analysis, as shown in Fig. 26, confirmed the results from the approximate analysis. None of the stresses (or strains) exceeded the material strengths.

Special attention has also been paid to the stresses in the glass fiber. This is

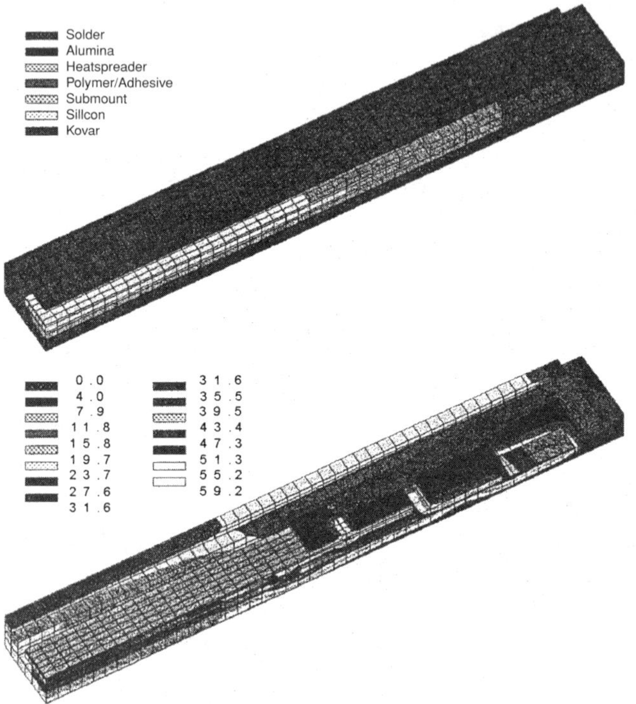

Figure 26 Structure and materials of quarter model of a package of a 1 × 8 BeamBox™ solid-state optical switch (upper). Result of an FE analysis showing the von Mises stress (in MPa) that occurs at a temperature sweep of 111 K.

necessary since the thermal expansion coefficient of the silica fiber is about one order of magnitude lower than that of the housing. Further analysis and solutions for this problem are presented in Refs. 80–82.

The durability of our housing technology was also experimentally verified by the following tests: heat storage (85°C), thermal shocks (30 shocks), 110 thermal cycles between −40°C and +75°C. All the packages (four per test) passed the hermeticity requirement. The moisture level was verified by a residual gas analysis test. This test showed not only that the moisture level was within the requirement, but also that there are no gases or vapors present in the package cavity that might have a negative effect on the reliability of the parts inside.

3.4.4.3 Thermal Management. As mentioned earlier, t.o. switches are by definition heat-dissipating components. The heat is dissipated by a heater element that creates a thermal gradient in the waveguide polymer, which in turn results

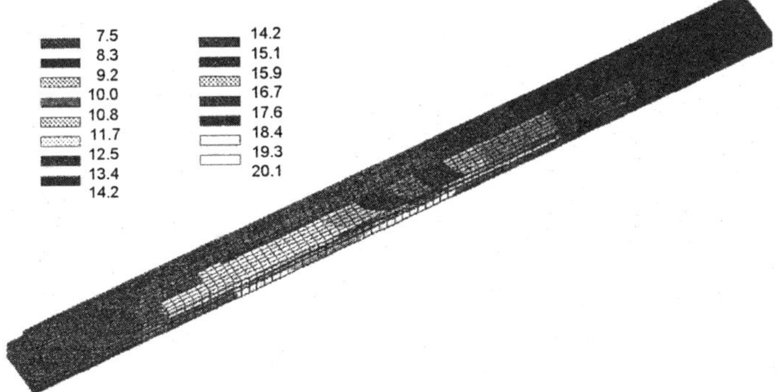

Figure 27 FE analysis showing the distribution of the temperature rise (in K) over the Si chip surface and other parts in the package of a 1 × 8 BeamBox™ solid-state optical switch under nominal operation and free convection.

in a refractive index change and optical switching. The required temperature difference is of the order of 10°C. We showed that many failure mechanisms are temperature dependent. Therefore it is advantageous to keep the temperature on the component as low as possible in order to improve reliability further [83].

During the design verification phase we optimized the thermal design of BeamBox™ solid-state optical switches in various ways. First, we optimized the t.o. design so that the maximum thermal gradient between the waveguides branches could be obtained with the lowest amount of heat dissipation. This resulted in a reduction of power dissipation of more than 100%. Second, the submount and package materials were chosen so that the thermal resistance between the optochip and the environment was considerably reduced. The use of an additional heat spreader also reduced thermal gradients in the package.

The design was again theoretically and experimentally verified. Experimental verification was done by temperature measurements outside and inside the package. The latter was done by monitoring the resistance change of heater elements when the component was switched on. In a 1 × 8 switch with a total dissipation of 1 W, this results in a maximum chip temperature rise of 20 K, as confirmed by FE analysis (see Fig. 27).

3.5 Testing of Optochip Life

During life testing, statistical data are gathered that can be used for estimation of device failure rate. Since the failure rate is almost constant during the useful life of a component, the maximum likelihood estimator for the failure rate is the

Table 8 χ_n^2 Values for 0, 1, and 2 Failures During a Life Test

Number of Failures, r	$\nu = 2r + 2$	χ_n^2 (60%)	χ_n^2 (90%)
0	2	1.83	4.61
1	4	4.04	7.78
2	6	6.21	10.6

number of failures that occurred during the test, divided by the total of accumulated operating hours:

$$\lambda_{\text{maxlik}} = \frac{\#\text{failures}}{T} \tag{48}$$

There is, however, no indication of how good the estimated value approximates the real value. For example: When 10^6 operating hours have been generated without failure, the estimated value for the failure rate with the maximum probability is 0 FIT, which is unlikely. Therefore it is better to give an upper limit for the failure rate that is combined with a statistical confidence level. In the case of exponential failure distributions, an upper limit for the failure rate with a confidence level of $a\%$ is given by the expression

$$\lambda_{UCL} = \frac{\chi_n^2}{2T} \tag{49}$$

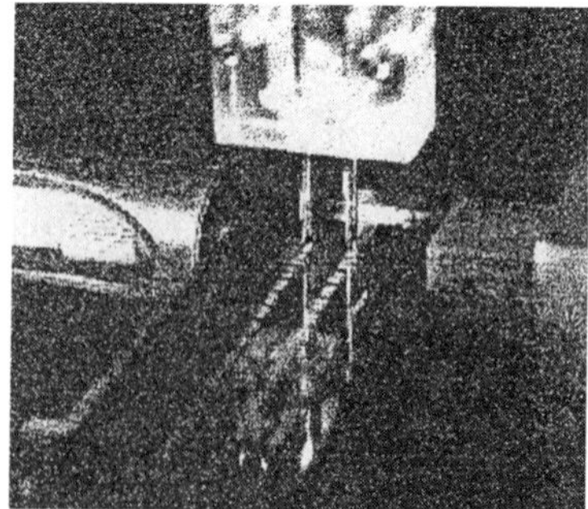

Figure 28 (a) Life testing of BeamBox™ optochips (not pigtailed). (b) Measurement of optical performance of the BeamBox™ optochips in a dedicated setup.

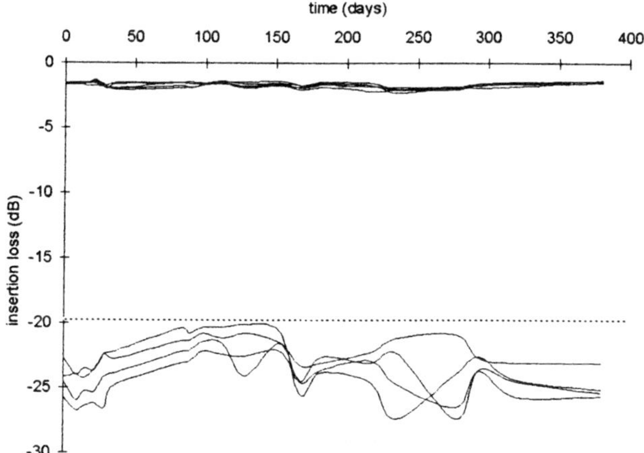

Figure 29 Typical evolution of the insertion losses of both *on* and *off* channels of five 1×2 optochips during the life test. The variations seen in the graph are caused by measurement inaccuracies (the chips are not pigtailed).

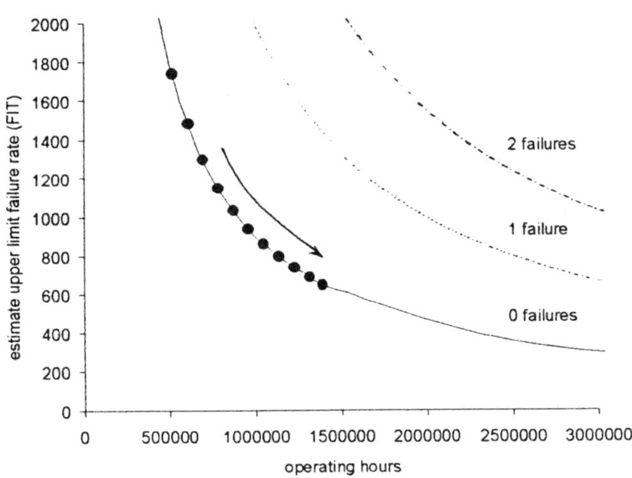

Figure 30 Upper limit of the failure rate with 60% confidence level for 0, 1, and 2 occurred failures during the life test. The dots show how our experiment follows the curve as more operating hours are accumulated.

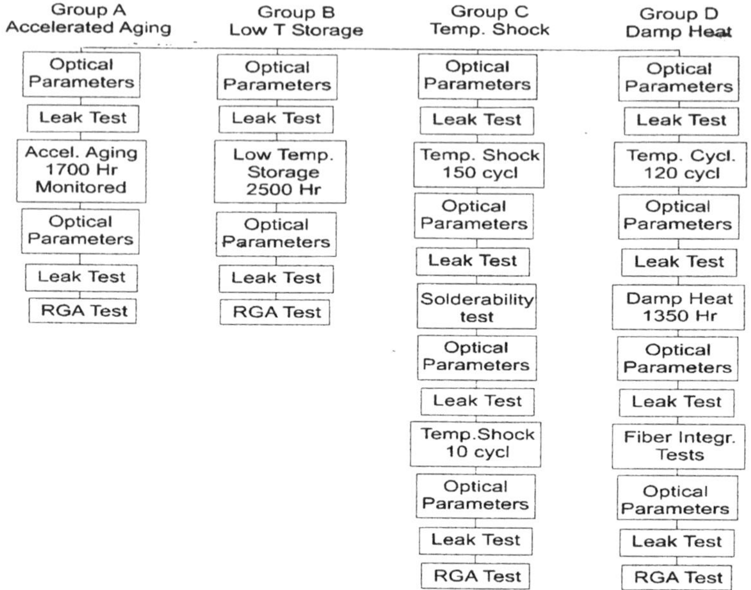

Figure 31 Overview of the different durability tests performed on 1 × 8 BeamBox™ solid-state optical switches.

where χ_n^2 is the *a*th percentile of the χ^2 distribution with $2r + 2$ degrees of freedom. T is the total amount of accumulated operating hours (a component that failed during the test does not accumulate operating hours after it failed). For our example, this yields an upper limit for the failure rate of 915 FIT. Table 8 gives the χ_n^2 values for 0, 1, and 2 failures during a life test.

We started life testing of BeamBox™ 1 × 2 optochips more than 1.5 years ago and up to now have obtained 1.4×10^6 faultless operational hours. According to the preceding equations, this yields an upper limit for the failure rate of less than 750 FIT with a 60% confidence level. Figure 28(a) shows a number of optochip arrays under test in an oven at 70°C, and Fig. 28(b) shows the (periodical) optical and electrical evaluation of these chips. In Fig. 29 a typical evolution of the insertion losses of both *on* and *off* channels of five 1 × 2 optochips can be seen.

The graph in Fig. 30 shows the trajectory in the λ_{UCL}-versus-time diagram, which we followed during this life-test experiment.

3.6 Durability Tests on Complete Components

Besides life tests on optochips, a considerable number of tests have also been performed on fully packaged BeamBox™ components that gave us further confidence that the product can be used in practical applications. An overview of the test scheme is given in Fig. 31. The standards to which the tests were performed are given in Table 9. The first group of tests included an accelerated-aging test at 70°C; the second group was exposed to a low-temperature storage test at −40°C. The next group of tests included 150 thermal shocks between −40°C and +70°C, followed by a solderability test, where the component was dipped into a molten solder bath for 10 seconds. Finally, another thermal shock test was performed in which the component was immersed alternately in boiling water and in ice water. The last group was exposed to a slow thermal-cycling test, a damp heat exposure (40°C, 95% RH), and a number of fiber integrity tests (fiber retention, fiber flexing, and fiber twisting). Before and after each test, the optical parameters and hermeticity were verified. At the end of each test group a residual gas analysis (RGA) was performed in order to investigate any outgasing products inside the package. No products with a detrimental effect on the switch could be found in any of the test groups.

Table 9 Durability and Aging Tests Performed on 1 × 8 BeamBox™ Solid-State Optical Switches

Test	Condition	Reference document
Mechanical Integrity		
Fiber pull	1 kg, 10s 3 times	EIA/TIA-455-6a
Fiber twist	1 lb, 10 times	EIA/TIA-455-36
Fiber flex	1 lb, 100 times	EIA/TIA-455-1A
Vibration	20g, 20–2000 Hz, 4 min/ cycle, 4 cycles/axis	MIL-STD-883D, Method 2007.2
Thermal shock	$\Delta T = 100°C$	MIL-STD-883D, Method 2031
Solderability	260°C, 10 s	MIL-STD-750, Method 2031
Endurance		
Accelerated Aging	70°C	Bellcore TR-NWT-000468
Thermal Cycling	−40°C–+70°C	MIL-STD-883, Method 1010
Low Temperature Storage	−40°C	EIA/TIA-455-4A
Moisture	+40°C/95% RH	MIL-STD-202, Method 103
Other		
Moisture content	< 5000 ppm	MIL-STD-883D, Method 1018, Procedure 1
Hermeticity	< 10^{-6} atm·cc air/s	MIL-STD-883D, Method 1014, Cond. C, A2

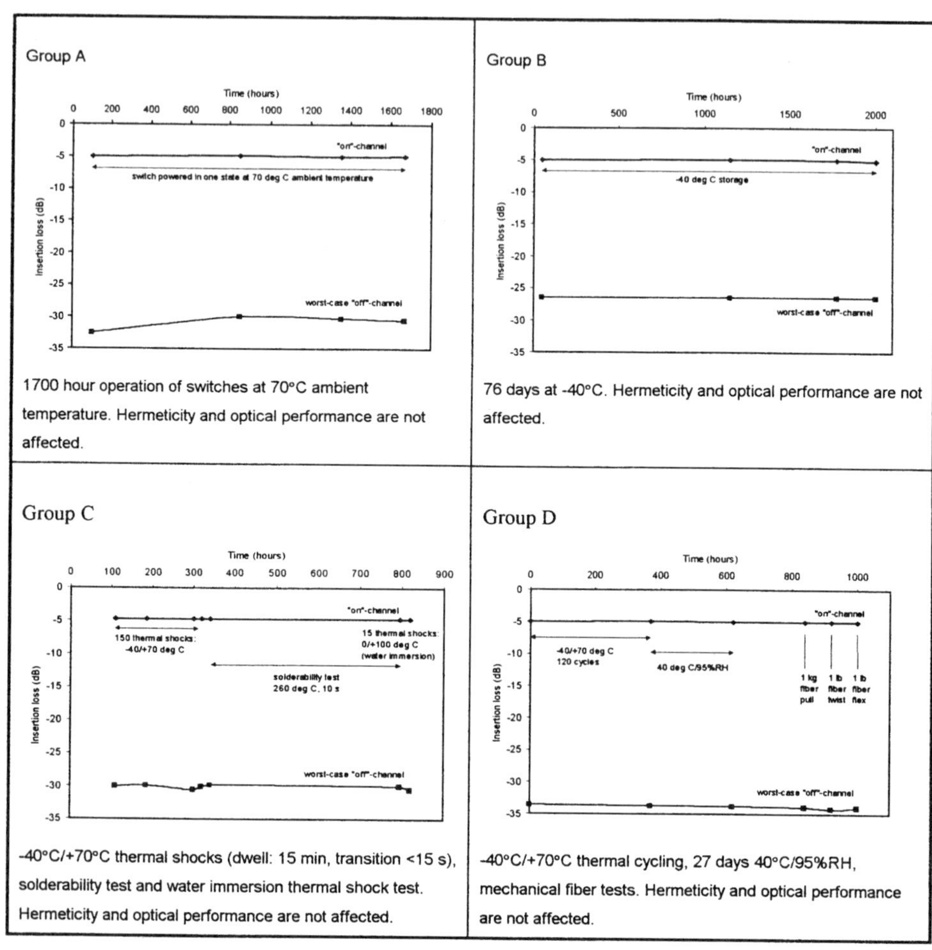

Figure 32 Summary of the results obtained during the robustness and aging tests performed on hermetically packaged BeamBox™ solid-state optical switches (1 × 8).

Figure 32 summarizes the results obtained during the robustness and aging tests performed on hermetically packaged BeamBox™ solid-state optical switches (1 × 8's).

3.7 Conclusions

Reliability study of integrated optic components has become a major research topic because of the coming massive deployment of these devices in telecommunication networks and the requirement of high reliability. Because of the many new materials and processes used to fabricate such devices, new failure mechanisms must be investigated and modeled. These studies will ultimately result in usable reliability models for these devices and give telecommunication system makers the necessary information for the development of high-availability telecommunications networks.

Our work on optimization and verification of the reliability and stability of BeamBox™ solid-state optical switches has been described. From the results it can be concluded that these components can meet the stringent reliability and stability requirements for the applications in optical fiber systems.

4 OPTICAL SWITCHING IN THE OPTICAL LAYER

4.1 Bandwidth Demand

The demand for bandwidth has exploded over the past few years, rapidly saturating the capacity of the telecom and datacom networks. The demand for bandwidth is most pronounced in the United States. Over the next several years, voice and fax traffic is expected to grow at a rate of 10% per year, data at 60% per year, and internet traffic at 100% per year. In other parts of the world, a similar growth is expected quickly to exhaust existing network capacities.

These growth rates are a nightmare to network planners. Newly installed equipment could be obsolete by the time it is up and running. Upgradability of equipment is therefore of utmost importance for network operators. This is one of the reasons that dense wavelength-division multiplexing (DWDM) has been such a success, enabling operators to increase the total capacity of their network by adding more wavelengths step by step. The day-one investments can be kept within reasonable limits this way, and future bandwidth demand can be quickly responded to. Deployment of DWDM is taking place worldwide.

Next to wavelength-division multiplexing, a growth path to higher-bandwidth systems using time-division multiplexing (TDM) is provided by system vendors. OC-48 (synchronous optical network—SONET) and STM16 (synchronous digital hierarchy—SDH), operating at a bit rate of 2.5 Gbit/s, has been succeeded now by OC-192 and STM64 at a bit rate of 10 Gbit/s. A significant number of fiber-optic links are currently provided with OC-192 equipment in the United States. Dense wave length-division multiplexing (up to 96 wavelengths at 2.5 Gbit/s), TDM (up to 10 Gbit/s), and the combination of DWDM and TDM (up to 40 wavelengths at 10 Gbit/s) make sure that the operators can handle future bandwidth demands when capacity bottlenecks at the access level are solved.

4.2 All-Optical Layer: Flexibility and Reliability

In addition to providing increased bandwidth rapidly, the emergence of DWDM has added another dimension to optical telecommunications. Dense wavelength-division multiplexing will accelerate the addition of an all-optical layer to the existing telecom infrastructure [84]. In the all-optical layer, high-bandwidth information carried by different wavelengths can be transferred to the proper location without leaving the optical domain for the electrical domain. This means that no optical-to-electrical or electrical-to-optical conversion is needed for this high-bit-rate traffic.

The all-optical layer will increase the reliability and flexibility of service delivery. Increased reliability is required because the impact of failures in the fiber trunk network rises in parallel with the increase of bit rates. Imagine a fiber break in a transmission line carrying 16 wavelengths at 2.5 Gbit/s. The equivalent of hundreds of millions 64-kbit/s phone or fax connections could be down for some time. Downtime can be reduced significantly in the all-optical layer if the proper optical protection hardware and management is installed.

Furthermore, flexibility is also improved by the optical network layer. Bandwidth can be allocated on demand, improving service to the customer and optimizing the installed capacity of the network system. The flexibility of the optical network is also increased by the transparent nature of the all-optical network. In principle, network elements of the all-optical network are transparent to signal format and protocol. This means that a wide variety of bit rates and protocols (asynchronous transfer mode—ATM, Internet protocol—IP, SONET/SDH, etc.) can be transmitted over such a system. This transparent and open nature of the all-optical network, however, is a complicating factor for system management. Issues like how to monitor the performance of a transparent network and how to address a basic functionality like protection in a multiprotocol environment have to be worked out to bring the all-optical network to maturity.

In order to enable operators to benefit from the increase in flexibility and reliability of their networks, the functionality of optical space switching is needed. Optical space switching is often referred to as *optical routing*, which is defined as the rearrangement of the optical paths in a network. Connections can be changed by means of optical switching. This functionality enables the development of different applications that will be part of the all-optical layer:

Optical line protection switching
Equipment redundancy switching
Flexible optical add/drop and access switching
Optical cross-connect switching

A mature optical space switching technology is one of the basic requirements for the emergence of a reliable and flexible optical network.

4.2.1 Switching for Optical Line Protection

Figure 33 represents the use of 1×2 switches for line protection applications. The switches can be used to rearrange the optical path to the backup fiber in case

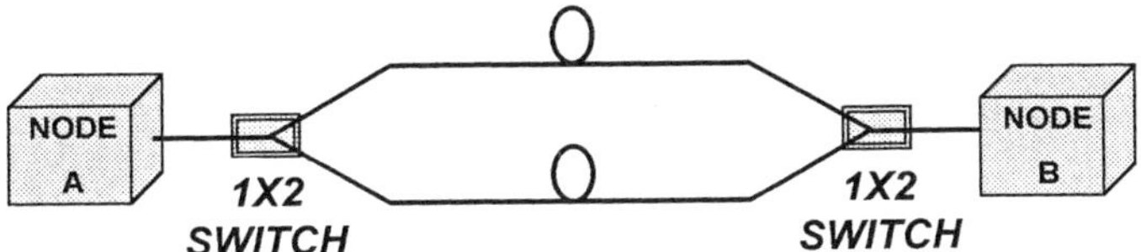

Figure 33 Optical line protection architecture using two 1×2 switches.

of fiber failure in the primary fiber. For this type of application, small port-size switches, such as the 1×2 and 2×2, are often used. Fiber-optic links most often consist of a bundle of fibers. For these links, arrays of optical protection switches can be used. As an example, the integrated 1×2 4-array solid-state optical switch, which consists of four individually controllable 1×2 switches in a single, compact package, is often used for this application. Optical line protection switching in the optical domain will exist next to the protection capability provided by the SONET/SDH layer. However, optical protection switching will be available at a lower cost and at a higher speed. It also provides protection capabilities for transmission protocols that lack protection features.

4.2.2 Equipment Redundancy Switching

High-bit-rate transmitters, receivers, and optical amplifiers dominate the costs of line terminal equipment. Optical switches are used to protect against failures of this line terminal equipment. If switches are employed for this application, significant savings can be accomplished on the total equipment costs. For this application, e.g., 1×2 and 1×4 switches are used, depending on the number of equipment units to be protected by a single switch. Figure 34 shows an example of a transmitter/receiver redundancy protection switching architecture. The transmitter/receiver connected to the 1×4 switch can be switched into any of the other lines in case one of the other transmitters or receivers fails. This way, the number of backup equipment elements can be reduced.

4.2.3 Flexible Optical Add/Drop and Access Switching

For most DWDM systems, it is possible to drop several fixed wavelengths at a specific location. This enables operators to assign fixed wavelengths to specific customers. A drawback of this configuration is that it provides only limited flexibility in terms of the particular wavelength of the carrier and the number of wavelengths to be dropped at the optical add/drop location.

The combination of wavelength (de-)multiplexers and optical switches provides flexible optical add/drop possibilities that can be used to create the required flexibility at the optical add/drop node. This way, rings can be tied together (e.g., in DWDM systems for metro applications) and service can be provided to new

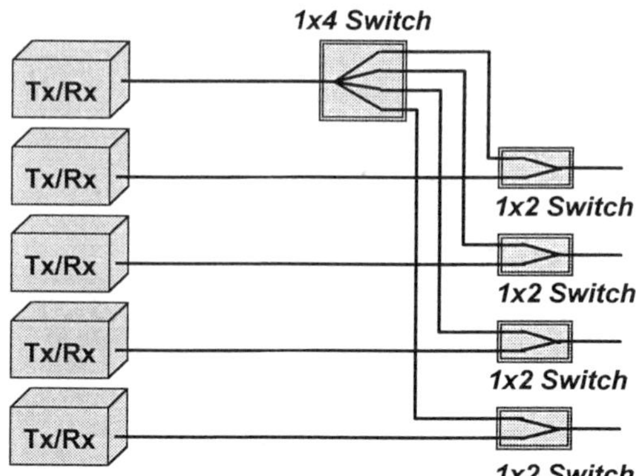

Figure 34 Equipment redundancy architecture using a combination of a 1 × 4 and four 1 × 2 switches. The four 1 × 2 switches can be replaced by a single and compact 4-array of 1 × 2 switches.

customers when required. Figure 35 shows 2 × 2 switches in combination with wavelength (de-)multiplexers. In the bar state (top switch), optical traffic passes the node. In the cross state of the 2 × 2 switch (bottom switch), one of the wavelengths is dropped. New traffic can be added to this same switch and can be launched into the system. The 4-array of BeamBox™ 2 × 2 switches is very well suited for this application.

The optical add/drop switch can also be used as a switch to get access to the (DWDM) optical network. Due to the transparency of the optical switch, a tool

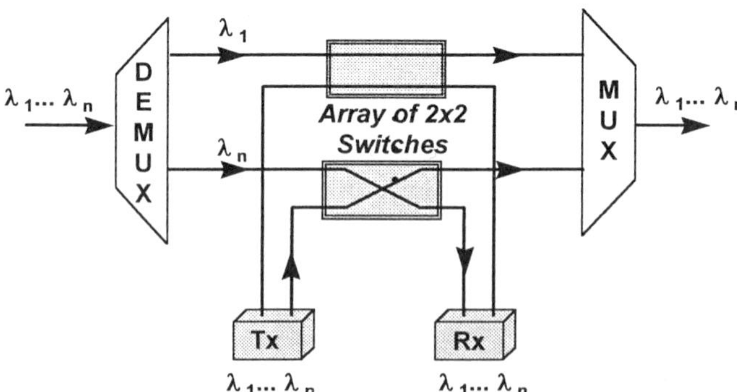

Figure 35 Flexible optical add/drop node. Wavelength (de-)multiplexers and a series of 2 × 2 switches allow adding and dropping of any wavelength at the node.

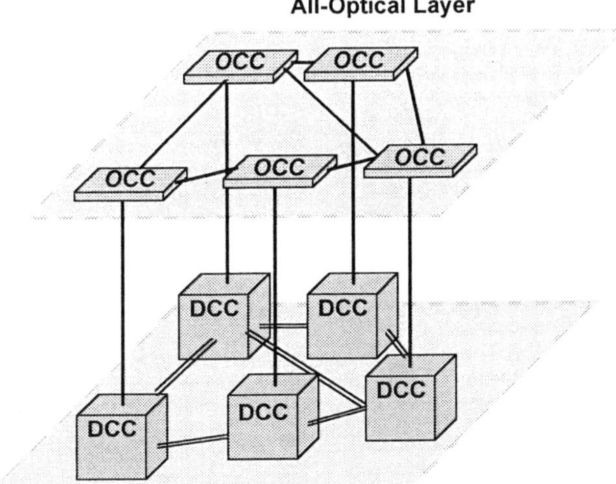

Figure 36 An all-optical reconfigurable transmission layer of optical cross-connects (OCCs) that allows bypassing of the digital cross-connect (DCC).

to get access for a wide variety of transmission protocols is provided by the switch.

4.2 Optical Cross-Connect Switching

The optical cross-connect (OCC) will be located at the major nodes of the all-optical network. Major United States carriers like AT&T and MCI WorldCom have recently done extensive successful trials utilizing OCCs in an all-optical network environment. Other operators are planning trials shortly. The OCC can be used for:

Fast, millisecond protection switching in the optical domain
Network restoration
Network reconfiguration to allocate bandwidth on demand
Bypassing of digital cross-connect (DCC) nodes
Optical wavelength management

The optical cross-connect is expected to be implemented next to the digital cross-connect. Traffic that is passing through a node (and that therefore does not have to be terminated) can be routed directly by the optical cross-connect. This way, the number and size of digital cross-connects can be reduced, providing significant cost savings.

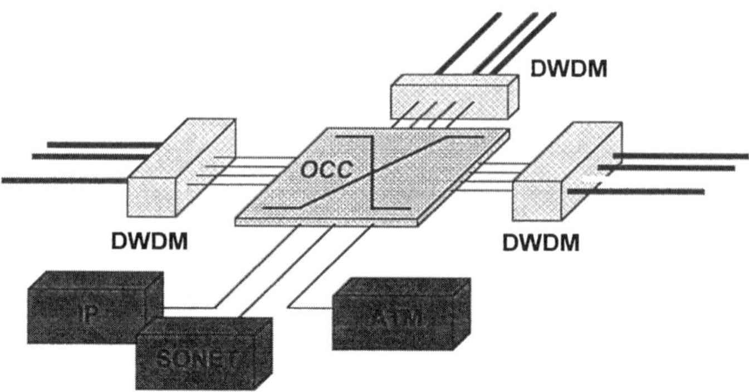

Figure 37 The optical cross-connect (OCC) in the opaque network.

The size of an optical cross-connect at the moment is typically on the order of 8×8, 16×16, or even 32×32. System vendors are asking for larger matrix switches as well, to deal with the increasing number of wavelengths and lines of a system.

Figure 36 shows the situation for a transmission network having a transparent all-optical layer. Some telecom operators, like AT&T, envision a role for optical cross-connects in an *opaque* optical network. In the opaque network, signals are all terminated and demultiplexed at the receiver end in the (DWDM) line terminals. These line terminals are connected via transponders to an optical cross-connect. The signals of the transponders are routed through the optical cross-connect to the transponders of the (DWDM) transmission equipment. One of the benefits of this architecture is that it offers good management possibilities provided by the return to the electrical domain. The performance of the cross-connect for this application has to be optimized for short-distance transmission. Figure 37 shows the setup for this opaque network architecture. The optical cross-connect in this figure also offers access for SONET, IP, and ATM equipment.

4.3. Optical Switch Requirements

There is no such thing as a generic optical switch requirement. Specific requirements are dependent on each application. There are some aspects that are of importance for each optical switch, such as:

Reliability The lifetime of equipment has to be at least 20 years. Since optical switches are often used to enhance the reliability of fiber-optic networks, the switches have to be intrinsically very reliable as well. For this reason, a solid-state technology without moving parts and a hermetically packaged component is highly preferred. The solid-state technology allows unlimited switching between states without wear and tear. In Sec. 2.7 it was shown that a 1 × 2 switch has been switched up to 10 million times. No optical or electrical degradation has been observed during this test. In Sec. 3 it was concluded that the hermetically packaged BeamBox™ solid-state optical switches can meet the reliability requirements for the applications in fiber-optic networks.

Optical performance The insertion loss (preferably zero), the isolation, and the return loss have to be such that error-free traffic transmission can be provided.

Transparency The optical switch and the optical cross-connect have to be transparent to bit rate, wavelength, and polarization. The all-optical layer has to be transparent to different protocols (ATM, IP, SDH/SONET, etc.) at all bit rates and wavelengths in order to avoid a return to the electrical domain for regeneration purposes. The wavelength transparency in both the 1310- and 1550-nm windows for a 1 × 8 switch is shown in Fig. 38. BeamBox™ solid-state optical switches can be used simultaneously in both windows.

Speed Speed is especially important for optical protection applications. SONET/SDH protection switching takes over after 50 ms. Fault location detection and optical protection switching have to be done within this timeframe. The fault location detection consumes most of this time leaving only milliseconds to the optical switch. In Sec. 2.9 it was shown that a fiber-compatible polymeric DOS has a switching time in the range 0.5–2 ms. A switching time of 1 ms for the BeamBox™ solid-state optical switch can be observed in Fig. 39.

Scalability (mainly for OCCs) The switching technology has to provide possibilities to scale up to larger port-count switching matrices. The growth path

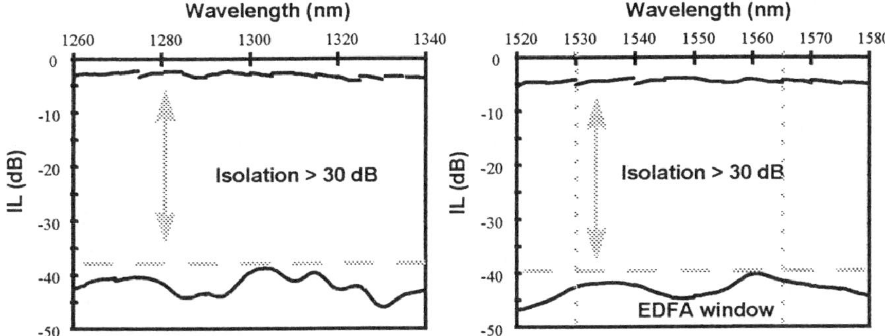

Figure 38 Insertion loss in both the 1310- and 1550-nm windows for a BeamBox™ 1 × 8 switch. The top line represents the insertion loss (IL) of the *on* channel; the bottom line shows the insertion loss for one of the *off* channels.

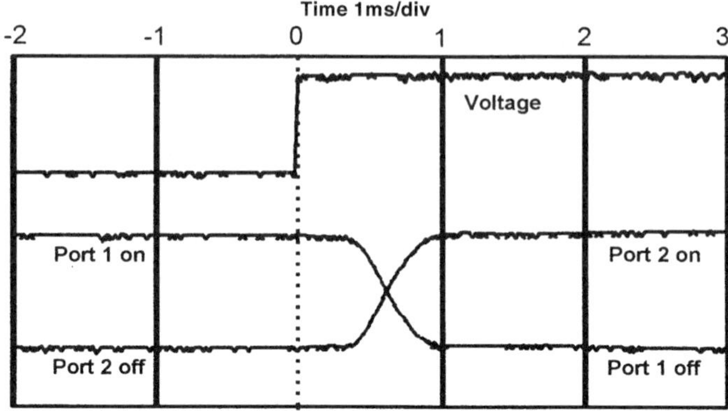

Figure 39 The top line shows the control voltage of 8 V of a BeamBox™ 1 × 2 switch. The bottom lines show the response of both outputs of the switch during the switching action at $t = 0$.

of DWDM systems is to add wavelengths. For optical cross-connects, a growth path in terms of port-count has to be provided in order to scale the system up if required by the operator. The solid-state technology, which is based on a single integrated chip approach, offers the potential to grow to larger-switch-size matrices. Single-chip switching matrices up to the 8 × 8 cross-connect have been realized.

4.4. BeamBox™ Solid-State Optical Switches

BeamBox™ solid-state optical switches are based on a proprietary planar waveguide technology, which has been developed by Akzo Nobel Photonics over the past 10 years. Based on a deep understanding and control of specialty photonic polymers, an integrated thermo-optic technology has been developed to produce fiber-compatible DOS's.

A wide variety of switch configurations can be integrated on a single (opto)chip. Switch sizes vary from the 1 × 2 switch to both the 4 × 4 and 1 × 16 switches on a single chip. One of the examples of the potential for integration is the realization of switch arrays. An array contains individually controllable switches in a single, compact package. Figure 40 shows an integrated BeamBox™ 1 × 8 switch on a single optochip. The optochip is provided with single-mode fibers and electrical connections, after which the assembly is placed in a hermetically sealed package. The compact package can then be mounted on circuit boards to fit into a wide variety of equipment racks. Figure 41 shows a picture of the BeamBox™ solid-state optical switches product line. These BeamBox™ solid-state optical switches:

Figure 40 An integrated BeamBox™ 1 × 8 switch on a single optochip.

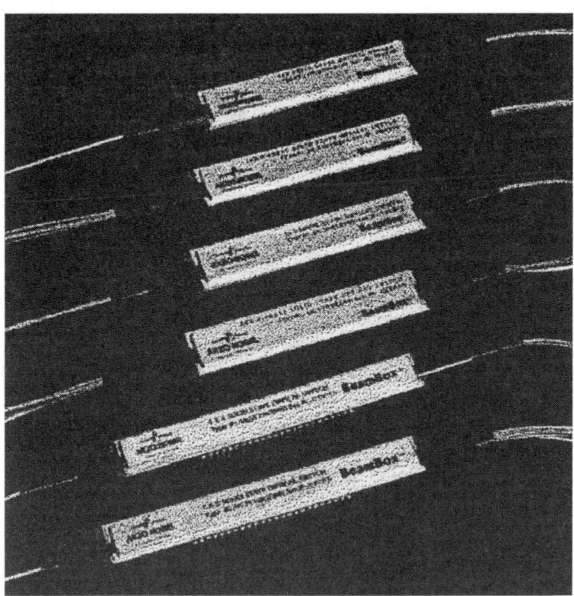

Figure 41 The BeamBox™ solid-state optical switches product line, consisting of a 1 × 2, 1 × 2 4-array, 2 × 2, 2 × 2 4-array, 1 × 4, and 1 × 8 switch.

Contain no moving parts
Are packaged in a hermetically sealed protective package
Allow switching of the optical path within 2 ms
Enable integration and scale-up on the chip level
Have a robust digital switching performance
Are transparent for bit rate, wavelength, and polarization

4.5. Conclusions

Solid-state optical switches take care of one of the basic functionalities in the all-optical layer of the network, i.e., optical switching. Together with the capability to (de-)multiplex optical wavelengths, this will provide the first building blocks for the all-optical layer. The possibility is there to get wavelengths all the way through the networks without conversion to the electrical domain. Optical switching will make this work.

BeamBox™ solid-state optical switches meet the requirements for successful implementation in the all-optical layer. System vendors have started exploiting the benefits of this technology by designing and realizing modules based on this type of switch. These modules have been tested and appear now as products on the market.

REFERENCES

1. Y. Silverberg, P. Perlmutter, and J. E. Baran. Appl. Phys. Lett. 51:1230–1232, 1987.
2. M. Haruna and J. Koyama. Electron Lett. 17:842–844, 1981.
3. M. Haruna and J. Koyama. Appl. Opt. 21:3461–3465, 1982.
4. M. Haruna and J. Koyama. Proc. Second European Conf. on Integrated Optics IEE, London, 1983, pp. 129–131.
5. N. Takato, K. Jinguji, M. Yasu, H. Toba, and M. Kawachi. J. Lightwave Technol. 6:1003–1010, 1988.
6. T. Goh, M. Yasu, K. Hattori, A. Himeno, and Y. Ohmori. Proc. APCC '97, 1997, pp. 232–238.
7. NTT Electronics Technology Corp. Technical Data Sheet TOS3-8M, 1995.
8. M. B. J. Diemeer, J. J. Brons, and E. S. Trommel. J. Lightwave Technol. 7:449–453, 1989.
9. G. R. Möhlmann, W. H. G. Horsthuis, J. W. Mertens, M. B. J. Diemeer, F. M. M. Suijten, B. Hendriksen, C. Duchet, P. Fabre, C. Brot, J. M. Copeland, J. R. Mellor, E. Van Tomme, P. Van Daele, and R. Baets. Proc. SPIE 1560:426–433, 1991.
10. F. M. M. Suijten, M. B. J. Diemeer, and B. Hendriksen. Proc. SPIE 2025:479–487, 1993.
11. M. B. J. Diemeer, F. M. M. Suijten, E. S. Trommel, A. Mc Donach, J. M. Copeland, L. W. Jenneskens, and W. H. G. Horsthuis. Electron. Lett. 26:379–380, 1990.
12. Y. Hida, H. Onose, and S. Imamura. IEEE Photon. Technol. Lett. 5:782–784, 1993.

13. H. Lausen, T. Klein, L. Bersiner, M. M. Klein Koerkamp, M. C. Donckers, B. H. M. Hams, and W. H. G. Horsthuis. Proc. EFOC&N '94, Heidelberg, 1994, pp. 99–101.
14. M. M. Klein Koerkamp, M. C. Donckers, B. H. M. Hams, W. H. G. Horsthuis, H. Lausen, T. Klein, and L. Bersiner. Proc. Integrated Photon. Res. 3(paper FG3-1): 274–276, San Francisco, 1994.
15. N. Keil, H. H. Yao, C. Zawazdki, and B. Strebel. Electron Lett. 30:630–640, 1994.
16. W. Horsthuis, B. Hendriksen, M. Diemeer, M. Donckers, T. Hoekstra, M. Klein Koerkamp, F. Lipscomb, J. Thackara, T. Ticknor, and R. Lytel. Proc. ECOC '95, paper Th.L.3.4., Brussels, 1995, pp. 1059–1062.
17. R. Moosburger, G. Fishbeck, C. Kostrzewa, B. Schuppert, and K. Petermann, Proc. ECOC '95, paper Th.L.3.5, Brussels, 1995, pp. 1063–1066.
18. R. Moosburger and K. Petermann. IEEE Photon. Technol. Lett. 10:684–686, 1998.
19. N. Ooba, S. Imamura, R. Yoshimura, A, Kaneko, T. Kurihara, M. Hikita, Y. Hida, and Y. Hibino. Proc. ACS PMSE 75:362–363, Orlando, 1996.
20. Y. Hida, N. Ooba, R. Yoshimura, T. Watanabe, M. Hikita, and T. Kurihara. Electron. Lett. 33:626–627, 1997.
21. T. Watanabe, N. Ooba, Y. Hida, S. Hayashida, T. Kurihara, and S. Imamura. Jpn. J. Appl. Phys. 36:L1672–L1674, 1997.
22. T. Ido, M. Koizumi, and H. Inoue. OFC '98 Tech. Digest, paper WH6, pp. 148–149, San Jose, 1990.
23. A. Borreman, T. Hoekstra, M. Diemeer, H. Hoekstra, and P. Lambeck. Proc. ECOC '96, paper THD.3.2, Oslo, 1996, pp. 5.59–5.62.
24. K. Propstra, T. Hoekstra, A. Borreman, M. Diemeer, H. Hoekstra, and P. Lambeck. Post-deadline papers ECIO '97, paper PD3, Stockholm, 1997, pp. PD3-1–PD3-4.
25. H. Nishihara, M. Haruna, and T. Suhara. Optical Integrated Circuits. New York: McGraw-Hill, 1989.
26. T. A. Tumolillo, M. Donckers, and W. H. G. Horsthuis. IEEE Commun. Mag., Feb. 1997, pp. 124–130.
27. M. Kawachi. Opt. Quantum Electron. 22:391, 1990.
28. S. Imamura, T. Yoshimura, and T. Izawa. Electron. Lett. 28:2135, 1992.
29. M. Usui, M. Hikita, T. Watanabe, M. Amano, S. Sugawara, S. Hayashida, and S. Imamura. J. Lightwave Technol. 14:2338–2343, 1996.
30. M. B. J. Diemeer, T. Boonstra, M. C. J. M. Donckers, A. M. van Haperen, B. H. M. Hams, T. H. Hoekstra, J. W. Hofstraat, J. C. Lamers, W. Y. Mertens, R. Ramsamoedj, M. van Rheede, F. M. M. Suijten, U. Wiersum, R. H. Woudenberg, B. Hendriksen, W. H. G. Horsthuis, M. M. Klein Koerkamp, et al. Proc. SPIE 2527:411–416, San Diego, 1995.
31. D. W. Van Krevelen. Properties of Polymers. Elsevier, 1990.
32. Y. Hida, H. Onose, and S. Imamura. IEEE Photon. Technol. Lett. 5:782–784, 1993.
33. H. Mueller. Physics 6:179, 1935.
34. R. M. Waxler D. Horowitz, and A. Feldman. Appl. Opt. 18:101–104, 1997.
35. R. M. Waxler and C. E. Weir. J. Res. Natl. Bur Stand. Sect. A 69:325, 1965.
36. S. Timoshenko. Theory of Elasticity. New York: McGraw-Hill, 1934.
37. Y. Inoue, K. Katoh, and M. Kawachi. IEEE Photon. Technol. Lett. 4:36–38, 1992.

38. H. S. Carslaw and J. C. Jaeger. Conduction of Heat in Solids, 2nd ed. Oxford: Clarendon Press, 1960.
39. B. A. Møller, L. Jensen, C. Laurent-Lund, and C. Thirstrup. IEEE Photon Technol. Lett. 5:1415, 1993.
40. R. Moosburger, C. Kostrzewa, G. Fishbeck, and K. Petermann. IEEE Photon. Technol. Lett. 9:1484–1486, 1997.
41. A. A. Bilotti. IEEE Trans. Electron. Devices ED-21:217–226, 1974.
42. J. L. Spencer and D. S. Kobayashi. Establishing reliability and availability criteria for fiber-in-the-loop systems. IEEE Communications Magazine, pp. 84–90, March 1991.
43. L. Wosinska. Reliability study of fault-tolerant multiwavelength nonblocking optical cross connect based on InGaAsP/InP laser-amplifier gate-switch arrays. IEEE Photon. Technol. Lett. 5:1206–1209, 1993.
44. N. Wauters, N. Lemaitre, J. P. Bataille, and M. Groisman. Protection in long distance single and multiwavelength transport networks using optical switch technology. Proc. First Int. Workshop on the Design of Reliable Communication Networks, DRCN '98, Brugges, Belgium, May 17–20, 1998.
45. F. Jensen. Electronic Component Reliability. New York: Wiley, 1995.
46. P. Tobias and D. C. Trindade. Applied Reliability, 2nd ed. New York: Van Nostrand Reinhold, 1995.
47. Bellcore Technical Reference TR-NWT-001073. Generic requirements for fiber optic switches. Issue 1, January 1994.
48. G. Di Giacomo. Reliability of Electronic Packages and Semiconductor Devices. New York: McGraw-Hill, 1997.
49. M. Pecht. Integrated Circuit, Hybrid and Multichip Module Package Design Guidelines. New York: Wiley, 1994.
50. D. A. Jeannotte, L. S. Goldmann, and R. T. Howard. Package reliability. In: R. R. Tummala and E. J. Rymaszewski, eds. Microelectronics Packaging Handbook. New York: Van Nostrand Reinhold, 1988.
51. J. W. McPherson. Accelerated Testing. In: Electronic Materials Handbook, Volume 1: Packaging 1989, pp. 887–893.
52. M. L. Williams, R. F. Landel, and J. D. Ferry. The temperature dependence of relaxation mechanisms in amorphous polymers and other glass-forming liquids. J. Am. Chem. Soc. 77:3701–3707, 1955.
53. MIL-HDBK-217F, Reliability Prediction of Electronic Equipment. Washington, DC:DOD, 1991.
54. CNET. Handbook of Reliability Data for Electronic Components. France Telecom, 1983.
55. ITU-T Recommendation G.911. Parameters and calculation methodologies for reliability and availability of fibre optic systems. April 1997.
56. L. Wosinska, and L. Thylen. Reliability performance of optical cross-connect switches—requirements and practice. Proceedings OFC '98, San Jose, Feb. 22–27, 1998, pp. 28–29.
57. R. Bose, M. R. Garzia, D. A. Hoeflin, D. R. Jeske, and W. B. Paulson. Availability measures for switching systems. Proceedings International Switching Symposium ISS'92, Yokahama, 1992, Vol. 2, pp. 364–368.

58. European Telecommunications Standards Institute. Equipment Engineering (EE); Environmental engineering; Guidance and terminology, ETR 035, July 1992.
59. Bellcore Generic Requirements GR-63-CORE. Network Equipment-Building System (NEBS) Requirements: Physical Protection, October 1995.
60. A. K. Agarwal. Fibre optic components for networks and the subscriber loop. Proceedings NOC '96, Heidelberg, Germany, June 25–28, 1996.
61. M. G. Pecht, A. A. Shukla, N. Kelkar, and J. Pecht. Criteria for the assessment of reliability models. IEEE Transac Components, Packaging Manufacturing Technol Part B 20: pp. 229–234, August 1997.
62. Bellcore Technical Advisory TR-NWT-000357. Generic requirements for assuring the reliability of components used in telecommunication equipment. Issue 3, October 1992.
63. Bellcore Generic Requirements GR-1221-CORE. Generic reliability assurance requirements for fiber optic branching components. Issue 1, December 1994.
64. Bellcore Technical Reference TR-NWT-00468. Reliability assurance practices for optoelectronic devices in interoffice application. Issue 1, December 1991.
65. M. Mortazavi, K. Song, H. Yoon, and I. McCulloch. ACS Preprints 208:198–199, 1994.
66. M. C. Flipse, C. P. J. M. van der Vorst, J. W. Hofstraat, R. H. Woudenberg, R. A. P. van Gassel, J. C. Lamers, E. G. M. van der Linden, W. J. Veenis, M. B. J. Diemeer, and M. C. J. M. Donckers. Recent progress in polymer based electro-optic modulators: materials and technology. NATO ASI Series 3: Photoactive Organic Materials, pp. 227–246, 1996.
67. P. S. Ho and T. Kwok. Electromigration in metals. Rep. Prog. Phys. 52:301–348, 1989.
68. A. Scorzoni, B. Neri, C. Caprile, and F. Fantini. Electromigration in thin-film interconnection lines: models, methods and results. Materials Sci. Rep. 7:143–220, 1991.
69. J. R. Black. Electromigration—A brief survey and some recent results. IEEE Transac Electron. Devices ED-16: pp. 338–347, 1969.
70. I. M. Plitz, O. S. Gebizlioglu, and M. P. Dugan. Reliability characterization of UV-curable adhesives used in optical devices. Proceedings SPIE 2290:150–156, 1994.
71. H. Nagata, M. Shiroishi, Y. Miyama, N. Mitsugi, and N. Miyamoto. Evaluation of new UV-curable adhesive material for stable bonding between optical fibers and waveguide devices: problems in device packaging. Optical Fiber Technol. 1:283–288, 1995.
72. T. Maruno and N. Murata. Properties of a UV-curable, durable precision adhesive. J. Adhesion Sci. Technol. 9:1343–1355, 1995.
73. Y. Hibino, F. Hanawa, H. Nakagome, M. Ishii, and N. Takato. High reliability optical splitters composed of silica-based planar lightwave circuits. J. Lightwave Technol. 13:1728–1735, 1995.
74. M. G. McMaster and D. S. Soane. Water sorption in epoxy thin films. IEEE Trans. Components, Hybrids Manufac Technol. 12:373–386, 1989.
75. H. Hanafusa, F. Hanawa, Y. Hibino, and T. Nozawa. Reliability estimation for PLC-type optical splitters. Electron. Lett. 33:238–239, 1997.
76. J. Yoshida, M. Yamada, and H. Terui. Packaging and reliability of photonic compo-

nents for subscriber network systems. IEEE Trans. Components, Hybrids Manufac. Technol. 16:778–781, 1993.

77. N. Fabricius. Reliability of integrated optical components. Proceedings MOC/GRIN '93, Kawasaki, Japan, 1993, p. 92.
78. R. K. Traeger. Hermeticity of polymeric lid sealants. Proceedings 25th Electronic Components Conference, Delft, April 1976, pp. 361–367.
79. MIL-STD-883D. Test Methods and Procedures for Microelectronics. Washington, DC:DOD, 1991.
80. H. Nagata and N. Mitsugi. Mechanical reliability of $LiNbO_3$ optical modulators hermetically sealed in stainless steel packages. Optical Fiber Technol. 2:216–224, 1996.
81. N. Mitsugi, H. Nagata, M. Shiroishi, N. Miyamoto, and R. Kaizu. Optical fiber breaks due to buckling: problems in device packaging. Optical Fiber Technol. 1: 278–282, 1995.
82. A. O'Donnell. Packaging and reliability of active integrated optical components. Proceedings ECIO '95, Delft, 1995, p. 585.
83. Thermal Management Concepts in Microelectronic Packaging: from Component to System (Project Coordinator, Robert T. Howard, Eds. Stephen S. Furkay, Richard F. Kilburn, Gabriel Monti, Jr.), Silver Spring, MD: International Society for Hybrid Microelectronics, 1984.
84. IEEE Communications Magazine 36:40–78, (February issue) 1998.

9

Hybrid Integration of Optical Devices on Silicon

Robert G. Peall

Nortel Networks
Harlow, England

1 INTRODUCTION

Optical hybridization describes a range of techniques for copackaging optical, optoelectronic, and electronic components. It is distinguished from alternative packaging and integration schemes by the fact that all the components are mounted on a single substrate. The substrate provides the mechanical alignment features to allow optical coupling between the components; it is frequently used to allow electrical interconnection as well. For reasons that will be discussed in detail later, silicon is by far the most researched and commercially exploited substrate material, and it is on silicon-based optical hybrids that this chapter will focus.

Silicon optical hybrids are referred to in various ways: silicon waferboards, silicon motherboards, optohybrids, silicon V-grooves, silicon optical benches. Although the approaches differ in their details, they rely on a few common underlying technologies: silicon micromachining, electrical interconnects, optical waveguides, and flip-chip alignment.

A generalized silicon optical hybrid is illustrated in Fig. 1. A precision groove is etched into the substrate, allowing the fiber to be passively aligned during assembly; this saves the time and cost associated with active fiber alignment. There is a pit etched behind the laser, formed at the same time as the fiber alignment groove, which provides coupling into a rear monitor diode. This simplifies assembly by eliminating any additional optical components. The silicon substrate is patterned with an electrical interconnect, which allows connection to the opto-

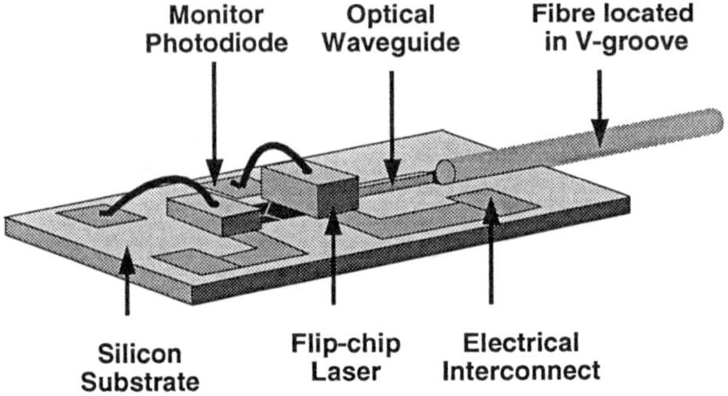

Figure 1 Illustration of a generic silicon optical hybrid (Copyright © Nortel, 1998).

electronic components; these lines can be impedance matched, allowing very high-speed operation. Flip-chip (or solder bump) technology provides both the physical alignment and electrical connection to the optoelectronic components. An optical waveguide is used to couple the light from the front facet of the laser into the optical fiber.

Figure 2 shows a real silicon hybrid substrate. It has some, but not all, of the illustrative features: there is a solder-bump-mounted laser and an etched V-groove for fiber location; a monitor photodiode is mounted behind the laser for rear facet control and a groove is etched between the laser and the photodiode

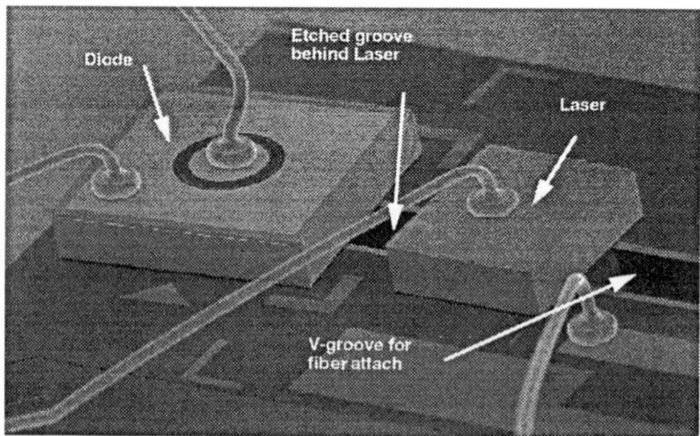

Figure 2 Scanning electron micrograph of a silicon hybrid submount showing laser, monitor diode, and etched grooves. (From Ref. 12. Copyright © Nortel, 1997)

to provide optical coupling; wire bond connections are made between the components, the substrate, and the package. This particular hybrid forms the basis of a commercially available, low-cost optical transmitter: optical hybrid technology is not just a research topic, it is an industry standard.

The most important driver for the development of hybrid optoelectronic integration has been the commercial pressure to reduce the cost of basic telecommunication network elements such as transmitters and receivers. Conventionally, these devices have been housed in metal packages and assembled using multiple ceramic- or diamond-piece parts that are actively aligned and laser welded. The high cost associated with this used to be acceptable because the volumes were relatively low, and reliability was more important than cost. This is no longer true; see Ref. 1, for example. As the optical network has continued to expand and competition has increased among both carriers and component suppliers, the price of basic optoelectronic components has come under great pressure. As a result, lower-cost assembly techniques, such as hybrid integration, have proved essential.

One reason that hybrid optoelectronic integration has been successful is that the goal of monolithic integration has continued to prove elusive. Although there have been numerous laboratory demonstrations of, for example, lasers integrated with rear facet monitors, even these ''simple'' forms of integration have proved unattractive for exploitation; it is hard to achieve high-yield, uniform performance and good reliability in monolithic devices because of the complex growth and processing required. At a more fundamental level, monolithic integration does not address the problem of aligning the completed device to an optical fiber. Hybrid integration allows the individual components to be optimized and offers a solution to the problem of fiber alignment.

Hybrid integration has recently been advanced by some significant improvements in laser technology. One of these is the development of lasers able to operate reliably at high temperatures without cooling and with very reduced thresholds. One result of this has been that thermoelectric coolers, which have generally been a significant proportion of the device cost, have been eliminated from many laser packages. More importantly for hybrid integration, these improved lasers are more tolerant of the thermal impedance between the laser junction and the heat sink. This means that diamond heat sinks are no longer mandatory, as they frequently were in the past, and silicon can be used as the substrate material.

Another improvement in laser technology has been the development of beam-expanded devices [2–4]. By matching the laser output field more closely to that of an optical fiber beam expanders allow both a significant reduction in the required alignment tolerances and the elimination of additional lenses in the optical train. This opens the way to nonhermetic packaging techniques [5,6]; these require that the space between the fiber and the laser be filled with a silicone or similar encapsulant to prevent water condensation. Because of the index-matching effect of the silicone, conventional lenses cannot be used; this is not the case with beam-expanded lasers, which are not significantly affected. The commercial implications of nonhermetic packaging are important; devices in a plastic package may prove to be extremely low cost, because the assembly can be automated and the raw materials are very cheap.

Although the drivers for developing hybrid optical integration are clear, the technical challenges associated with it are not insignificant. The basic underlying technologies are well established: anisotropic etching of grooves, deposition of solders for component connection, definition of electrical interconnect patterns, the formation of optical waveguides, etc. The challenge is to integrate these technologies while maintaining the required positional tolerances. The dimensions of optohybrid structures (typically 10–100 microns) are generally an order of magnitude greater than those of integrated circuits, for example, fiber-holding grooves can be up to 100 microns deep. Unfortunately, the tolerances required are similar to those for integrated circuits (0.1–1 microns). As a result of the large feature sizes, fine-tolerance photolithography is often compromised, especially when the substrates have, for example, deep-etched grooves for fiber placement or when patterning is required close to large ridges.

One of the widespread underlying beliefs behind optical hybridization has been that the cost associated with precision active alignment dominates the cost of optoelectronic packaging. Hence, simplifying the alignment process at the expense of more complex device and substrate processing will lead to an overall reduced cost. In this chapter the various ways in which this simplification has been attempted will be examined. The problem has generally been separated into two questions: (1) How do I align the optoelectronic component, for example, the laser or photodiode? (2) How do I align the optical component, for example, a fiber.

2 SILICON AS A SUBSTRATE FOR OPTICAL HYBRIDIZATION

Silicon is an excellent substrate material for manufacturing optical hybrids [7,8]. Due to its use in the semiconductor industry, it is widely available as 4-, 6-, and 8-inch wafers. These wafers are robust and relatively cheap; they are also well specified in terms of doping level, crystal cut, flatness, etc. Equipment for processing silicon wafers in high volumes can also be obtained easily: machines for metal and dielectric deposition, photolithography, and etching.

Thermal conductivity and expansion coefficient are important parameters for any substrate material. Table 1 gives a comparison between silicon and some other common substrates: diamond, silicon carbide (SiC), alumina (Al_2O_3), and

Table 1 Thermal Conductivity and Expansion Coefficient of Various Materials

Material	Thermal conductivity, W/cm°C	Expansion coefficient, 10^{-6}/°C
Si	1.57	2.33
Diamond	20	1.0
SiC	3.5	3.3
Al_2O_3	0.5	5.4
AlN	0.7	4.0
SiO_2	0.014	0.55
InP	0.68	4.5
GaAs	0.54	6.8

Source: Refs. 8 and 9.

aluminum nitride (AlN). Also shown are some of the common device materials: silica (SiO_2), indium phosphide (InP), and gallium arsenide (GaAs).

Although having a relatively high thermal conductivity, silicon is inferior to diamond and silicon carbide. However, these materials are expensive and not amenable to wafer-scale processing or micromachining. Both alumina and aluminum nitride are also frequently used as substrate materials, because of their low cost and widespread use as thick-film electrical substrates. These ceramics have poor thermal conductivity in comparison with silicon, which makes them unattractive for mounting lasers, and they are not suitable for very high-precision micromachining.

In order for hybrid assemblies to be reliable under thermal cycling it is important that the expansion coefficient of the substrate and that of the mounted components match closely. Silicon has an expansion coefficient intermediate between InP and silica, and so any stresses generated are shared between the optoelectronic components (typically InP) and the optical components (typically silica). The ceramics, though closely matched to InP, are poorly matched to silica. At the other extreme, diamond, because of its low expansion coefficient, can generate high stresses in InP components.

Taken together, low cost, high availability, good thermal conductivity and expansion match to other materials are sufficient to make silicon of significant interest as a substrate material. However, for optical hybrid integration, some of its other properties are perhaps more significant: it is amenable to a range of micromachining techniques, including anisotropic etching; as an electronic substrate it can be used for flip-chip mounting and for forming matched-impedance interconnects; it is an excellent substrate for low-loss optical waveguides. Each of these properties is important and will be described in detail in the following sections.

2.1 Silicon Micromachining

Silicon micromachining is well established as a technology for fabricating a range of devices, including pressure sensors, transducers, and accelerometers [8]. It has been apparent for some time that the same machining techniques could be used to fabricate precise ''optical bench'' structures [10,11]. In particular, one of the most attractive features of a silicon substrate is its suitability for forming high-precision grooves for aligning optical components. These features make use of the anisotropic etch properties of silicon; additionally, a number of other useful structures can be fabricated using reactive ion etching (RIE).

2.1.1 Anisotropic Etching

Precision V-grooves are generally produced using an anisotropic wet etch. The principle of this is illustrated in Fig. 3. The silicon is masked with a material such as silicon nitride in which windows are opened up. When exposed to the etchant, V-shaped grooves are formed, with a width determined by the window opening and side walls determined by the crystal planes.

There are a number of anisotropic etchants for silicon [8]; of these, two are most significant: ethylene diamine, pyrochatechol and water (EDP), and potassium hydroxide and water (KOH). Both of these etchants can be masked by a variety of materials, including SiO_2, Si_3N_4, and Au, and they show high (110)-

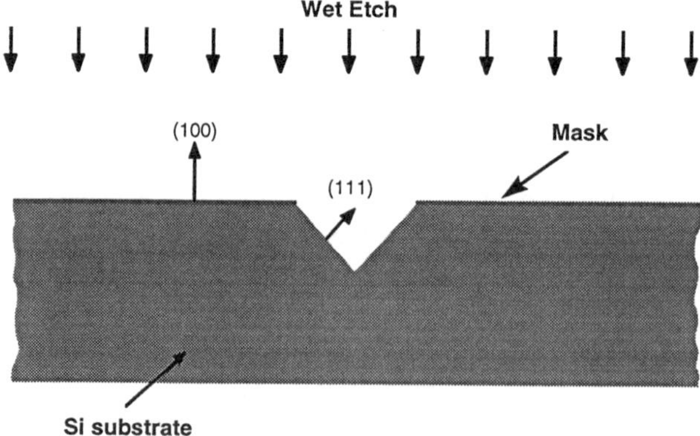

Figure 3 Principle of anisotropic V-groove etching (Copyright © Nortel, 1987).

to-(111) etch-rate ratios (typically 35:1). These high differential etch rates make them particularly useful for etching grooves with good dimensional control in (100)-oriented silicon; there is little undercutting of the mask material. EDP and KOH each have their own advantages and disadvantages. EDP is highly selective, showing minimal attack of SiO_2 layers, and the quality of the etched facets produced are excellent; however, it is expensive in comparison with KOH, has significant health and safety implications, and requires greater preparation before use. The main disadvantage of KOH is that it etches SiO_2, which can be important when silica interfaces are present, for example, at the overgrowth points where an underlying interconnect has been passivated.

The (100) silicon surface orientation is most commonly used in optical hybrid applications. The alternative, (110) orientation leads to vertically sided structures, which are not as suitable for accurately mounting optical components such as fibers. Because of the anisotropic nature of the etch, care needs to be taken when laying out structures so that they lie along the crystal axes. If the structures are misaligned, then rapid undercutting of the mask will occur to reveal (100) facets. This undercutting does not occur uniformly but tends to follow other etch planes, such as the (331). Care also needs to be taken when designing complex structures such as those with convex corners, for these are also rapidly undercut, with the etch time and the precise etch rates of the higher-order planes determining the final geometry.

2.1.2 Fiber Placement: Geometry and Tolerancing

Where the V-groove is being used for fiber placement, the height of the fiber core above the substrate is determined by the V-groove width, the fiber diameter, and the undercut. The depth of the groove is not important, because the fiber rests on the sides of the groove and not on the bottom.

Simple geometrical relationships govern the position of the fiber core with respect to the silicon surface. These are determined mainly by the fixed angle of the silicon crystal planes. This relationship is shown in Fig. 4, where

θ = angle of V-groove, 54.74°
d = etch depth
m = height of fiber axis above the surface of the silicon
r = radius of fiber/lens
u = undercut of mask material
v = final V-groove width
w = width of mask opening

The V-groove width is related to the mask opening by:

$$v = w + 2u$$

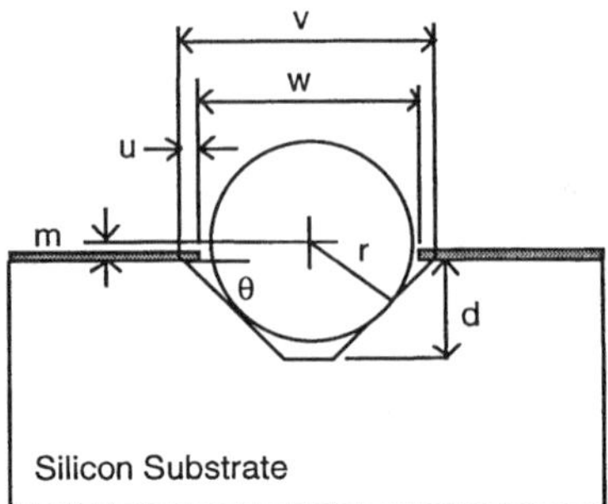

Figure 4 The geometry of a fiber aligned in an etched V-groove (Copyright © Nortel, 1998).

The height of the fiber axis above the silicon surface is given by

$$m = \frac{2r - v \sin \theta}{2 \cos \theta}$$

The width of the mask opening for a given axis height is

$$w = \frac{2(r - m \cos \theta)}{\sin \theta - 2u}$$

A number of factors can contribute to misalignment of the fiber in the groove. One of the most significant contributions comes from variations in the width and undercut rate of the V-grooves; these can both be controlled to submicron accuracies by careful processing. One way of presenting this tolerance is to measure the statistical variation in V-groove widths and to convert this to an equivalent variation in fiber axis height. The results achieved by Nortel using their process are shown graphically in Fig. 5 [12]. This shows that a 3σ tolerance of 1.6 μm is achieved, easily good enough to allow coupling of single-mode fibers.

Other significant contributions to fiber misalignment come from variations in the absolute fiber diameter and in the concentricity of the fiber core and cladding; these contribute to both lateral and vertical offsets. For a well-controlled process, the diameter can be controlled to better than 0.5 μm and the core concentricity to 0.3 μm. Because of their statistical nature, these variations set one of the fundamental yield limits on passive laser-to-fiber coupling.

2.1.3 Balls and Mirrors

Although V-grooves provide excellent fiber alignment structures, etched pits can be used to place ball lenses to very high precision. One significant application of these structures is in the manufacture of low-cost headers for coaxial lasers

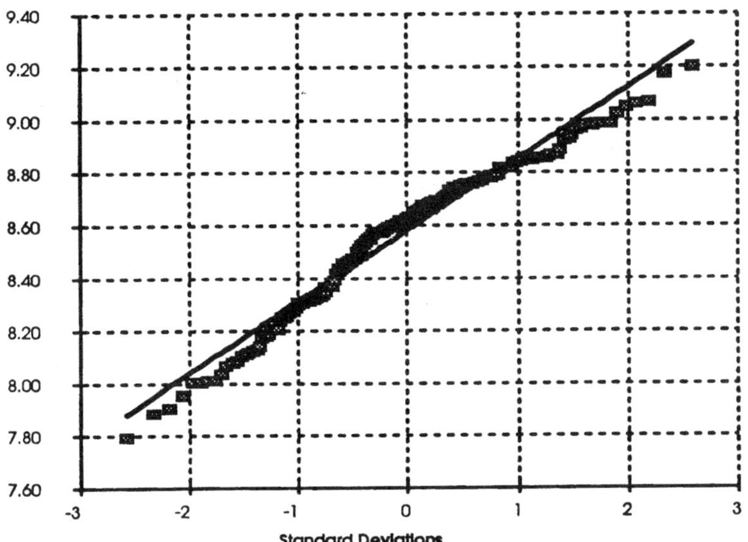

Figure 5 V-groove width control, represented as a linear-normal distribution of fiber heights that result from variations in the V-groove width. (From Ref. 12. Copyright © Nortel, 1997)

[13]. The etch pit provides a three-point contact for the ball lens, and its distance from the laser front facet can be set very precisely. More complex structures can also be formed in this manner where, for example, ball, drum, or GRIN lenses are used to perform expanded beam coupling of a laser to a fiber [14,15].

The previous paragraphs have dealt with the use of etched structures as alignment features for optical components. Another useful function they perform is as reflective elements. One application is illustrated in Fig. 6(a), where light is coupled from an optical fiber into a photodetector. Unlike laser-to-fiber coupling, where the core of the fiber lies above the substrate surface, wide grooves are used for photodiode coupling so that the fiber lies beneath the plane of the silicon. In the 1300- and 1500-nm regions, silicon is partially transparent, so to minimize the interface loss, the facets are metalized with Au or Al [16,17]. This metalization can be patterned in a number of ways, including etching and deposition through a thick-film resist mask. As a result of the metalization, near lossless interfaces can be formed. Another important application of a metalized groove is as a reflector for monitoring the output power from the rear facet of a laser. This is illustrated in Fig. 6(b). In these applications a relatively poor coupling efficiency (of the order of a few percent) is acceptable, and so no lens is used to collimate the signal. This type of approach is being exploited by a number of

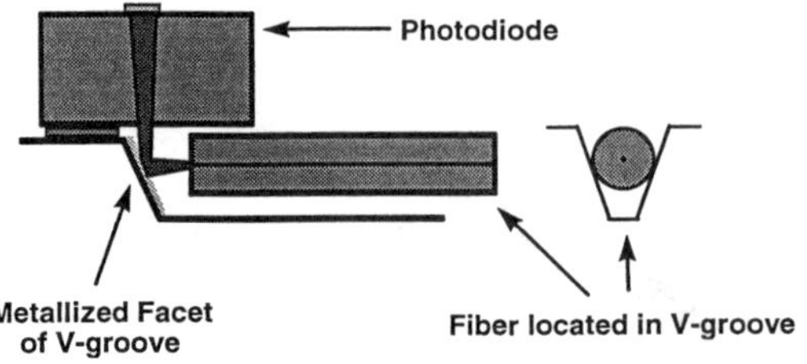

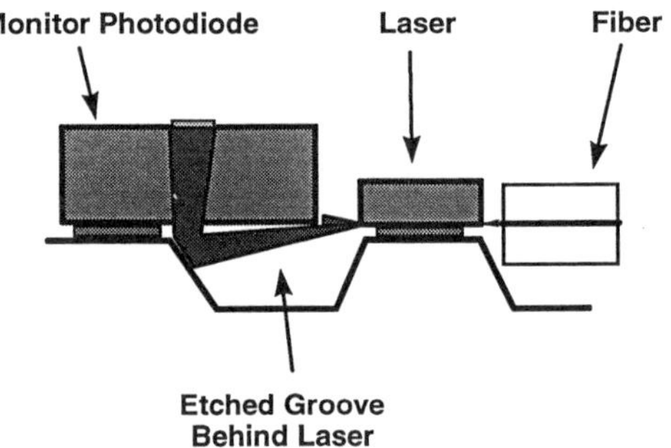

Figure 6 Use of etched grooves as reflectors (a) to couple light from a fiber to a photodiode, (b) to couple light from a laser rear facet to a photodiode. (Copyright © Nortel, 1997)

companies [12,18]. The use of these hybrid integrated monitor diodes is important, because it significantly simplifies both assembly and "burn in" compared with more conventional monitor-on-block approaches.

2.1.4 Process Routes

A fundamental challenge of optohybrid technology is how to integrate the V-groove etching, which inevitably results in a nonplanar substrate, with the formation and patterning of the other layers. Numerous different ways of doing this have been tried, with varying degrees of success.

One way of proceeding is to etch the V-grooves at the start of the process [14,19,20]. The advantage of this is that there is then no concern over the incompatibility of the etchants with any of the other materials. However, once deep grooves have been etched, the resulting substrate is not amenable to conventional patterning; spun photoresist tends to thin at the top edges of grooves and is much thicker at the bottom, and a wake is left around densely patterned features. A number of alternative resist systems can be used, including a planarizing resist, which fills the grooves, and a dry film resist, which "tents" over the V-groove area. Electrophoretic resists have been used successfully by Lucent Technologies. These are electrically deposited resists that have the ability to uniformly coat electrically conductive surfaces [19]. Complete optical bench structures, including interconnect, bond pads, mirror structures, and solder layers, have been formed in this way. An intriguing variation on the etch-first approach is the planarizing process reported by IMC [21]. They use a sacrificial polysilicon layer to define the V-groove area, which is overlaid by a silicon nitride layer into which holes are etched. When exposed to the V-groove etchant, the polysilicon is rapidly removed, leaving the silicon nitride as a membrane over the etched area. The holes in the membrane are subsequently sealed by a thicker silica layer, leaving a planar substrate for further processing. The membrane is removed at the end of the process to reveal the V-grooves.

An alternative way to proceed is to etch the V-grooves toward the end of the process [13,22]. This allows the interconnects, solder-wettable pads, and dielectrics to be deposited and patterned using conventional planar techniques. For example, standard photoresists can be used, and these can be spun on using wafer-track equipment. The advantage of the V-groove-last method is that excellent control of the feature sizes can be achieved without investment in nonstandard equipment or processes. The major disadvantage is that all the materials deposited on the substrate need to able to withstand the aggressive V-groove etchants. The silicon etchants are likely to attack metals such as aluminum; they will also attack weaknesses in any silica layers, such as at overgrowth points. The challenge with this route is to develop a material system inert to the V-groove etchant.

2.1.5 Reactive Ion Etching (RIE)

There are limitations to the use of the anisotropically wet-etched structures described in the previous section [23]. One is that the orientation of the grooves or pits is set by the crystal axis. This means that structures that would benefit from off-axis alignment are difficult to fabricate; an example of this is where an angled optical train is needed for reflection suppression. The second disadvantage is that the endfaces of the V-grooves frequently prevent sufficient proximity between the fibers and the optoelectronic components; it is common practice for this facet to be removed, with a shallow trench produced, by a wafer-dicing saw. In many cases, however, a sawn trench is undesirable, for it limits the possible layout of the substrate.

An attractive technique that does not impose these limitations is reactive ion etching. Reactive ion etching does not rely on the silicon orientation, and can produce near-vertical side walls, allowing components to be brought into near contact. Complete optical benches, including fiber-holding grooves, have been fabricated using this technique, by, for example, Schmidt et al. [24]. They produced deep, U-shaped grooves for fiber alignment by using a highly anisotropic etch, which due to side-wall passivation, gave an etch-rate ratio of 10:1 (vertical:

horizontal). Their structures also included structures for self-aligning lasers and trenches for monitoring the laser output.

2.2 Optoelectronic Component Mounting and Alignment

The previous section dealt with the alignment of optical components to the substrate. The next important consideration is the mounting and alignment of the optoelectronic components. This carries a further level of complexity, in that electrical connection must be made to the component in addition to high-precision alignment. In the same way that micromachining techniques were ''borrowed'' from more mature applications, component-mounting and interconnect techniques have also been ''borrowed.'' The most notable of these is flip-chip mounting.

2.2.1 Flip Chip

Flip-chip, or solder bump, technology was first developed by IBM in the late 1960s [25]. It is a method of forming numerous electrical connections simultaneously between a component—for example, an integrated circuit—and a substrate.

The assembly process is outlined in Fig. 7. Matching solderable pads are formed on the component and the substrate during their fabrication. The solder on one of the sets of pads is then preflowed to form bumps. During bonding, the two are brought into contact and a tack bond is formed. When the solder is again reflowed, it wets onto the other pad, forming a permanent solder joint. Under the right conditions, the surface tension of the molten solder will pull the component and substrate into accurate alignment to form a self-aligned solder bond.

Within the microelectronics industry, flip-chip mounting has frequently been applied to devices with large numbers of contacts, where the ability to form all the connections simultaneously has proved particularly attractive [26]. The electrical properties of the solder bump bonds are also attractive [27,28]. Unlike wire bonds, solder bumps have very low inductance, which is particularly important for current-driven devices such as lasers. Also, unlike wire bond pads, which

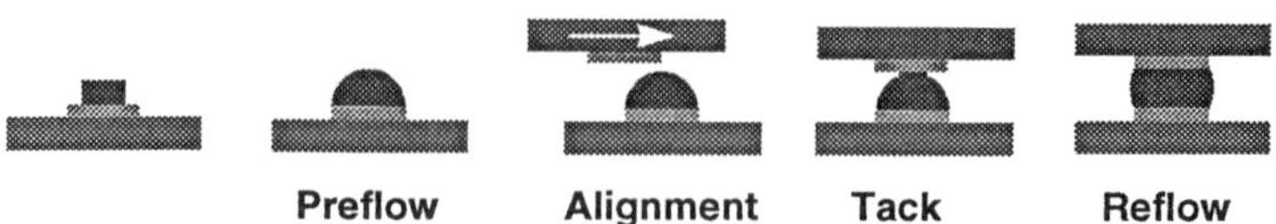

Figure 7 The principle of the solder bump alignment process (Copyright © Nortel, 1997).

need to be physically large (greater than 70-micron diameter), solder bump pads can be made very small (20–50 microns in diameter) and hence can have low capacitance; this is important for devices such as photodiodes.

Other properties of solder bump joints, namely, lateral self-alignment and high-accuracy vertical positioning, have proved very important to hybrid integration. One implementation, similar to that employed by IBM, is to use near-hemispherical bumps and to rely on the self-aligning forces to provide angular and lateral positioning and on the control of the solder volume and bump geometry for the vertical positioning. The tolerancing of these types of joint has been widely studied [29,30]. In their application [31,32], GPT (formally Plessey Research and Technology) used a Cr-Cu-Au wettable-pad structure, and PbSn solder formed by codeposition in an electron beam evaporator. Bumps produced in this manner have been applied to a range of devices, including modulators, lasers, and lithium niobate waveguide devices [33–35]. The assembly process has both been high yielding and demonstrated good alignment accuracies (horizontal and vertical) of around 1 micron.

National Equipment Corp. (NEC) has also reported [36–38] the use of hemispherical bumps, this time fabricated in AuSn solder. AuSn solder, unlike PbSn-based solders, is attractive because it is hard, has low creep, and can be used without flux. In order to achieve the thickness of solder required, they used preforms punched from a AuSn alloy ribbon. This avoided the need to evaporate or electroplate the AuSn, which can prove difficult to control in terms of composition and volume for thick layers. The bumps were reflowed in an N_2 atmosphere, and self-alignment was demonstrated to absolute accuracies of ± 2 microns.

An alternative approach is to use very low-profile bumps. These bumps have less tendency to self-align and so the lateral positioning relies more heavily on the accuracy that can be achieved by the flip-chip aligner bonder; this route has been followed by Nortel [22,39,12] and Fujitsu [40,41]. In their application, Nortel uses AuSn solder bumps formed by electron beam evaporation, and reflow is carried out without flux. Because of their small volume, these bumps are tolerant of the absolute bump geometry and the solder volume, and, as shown in Fig. 8(a), give excellent control of the vertical position. The lateral placement accuracy, shown in Fig. 8(b), is determine mainly by the performance of the flip-chip bonder; machines capable of placing devices to micron accuracies are commercially available.

An alternative to solder bump bonds is the use of thermocompression joints [20]. As with the low-profile solder joints, this approach gives good vertical control of the height positioning. It also relies on the accuracy that can be achieved with the flip-chip aligner bonder.

In comparison with other alignment techniques, such as the use of standoffs, the flip-chip approach is relatively simple; it requires little additional processing of the laser or photodiode beyond what it would normally receive. There are

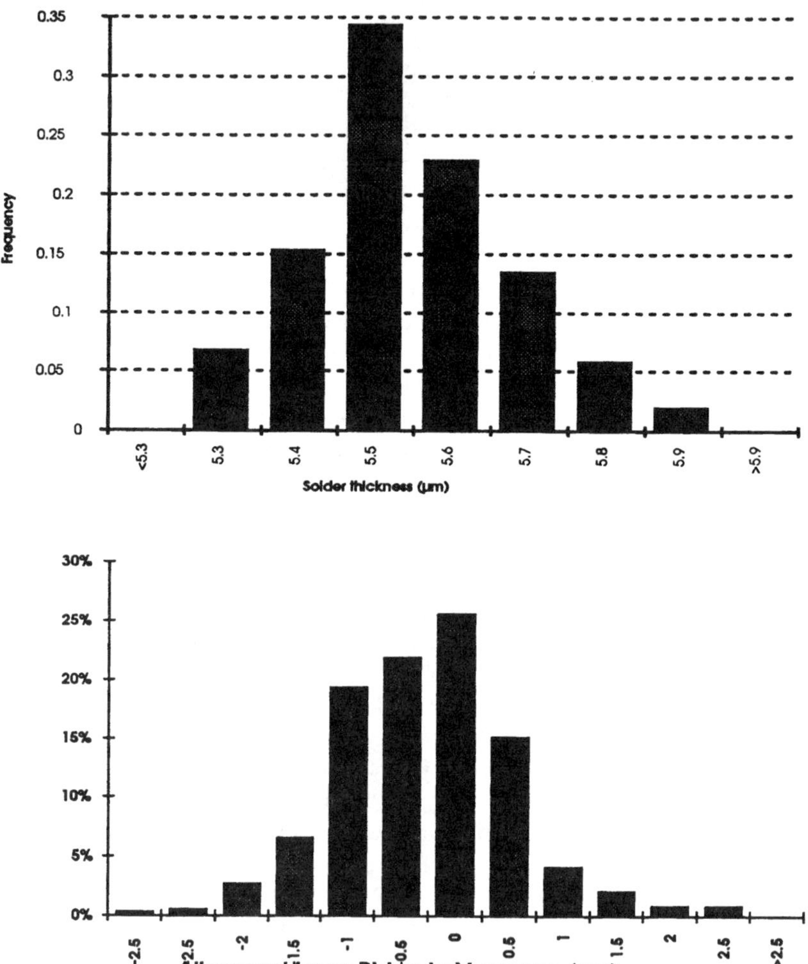

Figure 8 Graph of solder bump alignment accuracy. (a) Distribution of lateral alignment offset of laser ridge to V-groove. (b) Distribution of solder thickness of bonded lasers. (From Ref 12. Copyright © Nortel, 1997)

some disadvantages. One is that the final device position is controlled by statistical processes: the placement accuracy, the degree of self-alignment, the solder bond geometry, and the solder volume. Another is that the metal pattern on the component is rarely self-aligned to the feature of interest (the laser ridge or diode active area). This introduces an additional fabrication tolerance that needs to be well controlled.

2.2.2 Standoffs and Other Alignment Techniques

With flip-chip solder bump technology, the vertical and lateral device position is determined by the placement procedure and the solder. Alternative approaches do not make use of the solder to determine the positioning, but rely on mechanical standoffs or grooves formed on the substrate that are then aligned to physical features (such as grooves, ridges, or precision cleave facets) produced on the device.

General Telephone and Electric (GTE) has made extensive use of standoffs to assemble transmitter and receiver arrays [42,43]. A key aspect of their technology is the ability to form precise notches in the InP material from which the lasers and photodiodes are made. The position of these notches is determined in the first photolithography stage of processing, and so their positional accuracy relative to the active part of the device is a few tenths of a micron. The notches are 13–15 μm deep and wet etched with very low undercut (less than 0.2 μm) by using preferential chemical etchants [44]. During assembly the notches are butted against a set of silicon ''alignment pedestals'' that are themselves well aligned with respect to the fiber placement grooves. The vertical positioning is determined by another set of standoffs, this time formed of polyimide. The devices are held in place using an indium solder.

A similar concept has been reported by BT [17,45], which has used standoffs to align lasers and detectors to arsenic-doped glass (AsG) waveguides. In this case, the solder bond pads are recessed below the surface of the silicon in shallow, anisotropically etched grooves. During assembly the device is brought directly into contact with the silicon surface, and this provides the vertical positioning. Fine lateral positioning is achieved by bringing a ridge on the laser into contact with the wall of a narrow groove etched into the substrate. The ridge is formed at the same time as the laser stripe and is self-aligned to it. Poly silicon standoffs are used to fix the separation between the waveguide endface and the laser or photodiode.

IBM has also made use of standoffs, in their case to align laser and detector arrays to multimode optical waveguides [46,47]. The principle of this alignment scheme can be clearly seen form a scanning electron micrograph (SEM) of one of their devices, which is shown in Fig. 9. The waveguides are fabricated using a CVD flame-hydrolysis process [48], and the standoffs are formed at the same time. In order to provide alignment features, 20-μm-deep trenches are etched

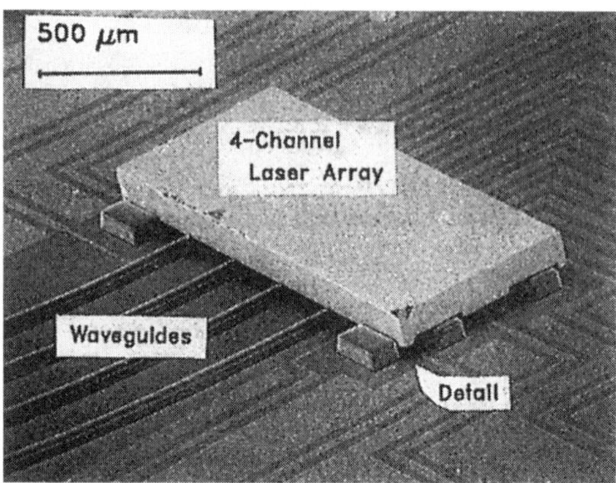

Figure 9 SEM of a laser array aligned to a silicon motherboard using etched standoffs. (From Ref. 46. Copyright © IBM, 1992)

into the components using chemical-assisted ion beam etching (CAIBE). During assembly the trenches interlock with the standoffs to determine the final device position, which has been found to be ±1.75 μm vertically and ±2.0 μm laterally. An interesting feature of this technology is the use of PbSn solder bumps to pull the components into alignment. Unlike conventional flip-chip alignment, the pads on the substrate and device are offset so that throughout the bond cycle the solder pulls the component against the standoff.

A technique for obtaining self-aligned coupling between single-mode optical fibers and a range of optoelectronic devices has been reported by ETH [49–51]. This differs from the techniques described earlier in that no additional features are formed on the substrate; instead, alignment ridges on the device are brought into contact with the edges of the fiber alignment groove. The technique is self-aligning, in that any process variations in the V-groove width cause the fiber and the device to move by the same amount.

The use of mechanical alignment features has proved an attractive alternative to solder-based alignment techniques. Standoffs avoid the statistical variations of solder deposition and reflow by providing a physical reference between the features of interest. Also, complex and high-precision (and expensive) flip-chip bonding equipment is not required. However, more complex motherboard and device processing are frequently needed to deposit and pattern the standoffs, and their dimensions are subject to variations during volume manufacture.

2.2.3 Electrical Interconnect

A key question in the fabrication of complex optoelectronic modules is how to integrate the electrical interconnect with the optoelectronic interface. More specifically, to what degree is it appropriate to integrate the electronic circuitry into the silicon motherboard?

The parasitic effects that arise from the conducting silicon substrate have been known for some time [52,53]. Even with the highest available resistivity (5–10 kΩcm), these effects are significant at the frequencies of interest within telecommunication systems. Hence, the silicon substrate is not suitable for use as the dielectric material in microstrip line structures without significant high-frequency loss.

One way around the problem of silicon resistivity is to form coplanar line structures on the silicon surface with a thick dielectric layer to isolate the fields from the underlying conductive silicon [20,54]. As will be discussed later, this thick dielectric may be provided by the undercladding of an optical waveguide [66]. Coplanar structures, because they require only a single interconnect layer, are relatively simple to fabricate and have been used to manufacture optoelectronic modules operating at line rates up to 2.5 Gbit/s [55,20].

Microstrip lines are more attractive than coplanar lines when complex interconnect structures are required, because they can be fabricated on a fine pitch without requiring the wide areas of ground plane needed to prevent inductive effects in the signal return path. As we have said, because of its finite resistivity, silicon is not suitable for forming the dielectric in a microstrip line structure. The silicon hybrid industry has circumvented this problem by fabricating full microstrip structures (ground plane, dielectric, and interconnect) on the silicon surface; the substrate plays no part in the electrical circuit. In order to obtain 50-Ω microstrip lines of sufficient width to allow significant current-carrying capacity, these circuits require dielectrics of thicknesses greater than 20 microns. Silica dielectrics of this thickness can be difficult to deposit onto a metal ground plane and take a long time to etch. One way around this is glass microwave integrated circuits (GMICs). These are formed by physically bonding a 200-micron-thick layer of silica onto a conventional silicon substrate [56,57]. Such GMICs are attractive because they offer a high level of integration of passive electronic components.

Although 50-Ω microstrip lines require relatively thick silica layers, this is not true of lower-impedance (5–10 Ω) lines. Indeed, a high current-carrying, low-loss microstrip line of around 10 Ω can be formed with a 3-micron silica layer and a 200-micron track width. These geometries are easily coped with by conventional deposition and photolithographic processes. This is very significant for driving directly modulated semiconductor laser devices, which have an impedance around this value. There are two circumstances where a low-impedance laser drive can be beneficial. One is at the interface to a laser array, where terminating resistors cannot be placed on a sufficiently fine pitch. The other is when the laser driver is to be copackaged with the laser. In this instance the intervening resistor, usually used to reduce the effect of parasitics at the package wall, is of no benefit. An example of where both of these features were combined is the 1×8 8-Gbit/s transmitter module [58]. This module (shown in Fig. 10) consisted of a silicon V-groove submount copackaged with a custom laser driver integrated circuit (IC). The laser array was solder bump mounted onto the V-groove carrier, p-side down, and connected to the laser driver via an array of low-impedance lines. The lasers were driven by direct modulation. The device operated at up to 8 Gbit/s, with the speed limited by the self-resonant frequency of the laser rather than the interconnect or the line parasitics.

Electrical multichip modules have generally used polyimide as the dielectric material rather than silica, because it can be deposited using conventional spinning techniques and can be etched rapidly in an oxygen plasma. Although it has been successfully demonstrated in the laboratory [59,39], the ''monolithic'' integration of polyimide electrical interconnects with etched fiber alignment structures has proved difficult. The reason for this is that polyimide structures are difficult to protect during the V-groove etch process. A much more successful

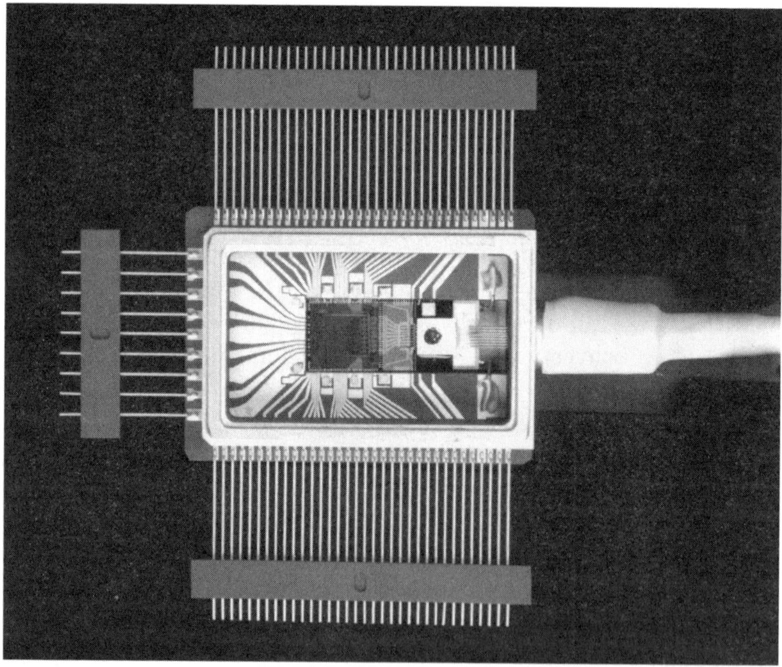

Figure 10 Photography of a 1×8 8-Gbit/s transmitter module. (From Ref. 58. Copyright © Nortel, 1997)

integration of polyimide-based interconnects has been a hybrid approach in which small silicon carriers taking fibers and optoelectronic components have been hybrid integrated with driver circuitry onto an underlying silicon hybrid circuit. Using this technique, a number of multichannel modules have been successfully assembled. One example of this is a 12-channel 2.5-Gbit/s receiver array module [60], shown in Fig. 11. This consisted of a 12-channel receiver array IC and a photodiode carrier mounted onto a silicon hybrid motherboard. The motherboard carried a two-level fine-pitch aluminium microstrip line structure with a polyimide interlevel dielectric. The motherboard provided fan-out of the high-speed signals from the IC to the package wall, and carried front-line decoupling of the IC power supply. This higher-level form of hybrid integration, in which a carrier taking the optoelectronic component and the fiber alignment grooves is itself mounted onto a true electronic hybrid, has been used widely [61–63].

In summary, electrical interconnects operating at rates up to and beyond 10 Gbit/s can be fabricated on optical hybrid substrates. However, care is needed in their design to avoid the parasitic effects arising for the substrate resistivity. One way to avoid these effects is, in a fully integrated fashion, using thick polyim-

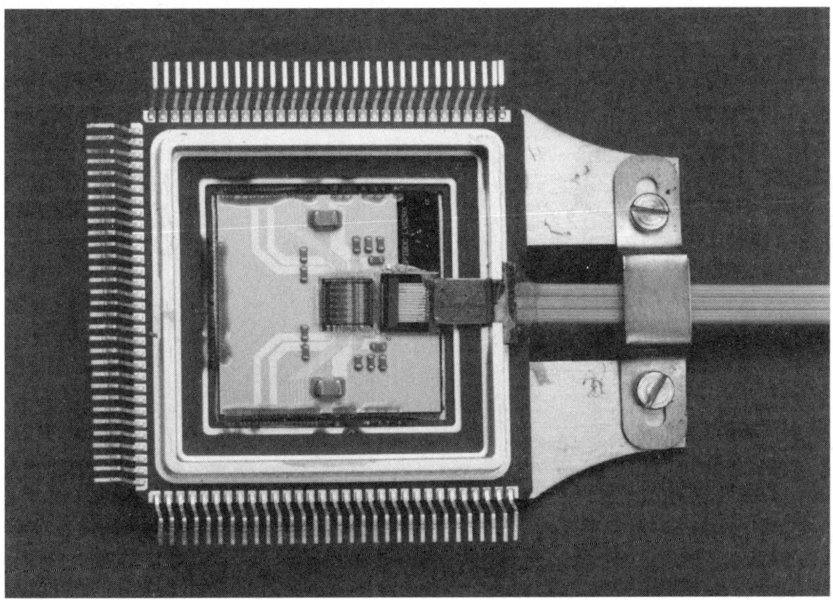

Figure 11 Photograph of a complete 12-channel 2.5-Gbit/s receiver module. (From Ref. 60. Copyright © Nortel, 1996)

ide or silica dielectrics. The alternative is to partition the circuitry and adopt a hybrid approach in which an optoelectronic carrier is mounted on a more complex electronic hybrid.

2.3 Integration of Optical Waveguides

The difficulties associated with integrating optical waveguides into the substrate of a hybrid optical bench should not be understated. In fact, there have been few successful demonstrations of the integration of V-groove fiber alignment features with optical waveguides and optoelectronic component mounting sites. One full-integration scheme that does include V-grooves, waveguides, and component bond sites has been reported by Lucent Technologies [64]. Their optical bench structure is fabricated on a 14-mask-level process that provides waveguides, fiber alignment V-grooves, and etched fiducial masks for alignment of the lasers to the waveguides. More commonly, partial integration has been achieved by omitting fiber alignment grooves and relying on a butt-aligned fiber joint in which the fiber is held in a polished silicon carrier and glued directly to the waveguide interface.

The integration of multimode waveguides with laser and detector arrays achieved by IBM was described earlier [46]. They used silica waveguides formed by a CVD flame hydrolysis process, and the patterning of the waveguides also provided the optoelectronic component alignment features. Interestingly, the waveguide features also provided the alignment for the guide pins of an optical connector.

Single-mode waveguide structures are essential for more complex module functions, such as the multiplexing and demultiplexing required for a wavelength-division multiplexed (WDM) transceiver. One of the most successful integration schemes for single-mode waveguide structures has been the silica-on-terraced-silicon platform advocated by NTT [65–67] and adopted more recently by Fujitsu [41]. The device structure is illustrated in Fig. 12. The silicon is anisotropically etched over a large area to leave a raised terrace onto which the laser is to be mounted. A silica undercladding layer is then deposited by flame hydrolysis so that it is planar with the surface of the silicon terrace. The undercladding provides both a buffer layer for the optical waveguides and high isolation of the coplanar electrical interconnects. Optical waveguides are formed on the undercladding in the conventional way, and the interconnect metalization is deposited. The silicon terrace ensures that the laser spot is at the same height as the waveguide core; it also ensures that the laser is properly heat-sunk into the bulk of the silicon substrate below.

The technique of waveguide integration into the optohybrid substrate has not yet been commercially exploited. There may be two interlinked reasons for this. One is that the target application, transceivers for broadband full-service access

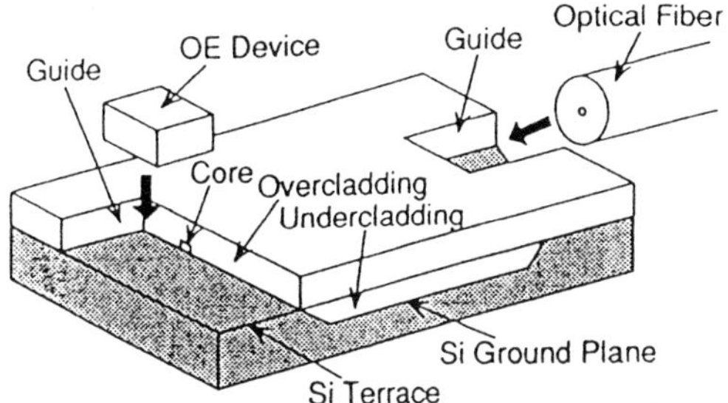

Figure 12 Configuration of silica waveguide on terraced silicon substrate. (From Ref. 65. Copyright © NTT, 1993)

networks, has not yet emerged as a volume market. The second is that the process for waveguide integration is complex, and only a high volume market justifies the manufacturing investment in such a process.

3 COMMERCIAL APPLICATIONS

The most significant commercial application of silicon hybridization has been in the manufacture of low-cost transmitters and receivers. Conventional dual in-line laser packages have been unable to compete effectively on price with coaxial laser products. In fact, the coaxial products themselves have also neared the bottom of their price evolution. In order to drive the price of dual in-line products down, a number of manufacturers, including Nortel, Hitachi, and Lucent Technologies, have standardized on a small ceramic package—the eight-pin Mini-Dil [68]. In order to use the Mini-Dil package, compact optical submounts are essential, and these are provided by silicon optical bench structures [18,12,6]. A number of other manufacturers, even if not following the Mini-Dil standard, have followed the same trend toward a compact ceramic package containing a silicon platform [69,70].

The contents of the Mini-Dil is common between the various manufacturers: in the transmitter are a laser, a rear monitor, and a fiber alignment structure; in the receiver are a pin diode and a preamplifier circuit. A schematic of the contents of the Min-Dil is shown in Fig. 13. Although the concept is similar, the actual implementations differ somewhat in their details between manufacturers; for example, where Nortel uses a V-groove for the alignment [12], Lucent has chosen to solder the fiber directly to the silicon surface [64].

The provision of optical components for the local loop has been a major driver toward reduced cost and higher volume. It has also driven the reliability requirements, which has been reflected in specifications such as Bellcore 983 [71], which

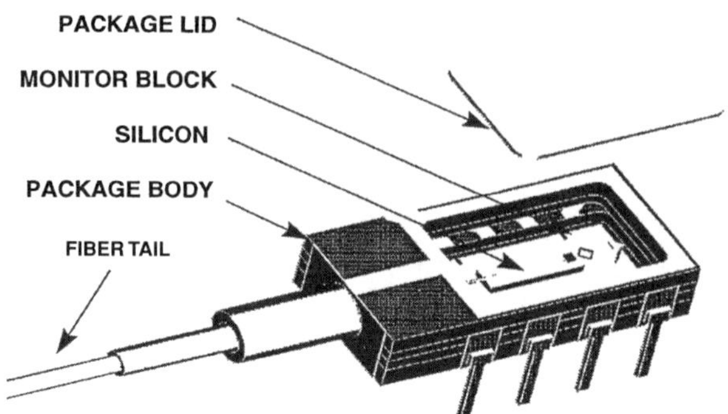

Figure 13 Content of the Mini-Dil package. (Copyright © Nortel, 1996)

stipulates 500 temperature cycles between −40 and +85°C. One of the key performance differentiators of alignment using a silicon platform is the stability achieved during long-term storage and temperature cycling. This is brought about by the short mechanical path between the fiber tip and the laser (or detector) and the robust fixing of the fiber. As an example, the following results were achieved by Nortel using their silicon V-groove technology in a Mini-Dil package [72,73]. The results were achieved in a series of overstress tests, using devices without any form of prescreening, and they show the inherent capability of the technology. Figure 14(a) and (b) are plots of the percentage change in monitor current required to achieve the original reference ex-fiber power at high and low temperatures. The periods have been extended well beyond the specified 2000 hours, and they show that the alignment remains stable to better than ±10% up to 7000 hours at maximum and minimum temperatures. Similar plots of the change in monitor current to achieve the original ex-fiber power are shown in Fig. 14(c). These show changes as a result of thermal cycling. This test has also been extended beyond the 500 cycles specified by Bellcore to 7000 cycles from −40 to 85°C, with no indication of any change in alignment.

The next generation of cost reduction may well come through the adoption of nonhermetic packaging techniques. The silicon platform will continue to provide robust and stable alignment between the components. However, the ceramic package will be replaced by a plastic housing either molded around a complete fiber assembly or into which the assembly is placed. In order to prevent condensation in the optical train, a silicone encapsulant is likely to be used. British Telecom has advocated this approach for some time [5], and many of the major manufacturers have published in this area [6,70]. The major challenge is to prove the reliability of these components, particularly under conditions of high temperature and humidity.

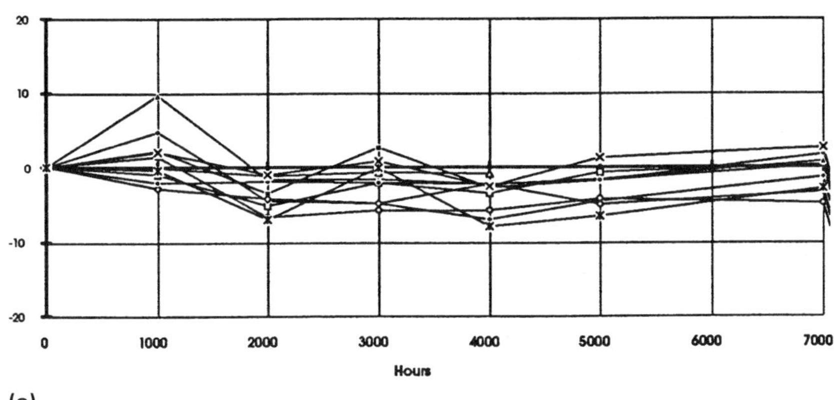

(a)

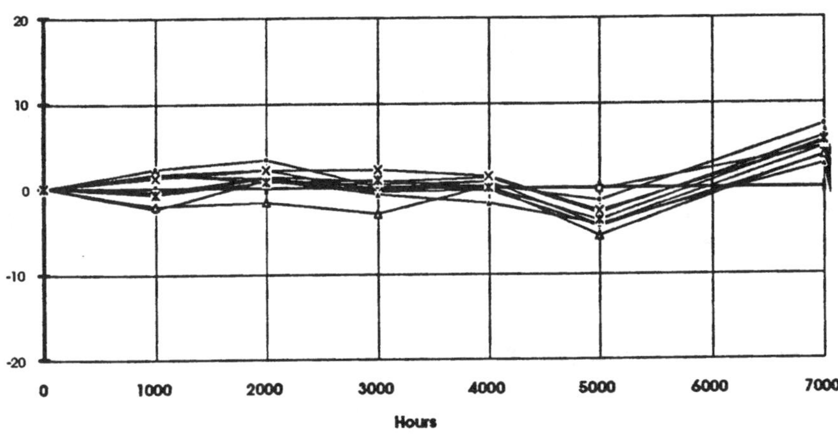

(b)

Figure 14 Reliability of Mini-Dil Transmitter. (a) Storage at 85°C for periods up to 7000 hr (1000–2000 hrs at 100°C). (b) Storage at −55°C for periods up to 7000 hr (0–1000 hr at −40°C). (c) Temperature cycling from −40 to 85°C for up to 7000 cycles. (From Ref. 12. Copyright © Nortel, 1997)

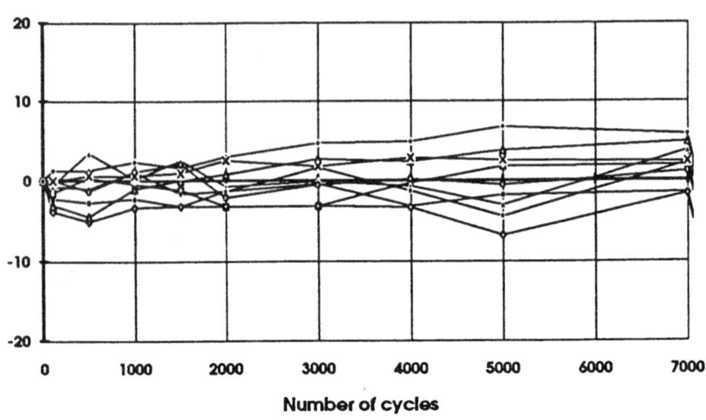

(c)

Figure 14 Continued

Having introduced silicon platforms into low-cost products, manufacturers are applying the same techniques to a range of other devices. For example, Lucent Technologies has concentrated on producing an optical subassembly (OSA) as a generic platform for their cooled products [13,18,74]. Here, they have concentrated on process automation and the ability to produce a low-cost common platform for a range of devices. Similarly, Nortel has extended its range to allow for active alignment to uncooled DFB devices, where the losses associated with passive alignment are unacceptable [75].

4 FUTURE DEVELOPMENTS

Over the last 10 years, optical hybridization has developed from an interesting laboratory concept to a commercial reality. Two factors will continue to play a vital role in the future development of the optohybrid technology. One is continued pressure on the price of existing components; the result of this will be the inexorable replacement of devices assembled using conventional techniques with their silicon optical bench equivalent. This has happened at the low-cost end of the market and will slowly extend to high-end transmission components. The other factor is the drive toward new functions, particularly as WDM networking spreads. The new functions include multifiber interconnects and transceiver devices.

4.1 Multifiber Interconnects

A very attractive application of the silicon hybrid technology is in the manufacture of multifiber array components. In fact, multifiber interconnects have been the prime motivator for the development of this technology for many manufacturers [10,76,47,43,20,61]. To date, multifiber links have been targeted mainly at board-to-board and rack-to-rack interconnects. These naturally lend themselves to the use of multiple fibers to achieve very high aggregate data rates either using the ribbon fiber interconnect as a parallel bus (bit parallel operation) or as multiple independent links.

The technology development associated with multifiber components has been very successful; this can be seen in the number of demonstrations that have been published [62,77–79]. However, despite great technical success, full commercialization of multifiber interconnects has been slow. This may be because the requirement for transporting very large quantities of data (5–30 Gbit/s) has not yet emerged. Another contributory factor may be that the technical challenges associated with demonstrating high yield and reliability of the optoelectronic subcomponents are severe. To date, the main challenge in multifiber interconnects has been perceived to be the packaging technology. However, as the silicon bench technology has matured, this is no longer the case, and the focus will inevitably shift toward proving the yield and the reliability of arrays of lasers and photodiodes. This proof is expensive to obtain, and may not be tackled until the application appears for which array technology is essential.

4.2 Transceivers and Waveguide Devices

As discussed earlier, one of the thrusts of optohybrid integration is toward the incorporation of waveguide devices into the substrate. To date, a prime motivation for this is the fabrication of transceivers for use in broadband full-service access networks [80,81,64]. This market will ultimately require transceivers at extremely low cost, and whether a substrate with an integrated waveguide can meet this challenge remains to be seen. However, even if this market does not emerge, there is a large number of applications within the optical network that could benefit substantially from the integration of waveguides and optoelectronic devices. In particular, the combination of the waveguide tap and a photodiode is ideally suited for nonintrusive monitoring of signals at key points in an all-optical network, for example, within a WDM router, where the signal could be used to provide channel equalization. The silicon hybrid technology is ideally suited to provide this sort of function.

5 CONCLUSIONS

Silicon optohybrid technology is itself a hybrid of a number of technologies. Without the understanding of topics such as anisotropic etching and flip-chip mounting that have been developed in other fields, this technology would not have reached commercial reality so quickly or so successfully. It has yet to be seen whether the silicon hybrid platform will be the final stage in optoelectronic packaging. One possibility is that injection-molded plastic materials will ultimately prove to have sufficient mechanical stability and low enough cost to supplant it. Nevertheless, for the foreseeable future, silicon-based hybridization will be the industry standard.

REFERENCES

1. Matthews M., Macdonald B., and Preston K. Optical components—the new challenge in packaging. IEEE Trans. Comp. Hybrids Manuf. Tech. CHMT-13:798–806, 1990.
2. Koch et al. Tapered waveguide InGaAs/InGaAsP multiple-quantum-lasers. IEEE Photon. Technol. Lett. 2:88–90, 1990.
3. Kasaya et al. IEEE Electron. Lett. 29:2067–2068, 1993.
4. Lealman I. F., Rivers L. J., Harlow M. J., and Perrin S. D. InGaAsP/InP tapered active layer multiquantum well laser with 1.8 dB coupling loss to cleaved singlemode fiber. IEEE Electron. Lett. 30:1685, 1994.
5. Hall I. P. Non-hermetic Encapsulation and Assembly Techniques for Optoelectronic Applications Proc. 10th European Microelectronics Conf, Copenhagen, 1995, pp. 62–69.
6. Tatsuno K., Kato T., Hirataka T., Yoshida K., Ishii T., Kono T., Miura T., Sudo R., and Shimaoka M. High performance and low-cost plastic optical modules for the access network applications. OFC 97, Dallas, 1997, pp. 111–112.
7. Henry C. H., Blonder G. E., and Kazarinov R. F. Glass waveguides on silicon for hybrid optical packaging. IEEE J-LT 7:1530–1539, 1996.
8. Petersen K. E. Silicon as a mechanical material. Proc. IEEE 70:419–457, 1982.
9. King J. A. Materials Handbook for Hybrid Microelectronics. Boston: Artech House, 1988.
10. Crow J. D., Comerford L. D., Laff R. A., Brady M. J., and Harper J. S. GaAs laser array source package. Optics Lett. 1:40–42, 1977.
11. Sumida S., Yamada Y., Yasu M., and Kawachi M. High silica guided-wave hybrid optical transmitting–receiving module. Trans. IECE Japan E69:352–354, 1986.
12. Harrison P. M., Cann R., and Spear D. Silicon V-groove technology in the volume production of optical devices. 10th Annual Meeting IEEE Lasers and Electro-optics Society, LEOS '97, San Francisco, 1997, pp. 134–135.
13. Dautartas M. F., Blonder G. E., Wong Y., and Chen Y. C. A self-aligned optical subassembly for multimode devices. IEEE Trans. Comps. Pack. Manuf. Tech. CHMT B 18:552–557, 1995.
14. Nakagawa G., Miura K., Tanaka K., and Yano M. Lens-coupled laser-diode module integrated on silicon platform. SPIE 2610:59–64, 1996.
15. Nakagawa G., Miura K., Tanaka K., and Yano M. Lens-coupled laser-diode module integrated on silicon platform. IEEE J-LT 14:1519–1523, 1996.

16. Silletti A., Tabasky M., Hill P. M., Haugsjaa P. O., and Urban M. A silicon waferboard detector module for an optical heterodyne balanced receiver with integrated bypass capacitor. Proc. SPIE 1794:293–298, 1992.
17. Godfrey D., Bailey S., Cooper K., Nield M., Hill J., and Welbourn D. Fully integrated silicon based optical motherboards. IEEE Multi-Chip Module Conference MCMC-92, Santa Cruz, 1992, pp. 146–149.
18. Gates J. V., Henein G., Shmulovich J., Muehler D. J., MacDonald W. M., and Scotti R. E. Uncooled laser packaging based on silicon optical bench technology. SPIE 2610:127–137, 1995.
19. Shmulovich J., Gates J. V., Kane C. F., Cappuzzo M. A., and Szalkowski J. M. Successful development on non-planar lithography for micro-machining applications. Opt. Soc. America Proc. of Integrated Photonics Res., 1996, Tech. Digest, 6, pp. 354–357, 1996.
20. Ambrosy A., Richter H., Hehmann J., and Ferling D. Silicon motherboards for multichannel optical modules. IEEE Trans. Comps. Pack. Manuf. Tech. CHMT A 19: 34–40, 1996.
21. Vieider C., Holm J., Forssen L., Elderstig H., Lindgren S., and Ahlfeldt H. A new process for combining anisotropic bulk etching with subsequent precision lithography. Proc. Int. Conf. on Solid State Sensors and Actuators, Transducers '97, Chicago June 1997, pp. 679–682.
22. Harrison P. M., Parker J. W., Peall R. G., Geear M. C., Cureton C. G., H. F. M. Priddle, Ayliffe P. J., Wood S. A., Buckley R. A., Spear D. A. H., Lambert N. J., and Postlethwaite R. Low cost transmitter and receiver modules fabricated in opto-hybrid technology. IEEE Colloquium on Planar Silicon Hybrid Optoelectronics, Digest No. 1996/198, 1994, pp. 14/1–6.
23. Johnson D. J., and DeVre M. W. Recent advances in the use of plasma etching for micro-machining. Advances Electron. Packaging 1997 19:413–419, 1997.
24. Schmidt J. P., Cordes A., Muller J., and Burkhardt H. Laser-fiber coupling by means of a silicon micro-optical bench and a self-aligned soldering process. Proc. SPIE 2449:176–183, 1995.
25. Miller L. F. Controlled collapse reflow chip joining. IBM J. Res. Dev. 13:226–238, 1969.
26. Pedder D. J. Flip chip solder bonding for microelectronic applications Hybrid Circuits 15:4–7, 1988.
27. Lindgren S. 24-GHz modulation bandwidth and passive alignment of flip-chip mounted DFB laser diodes. IEEE Photon. Lett. 9:306–308, 1997.
28. Katsura K., Hayashi T., Ohira F., Hata S., and Iwashita K. A novel flip-chip interconnection technique using solder bumps for high-speed photoreceivers J. Lightwave Technol. 8:1323–1327, 1990.
29. Bache R. A. C., Burdett P. A., Pickering K. L., Parsons A. D., and Pedder D. J. Bond design and alignment in flip chip solder bonding. Proc. 8th International Electronics Packaging Conference, Dallas, 1988, pp. 830–841.
30. Kallmayer C., Oppermann H., Engelmann G., Zakel E., and Reichl H. Self-aligning flip-chip assembly using eutectic gold/tin solder in different atmospheres. Proc. IEEE/CPMT Int. Elect. Manuf. Tech. Symp., Austin, 1996, pp. 18–25.
31. Wale M. J., and Edge C. Self-aligned flip-chip assembly of photonic devices with

electrical and optical connections. IEEE Trans. Comps. Pack. Manuf. Tech. CHMT 13:780–786, 1990.
32. Pedder D. J. Flip-chip solder bonding for advanced device structures. Plessey Res. Rev. Vol. No.:69–81, 1989.
33. Goodwin M. J., Moseley A. J., Robbins D. J., Kearley M. Q., Thompson J., Clewitt D., Goodfellow R. C., and Bennion I. Hybridized optoelectronic modulator arrays for chip-to-chip optical interconnection. SPIE 1281:207–212, 1990.
34. Edge C., Ash R. M., Jones C. G., and Goodwin M. J. Flip-chip solder bond mounting of laser diodes. IEEE Electron. Lett. 27:499, 1991.
35. Wale M. J., Edge C., Randle F. A., and Pedder D. J. A new self-aligned technique for the assembly of integrated optical devices with optical fiber and electrical interfaces. Proc. ECOC '89, Gothenburg, 1989, pp. 368–371.
36. Itoh M., Yoneda I., Sasaki J., Honmou H., Fukushima K., Nagahori T., and Kawasaki J. Self-aligned packaging of multi-channel photodiode array module using AuSn solder bump flip-chip bonding. Deutscher Verlag Proc. of EuPac '96, 2nd Euro. Conf. on Electronic Packaging Tech., Essen, 1996, pp. 75–77.
37. Itoh M., Sasaki J., Uda A., Yoneda I., Honmou H., and Fukushima K. Use of AuSn solder bumps in three-dimensional passive aligned packaging of LD/PD arrays on Si optical benches. Proc. 46th Electronic Components and Technology Conference, May 1996, pp. 1–7.
38. Itoh M., Sasaki J., and Uda A. Passive alignment packaging on Si optical bench using AuSn solder bumps LEOS '97, San Francisco, 1997, pp. 126–127.
39. Geear M. C., Parker J. W., Ayliffe P. J., Harrison P. M., and Peall R. G. High bit rate transmitter module realized in opto-hybrid technology. IEE Colloquium Digest, "The DTI optoelectronic systems LINK program," London, May 1994, pp. 1–5.
40. Sasaki S., Nakagawa G., Tanaka K., Miura K., and Yano M. Marker alignment method for passive coupling on silicon waferboard. IEICE Trans. Commun. E79-B:939–942, 1996.
41. Nakagawa G., Yamamoto T., Sasaki S., Norimatsu M., Yamamoto N., Nosaka T., Terada K., Tanaka K., Miura K., and Yano M. High power and high sensitivity planar lightwave circuit module incorporating a novel passive alignment method. IEEE J-LT 16:66–71, 1998.
42. Armiento C. A., Tabasky M., Jagannath C., Fitzgerald T. W., Shieh C. L., Barry V., Rothman M. A., Haugsjaa P. O., Holstrom R., Meland E., and Powazinik W. Passive coupling of an InGaAsP/InP laser array and single-mode fibers with silicon waferboard. Proc. OFC '91, 1991, p. 124.
43. Armiento C. A., Tabasky M. J., Chirravuri J., Rothman M. A., Choudhury A. N. M. M., Negri A. J., Budman A. J., Fitzgerald T. W., Barry V. J., and Haugsjaa P. O. Hybrid optoelectronic integration of transmitter arrays on silicon waferboard. SPIE 1582:112–119, 1991.
44. Armiento C. A., Negri A. J., Tabasky M. J., Boudreau R. A., Rothman M. A., Fitzgerald T. W., and Haugsjaa P. O. Four-channel, long wavelength transmitter arrays incorporating passive laser/single-mode-fiber alignment on silicon waferboard. Proc. 42nd Electronic Components and Technology Conf, IEEE, San Diego, 1992, pp. 108–114.
45. Jones C. A., Nield M., Cooper K., Waller R., Rush J., Fiddyment P., and Collins

J. An optical transceiver on a silicon motherboard. Proc. 7th Eur. Conf. on Int. Opt. (ECIO '95), Delft, 1995, pp. 591–594.

46. Jackson K. P., Flint E. B., Cina M. F., Lacey D., Trewhella J. M., Buchmann P., Harder C., and Vettiger P. Flip-chip, self-aligned, optoelectronic transceiver module. Proc. ECOC, Berlin, 1992, pp. 329–332.
47. Jackson K. P., Flint E. B., Cina M. F., Lacey D., Trewhella J. M., Caulfield T., and Sibley S. A compact multichannel transceiver module using planar-processed optical waveguides and flip-chip optoelectronic components. Proc. 42nd ECTC '92, San Diego, 1992, pp. 93–97.
48. Miyashita T., Kawachi M., and Kobayashi M. Silica-based planar waveguides for passive components. Proc. OFCC '88 4:173, 1988.
49. Hunziker W., Vogt W., and Melchior H. Self-aligned optical flip-chip OEIC packaging technologies. Proc. ECOC '93, Montreau, 1993, pp. 84–91.
50. Kraehenbuehl R., Bachmann M., Vogt W., Brenner T., Duran H., Bauknecht R., Hunziker W., Kyburz R., Holtmann C., Gini E., and Melchior H. High-speed low-loss InP space switch matrix for optical communication systems, fully packaged with electronic drivers and single-mode fibers. Proc. ECOC '94, Florence, 1994, pp. 511–514.
51. Hunziker W., Vogt W., Melchior H., Leclerc D., Brosson P., Pommereau F., Ngo R., Doussiere P., Mallecot F., Fillion T., Wamsler I., and Laube G. Self-aligned flip-chip packaging of tilted semiconductor optical amplifier arrays on Si motherboard. IEEE Electron. Lett. 31:488–489, 1995.
52. Hasegawa H., Furukawa M., and Yanai H. Properties of microstrip line on Si–SiO_2 system. IEEE Trans. MTT 19:869–881, 1971.
53. Geller B., Tyler J., Holdeman L., Phelleps F., Cline P., and Laird G. Integration and packaging of GaAs MMIC subsystems using silicon motherboards. GaAs IC Symposium, New Orleans, 1990, pp. 85–87.
54. Yamada Y. OE-device hybrid integration on PLC platform. LEOS '97, San Francisco, 1997, pp. 301–302.
55. Jones C. A., Cooper K., Nield M. W., Rush J. D., Thurlow A. R., Waller R. G., Ayliffe P. J., and Harrison P. M. A 2.4-Gbit/s transceiver using silica waveguides on a silicon optical motherboard. IEEE Electron. Lett. 31:2208–2210, 1995.
56. Iezekiel S., Soshea E. A., O'Keefe M. F. J., and Snowden C. M. Glass–silicon substrate for hybrid optoelectronic packaging. IEE Colloquium on planar silicon hybrid optoelectronics. Digest 1994/198, 1994, London, pp. 13/1–7.
57. Iezekiel S. Microwave photonic multichip modules. IEE Electron. Commun. J. 9: 156–164, 1997.
58. Peall R. G., Shaw B. J., Ayliffe P. J., Priddle H. F. M., Bricheno T., and Gurton P. 1×8 8-Gbit/s transmitter module for optical space switch applications. IEEE Electron. Lett. 33:1250–1252, 1997.
59. Harcourt R. W. A 2.4-Gbit/single and multifiber electro-optic transmitter. IEE Colloquium on planar silicon hybrid optoelectronics. Digest 1994/198, London, 1994, pp. 3/1–4.
60. Peall R. G., Priddle H. F., Geear M. C., Shaw B., Briggs A., Harrison P. M., Schmid A., Bitter M., Wieland J., Zorba O., and Harcourt R. 12×2.5-Gbit/s receiver array module. IEEE Electron. Lett. 32:682–683, 1996.

61. Arai Y., Takahara H., Koyabu K., Fujita S., Akahori Y., and Nishikido J. Multigigabit multichannel optical interconnection modules for asynchronous transfer mode switching systems. IEEE Trans. Comps. Pack. and Manuf. Tech. CHMT B 18:558–564, 1995.
62. Wong Y. M. et al. Technology development of a high-density 32-channel 16-Gb/s optical data link for optical interconnection applications for the optoelectronic technology consortium (OETC). IEEE J-LT 13:995–1016, 1995.
63. Ewen J. F., Jackson K. P., Bates R. J. S., and Flint E. B. GaAs Fiber-optic modules for optical data processing networks. IEEE J-LT 9:1755–1763, 1991.
64. Henein G. E., Muehlner D. J., Shmulovich J., Comez L., Capuzzo M. A., Laskowski E. J., Yang R., and Gates J. V. Hybrid integration for low-cost OE packaging PLC transceiver LEOS '97, San Francisco, 1997, pp. 297–298.
65. Yamada Y., Takagi A., Ogawa I., Kawachi M., and Kobayasi M. Silica-based optical waveguide on terraced silicon substrate as hybrid integration platform. IEEE Electron. Lett 29:444–445, 1993.
66. Mino S., Yoshino K., Yamada Y., Yasu M., and Moriwaki K. Opto-electronic hybrid integrated laser diode module using silica-on-terraced-silicon platform. Proc. 7th IEEE Lasers and Electro-Optics Society Meeting (LEOS '94), Boston, 1994, pp. 271–272.
67. Mino S., Ohyama T., Akahori Y., Yamada Y., Yanagisawa M., Hashimoto T., and Itaya Y. 10-Gbit/s hybrid-integrated laser diode array module using a planar lightwave circuit (PLC)-platform. IEEE Electron Lett. 32:2232–2233, 1996.
68. Turley et al. High-stability low-cost optical module packaging for the access market. First Optoelectronics and Communications Conference (OECC '96), paper 18D3-4, Makuhari, Japan, 1996, pp. 408–40.
69. Han H., Roff R. W., Boudreau R. A., Bowen T. P., and Wilson R. B. Pigtailed CATV PIN module packaged by passive alignment technology. LEOS '97, San Francisco, 1997, pp. 64–65.
70. Kurata K., Yamauchi K., Kawatani A., Tanaka H., Honmou H., and Ishikawa S. A surface mount single-mode laser module using passive alignment. IEEE Trans. Comps. Pack. and Manuf. Tech. CHMT B 9:524–531, 1996.
71. Bellcore Technical Advisory. Reliability assurance practices for optoelectronic devices in loop applications. TA-NWT-000983, Issue 2, December 1993.
72. Cann R., Harrison P. M., and Spear D. Use of silicon V-groove technology in the design and volume manufacture of optical devices. Proc. Optoelectronics '96, San Jose, 1996, pp. 17–23.
73. Cann R., Harrison P. M., and Spear D. Use of silicon V-groove technology in the design and volume manufacture of optical devices. SPIE 9th Annual Symposium Optoelectronics '97. Conference 3004, Dallas, February 1997.
74. Henein G. E., Gates J. V., Mulligan L. J., Presby H. M., and de Jong J. F. Fiber to laser coupling on silicon optical bench platform. Opt. Soc. America, Integrated Photonics Res., 1996, Boston, 1996, pp. 396–39.
75. Bricheno T., Priddle H. F. M., Peall R. G., and Spear D. Packaging of optoelectronic devices with high coupling efficiency using silicon V-groove technology. Proc. OFC '98, San Jose, February 1998, pp. 349–350.
76. Parker J. W. Optical interconnection for advanced processor systems: a review of

the ESPRIT II OLIVES program. IEEE J. Lightwave Technol. 9:1764–1773, 1991.

77. Rehm W., Dütting K., Kaiser N. M., Vetter P., Artigue C., and Fernier B. Optical interconnects: the key to high bit-rate communication within telecommunication equipment. Electrical Communication 1st Quarter 1995, 1995 pp. 65–73.
78. Peall R. G. Developments in multi-channel optical interconnects under ESPRIT III SPIBOC. 8th Annual Meeting IEEE Lasers and Electro-Optics Society, Vol. 1, November 1995, pp. 222–223.
79. Cannell G. J. Standardized packaging and interconnection for inter- and intra-board optical communication (SPIBOC). Proc. ECOC '96, Oslo, Norway, 15–19 September 1996, pp. 249–252.
80. Bell J., Snyder D., Yanushefski M., Yang R., Osenbach J., and Asom M. O-E transmitter packaging for broadband access systems (FTTX). LEOS '97, San Francisco, 1997, pp. 299–300.
81. Boudreau R. A., Bowen T. P., Han H., Schramm J., Zhou P., Hoot T., Mathews J., Tan S., Drabenstadt C., Feldman M., Tekolste R., Armiento C., Radcliffe J., Hookier B., Lee Y. C., Stirk C., Delen N., Baliga A., Bowler D., Saroya M., and Wilgus J. Single-mode bi-directional links for fiber-in-the-loop and optical networks. LEOS '97, San Francisco, 1997, pp. 56–57.

10

Integrated Optics in Sensors: Advances Toward Miniaturized Systems for Chemical and Biochemical Sensing

Rino E. Kunz

Centre Suisse d'Electronique et de Microtechnique SA (CSEM)
Zurich, Switzerland

1 INTRODUCTION

With respect to sensing applications, integrated optics plays very different roles, depending on the features that are most beneficial for the application aimed at. In one type of application, the key role of integrated optics is to read out the signal from external transducers and to provide optical signal processing. Well-known examples are displacement and rotation sensors (the optical gyroscope) [1–4 (Vol.4,Chap.11)], where an integrated optical (IO) circuit, attached to an external transducer, is used to obtain a linear or angular position signal. Similarly, an IO chip can be used for reading out polarization-sensitive transducers or as (pre-)processing circuits for metrology measurement techniques based on interferometers [1,2,4 (Vol.2),5].

In a second type of application, the IO chip acts as the transducer only. Examples are refractometry and (bio-)chemical sensing by means of uniform input and output grating couplers and by means of interferometers, where a dielectric waveguide is used to convert the change of the optical properties in the region of the evanescent wave of a guided mode (the sensing region) into a change of the effective refractive index of one or several modes. This change is then determined by external means, used to measure the resonance angle for input or output coupling, or the phase shift corresponding to the changes of the measurand. These applications have been described in many papers and reviewed in Refs. 1, 4, 6, and 7.

In a third type of application, IO is playing several roles at the same time. Examples are IO pickups for data storage applications [3,8,9], miniature goniometers [10], and chips acting as "smart beam splitters" with a built-in high-resolution spectrometer function for applications such as spectral beam sampling, and for realizing advanced tunable or stabilized light sources [11,12]. Another example are miniaturized IO modules for chemical and biochemical sensing [13–23].

This chapter concentrates on the last-mentioned topic. Its aim is to present concepts, aspects, advances, applications, and opportunities that have been the result of progress in recent years, and its emphasis is on aspects relevant to realizing complete miniaturized systems. The reason for choosing this scope is that books already exist (for example, Refs. 1, 2, 4, 6, and 7) that adequately describe the efforts and achievements in many specific disciplines of integrated optical sensing, but none exist that are devoted to the subject of miniaturizing the whole sensor system. This is also true for most other publications, due mainly to the fact that only one or two areas of expertise (e.g., chemistry, physics, electronics, mechanics, optics, microfabrication, theoretical modeling) are typically available in an advanced state at one single location.

A combination of the most important expertise and technologies was available during the last decade at the former Paul Scherrer Institute, Zurich (PSIZ), which has been transferred to the Centre Suisse d'Electronique et de Microtechnique SA (CSEM) in 1997. Due to this fact, work of a more comprehensive nature, aimed at the miniaturization of the complete sensor system, achieving high sensitivities, and also considering aspects relevant to practical applications, has been performed at this place. This chapter's subject are the most important aspects and results of this work at CSEM/PSIZ, the focus being on accomplishing all of the following main goals:

1. Moving the complexity from the system to sensor chips with enhanced functionality
2. Lowering the price of the chips
3. Increasing the sensitivity of the system (exploitable for practical sensors, not just in the laboratory)
4. Reducing the size, complexity, and price of the complete system

To favor a concise and coherent presentation, the scope will be restricted to so-called label-free (refractometric) methods based on measuring the real part of the effective mode index; i.e., no sensors are considered that measure intensity changes due to absorption, fluorescence, or scattering. Nevertheless, many of the principles and methods presented here are also suitable for miniaturizing intensity-based sensor systems [13]. Very recently, results for combined refractometric and fluorescence measurements by means of a single IO chip have been reported [19].

Comparing typical *electronic* integrated circuits and integrated *optical* sensor circuits reveals a remarkable difference. While the electronic circuits are based on standardized basic elements, such as operational amplifiers, and make use of a high level of integration, leading to compact and small devices, analogous opportunities offered by integrated optics have not been exploited for commercial sensors so far.

A typical integrated optical sensor system is shown schematically in Fig. 1(a). Because the level of integration on the chip level is low, several peripheral mechanical, optical, and electronic components are needed to provide the appropriate input to the IO transducer chip and for converting the optochemical effect into a convenient output signal. This applies in particular to IO refractometric sensors, which are suited to measuring and monitoring a great variety of important chemical processes based on converting the value of the measurand into a corresponding value of the effective refractive index N of guided modes. The majority of the various sensor types and principles published make use of grating couplers (see, for example, Refs. 1, 23–29) or of interferometers (see, for example, Refs. 1, 2, 23, 27, 30–38).

As is illustrated by the grating coupler example of Fig. 1(a), these systems are very complex, and the total sensor size is several orders of magnitude larger

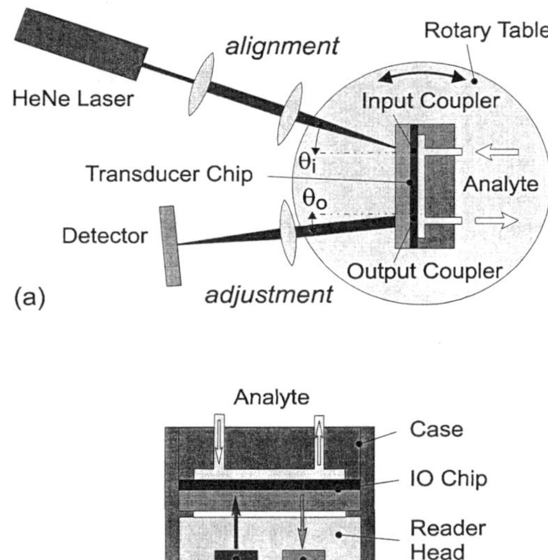

Figure 1 Schematic representation of two IO sensor system types: (a) typical conventional arrangement and (b) novel compact miniature sensor module.

than the chip size, mainly because the domain of integrated optics is limited to realizing the chemo-optical transducer only. If the sensing schemes are based on the use of external measuring variables, such as the angle of incidence θ_i or the outcoupling angle θ_o, peripheral equipment is needed to extract the value of the measurand via an external conversion of the angle into another variable (e.g., a position) suitable to be measured by photodetectors. This peripheral equipment needs space and a careful maintenance of adjustment. Often, the coupling to the waveguide also needs intensive alignment efforts of external focusing optics by means of precision mechanical parts, especially in the cases where end-face coupling schemes [30–32,35,37] or prism coupling schemes [39] are used.

The main purpose of this chapter is to work out the central thread providing a link between the principal ideas, new concepts, and practical issues that are important for realizing complete miniature sensor modules of the type shown in Fig. 1(b). The most important "ingredients" essential for achieving this goal are presented and discussed. In order to give concise guidance, only the most conspicuous end results of theoretical and experimental investigations are reported here, while the details on numerical and experimental procedures, parameters, and results are made available to the reader by citation of the original publications. It is thus hoped that this chapter will fulfill the needs of both expert readers for a methodical overview and for being referenced to the original work, as well as nonexperts interested in the more basic aspects or in the potential for realizing different types of miniature sensor modules. Emphasis is also on discussing many issues concerning important aspects and comparisons between different solutions which cannot appropriately be done in research papers focused on just one sensor type.

After presenting sensing pad fundamentals in Sec. 2, the most important key issues are discussed in the following sections concerning proper integrated optical sensing schemes for accomplishing on-chip measuring variables (Sec. 3), and sensor chip fabrication, characterization, experiments, modeling, new types of multilayer waveguides, on-chip referencing, and multicomponent sensing (Sec. 4). Finally, different compact sensor module types based on combining the aforementioned ingredients are presented and discussed in Sec. 5.

2 SENSING PAD FUNDAMENTALS

Figure 2 shows the principle tasks to be fulfilled by a miniature integrated optical sensor (MIOS) chip forming the basic building block for realizing compact, complete sensor systems of the type shown in Fig. 1(b). The most obvious tasks are to provide input ports (IPs) and output ports (OPs) for connecting the chip to the environment. The main purpose of the input port is to excite a guided mode u in a dielectric waveguide and to direct it into the sensing pad (SP) region, where its properties are changed by interaction with the measurand. The sensing pad

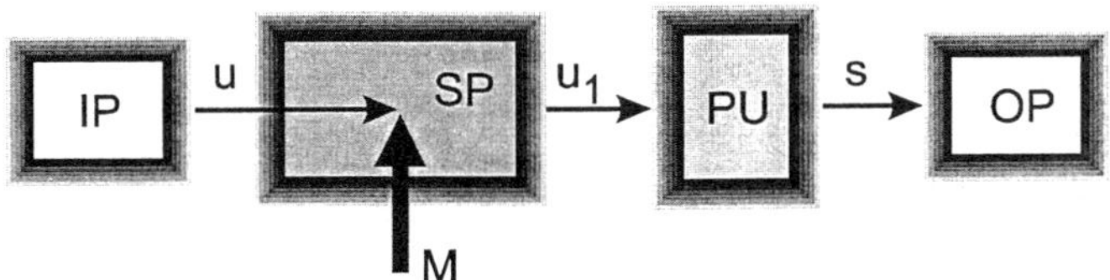

Figure 2 Basic functions to be performed by a MIOS chip.

provides the transducer function by "encoding" wave u with the measurand's value M, producing a modified wave u_1, which then enters the processing unit (PU). Finally, the processing unit "decodes" the information on the quantity to be measured from wave u_1 to produce a signal s in a form suitable for readout at the output port OP by means of simple, small, and standardized peripheral components. Since it is not easy to state precisely what a "suitable form" of an output signal is, I just give an example here: If the output signal is the *angle* of a beam coupled out from a waveguide, a considerable amount of effort and additional components are needed in order to decode the output beam further, i.e., to convert the *angle* into a *position* that can be measured directly by commercial detectors, such as position-sensitive devices or photodiode arrays. Hence, if an on-chip "detector-compatible" position signal is generated, the complexity of the whole system is significantly reduced. This and related aspects will be discussed further in the following sections.

Since many different possibilities for implementing the input and output ports as well as the sensing pad and the processing unit have been considered in Ref. 13, they are not presented here in detail. However, it is important to note that it is not necessary to implement the basic functions represented in Fig. 2 within different locations on a practical MIOS chip. In special cases, depending on the sensing principle and sensing schemes, all pads may be "collapsed" or "superimposed" into a single one; i.e., the functions of providing light input and output as well as chemo-optical conversion and signal processing can be fulfilled by a single pad. While it is feasible to create electrical output signals by means of on-chip optoelectronic processing units [13], the emphasis in the following is on purely optical (photonic) processing and output signals. One important way to implement the processing unit together with the sensing pad is that of using on-chip measuring variables (cf. Sec. 3). In any case, an important asset of the sensing schemes and principles presented in this chapter is their straightforward suitability for implementing multiple sensing pads on a single chip.

The novelty of the MIOS chip consists in the fact that it is an integral part of the whole sensor system, acting as a kind of "*smart planar optical transducer*" (SPOT), whose purpose is not only to provide the opto-chemical transducer function but also to fulfill all other essential tasks at the same time. This crucial point is best illustrated by comparison with some well-known examples of IO sensor types published in the literature. Many sensors just use the waveguide as a transducer, using prism [39] or end-face [30–32,35,37] coupling schemes. These arrangements lead to bulky and quite complex optomechanical systems, which is also the case for the type of grating coupler–based sensors depicted in Fig. 1(a) [24–29] and related types [33,34,36]. They typically need additional external components in order to convert the optical beams emerging from the opto-chemical transducer chip into a ''detector-compatible'' form for deriving the final sensor output signal. Since an overview of an extremely compact MIOS module type, namely, totally integrated, electrically connected IO chips comprising all components, including the light source and detector, has been given in Ref. 13, the rest of this chapter concentrates on miniature hybrid modules using MIOS chips with optical input and output ports.

Most of the considerations in the following are valid for a very wide class of IO sensor modules and sensor types, based on a great variety of effects, such as changes in the refractive index, absorption, fluorescence, or scattering properties of optical media used as waveguiding films or as substrates or located in the vicinity of a main waveguide. However, to keep our discussion of the main topics concise, only the limited range of evanescent wave sensing pad configurations shown in Fig. 3 will be considered. In addition, it is assumed that the value M of the measurand, i.e., the quantity to be measured or monitored, will be determined based on its effect on the effective refractive index

$$N = f_N \{n_s, n_f, n_\ell, n_c, h_f, h_\ell, \lambda, m, p\} \tag{1}$$

for the general type of chemical sensing pad shown in Fig. 3(a). Here, f_N denotes the functional dependence of the effective index N on the waveguide parameters, i.e., the refractive indices n_s, n_f, n_ℓ, n_c of substrate S, waveguiding film F, sensing layer SL, and cover medium C, respectively, and the respective thicknesses h_f and h_ℓ. The wavelength in air, the order of the waveguide mode, and the polarization (TE, TM) are denoted by λ, m, and p, respectively. Besides using commercial computer programs, some methods and useful expressions for calculating N can be found, e.g., in Refs. 24, 40, and 41.

In a practical sensor, the waveguiding film F, as well as the sensing layer SL, may actually consist of multiple layers. For conciseness, only single-layer configurations are assumed (except in Sec. 4.4). Three particularly simple yet practically relevant cases are shown in Fig. 3(b), (c), and (d).

The ''cover medium refractometer'' configuration of Fig. 3(b), obtained by setting $h_\ell \rightarrow 0$, provides access to measurands that affect the optical properties of the cover medium directly without the need for a separate sensing layer. For example, this sensing pad type is appropriate for measuring or monitoring changes in the bulk analyte refractive index, absorption, fluorescence, and scatter-

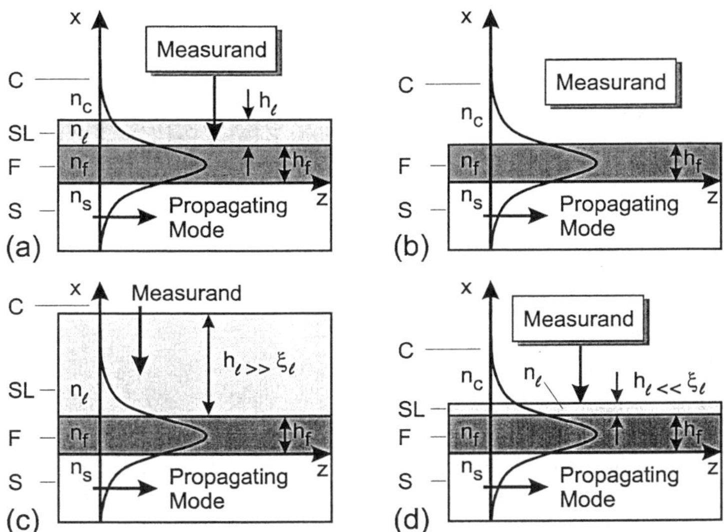

Figure 3 Typical sensing pad types: (a) general chemical sensing pad, (b) cover medium sensor, (c) optode membrane sensor, and (d) biosensor. The solid curves indicate a typical field distribution of the propagating mode.

ing. An almost identical situation [cf. Fig. 3(c)] is obtained if thickness h_ℓ is much larger than the $1/e$ field penetration depth ξ_ℓ of the propagating mode in the sensing layer (typically of the order of some 100 nm [42]), as is the case for chemical sensors making use of optode membranes with thicknesses in the micrometer range [43,44]. An important advantage of this configuration is the fact that the transducer output signal does depend only on the optical properties (n_ℓ) of the optode membrane and is affected neither by the refractive index n_c of the cover medium nor by the sensing layer thickness h_ℓ. Thus, the sensor is immune against disturbances originating from a bad quality of the sensing layer/cover interface or from excess absorption or scattering in the cover medium.

To simplify the discussion further, it is assumed that the measurand will affect only the properties of the sensing layer SL. Hence, if the thickness h_ℓ of the sensing layer is of the same order of magnitude as or smaller than the penetration depth of the propagating mode, the transducer effect consists in changing the refractive index n_ℓ and/or the thickness h_ℓ of the sensing layer upon changes of the measurand. A typical case of practical relevance is the biosensor example shown in Fig. 3(d), where a very thin sensing layer ($h_\ell \ll \xi_\ell$) is deposited on top of the waveguide. This configuration is also suitable for monitoring processes of thin-film growth, deposition, or adsorption on the waveguide's surface. If the refractive index n_ℓ remains constant during such processes, the optical transducer sensitivity, i.e., the changes of the effective index N upon changes of the layer thickness, can be expressed by the partial derivative

$$S_{N,h\ell} \equiv \frac{\partial N}{\partial h_\ell} = \frac{\partial f_N}{\partial h_\ell} \tag{2}$$

of the effective index function f_N with respect to the layer thickness h_ℓ.

Analogously, the sensitivity of the refractometric optode transducer pad corresponding to Fig. 3(c) is given by the partial derivative

$$S_{N,n\ell} \equiv \frac{\partial N}{\partial n_\ell} = \frac{\partial f_N}{\partial n_\ell} \tag{3}$$

of the effective index function f_N with respect to the optode refractive index n_ℓ.

Sensing pads with very high sensitivities $S_{N,h\ell}$, $S_{N,n\ell}$ and excellent stability can be obtained if hard, mechanically and chemically stable dielectric films with high refractive indices are used for fabricating the waveguides [14,15,17,19,22,44–50]. Materials such as Ta_2O_5, Nb_2O_5, TiO_2, ZrO_2, and Si_3N_4 with refractive indices $n_f > 2$ have been successfully used for fabricating waveguides. An additional advantage of using such stable high-index waveguiding films is that a very wide range of materials can be used for the substrates, sensing layers, and cover media with virtually any refractive indices n_s, n_ℓ, and n_c existing for practical substances.

The aforementioned sensing pad configurations are directly relevant for noncorrugated waveguide sensors such as various IO interferometer types [1,2,17, 20,22,27,30–37,51–54]. Another very important class of IO sensors makes use of grating couplers [1,14,15,17–19,24–29,42–50,55,56]. Figure 4 shows two configurations that are typical for sensor chips fabricated by depositing a waveguiding film F on structured substrates S [15,42,47,56]. A guided mode u_m is excited by illuminating the grating coupler with a collimated beam u_i at an angle of incidence θ_i measured in the ambient medium A with refractive index n_a. The grating is characterized by its length L, periodicity Λ, duty cycle κ, and the grating depths h_{fc} and h_{sf}. As has been shown in Ref. 42 for high-sensitivity waveguides, the effective index N is markedly influenced by the finite grating depth, even for geometrically shallow gratings. This leads to the fact that different effective indices apply for the corrugated (N_g) and noncorrugated (N_w) waveguide regions.

For simplicity, the thin-grating limit [46]

$$h_{fc} = h_{sf} \equiv h_g \to 0, \qquad N_g = N_w \equiv N \tag{4}$$

is assumed in the following. In this approximation, the effective index N is described by the very same function f_N [cf. Eq. (1)] for the configurations in Figs.

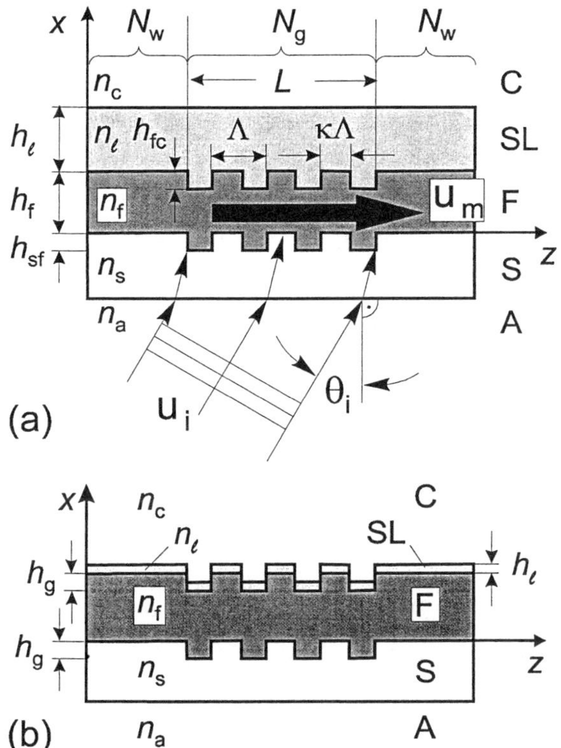

Figure 4 Grating coupler sensing pads (a) for thick and (b) thin sensing layers.

3 and 4. While the exact shape and profile depth of the grating structure must be taken into account for accurate modeling and reliable interpretation of the sensor output signals [42], they are neglected by using the thin-grating limit of Eq. (4).

Since the grating depth for high-sensitivity waveguides is typically only a few nanometers [14,15,42,45,47,55,56], thick sensing layers show a planarizing effect leading to the doubly stacked grating configuration according to Fig. 4(a). In contrast to this, very thin sensing layers give rise to the triply stacked configuration of Fig. 4(b). For such a configuration, with h_g and h_ℓ both on the order of 10 nm, immunosensor transducer pads have been demonstrated with very high resolutions and stabilities of about 1/4000 of a monolayer for a biochemical affinity assay [48].

In the thin-grating approximation, the grating coupler resonance condition for achieving maximum coupling efficiency is given by

$$N = n_a \sin \theta_i + \frac{m_g \lambda}{\Lambda} \tag{5}$$

where θ_i is the angle of incidence of the input beam, measured in the ambient medium with a refractive index n_a, and m_g is the grating diffraction order [46, p. 72].

An important aspect for designing and using sensor chips is the influence of the finite width and the light amplitude distribution of the incident beam, as well as the length L and the coupling strength of the grating coupler. Some practical aspects of this often-neglected topic have been pointed out [46,47], reporting results of theoretical and experimental investigations for high-sensitivity waveguides. Making use of a ''photonic picture'' for the action of a grating coupler as a ''momentum source'' during the interaction with the photons, the importance of taking into account several finite-length effects has been discussed, and approximate expressions for the width of different types of grating couplers and resonance phenomena have been given. Recently, the finite-grating-length effect for uniform-grating couplers has also been treated [57].

3 INTEGRATED OPTICAL SENSING SCHEMES

The choice of proper sensing schemes is crucial for exploiting an inherent advantage of integrated optics, that is, the possibility of obtaining arrays of sensing pads with an insignificant increase in complexity, fabrication effort, and cost. Careful consideration with respect to chip design and sensing schemes are also required to optimize the match between sensitivity, fabrication cost, and fulfilling the needs of practical applications.

Three different schemes are presented and discussed that demonstrate the importance of increasing the scale of integration on the chip level and of considering the chips as being an integral part of the whole sensor system in order to achieve SPOT chips that not only provide the task of conventional IO transducers, but also perform additional functions for which peripheral equipment is otherwise needed. Two important ingredients of these schemes are ''*on-chip measuring variables*'' and the use of ''*surface-emitting configurations*.''

3.1 Gradient Effective Index Sensors

Figure 5 shows a *smart planar optical transducer* example where an ''integrated optical light pointer'' IOLP is excited in a gradient effective index (GREFIN) waveguide [55]. An on-chip measuring variable, namely, the light pointer position $y_b(M)$, which depends on the value M of the measurand, is accomplished by introducing a spatial variation $h_f(y)$ of the waveguiding film thickness [see Fig 5(b)]. In this miniature sensor chip example, the light pointer is generated

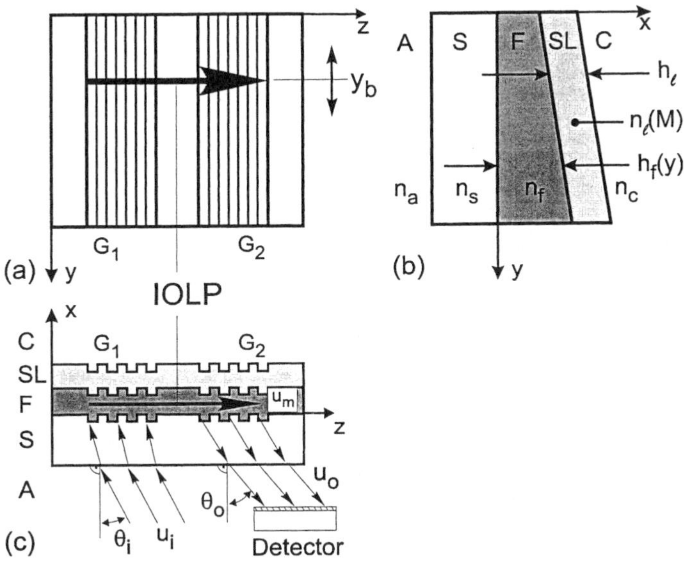

Figure 5 Miniature sensor based on a gradient effective index waveguide using uniform-grating couplers G_1 and G_2 in a tapered-thickness waveguide. (a) Top view and (b,c) cross-sections. The value M of the measurand is converted into the variable lateral position y_b of the integrated optical light pointer IOLP.

by illuminating the grating coupler sensing pad G_1 by a collimated laser beam incident at an angle θ_i measured in the ambient medium (air), as shown in Fig. 5(c). For deriving the sensor output signal, the IOLP beam is converted to the output beam u_o by the output coupler pad G_2, and the beam position y_b is determined by a position detector. This type of sensor chip has been shown to be suitable for measuring refractive index changes of the waveguiding film [55] and of the cover medium [17,46,58] with high sensitivity.

In order to be complementary to the theoretical expressions derived in Ref. 48 for a biosensor based on thin sensing layers, the most basic theoretical expressions are given here for modeling the performance of GREFIN sensors for refractometry of the cover medium. These applications are based on the sensing pad configurations shown in Figs. 3(b,c) and 4(a). The same expressions are valid for thick sensing layers (e.g., optode membranes) if $h_\ell \gg \xi_\ell$ and if n_c is replaced by n_ℓ (or by setting $n_\ell = n_c$, since the optode plays the role of the cover medium in this case). Assuming a linear taper of the waveguide thickness

$$h_f(y) = h_0 + g_h y \tag{6}$$

with a thickness gradient g_h and a thickness offset h_0, and introducing the position-dependent film thickness $h_f(y)$ as well as the measurand-dependent sensing layer refractive index $n_\ell(M) = n_c$ as waveguide parameters into Eq. (1) yields the modified functional dependence

$$N(y,n_c) = f_N\{h_f(y), n_c(M); n_s, n_f, \lambda, m, p\} \tag{7}$$

for the effective index $N(y,n_c)$ in the GREFIN sensing pad. For calculating the sensitivity, we follow the procedure given in Ref. 55, Sec. 3.3. The sensitivity S_{GREFIN} for converting the measurand-dependent refractive index n_c into the variable beam position y_b of the light pointer originates from two distinct contributions due to the shape of the thickness gradient ($S_{y,h}$) and the grating coupler resonance condition at constant effective index ($S_{h,n}$), and can be written as the product

$$S_{\mathrm{GREFIN}} \equiv \frac{\partial y}{\partial n_c} = S_{y,h} S_{h,n} = \frac{\partial y}{\partial h_f} \frac{\partial h_f}{\partial n_c} \tag{8}$$

of the partial sensitivities $S_{y,h}$ and $S_{h,n}$, which correspond to the respective partial derivatives shown in Eq. (8). $S_{y,h}$ corresponds to the inverse slope of the thickness taper and can easily be obtained from Eq. (6). Determining the partial sensitivity $S_{h,n}$ needs more effort, since the waveguide mode equation has to be solved. The final result is

$$S_{\mathrm{GREFIN}} = \frac{n_c}{g_h k \varepsilon_{fc} \sqrt{\varepsilon_{ec}}} \cdot \left[\frac{\varepsilon_{ec} + \varepsilon_e}{\varepsilon_{ec} + (\varepsilon_e / \varepsilon_{fc})} \right]^p \tag{9}$$

where

$$k = \frac{2\pi}{\lambda}; \qquad \varepsilon_{fc} \equiv n_f^2 - n_c^2; \qquad \varepsilon_{ec} \equiv N^2 - n_c^2; \qquad \varepsilon_e \equiv N^2 \tag{10}$$

and

$$p = \begin{cases} 0, & \text{for TE modes} \\ 1, & \text{for TM modes} \end{cases} \tag{11}$$

It was calculated analytically based on taking the appropriate derivatives in Ref. 46, Eq. (2).

The working principle of this MIOS chip makes use of the thickness gradient to *compensate* (“self-balance”) for the changes in the effective index N that would otherwise be induced by changes of the chemical-sensing layer. In other words, the photons always propagate at a fixed value of the effective index N, which is determined by Eq. (5). Another remarkable property of this sensor is its “surface-emitting configuration,” which provides the possibility of reading out the value of the measurand from the backside through the transparent substrate, as shown in Fig. 5(c). This is achieved by making use of grating pad G_1 as input pad, sensing pad, and optical processing unit at the same time (cf. Fig. 5.2) while pad G_2 is used as the output pad.

3.2 Chirped Grating Coupler Sensors

An interesting alternative to the GREFIN approach makes use of chirped grating couplers. Instead of the effective index gradient, a chirped grating coupler (CGC) with a spatially varying periodicity $\Lambda(y)$ is used for providing an additional on-chip degree of freedom. Figure 6 shows the corresponding sensing pad pattern with an input pad G_1 and an output pad G_2. While a uniform waveguiding film with constant thickness h_f can be used, more effort is required for fabricating the chirped grating couplers. The feasibility of fabricating these novel sensor chips by means of very low-cost processes such as replication and subsequent thin-film deposition has been demonstrated [15,56]. Experiments were successfully performed for replicated polycarbonate chips used in a flow cell to determine the refractive index of liquids. Due to the uniformity of the waveguide, the effective index function

$$N(M) = f_N \{n_\ell(M); n_s, n_f, n_c, h_f, h_\ell, \lambda, m, p\} \tag{12}$$

depends on the value M of the measurand only. Considering the general refractometric sensor example according to Fig. 3(a), and assuming a finite but constant

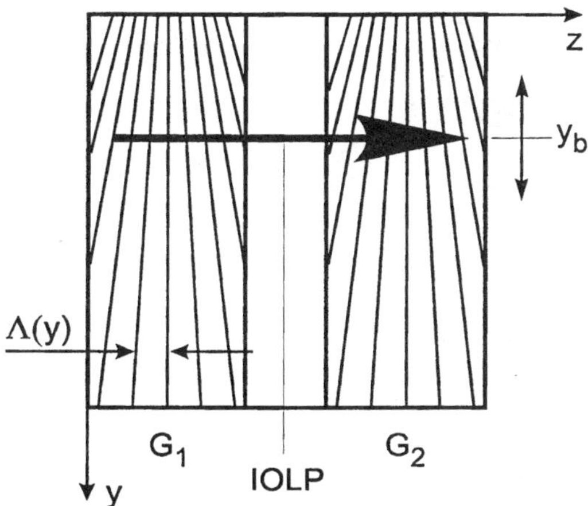

Figure 6 Miniature sensor chip based on a dual chirped grating pad structure (top view).

thickness h_ℓ for the sensing layer, the on-chip measuring variable is the light pointer position y_b, which is rendered dependent on the value M via the variable sensing-layer refractive index $n_\ell(M)$.

In the following, we assume a spatially varying grating periodicity

$$\Lambda(y) = \Lambda_0 + g_\Lambda y \tag{13}$$

with a linear chirp (constant chirp gradient g_Λ) and a periodicity Λ_0 at $y = 0$. Introducing the measurand-dependent effective index $N(M)$ and the "resonant grating periodicity" $\Lambda(y_b)$ at the IOLP position $y = y_b$ into the grating coupler resonance condition (5) yields the sensor characteristic

$$y_b(M) = \frac{\dfrac{m_g\lambda}{N(M) - n_a \sin\theta_i} - \Lambda_0}{g_\Lambda} \tag{14}$$

whose derivative

$$S_{CGC} \equiv \frac{\partial y_b}{\partial M} \tag{15}$$

is the overall CGC sensor sensitivity. More detailed expressions and results are given in Refs. 46–48, not only for the sensor characteristics for other types of sensors but also for the dependence of the IO light pointer width on various parameters, such as the effective length of the grating pad, the thickness taper for GREFIN sensor chips, and the periodicity chirp for CGC smart transducers.

As a consequence of the surface-emitting configuration, the very same optical readout arrangement as shown in Fig. 5(c) can also be used for the CGC chip! Both cases feature powerful intrinsic parameters for "programming" the shape (nonlinear or linear) of the sensor characteristic, for instance, using appropriate (nonlinear) gradients for the thickness taper (GREFIN) or the periodicity chirp (CGC). If pronounced nonlinearities are desired, which are useful for some monitoring applications, the CGC approach provides more flexibility, since each of the multiple grating pads on a single chip can easily be optimized for its specific purposes, and completely different grating patterns can be chosen for G_1 and G_2.

In both cases, it is not necessary to use separate input and output pads. Simpler single-pad configurations are obtained if the waveguide gratings are designed to work simultaneously at multiple diffraction orders or/and multiple waveguide modes. Another possibility is to superimpose grating G_2 on G_1, i.e., to use multiple periodic gratings [13,28,59–61]. However, the use of multiple separate grating pads is advantageous in many situations. An improved signal-to-noise ratio can be obtained if a separate output pad is imaged onto the detector, since in this case the background noise due to excess scattering in the illuminated input pad region can be suppressed, which is not the case if the input and output pads are superimposed. The gratings can also be designed to perform additional functions, such as imaging (e.g., focusing) onto a photodetector, including ''scan line corrections'' for obtaining an output beam u_o whose focus moves along a straight line on a linear detector array independent of changing analyte properties. Furthermore, in the arrangement of Fig. 6, it is possible to use both gratings G_1 and G_2 as active transducer pads simultaneously by illuminating each pad at different angles, wavelengths, or polarizations, using the other pad for outcoupling the respective guided mode(s). A very attractive possibility is to use such dual-pad configurations where one pad serves as a true on-chip reference for the other pad.

3.3 Interferometric Sensors

Using interferometry is an appropriate means to realize IO sensors for applications where very high resolution is of major importance. Many versions of IO interferometers have been proposed in the past [1,2,4,6,27,30–38]. However, most of the arrangements published so far do not fulfill the requirements needed for practical applications such as:

1. Simple and error-tolerant input/output coupling
2. Low number of peripheral optomechanical components
3. Small system size
4. On-chip measuring variables
5. Truly differential on-chip referencing capability
6. Suitability for realizing closely packed arrays
7. Very high sensitivity and stability
8. Low-cost fabrication

Work at CSEM/PSIZ has been done on realizing two interferometric module types aimed at fulfilling these requirements, namely, totally integrated monolithic versions [13,20–22,51–54,62,63,] and hybrid versions [16,17,48,51,64] of Mach–Zehnder interferometers (MZIs).

Figure 7(a) shows an example of a sensing pad suitable for realizing totally integrated monolithic sensors [13,20,22,48,52–54,62]. The sensing regions of the waveguide consist of sensor pads SP1 and SP2 implanted in the host IO circuit comprising the light source LS, channel waveguides WG, a Mach–Zehnder interferometer, including phase modulators PM1 and PM2, and a detector *D* [cf. Fig. 7(b)]. The waveguide cross section near the sensing regions depicted in Fig. 7(a) shows that the ''standard'' sensing pads according to Fig. 3 can also be used in this case. High sensitivities and a good stability can be obtained by depositing the sensing layer SL on a high-index waveguiding film *F* coated on a low-index buffer layer *B* with a thickness much greater than the penetration depth of the evanescent wave [42], and by placing the two sensing pads in close proximity.

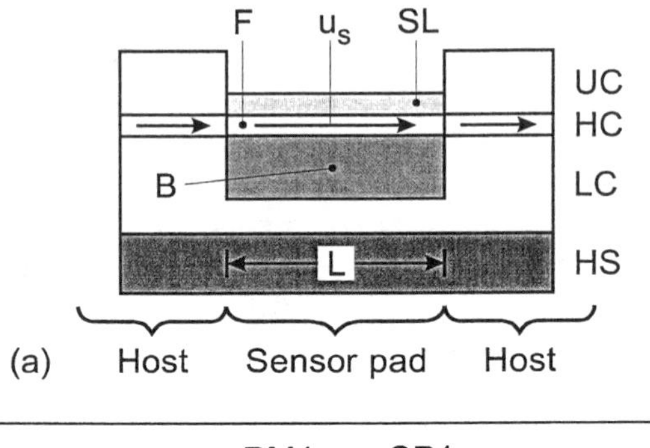

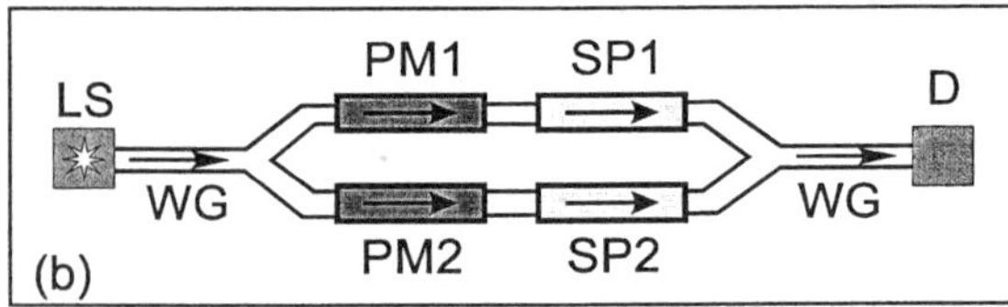

Figure 7 Totally integrated monolithic interferometer sensor. (a) Sensing pad cross section and (b) top view of a Mach–Zehnder interferometer with true differential chemical-sensing pads.

True differential operation is achieved by this highly symmetric arrangement (cf. Sec. 4.5). Working single-pad passive MZIs were recently demonstrated [20,22,52–54]. They have been fabricated by means of etching a sensor hole into a host layer structure consisting basically of an upper cladding UC, a host core HC, and a lower cladding LC grown on the host substrate HS.

In a biosensor example, the MZI converts the measurand-dependent sensing layer thickness $h_\ell(M)$ into a phase

$$\Phi_s(M) = 2\pi\, N(M)\frac{L}{\lambda} \tag{16}$$

accumulated by the guided mode u_s propagating through the sensing pad of length L with an effective index $N(M)$ [cf. Ref. 13, Eq. (11)]. Therefore, the basic interferometer sensitivity S_I, i.e., the change of the phase Φ_s induced by a change of the value M of the measurand, is given by

$$S_I = \frac{\partial \Phi_s}{\partial M} = \frac{2\pi L}{\lambda} S_{N,h\ell} S_{PT} \tag{17}$$

where S_{PT} and $S_{N,h\ell}$ are given by Eqs. (3.1.2) and (3.1.4) in Ref. 48, respectively. The advantages of using phase modulators have been outlined in Ref. 13.

For applications requiring disposable transducers, the use of replicated IO sensor chips is a very attractive possibility. Some configurations with appealing features have been proposed in Ref. 51, including theoretical results showing the feasibility of various novel IO interferometer types. One version of the transducer part is shown schematically in Fig. 8. In addition to the ''groove-type'' version of Fig. 8, a ''ridge-type'' MZI is readily obtained by using an inverted relief. Both types can be fabricated by just two low-cost full-area processes, namely, (1) embossing the surface relief into a plastic substrate, and (2) coating the whole structured substrate with a hard dielectric high-index waveguide. In order also to fulfill the requirement of simple and error-tolerant input/output coupling, preferentially in a surface-emitting configuration, a novel type of focusing grating coupler was developed that can be fabricated by the very same technology [16,17,64]. Figure 9(a) shows schematically such a coupler, consisting of a 12-nm-deep grating nanostructure located on a 3-μm-deep ridge waveguide microstructure, suitable for replication in a single step. A photograph of a working focusing coupler in a planar waveguide region is shown in Fig. 9(b). A collimated input beam ($\lambda = 633$ nm) was focused onto a 3-μm-wide spot within the waveguide. The intensity distribution of the focused TM_0 mode was derived from the stray light observed by means of a CCD camera attached to an optical microscope (cf. Refs. 17 and 64). Successful fabrication and light input/output coupling were

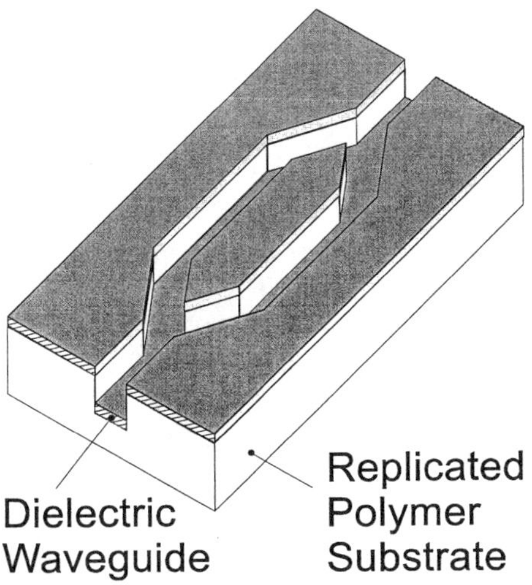

Figure 8 Replicated Mach–Zehnder interferometer structure suitable for low-cost mass production of miniature sensor chips.

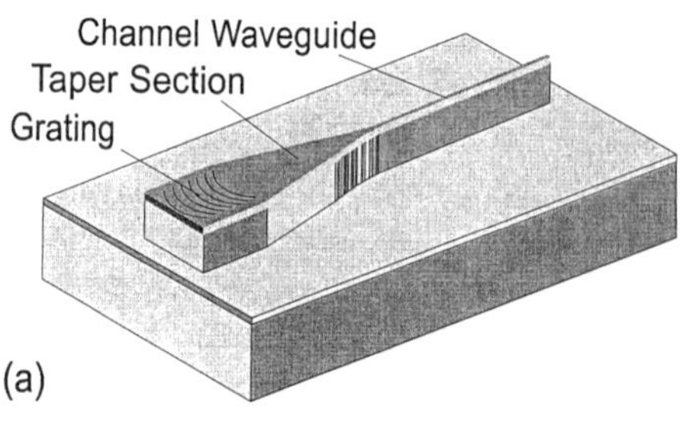

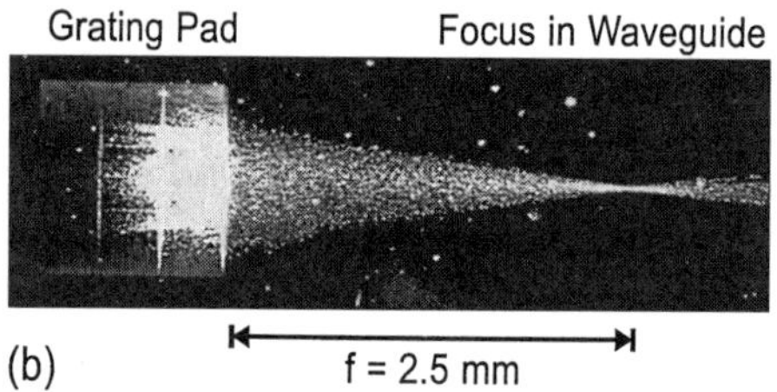

Figure 9 Replicated focusing grating coupler suitable as input and output pads for using the MZI of Fig. 8 in a surface-emitting configuration. (a) Schematic representation of replicated structure; (b) photograph of input coupler experiment.

also demonstrated for complete surface-emitting channel guide and MZI structures [16,17,64].

4 SENSOR CHIPS

4.1 Fabrication and Characterization

As was pointed out in Ref. 13, most of the technologies required to fabricate totally integrated measuring sensors based on III–V materials and silicon are already available, since they have been developed and are constantly being improved for opto- and microelectronic components used mainly for telecommunications applications. Some specific techniques suitable for fabricating the sensor chips are described in Refs. 20–22,47,62, and 63. A similar situation exists for implementing the sensing schemes based on hybrid arrangements, since light sources and photodetectors such as low-cost laser diodes, position-sensitive detectors, and photodiode arrays have been developed for consumer products such as compact disc players and photocopiers and for metrology applications.

For the fabrication of the MIOS chips, essential research and development work has successfully been performed in a joint effort in three distinct areas of technology, leading to: (1) methods for reproducibly fabricating micro- and nanostructures in glass and polymer substrates with a root mean square (r.m.s) surface roughness in the subnanometer range (at CSEM/PSIZ, Switzerland) [15,47,56,61,65–67], (2) procedures for the mass production of replicated chips by injection molding (at H. Weidmann AG Plastics Technology, Switzerland), and (3) a low-temperature process for coating the replicated substrates with high-quality waveguides consisting of hard dielectric films with a high refractive index (at Balzers Company, Lichtenstein) [15,45,47,56].

Figure 10 shows the fabrication steps for mass-producing replicated sensor chips. First, several steps are required in an ''origination'' sequence in order to produce a shim, which is then used to perform the replication sequence. Since

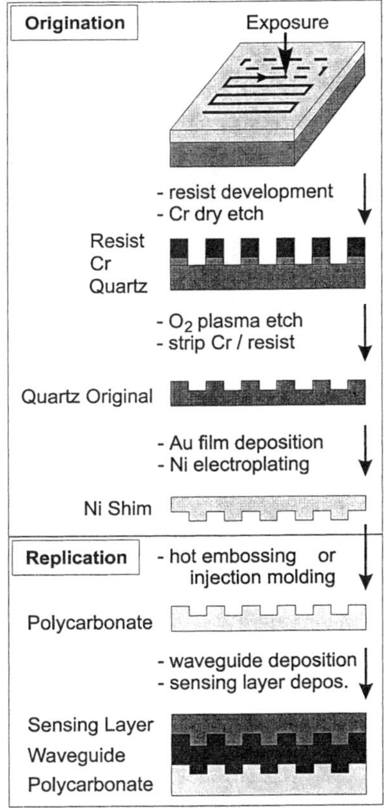

Figure 10 Chip fabrication: complete set of steps for origination and replication.

the origination sequence is required only once, it may be quite expensive, and the shims may bear a complex surface relief structure without significantly increasing the cost of mass-replicated MIOS chips.

Only a few important aspects of the fabrication process are outlined here. More details can be found in Refs. 15,17,47, and 56. While some advantages, such as a great flexibility, are offered if the original pattern is created by e-beam writing, other methods, for example, holographic exposure, are also being applied successfully. The process of transferring the resist pattern into a good-quality substrate, such as fused silica, is crucial, since the surface roughness can be controlled by the quality of the substrate material rather than that of the photo- or electron-resist film. For the pattern transfer, a careful choice from different (dry-etching) procedures, depending on the type of the relief, is important [15–17,47,56]. For depths in the nanometer range, O_2 plasma etching was chosen, since the very low etch rate is helpful in obtaining the desired depth with a high precision (on the order of 1 nm) just by timing, if the process parameters are reproducibly chosen and a proper calibration is made for the etching-time–to–geometrical-relief-depth ratio.

As is shown in Fig. 11, only two low-cost, full-area processes are required for chip mass production, namely, surface structuring (for example, by embossing or injection molding) and waveguide coating. An example of a replicated dual-chirped grating chip is also shown.

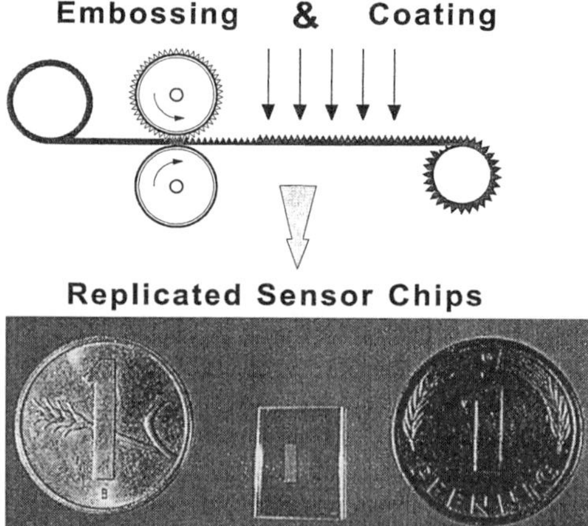

Figure 11 Schematic representation of the sensor chip replication process, and photograph of a chip based on a hot embossed polycarbonate substrate coated with a TiO_2 waveguiding film.

The characterization of the nano- and microstructures was performed optically and using an atomic force microscope (AFM). The optical procedures are based on measuring the first-order diffraction efficiency in the Littrow configuration [47] and on comparison with data computed using rigorous diffraction theory [68]. An optical setup for determining the thermal waveguide properties in situ during sensor experiments has been described in Ref. 69.

The surface relief shown in Fig. 12 demonstrates the state of the art in replication quality. The grating coupler nanorelief with a periodicity of 455 nm and a depth of 10 nm, was obtained by using a custom hot press in which the replication shim, polycarbonate foil (Röhm Europlex PC 99510, 250 μm thick), and a polished backing plate were hot-pressed together at a temperature of about 240°C for about 5 min. After cooling and separation, a high-fidelity copy of the shim surface relief was reproduced in the polycarbonate foil. The measured roughness of the polycarbonate surface was about 0.8 nm, both on top and at the bottom of the grating lines. Typical r.m.s. surface roughnesses achieved are 0.4 nm for structured fused silica and 0.8 nm for replicated polycarbonate foil. A low surface roughness is of utmost importance, not only for achieving good properties of the deposited film but also for minimizing optical scattering. After coating with TiO_2 waveguiding films, r.m.s. roughness values of 0.9 nm have been measured for waveguides on fused silica and about 1 nm for those on polycarbonate substrates. Work is currently in progress on optimizing the process parameters for a variety of techniques, including hot embossing and injection molding [15,70].

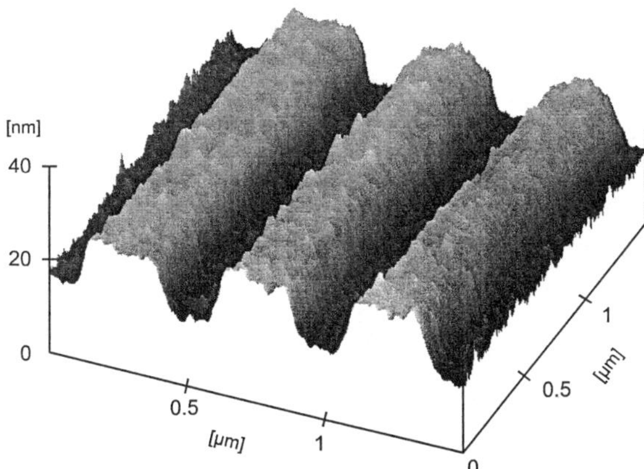

Figure 12 AFM picture of a 10-nm-deep grating coupler structure embossed into a polycarbonate substrate.

4.2 Sensing Pad Experiments

Three different types of experiments will be considered. The aim is to prove the feasibility of fabricating hard, compact dielectric waveguides with a high refractive index on structured substrates suitable for implementing high-sensitivity sensing and referencing pads in all the sensing schemes presented in this chapter. More specific experimental results will be presented in Secs. 4.4 and 4.5.

In a first series of experiments, chips in the ''bulk refractometric configuration'' according to Fig. 3(b) were used to demonstrate good performance of waveguides deposited on previously structured fused silica and plastic substrates [20,46,47,56,58]. The refractometric sensitivities on the order of 10^{-4}–10^{-5} achieved in these experiments can be further increased by several orders of magnitude, depending on the sensing schemes used and the efforts taken with respect to on-chip referencing, signal detection, and evaluation.

In a second type of experiment, the feasibility of realizing a novel kind of chemical optode sensor based on the configuration of Fig. 3(c) was demonstrated [43]. Figure 13 shows a top view of the sensor chip where an input grating coupler was used to excite guided modes u_s and u_s' at an optode pad and modes u_r and u_r' at a reference pad. The measurements were performed by scanning the collimated input beam from a laser diode over the two pads located side by side on the same grating coupler based on determining the grating coupler resonance angles. In the experiments where the plasticized polymer optode membranes provided the function of selective extraction and chemical recognition of the analyte, refractive index differences of $\sim 3 \times 10^{-3}$ per decade of molar Ca^{2+} concentration in buffered aqueous solutions have been observed. A typical overall resolution of 0.7–0.07% in units of Ca^{2+} concentration is expected if this type of sensing pad is used in miniature integrated optical devices. The experimental results demonstrated not only the IO optode sensor principle, but also the on-chip referencing capability in a practical application (cf. Sec. 4.5).

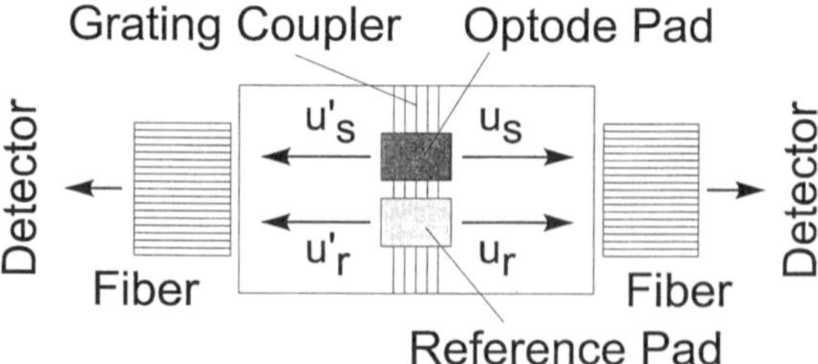

Figure 13 Integrated optical optode sensor with reference and optode membrane pads located side by side on the same grating coupler chip.

A third series of experiments was devoted to demonstrate the performance of "standardized" waveguides whose fabrication process was optimized to yield good sensing pads for a great variety of plastic, glass, or semiconductor substrates. The optochemical transducer consisted of a protein A layer immobilized on a compact TiO_2 waveguiding film fabricated by a low-temperature process [cf. Fig. 3(d)]. Results reported in Ref. 48 for a bioaffinity sensor chip at wavelengths of 633 nm and 785 nm showed that these sensing pads feature very high sensitivities and an excellent stability and that infrared laser diodes are adequate light sources. Resolutions and stabilities on the order of 1/4000 of an immunoglobulin (IgG) monolayer ($\sim$1 pg/mm^2) have been demonstrated by performing experiments in an input grating coupler configuration.

4.3 Sensing Pad Modeling

Figure 14 shows schematically the sensing pad structure that is typical for the case of biosensors where analyte molecules are being adsorbed to a sensing layer immobilized on a waveguide. An exact modeling of the real situation [cf. Fig. 14(a)] is not practical, since it would require a huge number of parameters, such as the three-dimensional refractive index distribution within the corrugated layers. While a multilayer approximation with thicknesses $h_1 \ldots h_{\max}$ and refractive indices $n_1 \ldots n_{\max}$ can easily be derived theoretically, it would require many parameters to be determined experimentally, which again is not practical.

An appropriate method for reporting results obtained for IO biosensors, where the measured quantity is the effective refractive index $N(M)$ of a guided mode, is to introduce a homogeneous "equivalent sensing layer" with refractive index n_ℓ whose variable "equivalent" thickness $h_\ell(M)$ depends on the value M of the measurand. As an example, if the sensor is used to determine the surface coverage

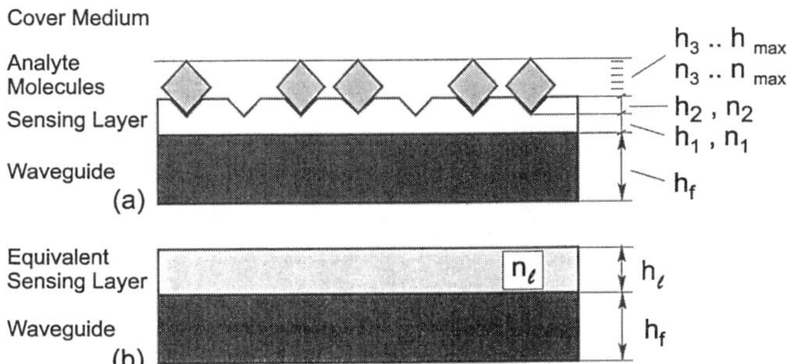

Figure 14 Typical (a) corrugated inhomogeneous sensing layer configuration for practical biosensor chips and (b) noncorrugated homogeneous equivalent sensing layer model.

$M = \Gamma$ of the analyte, the layer thickness $h_\ell(\Gamma)$ is *the numerical value that yields the same effective index* $N(\Gamma)$ *as observed in the experiment*. Despite the simplicity, no information is lost if the (low) number of variable parameters is matched to the (low) number of measured quantities, and the results can easily be corrected at any (later) time for more complex layer models, including, for example, the formation of multiple birefringent layers with variable thickness and refractive indices, if the correct layer data are available. An additional advantage of this model is its direct validity for applications where a ''pure'' layer thickness change is to be measured, as, for example, in layer growth studies. The practical usefulness of this approach is illustrated by looking at the results reported in Ref. 48, where a resolution $\partial h_\ell < 2$ pm of the ''equivalent sensing layer thickness'' was achieved. Based on the known density of the adsorbed material, an estimate of the corresponding surface mass coverage resolution $\partial\Gamma \approx 1$ pg/mm^2 is readily obtained from ∂h_ℓ, which is not the case if we had just reported the experimental resolution $\partial N \approx 9 \times 10^{-7}$ in the effective index determination.

However, accurate sensor pad modeling is crucial for a reliable determination of the properties of highly resonant waveguide configurations used for realizing IO sensors with high sensitivities and for reliably interpreting their output signals. This has been clearly demonstrated by comparing the results obtained from rigorous diffraction theory and by different approximations reported in Ref. 42. It concerns not only the finite-depth effects for grating coupler sensors, but also

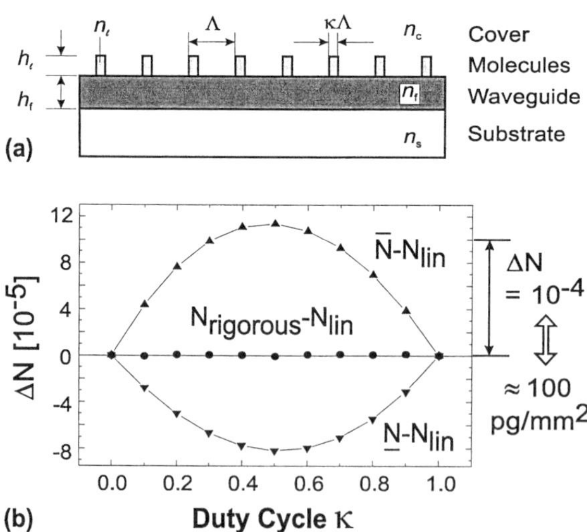

Figure 15 (a) Definition of the spatial duty cycle κ for a corrugated, thin sensing layer; (b) numerical results for the effective index effects obtained by different models.

the proper modeling of the corrugated sensing layers mentioned earlier. As an example, numerical results for the deviations ΔN of effective index changes originating from a variation of the duty cycle κ of a fractional monolayer of molecules directly adsorbed on a high-sensitivity waveguide are plotted in Fig. 15. For a parameter set with values close to the ones used for performing the immunosensor experiments reported in Ref. 48, deviations as high as $\Delta N \approx 10^{-4}$, corresponding to $\Delta\Gamma \approx 100$ pg/mm^2, result from using different approximations denoted by $\overline{N}$ and $\underline{N}$ instead of the rigorous theory [42]. These results show that by using rigorous theory, the experimental *resolution* of ~1 pg/mm^2 reported in Ref. 48 can be turned into an *accuracy* of the same order of magnitude.

4.4 Multilayer Waveguides and IO Chemical Benches

An important ingredient with respect to commercialization is the development of standardized processes for fabricating the transducer elements to be used in the sensor modules. Multilayer waveguides mentioned in Refs. 14 and 60 are a powerful means for accomplishing this goal. Much work has been done and is still needed for developing optochemical sensing layers with high sensitivity and stability. For a proper deposition/immobilization of these sensing layers, it is crucial to offer the possibility of using various types of chemical and physical surface properties in the sensing pad regions of MIOS chips. In the multilayer approach presented here, instead of trying to find a compromise between the often-conflicting chemical and optical requirements, separate layers are used to fulfill each of them.

Stable waveguide platforms with a high optical sensitivity are provided by a small set of standardized base waveguides with high refractive index (e.g., using compact Ta_2O_5 or TiO_2 films). A large variety of chemical and physical surface properties is then made available for implementing sensing layers with high chemical sensitivity and selectivity by means of additional "chemical interface layers" coated on the base waveguides. Sensitivity curves, i.e., the variation of the effective index N upon the adsorption of a 1-nm-thick protein layer ($n_\ell = 1.45$) on interface layers with thicknesses $h_i = 0, 20, 40$ nm, are shown in Fig. 16 as a function of the base waveguide thickness for three typical biosensor examples. A base waveguide consisting of a TiO_2 film on a polycarbonate (PC) substrate and an aqueous analyte were assumed.

The results show that not only high-index (Ta_2O_5, $n_i = 2.21$) and medium-index (SiN_x, $n_i = 1.85$) materials may be used as chemical interface layers, but low-index materials, such as SiO_2 ($n_i = 1.45$), can also be successfully applied without significantly reducing the maximum sensitivity if the layer thickness is kept below, say, 20 nm. Hence, it is possible to use a great variety of even low-index materials and still retain the very high sensitivity of TiO_2 waveguides. An example of practical importance would be to use multilayer sensing pads ac-

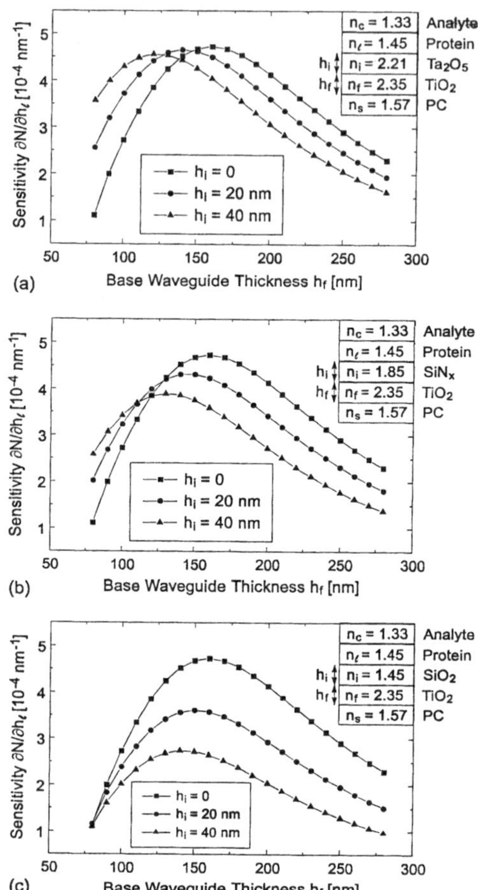

Figure 16 Calculated biosensor sensitivities for multilayer waveguide examples based on a TiO_2 base waveguide coated with (a) Ta_2O_5, (b) SiN_x, and (c) SiO_2 chemical interface layers. Results are for the TM_0 mode at $\lambda = 785$ nm.

cording to Fig. 16(b) for realizing sensing pad arrays based on advanced chemical-sensing-layer techniques [71], including surface immobilization of biomolecules by light on silicon nitride [72,73].

The multilayer approach is also suitable for optimizing many other types of sensing pads. As an illustration, sensitivity curves are shown in Fig. 17 for an optode sensor based on a plasticized PVC membrane [43], where the interface layer provides more flexibility with respect to adhesion and other interface prop-

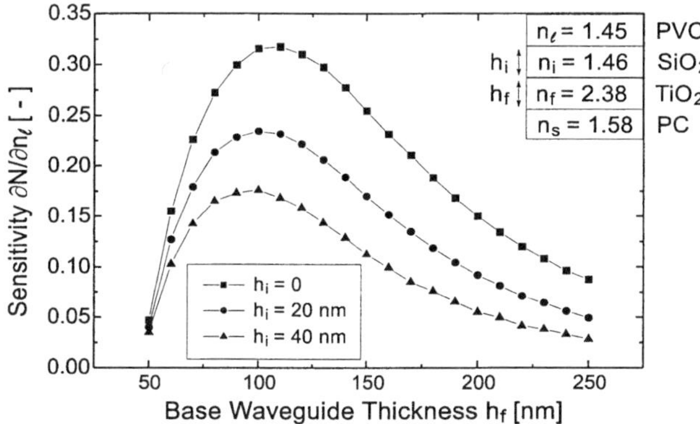

Figure 17 Calculated sensitivities for a multilayer waveguide chemical optode using the same TiO_2 base waveguide as for Fig. 16, coated with SiO_2 chemical interface layers. Results are for the TM_0 mode at $\lambda = 690$ nm.

erties. Also in this thick-sensing-layer case, high sensitivities $\partial N/\partial n_c > 0.2$ are available if the interface layer thickness is kept below ~20 nm.

These results clearly demonstrate the suitability of the multilayer waveguide approach for providing a kind of "IO chemical benches," consisting of MIOS chips fabricated by depositing different chemical interface layers to define different sensing pad regions on a single base waveguide. For applications requiring mass-produced disposable sensor chips, a particularly attractive version of a base waveguide are the compact TiO_2 waveguiding films on replicated polycarbonate foil described in Refs. 15 and 56.

4.5 On-Chip Referencing and Multicomponent Sensing

As a consequence of their high sensitivity, IO sensors based on thin, high-index waveguiding films are strongly affected by external disturbances originating, for example, from temperature variations, unspecific chemical effects, and instabilities of many (bio-)chemical sensing layers and fluctuations of the optical properties of the reagent solutions. By making use of the ingredients described in the previous sections, multiple sensing regions may be placed on a single MIOS chip without significantly increasing the cost. This opens up exceptional opportunities for realizing on-chip referencing as well as chemical multicomponent analysis.

To compensate or correct for unspecific effects, it is crucial to place the reference pads as close to the corresponding sensing pads as possible, since many disturbances result from local gradients of sensing layer and analyte properties. Results of detailed theoretical and experimental investigations on the temperature effect as an important example relevant for the majority of practical applications have been reported [19,50,69]. The experiments were conducted for high-sensitivity grating coupler chips based on fused silica and polycarbonate substrates in a dual-pad arrangement similar to the one shown in Fig. 13, but without chemical sensing layers. Analogous to the term *common-mode rejection ratio* (CMRR) used to describe the performance of electronic operational amplifiers, a *baseline-ripple rejection ratio*

$$\mathrm{BRRR} = \frac{\Delta O_n}{\Delta O_r} \tag{18}$$

is an adequate quantity to characterize the performance of MIOS chips with on-chip referencing. Here, ΔO_n and ΔO_r denote the baseline variations (ripple) due to unspecific effects for nonreferenced and referenced operation, respectively. Using this definition, the results obtained in Refs. 19, 50, and 69 can be summarized by stating that values of BRRR $\approx 30, \ldots, 350$ have been achieved for a reference-to-sensing pad distance of 2 mm. On-chip referencing does provide the possibility to compensate not only for unspecific chemical, but also for a much wider class of disturbing effects, including aging and changes in light source properties such as intensity, angle of incidence, and wavelength. By means of very detailed investigations, the benefit of on-chip referencing for practical sensing applications has been clearly demonstrated [19,49,50,69].

As an example, we consider the results shown in Fig. 18 for a series of repeated, consecutive immunoassay cycles performed with a replicated grating coupler sensor chip in the presence of a temperature variation between 24°C and >27°C, as given in Fig. 18(a) [19,50]. A TiO_2 waveguiding film with a refractive index $n_f = 2.349$ and a thickness $h_f = 150$ nm deposited on a replicated polycarbonate substrate was used to perform the experiments. Sensing and reference pads were formed by different chemical treatment of two adjacent grating regions, separated by about 2 mm. While the sensing region was incubated with protein A, the reference region was blocked by applying a bovine serum albumin (BSA) layer. Much more details are available in Refs. 19 and 50.

Figure 18(b) shows the apparent thickness variations $\Delta\tilde{h}_\ell^S$ and $\Delta\tilde{h}_\ell^R$ of the sensing and the reference pads, respectively, for six consecutive cycles of an immunoassay. The term *apparent* is used here to express the fact that these variations are affected by the unspecific temperature effect, which introduces a severe error into each of the signals, as is clearly demonstrated by the data points in Fig. 18(b). For $t = 0$, the apparent thickness variations $\Delta\tilde{h}_\ell^S$ and $\Delta\tilde{h}_\ell^R$ were set to zero. The figure shows six consecutive cycles out of a longer sequence, where a sample solution supply, containing 10^{-8} M r-IgG, resulted in an equivalent adsorbate layer increase of about 800 pg/mm^2 in each cycle, which was then removed in a regeneration step.

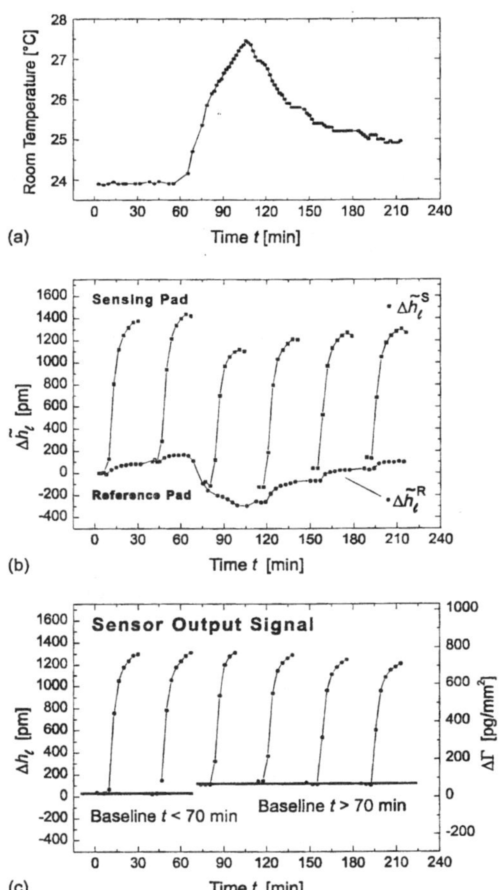

Figure 18 Variation of the room temperature (a) and resulting (apparent) sensing and reference layer thickness variations for a replicated polycarbonate sensor chip during six consecutive cycles of an immunoassay with regeneration step: (b) Apparent thickness variations $\Delta\tilde{h}_\ell^S$ (■) and $\Delta\tilde{h}_\ell^R$ (●) of the sensing and reference pad, respectively. (c) Calculated equivalent thickness variation $\Delta h_\ell = \Delta\tilde{h}_\ell^S - \Delta\tilde{h}_\ell^R$.

Making use of the response of the reference pad to compensate for the disturbance by the temperature effect, the corrected equivalent thickness variation plotted in Fig. 18(c) was obtained by just taking the difference $\Delta h_\ell = \Delta\tilde{h}_\ell^S - \Delta\tilde{h}_\ell^R$. As can be clearly seen, the baseline is now very stable during the first two assay cycles and during the last four cycles, in spite of the large temperature variation. What is also demonstrated is that on-chip referencing is more than just a means for temperature compensation. As a consequence of the markedly improved baseline stability, one can easily detect subtle changes and a different behavior of the biochemical layers forming the signal and reference pads. In the example shown in Fig. 18(c), such an effect takes place at time $t \approx 70$ min. Comparing the curves in Fig. 18(a–c), it can be derived that the effect is certainly not a simple direct temperature effect, which could not be concluded without on-chip referencing.

In spite of the fact that the main advantage of using mass-replicated chips occurs for applications such as the medical, where they will be used as disposables, they show a remarkable performance with respect to both sensitivity and stability. This is demonstrated by the set of 31 consecutive assay cycles plotted in Fig. 19 over a time period of 17 h for a similar chip as the one used to perform the experiments just described. A different chemical behavior of signal and reference pads against the regeneration procedure leads to the baseline variation for the very first three cycles. To demonstrate the long-term chip performance, r-IgG analyte concentrations were alternated between 10^{-8} M and 10^{-9} M after the second cycle. As can be seen from the inset, which is a magnification of the sensor response curves for the time window from 11 h to 17 h, the functionality of the sensing and reference pads is still excellent after this long measuring period. Even for the 10^{-9} M concentration, the transducer signal is more than two orders of magnitude above the detection limit.

With respect to temperature effects, the main advantage of on-chip referencing

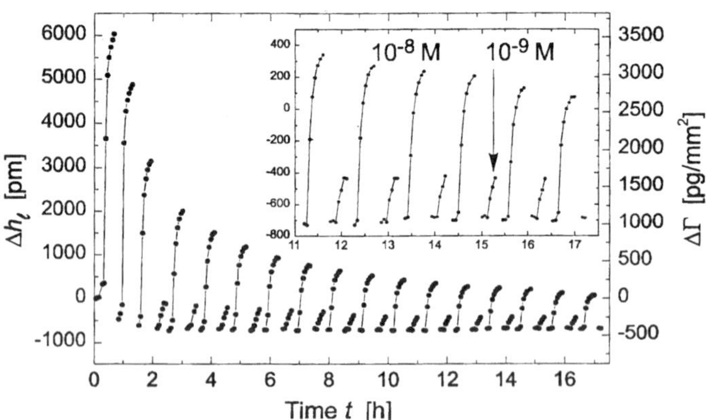

Figure 19 Sensor signal in terms of equivalent thickness variation Δh_l for a replicated polycarbonate sensor chip during 31 cycles of an immunoassay with regeneration steps. The inset shows a detailed view for $t = 11, \cdots, 17$ h.

is that the corrected sensor response is dependent no longer on the temperature itself, but just on temperature gradients. Therefore, the most effective means for achieving high performance are to realize a ''truly differential'' operation and to use *closely spaced* but still *separately accessible* reference and sensing pads in a highly symmetrical arrangement. A novel type of ''diagonal'' Mach–Zehnder interferometer designed for fulfilling these requirements was proposed in Ref. 51. As illustrated by the schematic representation in Fig. 20, it makes use of the easy access to the third dimension available in replicated structures, introducing a vertical rather than a lateral offset H for separating the interferometer arms with chemical pads 1 and 2. Calculations have shown that a vertical offset $H \approx 3\ \mu m$ is sufficient. A major advantage of this approach is that many separate sensing and referencing pads can be implemented in an extremely dense arrangement on one chip, since the waveguide width W is in the order of a few micrometers. As illustrated in Fig. 21, such interferometers are suitable for realizing multicomponent analysis chips, not only minimizing the required chip area and analyte volume, but also achieving a high BRRR due to true differential on-chip referencing. While the numerical results indicate the feasibility, this interferometer type has not yet been fabricated, in contrast to the replicated ''lateral'' MZI shown in Fig. 8, which has been demonstrated recently [16,17]. Both types are suited to realizing miniature modules based on single-chip sensor arrays, as will be discussed in the next section.

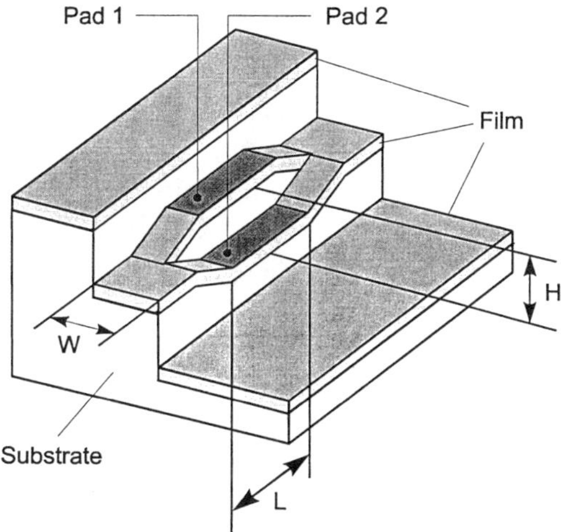

Figure 20 Novel ''diagonal'' Mach–Zehnder interferometer suitable for realizing replicated closely spaced sensing-pad array chips.

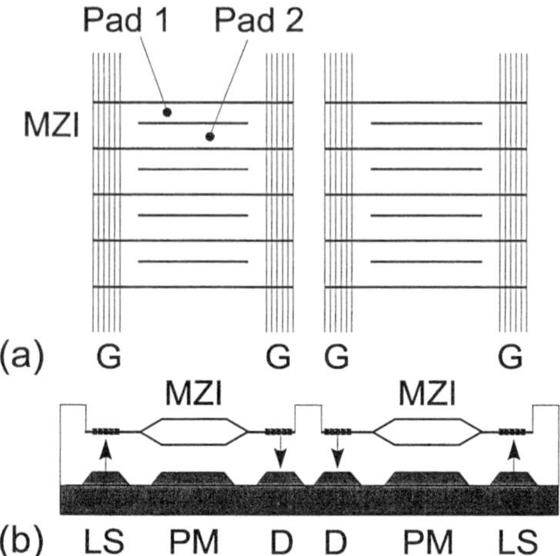

Figure 21 Example of a miniature single-chip sensor array based on diagonal interferometers (MZI) optically connected via grating couplers *G*. The lateral transducer pad arrangement is shown in the top view (a), while (b) is a cross section through the embossed structure on top of a miniature optoelectronic platform comprising light sources LS, phase modulators PM, and detectors *D*.

5 MINIATURE INTEGRATED OPTICAL SENSOR MODULES

General features and peculiarities are discussed for three basic types of *complete* miniature sensor modules that can be realized based on the ingredients just presented. While the work at CSEM/PSIZ in the past 10 years has concentrated on theoretically modeling and experimentally demonstrating the feasibility of all the ingredients, the succeeding years will be devoted to completing the sensor modules. The arrangements illustrate the potential for combining the ingredients for obtaining modules with different features able to achieve the goals stated in the introduction and facilitating the adaptation to different applications and production methods. For clarity, only one or two sensing pads may be shown; however, all sensing schemes are inherently suitable for accomplishing single-chip *arrays of transducer pads* to be fabricated at very low additional effort and cost.

5.1 Totally Integrated Monolithic Modules

The miniature sensor array chip shown in Fig. 21 is a good example to illustrate what is meant by a sensor module, since it is not just a transducer chip but represents a complete sensor system comprising all parts and providing all functions required to measure or monitor (bio-)chemical quantities in practical applications. The stacked monolithic ''multilevel'' arrangement, based on placing the planar optical transducer part in a distinct level above an optoelectronic integrated circuit [13–15,18,60], demonstrates the benefits of combining several of the ingredients discussed in the previous sections, in particular (1) close packing of true differential sensing pads, (2) low-cost mass production by using standard processes within each level, and (3) making use of the surface-emitting configuration to provide direct alignment-proof optical input/output coupling. Not only can the light sources LS and detectors D be included in the platform, but so can (electronic) circuits required for power supply, control, and signal evaluation.

In this example, the optochemical transducer part is realized by a single embossing step into a polymer layer, coated on the finished optoelectronic platform, and a single thin-film deposition step to form the waveguide. Work has been done successfully to demonstrate the fabrication of this type of waveguide based on polymer films coated on rigid substrates [15,65,66]. Two important features of this approach are the planarizing effect of the embossing step and the protection of the underlying optoelectronic platform by the polymer and waveguide layers. Other types of replicated sensor structures [16,56,64] may also be applied to realize such single-chip modules. Different technologies may be used and combined to fabricate the optoelectronic platform. One possibility is to realize a kind of ''optoelectronic motherboard'' with integrated vertical-cavity surface-emitting lasers (VCSELs) or light-emitting diodes (LEDs), together with silicon detectors and integrated circuits (ASICs). In an alternative module type already presented in Fig. 7, the transducer regions are directly fabricated within the optoelectronic platform itself in a ''single-level butt-coupling'' scheme [20,22,53]. Further variants of ultracompact, totally integrated modules making use of sensing pads directly integrated into a main waveguide or implemented by means of additional layers on top of the main waveguide (stacked monolithic modules) have been presented and discussed in Refs. 13, 14, 22, 52, 53, 54, and 60.

5.2 Stacked Hybrid Modules

For applications requiring disposable transducer elements, miniature modules can be obtained by using separate chips for fulfilling the different tasks. In the example shown in Fig. 22, a replicated SPOT chip is used to convert the chemical effect into a variable light pointer position (cf. Fig. 5). In this hybridly stacked multichip module, the transducer chip is removable and can be made reusable, exchangeable or disposable whatever best fits the application. In the very same replication step, alignment structures can also be produced which provide a ''snap-in''-type of positioning of the transducer chip with respect to a miniature optoelectronic platform (MOEP) or/and to a chemical analyte supply head (CASH).

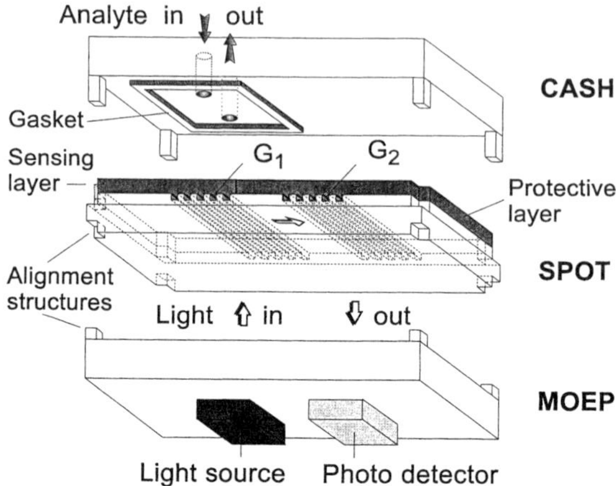

Figure 22 Triply stacked MIOS module consisting of a smart planar optical transducer (SPOT) supported mechanically, optically, and electronically by a miniature optoelectronic platform (MOEP) on one side and by a chemical analyte supply head (CASH) on the other side.

This approach provides an extreme flexibility for using mass-produced standardized components and still satisfying the needs of very different applications, since (1) different light sources and detectors may be incorporated that are fabricated for other mass market products, such as compact disc players and telecommunications equipment; (2) the transducer chips can be easily handled, since no other parts are connected to them; (3) on-chip referencing and sensor pad arrays can be accomplished by means of covering some pads with protective or "referencing" layers; and (4) only a small quantity of harmless material has to be disposed of after the usage.

Based on this concept, a compact integrated optical biosensor has been realized [18]. Figure 23 shows a photograph of the complete sensor module, including the detector and readout electronics in a box of about $10 \times 10 \times 10$ cm^3. The sensor chip (cf. Fig. 24), consisting of a TiO_2 waveguide on a replicated polycarbonate (PC) substrate, is read out from the back side in an arrangement similar to the one used for playing compact discs (CDs), while the chemical analytes are applied from the top side. Figure 24 shows how a sensing and reference channel are illuminated simultaneously by a single expanded laser beam u_i for simultaneously creating two guided light pointer beams. These beams lead to two separate intensity peaks in the output beam u_o, directed onto a single linear detector array by means of the output pads, without the need of any external optics. For a detector with 512 pixels, curves showing the adsorption kinetics

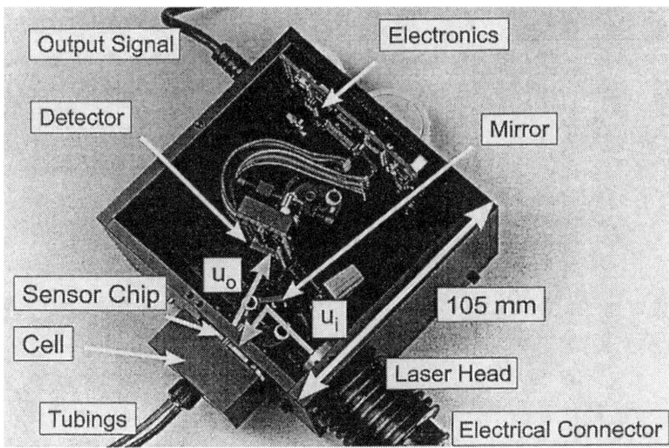

Figure 23 Compact integrated optical sensor module making use of IO light pointer chips.

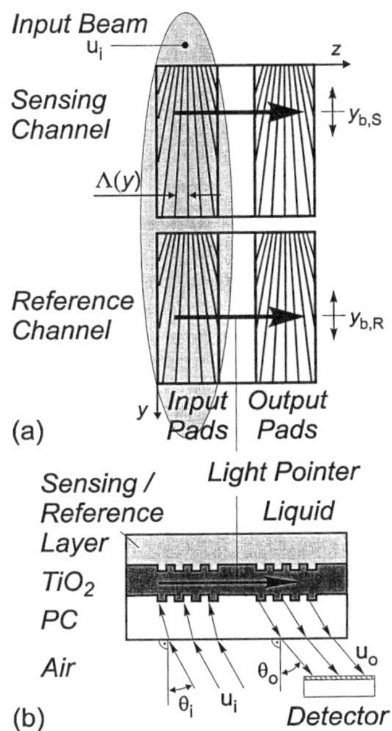

Figure 24 (a) Top view and (b) cross section of a dual-channel IO chirped grating pad (bio)chemical sensor.

have been recorded, with a data rate of a pair of signal and reference measurand values every 10 ms. The input and output grating pads were designed to have different spatially dependent periodicities $\Lambda(y)$ in order to get input and output angles θ_i and θ_o with values suitable for achieving a miniature arrangement with low background signal.

During a measurement, the two light pointer positions $y_{b,S}$ and $y_{b,R}$ indicate the values of the biochemical measurands at the location of the sensing and reference pads, respectively. Based on the known sensor chip sensitivities, the changes of the thicknesses $\Delta\tilde{h}_\ell$ and the surface mass coverages $\partial\Gamma$ are determined in real time from these positions for layers of analyte molecules being adsorbed on the signal and reference pads. For the example of an immunoassay, a typical sensor response is shown in Fig. 25. For this experiment, a polycarbonate-based replicated chip was used on which a protein A layer and a BSA layer were coated on the input pads to form the sensing and reference pads, respectively. The main parts of the assay consisted in flushing buffer solution over the chip (B, 0–9 min), applying analyte solution (A, 10^{-8} M r-IgG, 9–24 min), and flushing with buffer again (B, 24–30 min). In order to check the possibility of reusing the chip, a regeneration step (R, 30–36 min) and a buffer sequence (B, 36–46 min) were also performed. It can be clearly seen that the regeneration process practically did not affect the reference pad, while it established the former baseline for the sensing pad. Resolutions $\partial N \approx \pm 2 \times 10^{-6}$ and $\partial\Gamma \approx \pm 3$ pg/mm^2 have been achieved using this demonstrator module and replicated chips. Detailed informa-

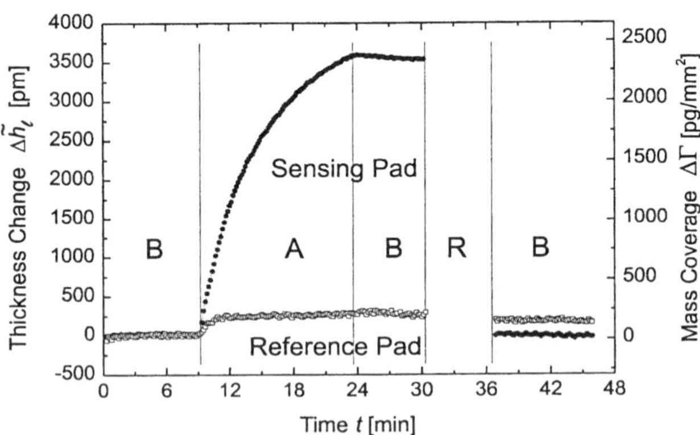

Figure 25 Apparent thickness variations $\Delta\tilde{h}_\lambda$ for the sensing pad (filled circles) and for the reference pad (open squares), and corresponding changes in surface mass coverage (right ordinate) during one cycle of an immunoassay, including a regeneration step. B: buffer solution (PBS + 1% BSA); A: analyte solution (10^{-8} M r-IgG in buffer); R: regeneration solution (glycine, pH 3.4).

tion on the compact module, chips, and assays is given in Refs. 18 and 19 and in the references cited therein.

Chemical sensing is another application area where miniature instrumentation and high resolution are important. As an example, we briefly mention pH sensing. Using the same compact module as just described, and the same replicated chips, but with a different chemical treatment of the input pads, a pH sensor was realized [44]. Figure 26 shows the sensor response in terms of changes of the refractive index n_m and in units of pH for a chip where the pH-sensitive swelling of a polymer membrane, deposited on the sensing pad, was detected by means of the compact module shown in Fig. 23. For a 300-nm-thick photopatterned hydrogel membrane, a response time of about 30 s and a resolution $\partial\text{pH} \approx 1 \times 10^{-4}$ in terms of the pH in the region of pH $\approx$ 7.5 were obtained. A resolution in the order of $\partial\text{pH} \approx 10^{-5}$ is likely to be achievable. This region is especially relevant for applications in microphysiometry, and the resolution is markedly higher than the 10^{-2}–10^{-3} of typical commercial instruments. Much more details on experiments and chip fabrication are given in Refs. 19 and 44.

Not only do these examples demonstrate the suitability of the approach for realizing miniature sensors for practical applications, but they also show the versatility of the concept and the possibility of using the same instrumentation and IO chips for very different applications, which is relevant for achieving high performance at a low cost.

For some applications it may be desirable to realize modules similar to the one shown in Fig. 27, where the transducer chip is not sandwiched between two other chips, but a kind of "optochemical reader head" is used for both supplying the chemical reagents and reading out the transducer signal from one side only.

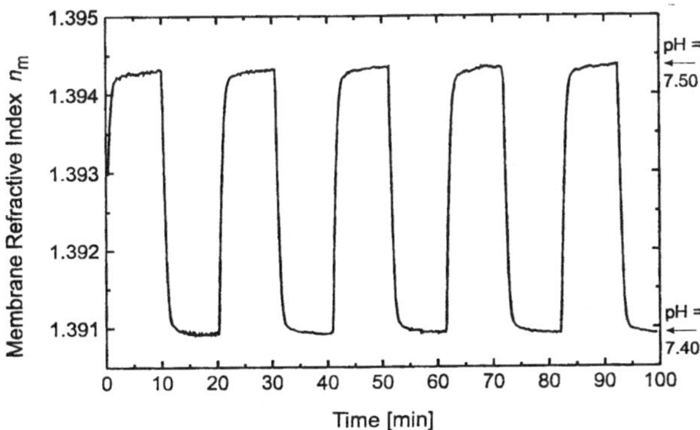

Figure 26 Sensor response, in terms of the membrane refractive index variation, upon repetitive pH switching between pH 7.40 and 7.50.

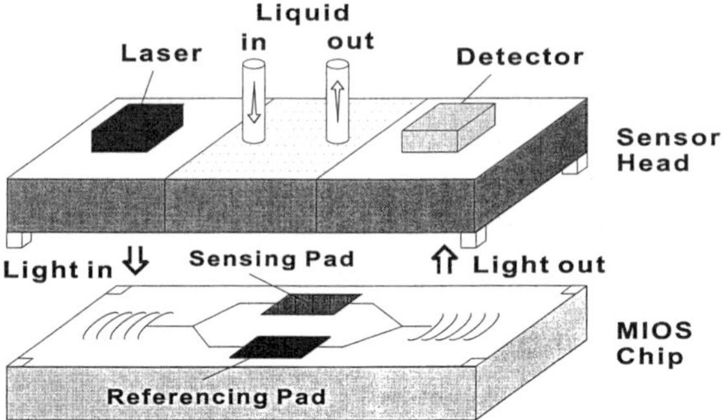

Figure 27 Doubly stacked sensor module consisting of a smart planar optical transducer (MIOS chip) based on a Mach–Zehnder interferometer and an optochemical sensor head.

A micropump [74] may be incorporated in the sensor head as well. This arrangement eases the feeding of transducer chips, for example, in batch processing applications.

One important aspect is how the tasks are distributed within the sensor system. As an example, the specific question of how much optics should be included in the MIOS chip will be discussed. Answering this question can be quite subtle, as in the case of designing the focusing grating coupler shown in Fig. 9 to be used in the module of Fig. 27. Two versions were considered: (1) using focusing grating couplers (FGCs) to focus the light emitted directly by a laser diode chip into the waveguide and (2) using FGCs for coupling collimated laser beams into the waveguide. Our approach was to develop FGCs for coupling collimated beams into and out of the waveguides [16], since, in this arrangement, errors in the lateral chip position have only negligible effect on the output signal in contrast to a ''point-to-point'' imaging. In order to get compact modules nevertheless, novel replicated planar micro-optical elements are being developed [75–77] that can be mass-produced and mounted close to the laser chips and detectors to form miniature illumination and detection submodules. Yet another possibility is to use optical fibers for providing the input/output connection [78].

Taking the modules depicted in Figs. 18 and 19 as representative examples, the main advantages with respect to conventional sensor schemes and principles can be summarized as follows:

1. Using surface-emitting configurations leads to (a) very compact arrangements with a low number of external components, (b) markedly relaxed optomechanical tolerances, and (c) low-cost chip fabrication since no endface preparation is required.

2. Optional readout via transparent substrates provides great flexibility of the arrangement.
3. On-chip measuring variables (a) enable the construction of very compact modules, (b) reduce the need, space, and adjustment effort for external components, and (c) increase the stability of the sensor system.
4. True differential on-chip referencing by means of multiple sensing pads that are separately accessible but closely spaced are essential for exploiting the very high sensitivities achieved by sensor chips based on high-index waveguides for practical applications.

Advantages 3 and 4 are of utmost importance for interferometric sensors in order to turn the high resolution into a high effective accuracy. If external components are used for the beam splitting and recombination, the effective accuracy and stability of the sensor output signal depends on the quality and stability of the optical connection to the chip. This also applies to the ''difference interferometer'' presented in Refs. 27, 31, 32, and 37, which neither uses an on-chip measuring variable nor provides a true differential referencing capability, since the interference is established by external components and two different modes are used that are differently affected by unspecific effects such as temperature, analyte, and sensing-layer properties. Based on calculations taking into account the importance of these aspects for biochemical sensors, surface coverage resolutions in the $<50\text{-fg/mm}^2$ range have been estimated for high-index sensing pads used in integrated optical interferometers with true differential on-chip referencing capability [48].

5.3 Remote Reader Head–based Modules

In the sensor modules just presented, an optical head attached to the MIOS chip is used to read out the information on the chemical measurand. For applications where smart optical transducers with large sensing pad arrays are desired, several solutions are adequate, for example, modules based on (1) large optoelectronic light source and detector arrays (cf. Fig. 21), (2) light source fan-out geometries [75], (3) scanning arrangements, or combinations thereof.

In this section, a type (3) sensor system is outlined where a remote optical reader head is scanning a planar optical transducer in a similar arrangement as is used for reading out the music information stored on the well-known compact disc (CD). Figure 28 shows an example where an array of sensing and reference pads is placed on a kind of ''chemical disc'' and a remote reader head is used to determine the measurand's value at the different pad locations. Examples of applications for which such chemical discs or cards are especially attractive are in situ gas sensing or the examination of MIOS array chips where the (bio-)chemical reactions have taken place in a previous step, possibly at a different location. This is comparable to the recording of a writable CD with any information at

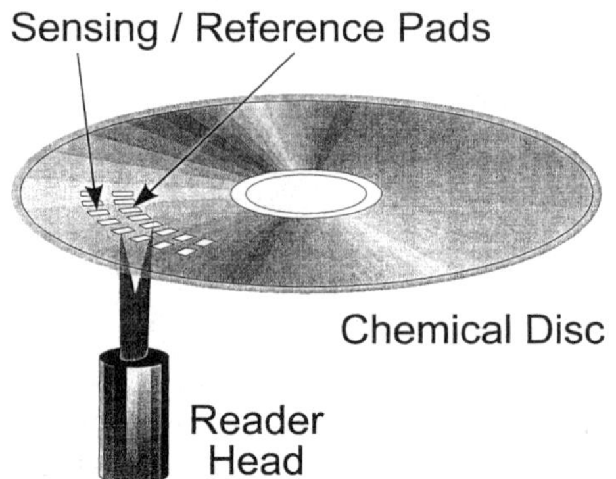

Figure 28 Schematic representation of a CD player type of arrangement where a dual-beam optical reader head is scanning a sensing pad array placed on a chemical disc.

one location and reading out the information at another location. To speed up the reading operation, reader heads with multiple beams are advantageous. Miniaturized versions can be accomplished by realizing micro-optical multiple-beam pickups based, for example, on using laser diode arrays and/or planar fan-out elements [75,78] or IO disc pickups [8,9]. In the example shown, a dual-beam head is used to scan simultaneously a pair of sensing and reference pads in order to provide true differential on-chip referencing [49,50,69].

The arrangement of Fig. 28 is also suitable for developing new chemicals or processes or for screening purposes, with the chemical disc playing the role of a ''removable IO chemical bench'' (cf. Sec. 4.4). Many different sensing schemes may actually be applied as well in such a scanning arrangement. In addition to refractometric sensing, all module types presented in this and the previous sections are suitable for incorporating means to perform other types of measurements based on quantities such as absorption, fluorescence, and scattering in hybrid or monolithic sensor arrangements.

CONCLUSIONS

An overview was given on the principal ideas, new concepts, and practical issues for accomplishing complete miniature sensor modules for chemical and biochemical applications, which are at the same time small, highly sensitive, and stable. It was shown that the size, cost, and complexity of the complete sensor system used for measuring, monitoring, or controlling (bio-)chemical processes can be reduced substantially by increasing the scale of integration on the chip level and by considering the chips as being an integral part of the whole sensor system.

Different types of *smart planar optical transducer* chips have been presented that not only handle the task of conventional integrated optical transducers, but also perform additional functions for which peripheral equipment is otherwise needed. The emphasis was on pointing out and discussing the most important aspects for accomplishing refractometric modules and on comparing different approaches and sensor types with respect to fulfilling the requirements for practical applications.

The following key issues were discussed:

1. The importance of a proper theoretical modeling of the sensor chips and the (bio-)chemical sensing layers
2. The availability of micro- and nanofabrication technologies suitable for realizing sensors based on a variety of substrate materials, such as glasses, polymers, and semiconductors
3. The possibility of mass-producing low-cost *smart planar optical transducer* chips
4. Producing *standardized* waveguides for the implementation of *miniature IO chemical benches*
5. Employing planar micro-optical elements for accomplishing illumination and detection submodules
6. The advantages of surface-emitting chip configurations with respect to miniaturization and adaptation to different applications
7. Schemes for single-chip multicomponent analysis based on sensor pad arrays
8. The significance of using on-chip measuring variables and true differential on-chip referencing for achieving high sensitivity and stability at the same time

Finally, examples of miniature integrated optical sensor modules have been presented and discussed that are based on monolithic integration schemes for achieving ultracompact systems, on stacked multichip configurations consisting of smart transducer chips in combination with miniature reader heads, and transducers in the form of discs or cards being scanned by remote reader heads. These examples illustrate the potential inherent in combining the aforementioned ingredients for realizing a few standardized miniature modules that can be adapted easily to a great variety of applications, reducing the cost by increasing the production volume of modules and chips.

ACKNOWLEDGMENTS

It is a pleasure for me to thank my colleagues at CSEM/PSIZ as well as external partners for the fruitful collaboration, discussions, support, and direct help in performing the work presented in this chapter. I gratefully acknowledge L. G. Baraldi, U. Bruhin, M. Derendinger, J. Dübendorfer, M. T. Gale, Th. Hessler,

L. U. Kempen, K. Knop, M. Kuhn, E. Meier, P. Metzler, R. H. Morf, J. Pedersen, M. Rossi, H. Schütz, J. Söchtig, R. Stutz, S. Westenhöfer, M. Wiki, A. Vonlanthen, H. P. Zappe, and P. Zeller, all of CSEM/PSIZ, Zurich; J. Edlinger and B. Maisenhölder of Balzers, Lichtenstein, for depositing the hard dielectric waveguides; P. Sixt of Photronics SA, Neuchâtel, for support with respect to e-beam writing; T. Callenbach and K. Eggmann of Weidmann AG, Rapperswil, for the cowork on injection molding; G. Duveneck and M. Ehrat of Novartis, Basel, for the collaboration on immunosensors; U. E. Spichiger, D. Freiner, D. Citterio, and M. Linnhoff of the Center for Chemical Sensors at ETH Zurich for the cowork on chemical sensors; and G. Jobst, I. Moser, and G. Urban of Albert Ludwigs University, Freiburg, Germany, for the collaboration on pH sensors. Acknowledgment is also made to the Swiss Kommission für Technologie und Innovation (KTI), Schweizerische Stiftung für mikrotechnische Forschung (FSRM), and Optique (SPP) for partial support of this work.

REFERENCES

1. H. Teichmann. Integrated optical sensors: new developments. In: H. Meixner and R. Jones, eds. Sensors, Vol. 8: Micro- and Nanosensor Technology/Trends in Sensor Markets. Weinheim: VCH Verlagsgesellschaft, 1995, Chap. 7.
2. R. Th. Kersten. Integrated Optics for Sensors. In: J. Dakin and B. Culshaw, eds. Optical Fiber Sensors, Vol. 1: Principles and Components. Boston: Artech House, 1988, Chap. 9.
3. H. Nishihara, M. Haruna, and T. Suhara. Optical Integrated Circuits. New York: McGraw-Hill, 1989.
4. J. Dakin and B. Culshaw, eds. Optical Fiber Sensors, Vols. 1–4. Boston: Artech House, 1988/89/96/97.
5. G. Voirin, F. Gradisnik, O. Parriaux, M. T. Gale, R. E. Kunz, B. J. Curtis, and H. W. Lehmann. Waveguide grating for polarization preprocessing circuits. Proc. SPIE 1126:50–56, 1989.
6. G. Boisdé and A. Harmer. Chemical and Biochemical Sensing with Optical Fibers and Waveguides. Boston: Artech House, 1996.
7. O. Parriaux. Guided wave electromagnetism and opto-chemical sensors. In: O. S. Wolfbeis, ed. Fiber Optic Chemical Sensors and Biosensors. Boca Raton, FL: CRC Press, 1991, Vol. 1, Chap. 4.
8. S. Nishiwaki, J. Asada, and S. Uchida. Optical head employing a concentric-circular focusing grating coupler. Appl. Opt. 33:1819–1827, 1994.
9. I. Kawakubo, J. Funazaki, K. Shirane, and A. Yoshizawa. Integrated optical-disk pickup that uses a focusing grating coupler with a high numerical aperture. Appl. Opt. 33:6855–6859, 1994.
10. R. E. Kunz and J. Dübendorfer. Novel miniature integrated optical goniometers. Sensors Actuators A 60:23–28, 1997.
11. R. E. Kunz and J. Dübendorfer. Miniature integrated optical wavelength analyzer chip. Opt. Lett. 20:2300–2303, 1995.

12. M. Wiki, J. Dübendorfer, and R. E. Kunz. Spectral beam sampling and control by a planar optical transducer. Sensors Actuators A 67/1–3:120–124, 1998.
13. R. E. Kunz. Totally integrated optical measuring sensors. Proc. SPIE 1587:98–113, 1992.
14. R. E. Kunz. Miniature integrated optical modules for chemical and biochemical sensing. Sensors Actuators B 38–39:13–28, 1997.
15. R. E. Kunz. Integrated optical sensors based on hard dielectric films on replicated plastic substrates. Proc. 8th Eur. Conf. Integrated Optics ECIO '97, Stockholm, Sweden, April 2–4, 1997, pp. 86–93.
16. L. U. Kempen and R. E. Kunz. Replicated Mach–Zehnder interferometers with focusing grating couplers for sensing applications. Sensors Actuators B 38–39:295–299, 1997.
17. L. U. Kempen. Integrated optical sensor modules. Ph.D. dissertation, University of Neuchâtel and Paul Scherrer Institute, Switzerland, July, 1996.
18. J. Dübendorfer and R. E. Kunz. Compact integrated optical immunosensor using replicated chirped grating coupler sensor chips. Appl. Opt. 37:1890–1894, 1998.
19. J. Dübendorfer. Replicated integrated optical sensors. Ph.D. dissertation, University of Fribourg and Paul Scherrer Institute, Switzerland, November, 1997.
20. B. Maisenhölder, H. P. Zappe, R. E. Kunz, P. Riel, M. Moser, and J. Edlinger. A GaAs/AlGaAs-based refractometer platform for integrated optical sensing applications. Sensors Actuators B 38–39:324–329, 1997.
21. M. T. Gale, R. E. Kunz, and H. P. Zappe. Polymer and III–V transducer platforms for integrated optical sensors. Opt. Eng. 34:2396–2406, 1995.
22. B. Maisenhölder. A monolithically integrated optical Mach–Zehnder interferometer for refractometric sensing applications. Ph.D. dissertation, University of Konstanz, Germany, and Paul Scherrer Institute, Switzerland, October, 1997.
23. R. E. Kunz and K. Cammann, eds. Proc. EUROPTRODE III, 3rd Europ. Conf. on Optical Chemical Sensors and Biosensors, Zurich, Switzerland, March 31–April 3, 1996; printed in Sensors Actuators B 38–39, 1997.
24. K. Tiefenthaler and W. Lukosz. Sensitivity of grating couplers as integrated-optical chemical sensors. J. Opt. Soc. Am. B 6:209–220, 1989.
25. W. Lukosz, D. Clerc, and Ph. M. Nellen. Input and output grating couplers as integrated optical biosensors. Sensors Actuators A 25–27:181–184, 1991.
26. E. M. Bowman, L. W. Burgess, S. W. Wenzel, and R. M. White. Optical and piezoelectric analysis of polymer films for chemical sensor characterization. Proc. SPIE 1587:147–157, 1992.
27. W. Lukosz. Integrated optical chemical and direct biochemical sensors. Sensors Actuators B 29:37–50, 1995.
28. Ch. Fattinger, C. Mangold, M. T. Gale, and H. Schütz. Bidiffractive grating coupler: universal transducer for optical interface analytics. Opt. Eng. 34:2744–2753, 1995.
29. A. Brandenburg, R. Polzius, F. Bier, U. Bilitewski, and E. Wagner. Direct observation of affinity reactions by reflected-mode operation of integrated optical grating coupler. Sensors Actuators B 30:55–59, 1996.
30. Y. Liu, P. Hering, and M. O. Scully. An integrated optical sensor for measuring glucose concentration. Appl. Phys. B 54:18–23, 1992.

31. Ch. Stamm and W. Lukosz. Integrated optical difference interferometer as refractometer and chemical sensor. Sensors Actuators B 11:177–181, 1993.
32. Ch. Fattinger, H. Koller, D. Schlatter, and P. Wehrli. The difference interferometer: a highly sensitive optical probe for quantification of molecular surface concentration. Biosensors Bioelectronics 8:99–107, 1993.
33. R. G. Heideman. Optical waveguide based evanescent field immunosensors. Ph.D. dissertation, University Twente, The Hague, 1993.
34. R. G. Heideman, R. P. H. Kooyman, and J. Greve. Performance of a highly sensitive optical waveguide Mach–Zehnder interferometer immunosensor. Sensors Actuators B 10:209–217, 1993.
35. I. Schanen Duport, P. Benech, and R. Rimet. New integrated-optics interferometer in planar technology. Appl. Opt. 33:5954–5958, 1994.
36. L. M. Lechuga, A. T. M. Lenferink, R. P. H. Kooyman, and J. Greve. Feasibility of evanescent wave interferometer immunosensors for pesticide detection: chemical aspects. Sensors Actuators B 24–25:762–765, 1995.
37. W. Lukosz, Ch. Stamm, H. R. Moser, R. Ryf, and J. Dübendorfer. Difference interferometer with new phase-measurement method as integrated-optical refractometer, humidity sensor and biosensor. Sensors Actuators B 38–39:316–323, 1997.
38. E. F. Schipper. Waveguide immunosensing of small molecules. Ph.D. dissertation, University of Twente, the Netherlands, November, 1997.
39. R. P. Podgorsek and H. Franke. Selective optical detection of *n*-heptane/iso-octane vapors by polyimide lightguides. Opt. Lett. 20:501–503, 1995.
40. D. W. Hewak and J. W. Y. Lit. Generalized dispersion properties of a four-layer thin-film waveguide. Appl. Opt. 26:833–841, 1987.
41. E. Anemogiannis, E. N. Glytsis, and T. K. Gaylord. Opimization of multilayer integrated optics waveguides. J. Lightwave Technol. 12:512–517, 1994.
42. R. E. Kunz, J. Dübendorfer, and R. H. Morf. Finite grating depth effects for integrated optical sensors with high sensitivity. Biosensors Bioelectronics 11:653–667, 1996.
43. D. Freiner, R. E. Kunz, D. Citterio, U. E. Spichiger, and M. T. Gale. Integrated optical sensors based on refractometry of ion-selective membranes. Sensors Actuators B 29:277–285, 1995.
44. J. Dübendorfer, R. E. Kunz, G. Jobst, I. Moser, and G. Urban. Integrated optical pH sensor using replicated chirped grating coupler sensor chips. Sensors Actuators B 50:210–219, 1998.
45. R. E. Kunz, C. L. Du, J. Edlinger, H. K. Pulker, and M. Seifert. Integrated optical sensors based on reactive low-voltage ion-plated films. Sensors Actuators A 25:155–159, 1991.
46. R. E. Kunz and L. U. Kempen. Miniature integrated optical sensors. Proc. SPIE 2068:69–86, 1994.
47. R. E. Kunz, J. Edlinger, B. J. Curtis, M. T. Gale, L. U. Kempen, H. Rudigier and H. Schütz. Grating couplers in tapered waveguides for integrated optical sensing. Proc. SPIE 2068:313–325, 1994.
48. R. E. Kunz, G. Duveneck, and M. Ehrat. Sensing pads for hybrid and monolithic integrated optical immunosensors. Proc. SPIE 2331:2–17, 1994.
49. J. Dübendorfer, R. E. Kunz, E. Mader, G. L. Duveneck, and M. Ehrat. Reference

and sensing pads for integrated optical immunosensors. Proc. SPIE 2928:90–97, 1996.

50. J. Dübendorfer, R. E. Kunz, E. Schürmann, G. L. Duveneck, and M. Ehrat. Sensing and reference pads for integrated optical immunosensors. J. Biomed. Opt. 2:391–400, 1997.
51. R. E. Kunz and J. S. Gu. Design of integrated optical couplers and interferometers suitable for low-cost mass production. Proc. ECIO '93, 1993, pp. 14–36/37.
52. B. Maisenhölder, H. P. Zappe, R. E. Kunz, P. Riel, M. Moser, and G. L. Duveneck. A GaAs/AlGaAs-based Mach–Zehnder interferometer as integrated optical immunosensor. Proc. SPIE 2928:144–152, 1996.
53. B. Maisenhölder, H. P. Zappe, M. Moser, P. Riel, R. E. Kunz, and J. Edlinger. Monolithically integrated optical interferometer for refractometry. Electron. Lett. 33: 986–988, 1997.
54. H. P. Zappe, D. Hofstetter, B. Maisenhölder, M. Moser, P. Riel, and R. E. Kunz. Physical and chemical sensing using monolithic semiconductor optical transducers. Conference on Micromachining and Microfabrication '97, Austin, TX, September 29–30. SPIE 3226:180–187, 1997.
55. R. E. Kunz. Gradient effective index waveguide sensors. Sensors Actuators B 11: 167–176, 1993.
56. R. E. Kunz, J. Edlinger, P. Sixt, and M.T. Gale. Replicated chirped waveguide gratings for optical sensing applications. Sensors Actuators A 47:482–486, 1995.
57. J. C. Brazas and L. Li. Analysis of input-grating couplers having finite lengths. Appl. Opt. 34:3786–3792, 1995.
58. L. U. Kempen and R. E. Kunz. Miniature integrated optical refractometer chip. Proc. SPIE 2208:124–129, 1994.
59. R. Zengerle. Light propagation in singly and doubly periodic planar waveguides. J. Modern Opt. 34:1589–1617, 1987.
60. R. E. Kunz. Process and device for determining measured quantities by means of an integrated optical sensor module. International patent No. PCT/CH92/00078; WO 92/19976.
61. B. J. Curtis, M. T. Gale, R. E. Kunz, and H. Schütz. The fabrication and characterization of shallow diffraction gratings for applications in integrated optical chemical sensors. Proc. 11th International Symposium on Plasma Chemistry, Loughborough, UK, Aug. 23–27, 3, 1993, pp. 885–890.
62. H. P. Zappe, H. E. G. Arnot, and R. E. Kunz. Technology and devices for hybrid and monolithic integrated optical sensors. Sensors Actuators A 41:141–144, 1994.
63. H. P. Zappe, H. E. G. Arnot, and R. E. Kunz. Technology for III–V-based integrated optical sensors. Sensors Materials 6, 1994.
64. L. U. Kempen, R. E. Kunz, and M. T. Gale. Micromolded structures for integrated optical sensors. Proc. SPIE 2639:278–285, 1995.
65. L. Baraldi, R. E. Kunz, and J. Meissner. High-precision molding of integrated optical structures. Proc. SPIE 1992:21–29, 1993.
66. L. G. Baraldi. Heissprägen in Polymeren für die Herstellung integriert-optischer Systemkomponenten. Ph.D. dissertation ETH Zurich, 1994.
67. M. T. Gale, L. G. Baraldi, and R. E. Kunz. Replicated microstructures for integrated optics. Proc. SPIE 2213:2–10, 1994.

68. R. H. Morf. Exponentially convergent and numerically efficient solution of Maxwell's equations for lamellar gratings. J. Opt. Soc. Am. A 12:1043–1056, 1995.
69. J. Dübendorfer and R. E. Kunz. Reference pads for miniature integrated optical sensors. Sensors Actuators B 38–39:116–121, 1997.
70. M. T. Gale, Th. Hessler, R. E. Kunz, and H. Teichmann. Fabrication of continuous-relief micro-optics: progress in laser writing and replication technology. In: Diffractive Optics and Micro-Optics, Vol. 5, OSA Tech. Digest Series. Washington DC: OSA, 1996, pp. 335–338.
71. W. Göpel. New materials and transducers for chemical sensors. Sensors Actuators B 18/19, 1–21, 1994.
72. H. Sigrist, A. Collioud, J.-F. Clémence, H. Gao, R. Luginbühl, M. Sänger, and G. Sundarababu. Surface immobilization of biomolecules by light. Opt. Eng. 34:2339–2348, 1995.
73. H. Gao, R. Luginbühl, and H. Sigrist. Bioengineering of silicon nitride. Sensors Actuators B 38–39:38–41, 1997.
74. B. H. van der Schoot, S. Jeanneret, A. van den Berg, and N.F. de Rooij. Microsystems for flow injection analysis. Anal. Methods Instrum. 1:38–42, 1993.
75. M. Rossi and R. E. Kunz. Focusing fan-out elements based on phase-matched Fresnel lenses. Opt. Commun. 112:258–264, 1994.
76. T. Hessler, M. Rossi, and R. E. Kunz. Diffractive optical elements for laser diodes. Tech. Digest Workshop on Diffractive Optics, Prague, August 21–23, 1995, pp. 52–53.
77. T. Hessler and R. E. Kunz. Relaxed fabrication tolerances for low-Fresnel-number lenses. J. Opt. Soc. Am. A 14:1599–1606, 1997.
78. M. Rossi, G. L. Bona, and R. E. Kunz. Arrays of anamorphic phase-matched Fresnel elements for diode-to-fiber coupling. Appl. Opt. 34:2483–2488, 1995.

11

Nonlinear Integrated Optical Devices

George I. Stegeman

University of Central Florida, Orlando, Florida

Gaetano Assanto

Terza University of Rome, Rome, Italy

I INTRODUCTION

The use of integrated optics waveguides to maintain the high intensities needed for efficient nonlinear optical interactions is not a new concept. It followed the birth of nonlinear optics in the 1960s [1]. The first efforts dealt with the second harmonic generation (SHG) in planar waveguides, which was the key phenomenon of interest at that time [2]. Progress in second-order nonlinear optics was essentially stymied and limited to birefringent phase-matching in $LiNbO_3$ Ti-indiffused channel waveguides for over 15 years [3–6]. Although the devices were essentially perfected, the small nonlinearities and difficult technology involved in making long waveguides led to serious questions about their applications. The breakthrough needed came from two laboratories in the late 1980s, at Stanford University and at the Institute of Optical Research in Sweden [7,8]. Based on earlier, pioneering work on crystals at Nanjing University they found that they could invert periodically with distance the ferroelectric domains in $LiNbO_3$ waveguides [9]. In this way, the nonlinearity could be inverted every half coherence length, leading to a form of phase-matching called ''quasi-phase-matching'' (QPM) [10,11]. This concept has led to a revolution in second-order nonlinear guided wave optics, and the current status of the field owes its sophistication to this QPM approach [12]. Although other materials and phase-matching techniques are still being investigated, QPM in ferroelectrics is currently acknowledged to be the best approach to efficient guided-wave frequency conversion.

The generation of new frequencies in channel waveguides is now an efficient process with many applications [12]. The original thrust was into blue sources from GaAs-based lasers for applications to data storage and xerography [13]. In fact, figures of merit over 1500%/W-cm^2 have now been reported, which translates into 15% conversion of a 10-mW input [14,15]. This same fabrication technology has been used for difference frequency generation to produce radiation for remote sensing and medical applications, to shift signals between the two communications bands at 1.3 and 1.55 μm, and wavelength shifting within the 1.55-μm band for wavelength-division multiplexed (WDM) systems [16–21]. Furthermore, as the losses of QPM channel waveguides, particularly in $LiNbO_3$, have been reduced, the threshold for waveguide optical parametric oscillators has been pushed into the hundreds of milliwatts range to give tunable sources [22]. A whole new area is the use of "cascading" as a mechanism for simulating very efficient third-order nonlinear effects with second-order nonlinearities [23]. The key feature of QPM systems is that the phase-matching condition can be moved into any convenient region of the spectrum, increasing the flexibility of parametric interactions to previously undreamed of levels and allowing a variety of applications to be tested in one waveguide system.

Research continues into potentially better waveguide systems than QPM ferroelectrics [24–26]. Very promising are semiconductors based on GaAs with their hundreds of pm/V nonlinearities [24]. Another option is to use polymers, which also offer a lot of versatility in terms of phase-matching, etc., and potentially offer much larger figures of merit [25]. However, because of absorption–nonlinearity trade-offs, polymeric devices will probably be limited to the 700–1800-nm region of the spectrum. Yet another promising approach, driven by the high nonlinearities available (thousands of pm/V), is single-crystal organics [27]. However, a great deal of material and waveguide engineering needs to be done to make these materials useful for this purpose.

Third-order phenomena, specifically the intensity-dependent refractive, have also led to a number of applications in integrated optics. In the 1970s the development of low-loss fibers first led to an intense interest in third-order fiber nonlinear optics [28]. It was not until the mid-1980s that the prospects of all-optical signal processing stimulated interest in third-order waveguide phenomena based on an intensity-dependent refractive index [29,30]. The problem to be overcome, as it is in most promising areas of technology, is one of materials [31]. In passive materials, i.e., ones in which there is no charge injection or gain, the key trade-offs are between the propagation losses (which limit the effective device length), the nonlinearity (which combined with losses determines the required device power), and the nonlinearity response time (usually the recovery time, i.e., the time it takes for the index change to decay away after the illumination is turned off) [31]. There has been a great deal of progress in this area in the last decade with semiconductors (with their rich variety of nonlinear mechanisms). Organic materials still appear promising. However, it took a recent breakthrough to give the field fresh perspective and stimulus and the chance to try many different device configurations. This was provided by the discovery of large nonlinearities in semiconductor amplifiers. They were introduced as signal-processing elements by researchers at British Telecom Labs and at NTT in 1993 [32]. Subsequently they were the impetus for the development of state-of-the-art devices [33]. Reported to date are applications to wavelength shifting, time domain demultiplexing, etc.

Nevertheless, materials scientists continue to investigate new materials for higher nonlinearities and quantum-confinement effects in existing materials, hopefully leading to lower device-operating powers. This work has led to the investigation of multiple quantum wells, quantum wires, quantum dots, etc. in semiconductors [29,34,35]. Research continues into organic materials, which already claim the highest nonresonant nonlinearities of any known materials system but are difficult to process into device structures [36].

It is interesting to note that these two apparently different aspects of nonlinear optics, frequency conversion, usually associated with $\chi^{(2)}$ nonlinearities, and all-optical effects, via $\chi^{(3)}$, have recently been merged, and roles have even interchanged with the advent of the concept of ''cascading'' [23]. This offers a completely different approach to all-optical processing devices.

There have been a number of review papers dealing with both second- and third-order nonlinear integrated optics [3,4,30,37]. They contain detailed explanations of device geometries, their theoretical analyses, and their operating characteristics. Such details will not be repeated here. This chapter is focused on defining the state of the art in nonlinear integrated optics, and specifically the impact of the breakthroughs on this status.

2 SECOND-ORDER EFFECTS AND DEVICES

This field was initially driven by the need for a compact blue source for a variety of applications [13]. These included data storage, where the surface storage density of a disk increases as λ^{-2}. Also, xerography requires short-wavelength beams, not just for optimum resolution but also because many xerographic materials have optimum sensitivity in the blue region of the spectrum. These and other applications led to the idea that a semiconductor laser in the 800–900-nm range could be doubled with sufficient efficiency in a waveguide to be useful.

There are essentially two interaction geometries that produce second harmonic generation in an integrated optics context [3,37]. They are shown schematically in Fig. 1. In the usual case (a), the fundamental and harmonic propagate in essentially the same direction. For the second, the input fundamental fields propagate colinearly but in opposite directions, and the harmonic signal is generated normal to the waveguide surfaces. The copropagating case (a) is clearly the most popular,

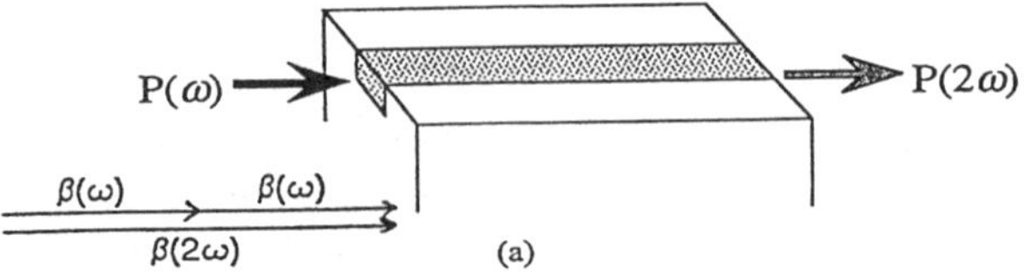

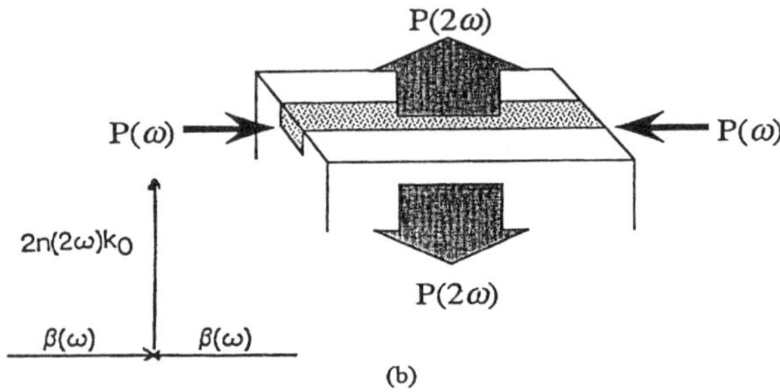

Figure 1 The two SHG geometries. (a) The fundamental and harmonic beams propagate as guided waves in the same direction. (b) The fundamental beams propagate in opposite directions and the harmonic is emitted normal to the waveguide surfaces.

and for a phase-matchable material gives the best conversion efficiency. However, there are materials with large resonant nonlinearities that cannot be phase-matched in the conventional sense but that can potentially be used in the counter-propagating geometry. Furthermore, the counter-propagating geometry is much more robust than the copropagating case, and this feature could make it useful, even though the conversion efficiencies are at least one to two orders of magnitude lower than for the conventional case.

2.1 Second Harmonic Generation

Most of the activity has been in codirectional SHG, and that case will be discussed first. In bulk crystals, Type II phase-matching with two orthogonally polarized input fundamental beams, sometimes called birefringent phase-matching, is the most popular because of dispersion in the refractive index in the visible and infrared regions of the spectrum [38]. In waveguides there are many additional options for phase-matching, and Type I phase-matching, which requires only one input fundamental beam polarization is not only feasible but also used most frequently [3,12]. The theory, at least in the low-depletion limit, is well-known for this case, and the figures of merit (FOMs) can be given in a simple form.

For Type I phase-matching, used for both QPM and modal phase-matching, one can write the incident and harmonic fields as:

$$\mathbf{E}(\mathbf{r},\mathbf{t}) = \frac{1}{2}\hat{\mathbf{e}}f_{\hat{e}}(\mathbf{r}_{\|};\Omega)a(z;\Omega)\exp[i\Omega t - \beta(\Omega)z] + c.c. \tag{1}$$

where $\hat{\mathbf{e}}$ is the electric field unit vector, $\mathbf{r}_{\|}$ identifies coordinates in the plane orthogonal to z, $f(\mathbf{r}_{\|};\Omega)$ is the transverse field distribution across the channel waveguide, Ω is the frequency, with $\Omega = \omega$ for the fundamental and $\Omega = 2\omega$ for the second harmonic, $\beta(\Omega)$ is the propagation wavevector for either the fundamental or the harmonic, and $a(z;\Omega)$ is a complex amplitude normalized so that the guided wave power is given by $P(\Omega) = |a(z;\Omega)|^2$. The second harmonic power $P(2\omega)$ can easily be calculated in the low fundamental beam depletion limit to be

$$P(2\omega) = \frac{2\omega^2}{n^3c^3\varepsilon_0}[d_{\text{eff}}^{(2)}L_{\text{eff}}]^2\ |F(\Delta\phi)|^2P(\omega)^2 \tag{2}$$

where $P(\omega)$ is the input fundamental power and L_{eff} is the effective length over which the second harmonic accumulates. L_{eff} can be limited by the physical length of the sample, propagation losses at the fundamental or harmonic, or imperfections in the channel waveguide manufacture. For the waveguide case, the effective second-order nonlinearity is given in terms of the material's nonlinear tensor $d_{ijk}^{(2)}(\mathbf{r})$ as

$$d_{\text{eff}}^{(2)}g(\mathbf{r}_{\|}) + \Delta d_{\text{eff}}^{(2)}\,\Delta g(\mathbf{r}_{\|})\sin(\kappa z) = [d_{ijk}^{(2)}(\mathbf{r}_{\|}) + \Delta d_{ijk}^{(2)}(\mathbf{r}_{\|})\sin(\kappa z)] \times f_i(\mathbf{r}_{\|};2\omega)f_j(\mathbf{r}_{\|};\omega)f_k(\mathbf{r}_{\|};\omega) \tag{3}$$

where the nonlinearity can be sinusoidally modulated along z with period $\Lambda = 2\pi/\kappa$ [10–12]. The product of the overlap integral with the cumulative phase-mismatch function is given by

$$F(\Delta\phi) = \int_0^{L_{\text{eff}}} \frac{Kd_{\text{eff}}^{(2)} + \Delta K\ \Delta d_{\text{eff}}^{(2)}\sin(\kappa z)}{d_{\text{eff}}^{(2)}L_{\text{eff}}}\exp[i[2\beta(\omega) - \beta(2\omega)]z]\,dz \tag{4}$$

Although the argument $\Delta\phi$ does not explicitly appear inside the equation, we use this definition to draw attention to its role in determining the cumulative phase-mismatch in SHG. Here the overlap integral has two possible contributions, K and ΔK, which describe the spatial overlap of the transverse field distributions of the fundamental and harmonic with the nonlinearity,

$$K + \Delta K = \int_{-\infty}^{\infty}\int_{-\infty}^{\infty}[g(\mathbf{r})_{\|} + \Delta g(\mathbf{r})_{\|}]\,d\mathbf{r}_{\|} \tag{5}$$

The details of all of the terms depend on the phase-matching technique and material system used. These will be discussed next.

2.2 Phase-Matching Techniques

Phase-matching is tied directly to the techniques used for optimizing the factor $F(\Delta\phi)$. For simplicity, consider the case where the nonlinearity is uniform along the propagation direction; i.e., $\Delta K = 0$ and $K \neq 0$. In this limit,

$$|F(\Delta\phi)|^2 = |K|^2 \mathrm{sinc}^2\, \Delta\phi \tag{6}$$

where $\Delta\phi = \frac{1}{2}[2\beta(\omega) - \beta(2\omega)]L_{\mathrm{eff}}$. This can be expressed in terms of the effective indices (N_{eff}) of the waveguide as [3,31]

$$\Delta\phi = k_{\mathrm{vac}}(\omega)[N_{\mathrm{eff}}(\omega) - N_{\mathrm{eff}}(2\omega)]L_{\mathrm{eff}}$$

Although this is somewhat more complicated in the QPM case and for birefringent phase-matching, the goal is essentially to ensure wavevector conservation between the modes and any other periodic structure present that modulates the optical properties. Note, also, that a large value of $|K|$ is necessary for efficient conversion.

With the exception of a few organic materials, the "diagonal" nonlinear coefficients $d_{il}^{(2)}$ (in the Voigt contracted notation, d_{iii} in full notation) have larger nonlinearities than the off-diagonal ones [39]. For example, $d_{15}/d_{33} = 1/7$ in $LiNbO_3$. Therefore there is a clear advantage to using such diagonal elements, since $P(2\omega) \propto |d_{\mathrm{eff}}|^2$, a factor of 49 in the $LiNbO_3$ case. However, there are problems with using diagonal elements, due to "normal" dispersion in the refractive index. For example, for a slab (1D confinement versus 2D confinement in channels) waveguide, the dispersion in the effective index does not allow phase-matching between the lowest-order modes of the same polarization in a waveguide; see Fig. 2, which shows the dispersion in the effective index with the normalized quantity $k_{\mathrm{vac}}h$, where h is the waveguide thickness. Clearly there is no intersection (which signifies wavevector conservation) between the $TE_0(\omega)$ and $TE_0(2\omega)$ dispersion curves. However, wavevector matching is possible between the $TE_0(\omega)$ and $TE_1(2\omega)$ modes. But, as will be discussed later, the SHG conversion efficiency can be greatly reduced due to the overlap integral K.

2.2.1 Quasi-Phase-Matching

This approach relies on (1) the small amount of free energy it takes to have two oppositely directed polarization domains adjacent to each other in a ferroelectric crystal like $LiNbO_3$ or $LiTaO_3$, and (2) the ease with which the domains can be inverted with an applied electric field. In the earliest experiments it was found that Ti in-diffusion through the $+c$-cut phase caused the domain to invert in the

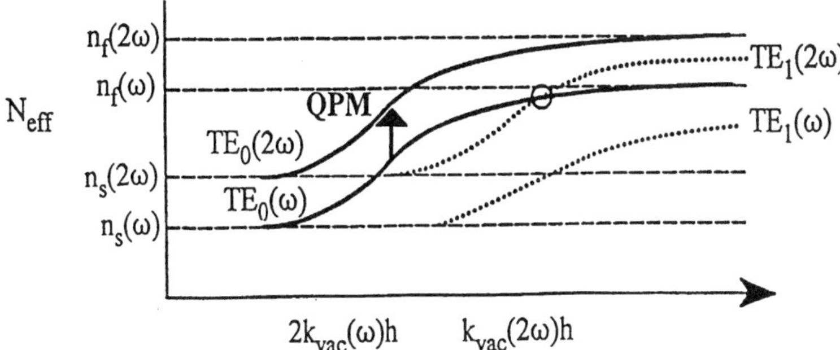

Figure 2 Typical dispersion curves for the effective index at the fundamental (ω) and harmonic (2ω) frequencies for the TE_0 and TE_1 modes of a slab waveguide. The intersection of the fundamental and harmonic curves corresponds to wavevector matching. The vertical arrow corresponds to the contribution of the grating to wavevector matching via QPM.

region into which Ti was in-diffused [7,8]. This led essentially to a triangular domain inversion (and therefore to reversal of $d_{33}^{(2)}$), as indicated in Fig. 3(a). Because the triangular regions were not a good match to the spatial depth dependence of the guided wave fields, as can be shown by evaluating ΔK, the conversion efficiency was far from optimum. The breakthrough came when it was found that when strong electric fields are applied along the *c*-axis, domains could be reversed all of the way through the sample [see Fig. 3(b)] [40]. This optimized the spatial overlap between the guided mode fields and the modulated nonlinearity. As a result, the harmonic conversion efficiency is now within a factor of 2 of theory. Furthermore, as the techniques improved over time, such periodically reversed structures could be maintained over sample lengths in excess of 2 cm, i.e., $L_{\text{eff}} > 2$ cm [22].

With electric field poling, the spatial modulation of the nonlinearity is square-wave-like and can be expanded in a Fourier series. In this case, the $F(\Delta\phi)$ contains a summation over the Fourier components (labeled p), and each term can be written as $F_p(\Delta\phi)$, with $\Delta d_{\text{eff}}^{(2)}\sin(\kappa z)$ replaced by $\Delta d_{p\text{eff}}^{(2)}\sin(p\kappa z)$, where $\Delta d_{p\text{eff}}^{(2)} = \Delta d_{\text{eff}}^{(2)} 2/\pi p$. Note that for domain reversal, $d_{\text{eff}}^{(2)} = 0$. The wavevector conservation condition in this case is $\Delta\beta = 2\beta(\omega) - \beta(2\omega) \pm p\kappa = 0$, or $N_{\text{eff}}(\omega) - N_{\text{eff}}(2\omega) \pm p\kappa/2k_{\text{vac}}(\omega) = 0$, where p is the grating order used. In terms of the phase-matching diagram (Fig. 2), the vertical arrow indicates the change in N_{eff} introduced by the grating. Clearly, wavevector matching can now be implemented at *any wavelength*, since the length of the arrow is proportional to Λ^{-1}, which is a free variable as long as the periodicity and domain inversion are technologically feasible.

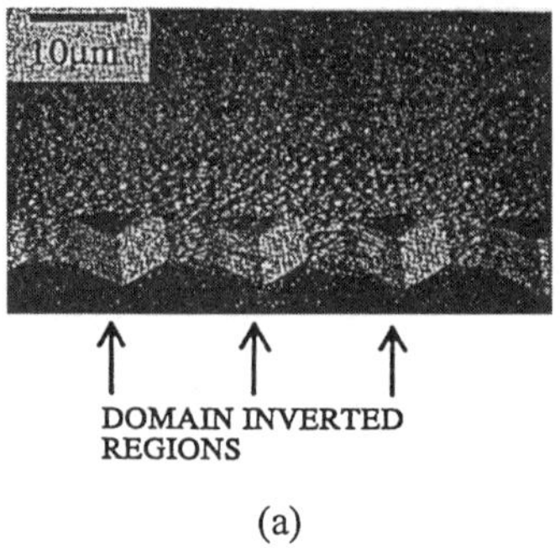

(a)

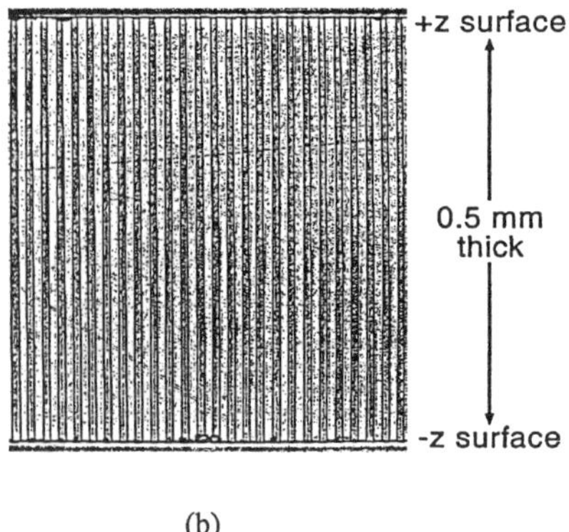

(b)

Figure 3 Domain inversion in QPM $LiNbO_3$ waveguides revealed by etching with HF acid. (a) The triangular regions correspond to inverted domains due to Ti in-diffusion at the surface of the sample. (b) The vertical stripes correspond to domains inverted by electric field poling. (From Ref. 101.)

The fabrication technology for QPM $LiNbO_3$ and $LiTaO_3$ devices is now excellent. It is useful to describe the low fundamental depletion case in terms of wavelength-dependent figures of merit by which different materials and waveguide systems can be compared. From Eq. (2),

$$\eta = \frac{P(2\omega)}{[P(\omega)L_{\text{eff}}]^2} \qquad \eta' = \frac{P(2\omega)}{P(\omega)^2} \tag{7}$$

The first figure of merit normalizes the conversion to a device of unit length, whereas the second one (η') gives the actual efficiency obtained in a given device. Both are important, because the first (η) measures the potential of a structure, and the second reflects the best technological application of that material, which includes the distance over which phase-matching has been achieved as well as low loss (which limit L_{eff}). Some state-of-the-art results are listed in Table 1 in terms of the two pertinent figures of merit.

A simple analysis of the wavelength dependence of the figures of merit shows that they vary as λ^{-4}. This explains the large difference between the results at 800 nm versus those obtained at 1550 nm. It is also useful to comment on the meaning of numbers that exceed 100%/W-cm^2, which obviously have no meaning for devices 1 cm long and watt input powers. It is important to recall that these values refer to the weak depletion limit and that the correct way to read this number is that 10 (20) mW of input power can produce 1 (4) mW of doubled power, etc. These are indeed very impressive and practically useful results. When the depletion of the fundamental is significant, the interaction needs to be analyzed in a coupled-mode context, and the perfectly phase-matched conversion efficiency (the best case) varies as $\tanh^2 \xi$, where $\xi \propto d_{eff}^{(2)} L P^{1/2}(\omega)$, asymptotically leveling the conversion efficiency at 100% [48].

In the last few years the large off-diagonal nonlinear coefficients of semiconductors of the zinc-blend class (e.g., GaAlAs) have been successfully phase-matched [24,43]. This has required innovative growth and processing techniques, and the results are already impressive and future improvements in the losses are expected. A typical structure is shown in Fig. 4 [24]. A key point is that if the losses are equal for both polarizations, the conversion efficiency is independent

Table 1 FOMs for Waveguide SHG for Cases in Which Both Fundamental and Harmonic Are Guided (η and η') and for the Cerenkov Case (η'')*

Phase-match method	Waveguide	η %/[W-cm^2]	η' %/W	λ (nm)
QPM	KTP[42]	800	115	830
QPM	$LiTaO_3$[15]	1500	960	830
QPM	$LiNbO_3$[14]	1600	150	1064
QPM	$LiNbO_3$[22]	44	250	1550
QPM	AlGaAs[43]	76	22	1550
Birefringent	$LiNbO_3$[6]	4	12	1064
Birefringent	$LiNbO_3$[5]	1.8	40	1091
Birefringent	$LiNbO_3$[5] (resonator)	45	1000	
MDPM	DR1[44] (polymer)	44	0.6	1550
MDPM	SiO_2/Ta_2O_5/KTP[45]	960	161	830
		η'' %/[W-cm]		
CPM	$LiNbO_3$[46]	42	12	830
CPM	DMNP core fiber[47]	40	20	830
CPM	SiO_2/Ta_2O_5/KTP[45]	210	95	830

* Superscripts indicate references.

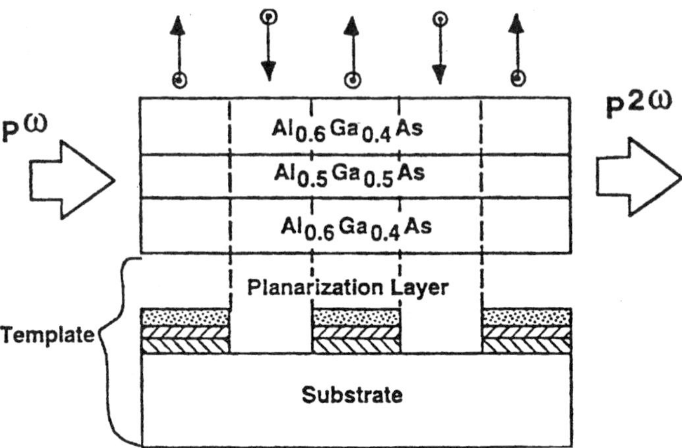

Figure 4 QPM inverted structure in GaAlAs waveguide structure fabricated by wafer bonding and regrowth techniques. The direction of the arrows corresponds to the sign of $d^{(2)}_{\mathrm{eff}}$. (From Ref. 24.)

of input polarization. As discussed later, this is useful for wavelength shifting of WDM signals.

2.2.2 Birefringent Phase-Matching

This approach merits attention because it is was the leading phase-matching technique right up to the 1990s, and there are some very recent developments that could prove it valuable again [3,4–6,26,49]. The waveguide implementation was initially a direct analog to bulk doublers, with the exception that the bulk refractive indices are replaced by waveguide effective indices in the phase-matching conditions. A birefringent material is needed so that the mixing of the fundamental beams involves one of each polarization. As a result, (1) off-diagonal (typically smaller) elements of the nonlinearity tensor are involved, and (2) it may be possible to find geometries in which all three guided wave modes belong to the lowest order (which optimizes the overlap integral K). Furthermore, labeling the two fundamental guided waves as 1 and 2, the phase-matching condition is $\beta_1(\omega) + \beta_2(\omega) - \beta(2\omega) = 0$. Although a quick perusal of Table 1, most specifically the figures of merit for some $LiNbO_3$ devices, shows large conversion efficiencies, these devices had 5-cm-long channels that were resonators at the fundamental and/or second harmonic [5]. The single-pass efficiency is small, which explains why this approach was essentially abandoned in $LiNbO_3$. A very recent result in $KNbO_3$ has produced blue light with an FOM of $25\%/\mathrm{W\text{-}cm}^2$ in 0.73-

cm-long waveguides, despite the relatively large losses (which can still be improved) at both the fundamental and harmonic wavelengths (3 dB/cm and 1.7 dB/cm, respectively) [50].

Normally, birefringent phase-matching is impossible in zinc-blend semiconductors, because their refractive index is isotropic, and hence in their transparent regime phase-matching between lowest-order modes is impossible. Recently a new technique has been reported that artificially introduces a large dispersion in the effective indices between the fundamental and harmonic modes. This dispersion has an opposite sign to the usual waveguide-induced dispersion and hence allows the lowest order to be phase-matched [26]. This is very promising, since the extra dispersion can be controlled, allowing phase-matching over a range of wavelengths.

2.2.3 Modal Dispersion Phase-Matching

This technique, which has no analogy in bulk media, relies on control of the nonlinearity across one of the transverse dimensions of a waveguide. It has both advantages and disadvantages. Its principal value is that it optimizes the overlap integral K and utilizes the diagonal element of the $\mathbf{d}^{(2)}$ tensor, despite the fact that the lowest-order fundamental mode is phase-matched to a higher-order harmonic guided mode. Its disadvantage is that the spatial structure of the output harmonic field is complicated and contains multiple maxima and minima. This limits this approach to applications in which the fundamental is the desired output, modified in phase or amplitude by the harmonic generation process. An example is "cascading," discussed in a later section.

The key here is the value of the overlap integral K for Type 1 phase-matching. This is given simply by

$$\frac{\int_{-\infty}^{\infty}\int_{-\infty}^{\infty} f_i^2(\mathbf{r}_\parallel;\omega) f_i^*(\mathbf{r}_\parallel 2;\omega) d_{iii}^{(2)}(\mathbf{r}_\parallel) d\mathbf{r}_\parallel}{d_{\text{eff}}^{(2)}} \tag{8}$$

If the nonlinearity is uniform across the waveguide, then $K \propto \int f^2(\mathbf{r}_\parallel,\omega) f^*(\mathbf{r}_\parallel,2\omega)\, d\mathbf{r}_\parallel$. Here, $f^2(\mathbf{r}_\parallel,\omega)$ is always positive, but $f^*(\mathbf{r}_\parallel,2\omega)$ reverses sign at least once for higher-order modes so that the value of the integral is reduced. (Recall that higher-order modes are necessary for phase-matching due to index dispersion, as discussed previously). However, if the harmonic mode is higher order in just one of the two transverse channel waveguide dimensions, and the direction of the nonlinearity can be reversed whenever the harmonic field changes sign, then the overlap integral is optimized, since the integrand is always positive. Alternatively, the nonlinearity can be nulled out whenever the field profile has a negative sign: This reduces the figure of merit by about a factor of 4 relative to the nonlinearity reversal case.

This technique was demonstrated with $d^{(2)}$-inactive layers in the early days of SHG in waveguides [51]. It has recently been revived using poled polymers for which the nonlinearity can actually be reversed within a waveguide core [44]. Modulation of the nonlinearity along one transverse direction has also been achieved by making alternate layers $d^{(2)}$-inactive. Better still, polymer stacks in which adjacent layers have been poled to have opposite nonlinearities. An example of a doubler constructed in this way is shown in Fig. 5. The early polymer results achieved are shown in Table 1. The η figure of merit is impressive in view of the small poling voltages used (90 V/μm versus maximum possible values of 360 V/μm), approaching that of $LiNbO_3$ around 1600 nm. However, the large losses at the harmonic frequency have limited L_{eff} to at best a few millimeters. Orders of magnitude improvement can be expected, due to both better poling and lower harmonic losses.

Modal dispersion phase-matching has also been implemented in the case where the substrate is nonlinear [45]. Although there are no interference effects here, the waveguide must be operated near cutoff for both modes in order to have

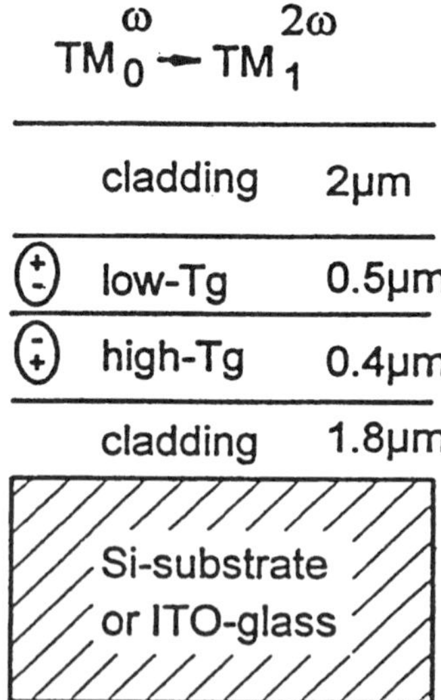

Figure 5 Structure of a poled polymer frequency doubler for modal dispersion phase-matching between a TE_0 and a TE_1 waveguide mode. The use of two polymers with different glass temperatures has allowed the nonlinearities to be reversed about halfway through the guiding region. (From Ref. 25.)

a large fraction of the power carried in the substrate for an efficient interaction. In this case, the output beam shape is strongly asymmetrical. This approach has been used very successfully, as shown in Table 1, for the case of a Ta_2O_5 channel on top of a KTP substrate [45].

2.2.4 Cerenkov

This approach to waveguide SHG historically offered the most promising results in the time interval between birefringent phase-matching and QPM, both in $LiNbO_3$. It is attractive because it reduces dramatically the constraints on phase-matching and still allows the large diagonal elements of $d^{(2)}$ to be used. The harmonic output is not guided but appears as a radiation field, as indicated in Fig. 6. This can happen only if the refractive index of the doubled beam in the substrate, n_s, is higher than the effective index of the fundamental guided beam. For example, in Fig. 2 the range available for Cerenkov SHG corresponds to the cutoff for $TE_0(\omega)$ to the point at which the $TE_0(\omega)$ dispersion curve intersects $n_s(2\omega)$ [46]. Wavevector is always conserved along the waveguide propagation direction so that the projection of the harmonic beam wavevector

$$n_s(2\omega)k_{\text{vac}}(2\omega)\cos\theta = 2k_{\text{vac}}(\omega)N_{\text{eff}}(\omega)$$

where θ is the radiation direction into the substrate. The smaller the radiation angle, the larger the conversion efficiency. The detailed analysis of this interaction shows that the $P(2\omega)$ varies linearly with the sample length L, in contrast to the usual all-guided SHG case, where it varies quadratically with length [52]. As a result, the normalized figure of merit η'' is defined as $\eta'' = P(2\omega)/P^2(\omega) - L$ in %/W-cm.

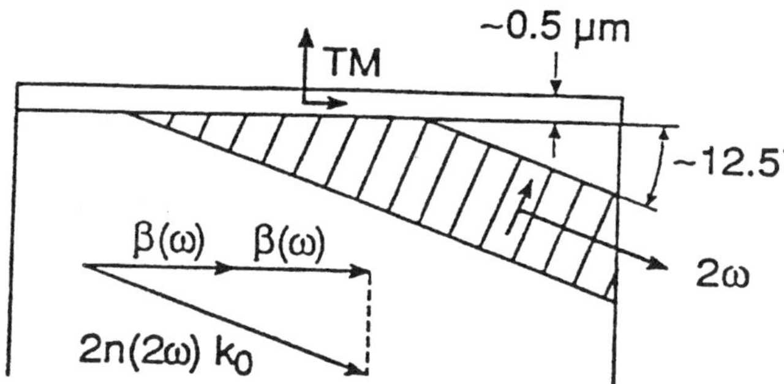

Figure 6 An example in $LiNbO_3$ of the geometry for Cerenkov SHG in which the harmonic is radiated into the substrate. Wavevector conservation parallel to the surface is also shown.

Cerenkov waveguide SHG was first reported in $LiNbO_3$ waveguides, and the best result for this case is given in Table 1, i.e., tens of %/W-cm [46]. The SHG output beam had a strongly asymmetrical beam shape, one of the disadvantages of this approach. Similar conversion results were obtained in fibers filled with single-crystal organics, where the output was a hollow circle [47]. Finally, for the Ta_2O_5 stripe on KTP, very large values of η'' were obtained just below cutoff, where the radiation angle is very small [45]. In this case the output beam was very asymmetrical. In general, this approach has been abandoned, due to the advantages available from the QPM approach.

2.2.5 Contra-Propagating Second Harmonic Generation

This unusual geometry corresponds to Fig. 1(b), with the mixing beams traveling in opposite directions and the harmonic radiated normal to the waveguide surface [3]. The product of the two beams leads to nonlinear polarization fields proportional to $g^2(r_\perp)\{\exp[2i\beta(2\omega)t] + \exp[2i\omega t]\}$ where $r_\perp$ is the coordinate normal to the waveguide surface. The term proportional to exp $[2i\omega t]$ has no spatial periodicity parallel to the waveguide surface and hence can radiate at 2ω only perpendicular to the surface. Therefore the harmonic can only grow over the waveguide cross-sectional dimension, and is emitted in the form of a line along the length of the waveguide. For an idealized channel waveguide of width W and length L,

$$P(2\omega) = A^{NL}\frac{L}{W}P_+(\omega)P_-(\omega) \tag{9}$$

Here, $A^{NL} \propto |d_{\text{eff}}|^2$. The first experiments were reported in lithium niobate waveguides, and this result has been improved upon both by improved waveguide design and by placing the waveguide in a resonator to enhance the signal [53,54]. Values for the parameter A^{NL} are listed in Table 2, giving, for example, a conversion efficiency of 10%/W in the poled polymer DANS [55]. Although this is clearly not competitive with copropagating SHG, there are advantages to this approach. For example, there is no phase-matching condition that needs to be precisely maintained, and in fact many wavelengths can be doubled simultaneously. In fact, this approach has not been fully tested by using resonant nonlinearities in materials that are not normally phase-matchable, like organics [27].

Table 2 Typical SHG Cross Sections for the Counter-Propagating Case*

Material	$LiNbO_3$[53]	GaAs[56]	DANS[55]
A^{NL} (W^{-1})	10^{-8}	10^{-5}	$\rightarrow 10^{-4}$

* Superscripts indicate references.

Because the two fundamental beams pass through each other, this geometry can be used for signal processing. For example, if one fundamental is a fixed frequency and the second contains a spectrum of frequencies, the sum frequency signal is radiated at an angle from the surface normal given by the difference between the fixed frequency and the signal frequency. That is, the device operates like a spectrum analyzer [57]. Also, if one of the inputs is a very short pulse, the signal envelope of a longer pulse can be captured on a detector array [58]. And so on.

2.3 Difference Frequency Generation

In this case there are two input beams of different frequency (ω_1 and ω_2) that mix to produce a smaller frequency ω_3 via $\omega_3 = \omega_1 - \omega_2$. The wavevector matching condition is $\beta_3 = \beta_1 - \beta_2 \pm \kappa$, where the possibility of QPM is included via the grating wavevector κ. All of the phase-matching techniques discussed previously for SHG can be used, and the figures of merit are essentially defined the same way. This process was first efficiently implemented in birefringence phase-matched, and later in QPM channel $LiNbO_3$ waveguides [16,17].

One application of this process has been to generate radiation around 2 μm and beyond. This wavelength region is important for both remote sensing of pollutants and medical applications, as is the mid-infrared [17]. Most of the state-of-the-art work has been performed with QPM lithium niobate channel waveguides. When the three wavelengths involved are separated, for example, by a factor of 2 or more, the minimum waveguide dimensions are determined by the cutoff thicknesses of the longest wavelength involved. Invariably this means that the waveguide is multimode at the shorter wavelengths, and the clean excitation of the lowest-order modes becomes a difficult problem. This problem has recently been solved by using segmented tapers to transition a waveguide that is single mode at a short wavelength to one that is single mode at long wavelengths [59]. To date the best figures of merit for generating 2-μm radiation is only 4%/W-cm^2 and 2%/W for η and η', respectively. However, the latest state-of-the-art waveguides have not been applied to this problem yet.

A second major application has been to shifting of signals between the 1.3-μm and 1.55-μm communications bands or within the 1.55-μm band for WDM [18–21]. The most recent results for interband conversion have had η''s of 450%/W in 3.3-cm-long waveguides, more than adequate for the purpose [19]. A laser of wavelength 710 nm was used as the short-wavelength source. A similar scheme was used to convert signals from one center wavelength to another within the erbium-fiber amplifier band, for example, for WDM systems [20,21]. Shown in Fig. 7 are some typical results. Because the shifts are small, at most 100 nm, the short-wavelength laser is centered near 775 nm, the degeneracy point for this interaction. This has the advantage of producing a very large bandwidth for the

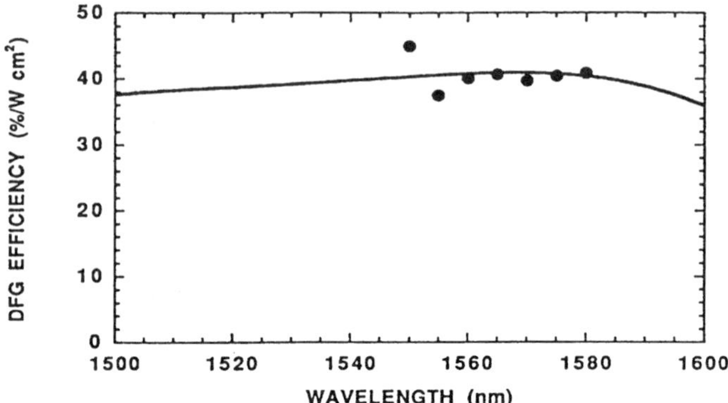

Figure 7 Shown is the FOM for the frequency-shifted signal inside the 1550-nm communications window obtained by tuning the pump laser. (From Ref. 20.)

wavelength conversion process; in fact, the 3-dB roll-off points span the full erbium gain range.

2.4 Optical Parametric Amplifiers and Oscillators

Both amplifiers and oscillators have been demonstrated, primarily in QPM $LiNbO_3$ (also in birefringent phase-matched [60]) channel waveguides [22]. In the amplifier configuration, both the pump signal and the small signal are injected into one end of the waveguide, and the signal is amplified via difference frequency generation, with the concomitant generation of an idler wave. More useful is an optical parametric oscillator, for which one or more of the input and/or generated waves are enhanced by enclosing the waveguide in a resonator. The end mirrors are coated for high reflectivity at one or more of the interacting wavelengths, usually the pump wavelength. The technology has been developed to contact the end mirrors optically right onto the endfaces of the waveguide [60,22].

Optical parametric oscillators (OPOs) based on birefringently phase-matched $LiNbO_3$ reached a very high level of sophistication, including enclosure in a doubly resonant cavity. A threshold value of 27 mW was achieved with a single frequency source around 600 nm, within a factor of 4 of theory. The doubly resonant figures of merit were $\eta = 45\%/\text{W-cm}^2$ and $\eta' = 1000\%/\text{W}$ for a 5-cm-long sample—very impressive. More recent results in QPM channel waveguides in a 2.4-cm-long sample produced a parametric $\eta' = 170\%/\text{W}$ (versus 40%/W for the shorter birefringent phase-matched sample) for single pass, and for the singly resonant case a threshold power of 1.6 W. The tuning range of a waveguide OPO is shown in Fig. 8. This spans a number of useful wavelength ranges,

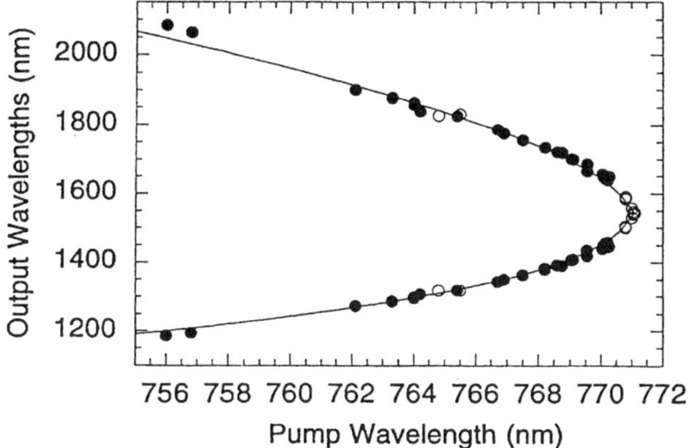

Figure 8 The tuning range for the signal and idler for a waveguide OPO in QPM $LiNbO_3$. (From Ref. 22.)

including the 1.55-μm communications band and the medical- and remote-sensing wavelengths. The key problem has been to lower the threshold for oscillation.

3 THIRD-ORDER EFFECTS AND DEVICES

3.1 Basic Concepts for All-Optical Effects and Devices

The concept of all-optical effects for signal processing is an attractive one. Progressively more information is being transferred in the form of optical pulses so that it is efficient to manipulate the information without first converting it into the electronics domain, processing it there, and then regenerating optical signals. Furthermore, in most electronics devices, the response time is limited by the transit time of electrons to cross some barrier, and it is believed that this will be of the order of 10 psec. This limits processing speeds to maybe a 100 Gb/s. One solution currently being pursued is to leave the information in the optical domain and to use the electro-optic effect to control routing electronically, etc. [61]. This again appears to be limited to perhaps 100 Gb/s due to the temporal response of traveling-wave electro-optic modulators and the electronic circuits needed to drive them efficiently. This approach works well if many very fast pulses in a sequence are routed to a common destination and the routing direction needs to be changed only occasionally on a >20-psec time scale.

Another concept, initiated in the late 1970s with studies on optical bistability, was to use light to manipulate light [62]. That is, if intense enough, an optical signal can change the optical properties of a medium, which then affects its own propagation or that of a second beam or subsequent optical beams. There are many physical phenomena that can modify either the refractive index or the attenuation (or gain) of a medium, each operating on different time scales. Typically the time it takes to create an all-optical effect depends on the input intensity or pulse energy, and it can be very fast. But the relaxation (turnoff) time depends on the detailed physical mechanism involved. The response time usually referred to is the turnoff time. For example, reorientation times in liquid crystals that produce large changes in the refractive index can operate on the second time scale, thermal effects that change index by changing the temperature have decay times of milli- to microseconds, and the nonresonant third-order response of a material can be just a few femtoseconds. In a typical device, more than one mechanism is usually operative, and thermal effects, which are cumulative, are always present and need to be controlled. To process information at very high bit rates, very fast response times are needed to reduce crosstalk between successive events; i.e., the next signal pulse should not be affected by the index change produced to control the previous pulse or the accumulated effects of a train of pulses. Fast-enough materials can be used to manipulate individual signals on time scales less than 10 ps, a niche not accessible with other approaches. Therefore the current thrust is into very fast phenomena, certainly on the time scale of picoseconds.

There are a number of discrete functions that have been or can be implemented all-optically:

1. Self-routing, in which the intensity or energy of a pulse determines its own routing, attenuation, etc.
2. Control beam routing, in which the presence of a control beam determines the routing of a signal beam. This is useful for routing, multiplexing, and demultiplexing (in the time domain).
3. Wavelength shifting, by which a signal's wavelength is changed.
4. Optical logic functions, such as AND and OR gates. These typically require multiple-pulse inputs and a logic-pulse output.

These operations are achieved by controlling all-optically the absorption (or gain), refractive index (usually an accumulated nonlinear phase shift), propagation direction, and/or polarization of signal beams in a medium.

There are multiple mechanisms in semiconductors that can be used—the most common being changes in the carrier density (and hence the refractive index) due to absorption of photons [29]. This nonlinear mechanism can be very fast, provided that the carriers can be removed from the optical beam transit region by some fast process, such as electric-field-induced sweep-out [63]. Typically the control optical pulse duration Δt is shorter than the decay time τ of the optically induced change in the carrier density. This leads to an index change $\Delta n(t)$

and an absorption (or gain) change $\Delta\alpha(t)$, both proportional to $\int\alpha_1(t')I(t')dt'$, which then decays on a longer time scale as exp $[-t/\tau]$. If the pulse duration is much longer than τ (effectively the cw case), then Δn follows the pulse temporal profile and one can define an $n_{2\text{eff}} = \Delta n/I$ that is "independent of time." A long decay time reduces the need for exact coincidence between two optical pulses, as long as the signal pulse follows the control pulse. On the other hand, too long a decay time means crosstalk between successive processes. This is the difficult trade-off encountered in all all-optical devices in which high-speed individual pulse processing is needed. It can be somewhat relieved by using interferometric configurations, as discussed later.

The ultrafast third-order nonlinearity $\chi^{(3)}$ associated with bound electrons and virtual states leads to an intensity-dependent refractive index change of the form $\Delta n = n_2 I$ and in some cases an intensity-dependent change in absorption of the form $\Delta\alpha = \alpha_2 I$, where I is the local intensity. In the case of nonresonant $\chi^{(3)}$, i.e., far from any absorption resonances, these changes typically decay on a time scale shorter than the pulse width because only virtual transitions are involved. This is an advantage in terms of crosstalk between successive processing events, but it is also a problem because the overlap of two pulses in time must be very precise for a control pulse to influence a signal pulse efficiently. For near-resonant processes, there is a finite (typically >10 ps) turnoff time, and the crosstalk-versus-pulse-timing trade-offs are similar to the semiconductor case just discussed. Note that semiconductors also have $\chi^{(3)}$ or ultrafast Kerr nonlinearities, but they are much smaller than those associated with carrier effects.

All of these mechanisms lead to an index and/or an absorption (or gain) change experienced by a beam as it traverses the device [30,33]. The index change is translated into a net induced phase change that can be used in interferometric geometries either to modulate the device transmission in various output channels or to rotate the plane of polarization. An absorption (or gain) change also modifies device transmission, usually in a single output channel, unless a second (usually linear) device is used.

Device throughput is a key factor, especially in cases where some kind of absorption mechanism, linear or nonlinear, is a factor. (Devices based on semiconductor amplifiers can experience net gain, which can be an attractive feature as long as excessive noise is not introduced on the output.) Absorption limits the distance over which a nonlinear phase change can effectively be accumulated, which is critical for interferometric devices that need a minimum nonlinear phase change for good device characteristics, such as complete switching [64]. Analysis of a number of devices has led to two figures of merit that need to be satisfied, one dealing with large linear absorption (α_1) and one for two-photon absorption (α_2). Respectively they are $W > 2\Delta n/\alpha_1\lambda_{\text{vac}} > p$ and $T^{-1} > 2n_2/\alpha_2\lambda_{\text{vac}} > p$, where p depends on the individual device characteristics and $\sim 3 \geq p \geq \sim 0.5$

[31]. (Usually, the more desirable the device characteristics, the larger the value of p [31]. These figures of merit put difficult constraints on the kind of materials that can be used.

In the next subsections we will explore how different materials systems have been used to demonstrate all-optical processes and how effective they have been.

3.2 Passive Semiconductor Effects and Devices

It proves convenient to classify the nonlinearities as passive (conduction band initially populated in thermal equilibrium with the valence band) and active, in which the population of the conduction band is initially out of thermal equilibrium due to electrical or optical pumping of electrons into the conduction band [29,30,33,65,66]. Some typical properties of passive nonlinearities are listed later. Note that there is a crucial trade-off here between loss and nonlinearity, as expressed by the W and T FOMs, and this must always be taken into account in evaluating device performance [30]. For example, it is known that if loss is excessive and $W < 1$, switching is incomplete [64].

There is always one "nonlinearity" that must be taken into account when operating in a region of strong linear absorption, i.e., near the bandgap. Sample heating can dominate the index change when the average energy absorbed over the thermal relaxation time is large, despite a small net absorption over any one signal pulse. It can dramatically affect (and dominate) the effective duty cycle of a device, even with femtosecond pulses. In fact, thermal effects are misinterpreted in many papers as carrier-related nonlinearities. The main characteristics of the thermal nonlinearities can be summarized as follows [29]:

Thermo-optic effect

Absorption $\Delta T \rightarrow \Delta n$(via dn/dT)
Large effective n_2 ($\sim\alpha/\sigma$, where σ is the conductivity), slow ($\tau \sim \mu$s)

3.2.1 Carrier Excitation and Devices [29]

Carrier excitation and exciton bleaching were the first semiconductor electronic nonlinearities to be used. The characteristics, which are strongly wavelength dependent and must be considered in the context of the FOM, are as follows.

1. Bandfilling

Associated with one-photon absorption $\rightarrow$ carrier generation.
All parameters depend on detuning from the bandgap.
$0 < |n_2| < 10^{-6}$ cm^2/W; $|\Delta n_{sat}| < 0.1$; $\tau \sim 10$ ns (bulk semiconductors)

2. Exciton bleaching

Associated with one-photon absorption.
Parameters depend on detuning from the exciton line.
$0 < |n_2| < 10^{-5}$ cm²/W; $|\Delta n_{sat}| < 0.1$; $\tau \sim 10$ ns (ends up in carrier generation)

Note that the relaxation time can be reduced by introducing defects, impurities, electric field sweep-out of carrier, low-temperature growth, etc. to the 10-ps range [63,67]. This also reduces the effective n_2. However, the pulse energy needed to make an all-optical device functional remains the same.

Carrier-related nonlinearities have been used from the earliest days of this field [68]. They have been used primarily in channel waveguides for nonlinear directional couplers (NLDCs), nonlinear Mach–Zehnder interferometers, and, very recently, nonlinear distributed-feedback gratings (for example, Ref. 68). The simplified device geometries and their response to cw excitation are shown in Fig. 9 for these three cases [30]. A number of functions have been demonstrated, mostly self-switching, which was typically the first operation demonstrated [30].

Potentially more useful has been the switching of signal beams by a control beam, both in NLDCs and Mach–Zehnder interferometers [63,68,69]. An example of the routing of a signal pulse by a control pulse is shown in Fig. 10 for a zero-gap NLDC, i.e., a mode mixer [63]. Note that the switching can be essen-

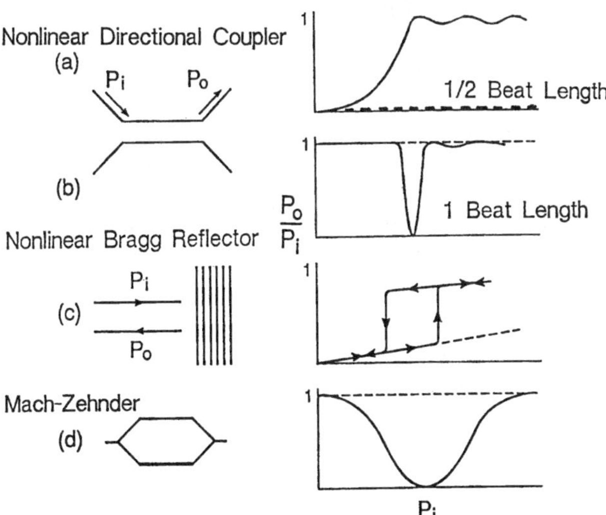

Figure 9 Three all-optical switching device geometries and their response to increasing cw intensity.

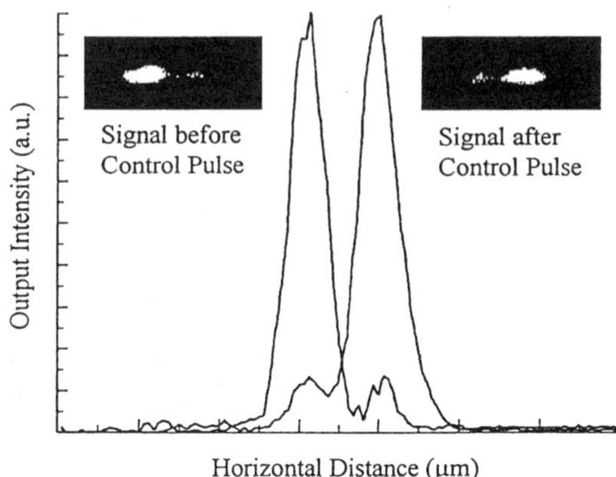

Figure 10 Output from an NLDC at high and low input powers. (From Ref. 63.)

tially complete, because the turnoff time for the index change is much longer than the signal pulse width. In fact, most of the work to date has been in devices in which the turnoff time was limited by carrier recombination to an order of nanoseconds. The most sophisticated implementation has included strong electric fields to sweep the generated carriers out of the waveguide region [63]. Typically, the switching energies were in the 1–10-picojoule range for the control pulse.

A novel approach to reducing the "effective" recovery rate of a device is based on an interferometric approach first introduced for implementing logic gates in $LiNbO_3$ waveguides [70]. The concept is similar to the TOAD and SLALOM devices (discussed later under semiconductor amplifiers, Sec. 3.3, which were introduced at about same time. The geometry is shown in Fig. 11 [69]. Two control pulses of one wavelength, which lies inside the semiconducting bandgap so that they are absorbed and efficiently excite carriers, are used to switch a signal

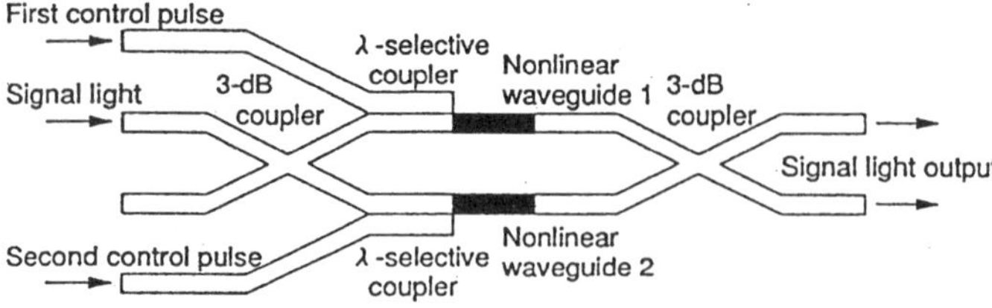

Figure 11 Schematic of a symmetric Mach–Zehnder nonlinear switching device. (From Ref. 69.)

pulse of a longer wavelength (chosen to have low loss). After 50:50 splitting, the input signal in each arm experiences a phase shift of π if a control signal has preceded it. If there are either no control pulses at either input or control pulses at both inputs, then the signal exits in the same output channel, say, A. However, if only one control pulse is present, then the signal exits from the alternate output channel (B). If there is a time delay between the two control pulses, the delay defines a time window in which a modulated version of the long-wavelength signal can appear in B. Time windows on the order of a picosecond have been demonstrated, so very high-bit-rate data streams can be processed, subject to carrier saturation effects [69]. Here, the longer the carrier relaxation time, the more complete the switching.

3.2.2 Kerr Nonlinearities

The Kerr nonlinearity arises due to the response of bound electrons with response times on the femtosecond time scale [71]. Near the semiconducting bandgap, this is also called the *optical Stark effect*, in which the band edge is shifted at high field intensities. This nonlinearity is enhanced near the bandgap, just as it would be near any resonance [71,72]. There, however, the trade-off with the linear absorption is not attractive for reasonable duty cycles. It is also enhanced near half the bandgap, which corresponds to the band edge for two-photon absorption [71]. In fact, for photon energies below half the bandgap, where two-photon absorption is forbidden, the Kerr nonlinearity is dominant, the FOMs are very attractive for devices, but the nonlinearity is not large enough for practical devices [65]. The properties of the nonlinearities can be summarized as follows.

1. Near-gap Kerr nonlinearities

Ultrafast optical Stark effect.
n_2 is negative and decreases with detuning from the bandgap.
$0 < |n_2| < 10^{-11}$ cm^2/W; $|\Delta n_{sat}| < 0.01$; $\tau \sim 10$ fs

2. Two-photon Kerr nonlinearities

Ultrafast two-photon Kerr nonlinearity
The sign of n_2 is positive and depends on detuning from half the bandgap.
$0 < |n_2| < 2 \times 10^{-13}$ cm^2/W; $|\Delta n_{sat}| < 0.01$; $\tau \sim 10$ fs

Some NLDCs have been reported in which the response was subpicosecond for photon energies just below the bandgap [73]. At high powers, the switching was accompanied by an increase in transmission, reminiscent in response to absorption bleaching. Carrier generation, perhaps due to two-photon absorption, spoiled the switching characteristics at high powers. As the detuning is increased to smaller photon energies, and hence to much-reduced one-photon absorption, two-photon absorption becomes the limiting factor, and $T > 1$ until below half the bandgap [30].

Another useful spectral region is for photon energies just below one half the semiconducting bandgap, where two-photon absorption rapidly goes to zero [65,71]. Here in the AlGaAs system with composition tuned for operating in the communications band at 1550 nm, $n_2 \approx 2 \times 10^{-13}$ cm^2/W with the FOM $W > 10$ and $T < 0.2$. Furthermore, for propagation in channel waveguides along [1,−1,0], the self- and cross-phase-matching coefficients and the nonlinearities for both polarizations are essentially equal [65]. To date the best results have been obtained in bulk AlGaAs samples. Demonstrated to date has been a variety of self-switching devices, including NLDCs, X-junctions, nonlinear Mach–Zehnder interferometers, polarization switches etc [74,75]. An example of an NLDC's response is shown in Fig. 12. Note that in contrast to the cw case or the integrating nonlinearity case (carrier excitation), very high intensities are needed for complete switching, because the nonlinearity responds on a time scale much shorter than the pulse width. As a result, each part of the pulse envelope essentially switches independently, leading to incomplete switching, in which the tails of the pulse are the last to switch with increasing intensity. This also leads to "breakup" of the pulse envelope. These are very undesirable features.

The switching of signal beams by more intense control beams has also been used to implement demultiplexing, as indicated in Fig. 13 [76]. Pulse breakup was avoided by using control pulses temporally longer than the signal pulses

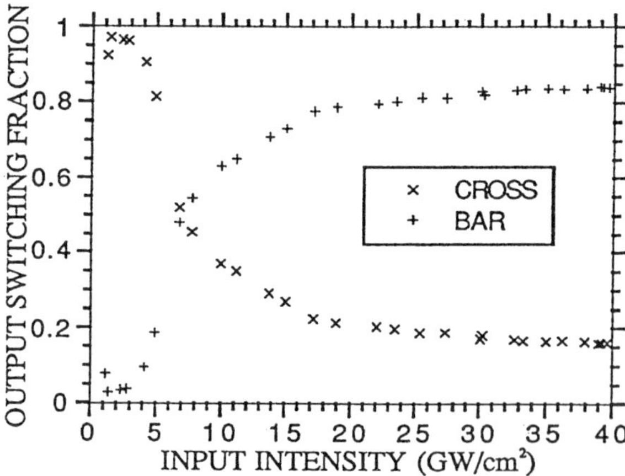

Figure 12 Fraction of power appearing in each of the two output channels for an NLDC made from AlGaAs (Kerr medium). (From Ref. 75.)

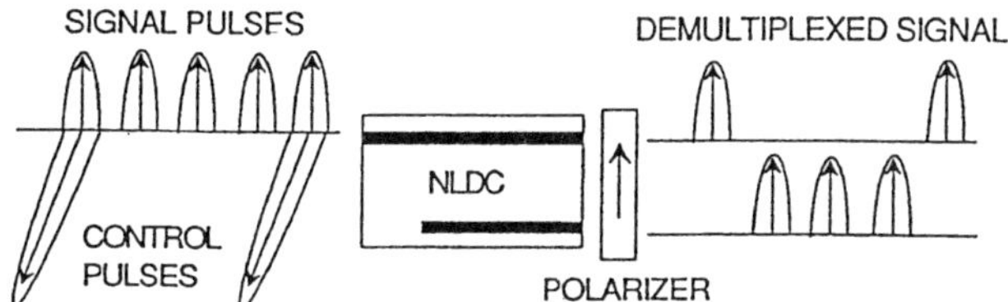

Figure 13 The beam geometry used for time-domain demultiplexing via NLDC. (From Ref. 76.)

[76]. As with all of these ultrafast operations, the timing of the control pulses is critical [76]. Even though the half-bandgap nonlinearities are too small for practical devices, this spectral region has proven useful as a testing ground for a rich variety of device concepts.

3.3 Semiconductor Amplifiers as Signal-Processing Devices

A very promising direction being vigorously pursued in the 1990s has been the use of semiconductor amplifiers as signal-processing elements. First of all, the potential for gain means that the paradigms regarding loss–nonlinearity tradeoffs are no longer relevant for some mechanisms. Second, ultrafast operations can still be implemented with relatively slow processes if the loss associated with the roll-off in frequency response can be compensated by gain. Third, these devices now allow modulation of gain instead of loss. And fourth, innovative ways of effectively decreasing the effective response time are possible.

3.3.1 Nonlinear Mechanisms

Active interband (between valence band and conduction band) nonlinearities have proven very useful for various communications functions. Electrons are ''pumped'' into the conduction band either by direct electrical injection (via electrodes) or by absorption of radiation. When in the gain regime, an incident photon can stimulate an electron transition from the conduction band to the valence band, producing device gain. The resulting change in population in the two bands induces an index change that has a sign opposite to that associated with interband absorption [33]. There are additional mechanisms that lead to index changes on a subpicosecond time scale due to the induced departure of the electron system from its initial equilibrium [77,78]. Finally, the Kerr effect is always present.

Active amplifier nonlinearities (electrical/photon pumping) [33,66]

1. Ultrafast nonlinearities near transparency:

   Two-photon Kerr nonlinearity
   Spectral hole burning
   Carrier heating

2. Carrier nonlinearities: Stimulated emission (gain)

The ultrafast nonlinearities in amplifiers are very interesting from a physics perspective. They are dominant only near and at the transparency point, i.e., the point at which stimulated emission and absorption are just balanced [33,77,78]. (Note that this is not necessarily a lossless case!) An incident photon flux ''burns'' a hole via stimulated emission into the equilibrium electron distribution in the conduction band. This occurs on a sub-100-fs time scale. Over a time period of hundreds of femtoseconds, the electron distribution redistributes to a new equilibrium distribution by ''carrier heating.'' This second intraband mechanism produces a relatively large index change, i.e., an effective n_2, comparable to the Kerr effect. The properties of the ultrafast mechanisms are summarized in Table 3 [77,66,78].

The dominant nonlinearity in amplifiers is carrier related; i.e., it corresponds to interband transitions [33]. When the population near the bottom of the conduction band is inverted with respect to the top of the valence band, this situation is far from equilibrium for the electron temperature. If left without further pumping, the nonlinear phase shift experienced by a transit beam would change by about 10–20π as the system returned to thermal equilibrium between the valence and conduction bands. This corresponds to a very large effective nonlinearity. This process can be enhanced by an incident optical beam if the system is still inverted and stimulated emission can occur. Typically, a 3-pJ input pulse energy is sufficient for creating a π-phase shift! However, the recovery time due to carrier

Table 3 Parameters of Ultrafast Nonlinearities in Semiconductor Optical Amplifiers

Mechanism	λ (μm)	n_2 (cm^2/W)	α (cm^{-1})	W	T
Ultrafast *Kerr* (40 fs)	1.5	-3×10^{-12}	40	0.5	4
Carrier heating (carriers thermalize in band, 100's of fs)	1.5	4.5×10^{-12}	40	0.75	3
Spectral hole burning (photons burn ''hole'' in conduction band, <100 fs)		Small			

recombination is in the nanosecond range. But if the system is continuously pumped near or at gain saturation, the population recovery to its initial (before the incident beam) inverted state can be as short as 10 ps. Thus, there is gain on the input signal, a nonlinear phase shift of π requires only 3 pJ of energy, and the recovery time is on the order of 10 ps, all very useful for ultrafast signal processing [79].

Another useful aspect of amplifier nonlinearities is saturation of the gain [33]. This is essentially a homogeneously broadened system. Therefore the gain across its full bandwidth is affected by changes in the saturation caused by beams at a specific wavelength. This gain modulation can be used to modulate an incident cw beam at a different wavelength than the one that modulates the saturation.

All of the nonlinear mechanisms have a spectral bandwidth associated with them [66]. Strong modulation at one wavelength modulates optical beams at other wavelengths. For slow-relaxing processes, the sensitivity rolls off with detuning from the 3-dB bandwidth. However, with gain present, the effective bandwidth is enhanced. This is a useful feature for processes such as nondegenerate four-wave mixing. In fact, for frequencies separated by THz, the carrier nonlinearity is still comparable in magnitude with the ultrafast ones: This feature has proven useful for frequency shifting for WDM.

3.3.2 Devices

A number of potentially very useful devices based on the nonlinearities just discussed have been implemented in semiconductor optical amplifiers (SOAs).

Self-Switching Devices Some of the earliest demonstrations of amplifiers involved self-switching [80]. The goal in general was to use the ultrafast mechanisms known near the transparency point to implement an NLDC and switching via nonlinear polarization rotation. Typically, switching energies of picojoules were found, although the devices were lossy.

Interferometric Devices Based on Semiconductor Amplifiers Semiconductor optical amplifiers have been used in conjunction with fibers for all-optical devices since the early 1990s [81]. In conjunction with nonlinear loop mirrors, they were the basis of TOAD and SLALOM demultiplexing devices [82]. In the mid-1990s these ideas were turned into devices based on Mach–Zehnder configurations and ultimately into all integrated devices [83,84]. As shown in Fig. 14, the final device is a four-port device, with two ports at each end. Note that the wavelengths of the control pulses lie within the gain region, whereas the signal pulses are at longer wavelengths so that the device is ''transparent'' in terms of losses. There is a phase shift of 0 or π between the two interferometer arms, introduced usually via the electro-optical effect or very precise fabrication control. One or more of the ports can be used as either an input or an output, making for a very versatile device. The two SOAs are displaced from the midpoint of each arm, one to each side.

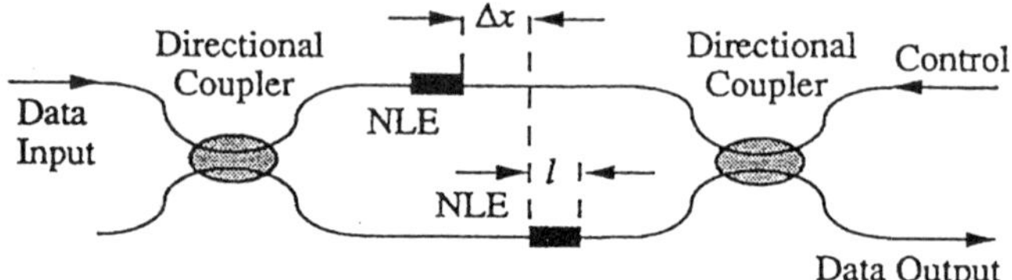

Figure 14 Integrated asymmetrical Mach–Zehnder interferometric SOA switching device in which the two SOAs are spatially offset in the two interferometer arms. (From Ref. 69.)

This powerful device can be used in various ways. If a control pulse is incident from, for example, the right, it arrives at each amplifier at different times. Assuming that the pulse changes the population inversion between the conduction and valence bands, then there is a refractive index change whose time evolution is shown in Fig. 15. If another, for example, cw, beam is present, then before the control pulses arrives at either amplifier, the phase change experienced in each arm differs by π and there is no cw beam output. During the time that one amplifier is excited and the other is not, an additional phase shift of π is produced on one beam and there is an output of the left-traveling beam. As soon as the second amplifier is excited, the phase shifts become equal again, leading to destructive interference and no left-traveling beam output. Thus a pulse is created whose duration is given primarily by the SOA offset. In this mode of operation, the pulse train of the control beam is imparted onto the signal beam. If they have different wavelengths, this is a wavelength shifter [85].

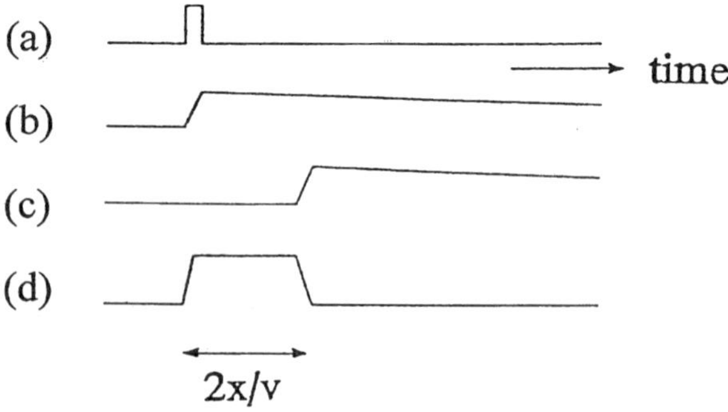

Figure 15 Time evolution of the index change induced in the two SOAs (a,b), (c) the time window created for the signal pulse, and (d) the modulated signal pulse output.

This device can also function as a demultiplexer. The control beam allows a pulse to be passed within the temporal window it creates, and a pulse that is in that window gets switched out. By offsetting the SOAs, the width of this window can be controlled, reducing the need for exact coincidence of the control and demultiplexed pulses by increasing the size of the temporal window, at the expense of processing speed. Demultiplexing in the 100-Gb/s range has been realized with picojoule switching energies [83,84].

Nondegenerate Four-Wave Mixing Devices The third approach utilizes near-degenerate four-wave mixing in active amplifiers (see Fig. 16 for the interaction geometry). This utilizes two input beams, one "pump" beam (ω_p), and one input signal (ω_i), which produces an output signal at $\omega_s = 2\omega_p - \omega_i$ via some effective $\chi^{(3)}$ in the medium. As in the preceding case, this concept was first implemented in fibers and later in SOAs [86]. For the THz frequency shifts needed for operating over the erbium-fiber amplifier bandwidth, the three mechanisms have comparable strengths [66]. Progress in this area has been steady, and the use of high-gain amplifiers with unsaturated gain in excess of 30 dB has led to frequency conversion over the full erbium-fiber bandwidth, with net gain [87].

Gain Saturation Devices Semiconductor amplifiers operating near their gain saturation point offer yet another method for implementing wavelength shifting [88]. It relies on modulating the gain saturation, which is then experienced by a guided wave beam of a different wavelength. One cw laser is used to maintain the amplifier in a state of saturated gain; i.e., it is a holding beam with a cw power of a few hundred milliwatts for recovery times of 13 ps. This is the λ_1 beam in Fig. 17. An input, modulated with information, is incident at a second wavelength λ_2. It causes a modulation in the gain saturation that is transferred to a third cw input beam (λ_3), which is output as a wavelength-shifted signal. The net result is a transfer of the modulation to a beam at a different frequency. The powers involved for λ_2 and λ_3 are typically 100 mW peak and a few milliwatts, respectively.

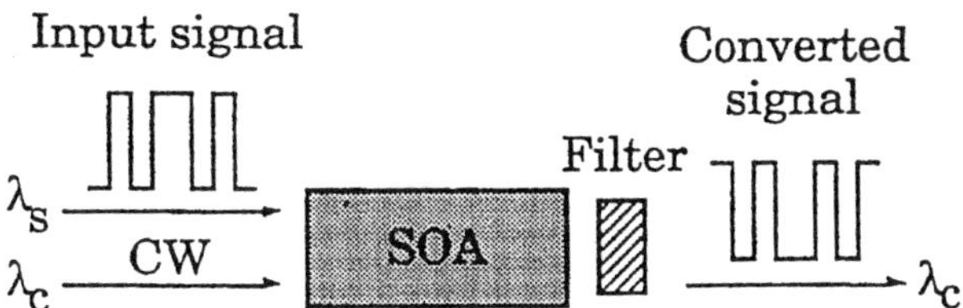

Figure 16 Geometry for nondegenerate four-wave mixing for frequency shifting in an SOA. The filter is used to isolate the input signal and other spurious signals. (From Ref. 87.)

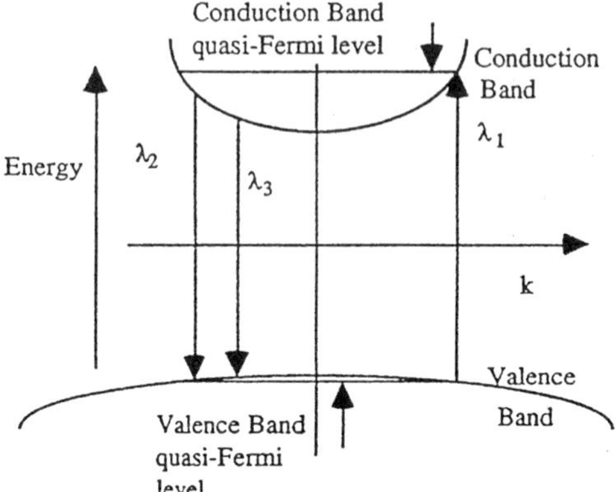

Figure 17 The wavelengths involved in wavelength shifting in an SOA pumped to saturation by a holding beam of wavelength λ_1. The λ_2 beam is the strong input signal, modulated with information. The λ_2 is a cw input beam, which is modulated by the change in the gain [88].

Miscellaneous There have been a number of other fascinating devices based on semiconductor amplifiers. A recent example is another approach to wavelength shifting that operates with subpicojoule energies with a single amplifier and an element that delays two orthogonal control beams (derived from a single control beam) relative to each other. The resulting time window allows a cw wavelength source at a different (longer) wavelength from the control beams to be pulse modulated [89]. Some other examples include optical sampling, tunable wavelength filters, all-optical AND gate, photonic packet switching and ATM switching, clock recovery, and time slot interchange [90].

4. Spatial Solitons

Spatial solitons are beams that propagate in a self-focusing medium ($n_2 > 0$) without spatial spreading. Diffraction in a slab waveguide is counteracted in a robust way by a self-induced lens action [91]. Solitons usually produce index changes equivalent to a waveguide; hence, other beams of opposite polarization and/or wavelength can be guided in them. Therefore, signal beams can be guided inside a soliton.

Bright spatial solitons have been reported in a number of slab waveguides in Kerr nonlinear media. The earliest work dealt with CS_2 and glasses as the core guiding medium [92]. They have also been reported in semiconductors, both in AlGaAs operated in the below-half-bandgap spectral region and in amplifying

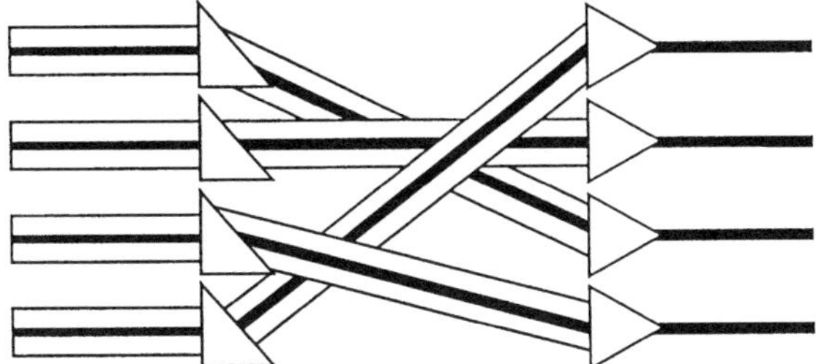

Figure 18 A concept for an $N \times N$ reconfigurable interconnect based on the steering of soliton beams and the signals carried by them.

media [93]. Most recently they have been observed in organic films, and in $LiNbO_3$ via the ''cascading'' $\chi^{(2)}$ mechanism [94,95].

There is a new approach to signal manipulation being investigated based on spatial soliton waveguide phenomena. Although it is still in the physics stage, it does offer some interesting possibilities for reconfigurable interconnects [95]. Solitons can be deflected by prisms induced in a medium either by the electro-optical effect or by charge injection. Thus the signal beams guided by the soliton are also deflected. The concept, as sketched in Fig. 18, is to produce an $N \times N$ interconnect that can be dynamically reconfigured via electrical control of the prisms.

The first steps have already been reported in AlGaAs slab waveguides at 1.55 μm for below-half-bandgap operation. Solitons have been generated, proven to guide beams of different polarization, found to be robust (not fall apart) for propagation through index-induced prisms, and electrically deflected by the prisms [97].

5 CASCADING

A new approach to producing an effective (but strange) n_2, or more precisely a nonlinear phase shift, has recently been investigated in media with $\chi^{(2)} \neq 0$ [23]. This effect was first predicted back in 1967, but it was not exploited until about 1990. Near a phase-matching condition, for example, during SHG, the phase of the fundamental beam is augmented in a nonlinear fashion.

The reason why a nonlinear phase shift exists during a second-order process like second harmonic generation is relatively easy to understand [23]. In non-phase-matched SHG, part of the input fundamental energy is periodically (with distance) converted to the harmonic and then converted back to the fundamental. Since the harmonic phase velocity is different from that of the fundamental, a phase difference accumulates between the fundamental and harmonic fields. Because the harmonic converts periodically back to the fundamental, the phase difference obtained during down conversion adds to the net phase of the fundamental beam. Thus the phase of the total fundamental is altered cumulatively with each periodic exchange. It is nonlinear because the higher the input fundamental power, the stronger the SHG and hence the larger the net phase shift.

The phase shift can be calculated from the coupled-mode equations that describe SHG [23]. The solutions have many interesting characteristics, such as a steplike response with distance. For large wavevector mismatch, there is a simple formula that gives the effective nonlinearity:

$$n_{2,\mathrm{eff}}(z) = \frac{2\pi[d_{\mathrm{eff}}^{(2)}]^2}{\varepsilon_0 c n^4 \lambda\ \Delta\beta}\,[1 - \cos(\Delta\beta z)] \tag{10}$$

where $\Delta\beta$ was defined previously. Note that the sign of the effective nonlinearity is determined by the sign of the detuning and that the magnitude is proportional to the usual figure of merit for SHG. The promise of this approach is that since $n_{2\mathrm{eff}} \propto |d_{\mathrm{eff}}^{(2)}|^2$, it may prove possible to phase-match some of the new organic materials with $d^{(2)}$'s > 100 pm/V, and hence provide effective nonlinearities larger than any available from $\chi^{(3)}$.

The existence of such phase shifts has now been verified experimentally in waveguides [98]. Normally some of the input fundamental is left converted to the harmonic, which acts like a loss for the fundamental. However, it has been found that by controlling the wavevector mismatch with distance it is possible to produce large phase shifts with minimal loss to the harmonic, even for pulses [98].

In analogy to third-order all-optical devices, it was predicted that the nonlinear phase shift could be used for all-optical switching devices in waveguide geometries [99]. Some all-optical devices have already been demonstrated using the "cascaded" phase shift, in direct analogy to the $\chi^{(3)}$ case. A mode mixer, an NLDC, and a nonlinear Mach–Zehnder interferometer were all designed for minimum loss of the fundamental and implemented in birefringently phase-matched $LiNbO_3$ channel waveguides [100]. These initial experiments were done with $d^{(2)}$'s of a few pm/V, and so the switching powers were large. Simple calculations show that QPM $LiNbO_3$ should give peak switching powers of only a few watts. There are organic materials available with nonlinearities of hundreds to thousands of pm/V, but it still remains a challenge to phase-match such materials.

6 CONCLUSIONS

The progress in nonlinear integrated optics has been spectacular over the last decade. In the area of second-order effects, the evolution has gone from rudimentary science to practical devices operating at useful input power levels. A key factor has been the development of QPM, which has almost reached its theoretical limits. There are other possibilities on the horizon, but at this time they are speculative.

In the area of third-order effects for signal processing, the discovery that semiconductor amplifier nonlinearities could be useful has resulted in a variety of practical devices. This has moved this field from basic science and demonstration devices to one in which realistic devices are being reported at a rapid rate. As is healthy for any field, there are also a number of new concepts being developed that could prove useful.

REFERENCES

1. P. A. Franken, A. E. Hill, C. W. Peters, and G. Weinreich. Phys. Rev. Lett. 7:118, 1961.
2. D. B. Anderson and J. T. Boyd. Appl. Phys. Lett. 19:266, 1971.
3. Reviewed in G. I. Stegeman, C. T. Seaton, and J. J. Burke. In: L. D. Hutcheson, ed. Integrated Optical Circuits and Components: Design and Application. New York: Marcel Dekker, 1987, Chap. 9, p. 317; G. I. Stegeman and C. T. Seaton. Appl. Phys. Rev. (J. Appl. Physics) 58:R57, 1985.
4. Early $LiNbO_3$ results reviewed in W. Sohler and H. Suche. In: L. D. Hutcheson and D. G. Hall, eds. Integrated Optics III. Proc. SPIE 408:163, 1983; W. Sohler. In: D. B. Ostrowsky and E. Spitz, eds. New Directions in Guided Wave and Coherent Optics Vol II. The Hague: 1984. NATO ASI Series, Vol. 79, pp. 449–479.
5. W. Sohler and H. Suche. Appl. Phys. Lett. 33:518, 1978; R. Regener and W. Sohler. J. Opt. Soc. Am. B 5:267, 1988; H. Hermann and W. Sohler. J. Opt. Soc. Am. B 5:278, 1988.
6. N. Uesugi and T. Kimura. Appl. Phys. Lett. 29:572, 1976.
7. E. J. Lim, M. M. Fejer, and R. L. Byer. Electron. Lett. 25:174, 1989; G. A. Magel, M. M. Fejer, and R. L. Byer. Appl. Phys. Lett. 56:108, 1990.
8. J. Wejborn, F. Laurell, and G. Arvidsson. J. Lightwave Technol. 7:1597, 1989; J. Webjorn, F. Laurell, and G. Arvidsson. IEEE Phot. Techn. Lett. 1:316, 1989.
9. D. Feng, N. B. Ming, J.-F. Hong, Y.-S. Yang, J.-S. Zhu, Z. Yang, and Y.-N. Wang. Appl. Phys. Lett. 37:607, 1980; Y. H. Xue, N. B. Ming, J. S. Zhu, and D. Feng. Chin. Phys. 4:554, 1983.
10. M. M. Fejer, G. A. Magel, D. H. Jundt, and R. L. Byer. IEEE J. Quant. Electron. 28:2631, 1992.
11. T. Suhara and H. Nishihara. IEEE J. Quant. Electron. 26:1265, 1990.
12. M. M. Fejer. In: F. Kajzar and R. Reinisch, eds. Beam Shaping and Control with Nonlinear Optics. New York: Plenum Press, 1998, Vol. 369, p. 375.
13. C. T. Whipple. Photonics Spectra 33:116, (1998).
14. K. Kiinata, M. Fujimura, T. Suhara, and H. Nishihara. J. Lightwave Technol. 14: 462, 1996.
15. S.-Y. Yi, S.-Y. Shin, Y.-S. Jin, and Y.-S. Son. Appl. Phys. Lett. 68:2493, 1996.

16. N. Uesugi. Appl. Phys. Lett. 36:178, 1980.
17. E. J. Lim, M. L. Bortz, and M. M. Fejer. Appl. Phys. Lett. 59:2207, 1991; S. Sanders, D. W. Nam, R. J. Lang, M. L. Bortz, and M. M. Fejer. Technical Digest of CLEO '94, Washington, DC: Opt. Soc. Am., 1994, 287.
18. C. Q. Xu, H. Okayama, and M. Kawahara. Electron. Lett. 30:2168, 1994.
19. M. H. Chou, K. R. Parameswaran, M. A. Arbore, J. Haudem, and M. M. Fejer. Technical Digest of CLEO '98, Washington, DC: Opt. Soc. Am., 1998, p. 475.
20. C. Q. Xu, H. Okayama, K. Shinozaki, K. Watanabe, and M. Kawahara. Appl. Phys. Lett. 63:1170, 1993; C. Q. Xu, H. Okayama, and M. Kawahara. Appl. Phys. Lett. 63:3559, 1993.
21. K. Gallo, G. Assanto, and G. I. Stegeman. Appl. Phys. Lett. 71:1020, 1997; C. Trevinio-Palacios and G. I. Stegeman. Electr. Lett. 34:2157, 1998.
22. M. L. Bortz, M. A. Arbore, and M. M. Fejer. Opt. Lett. 20:49, 1995; M. A. Arbore and M. M. Fejer. Opt. Lett. 22:151, 1997.
23. Reviewed in G. I. Stegeman, D. J. Hagan, and L. Torner. J. Opt. Quant. Electron. 28:1691, 1996.
24. S. J. B. Yoo, C. Caneau, R. Bhat, M. A. Koza, A. Rajhel, and N. Antoniades. Appl. Phys. Lett. 68:2609, 1996; S. J. B. Yoo. J. Lightwave Technol. 14:955, 1996.
25. M. Jager, G. I. Stegeman, M. Diemeer, C. Flipse, and G. Mohlmann. Appl. Phys. Lett. 69:4139, 1997; W. Wirges, S. Yilmaz, W. Brinker, S. Bauer-Gogonea, S. Bauer, M. Jager, G. I. Stegeman, M. Ahlheim, M. Stahelin, B. Zysset, F. Lehr, M. Diemeer, and R. Felipse, Appl. Phys. Lett. 70:3347, 1997.
26. A. Fiore, V. Berger, E. Rosencher, P. Bravetti, and J. Nagle. Nature 391:463, 1998.
27. G. Knopfle, R. Schlesser, R. Ducret, and P. Gunter. Nonlinear Optics 9:143, 1995; C. Bosshard, G. Knopfle, P. Prere, S. Follonier, C. Serbutoviez, and P. Gunter. Opt. Eng. 34:1951, 1995; Ch. Bosshard, K. Sutter, R. Schlesser, and P. Gunter. J. Opt. Soc. Am. B 10:867, 1993.
28. R. H. Stolen and C. Lin. Phys. Rev. A 17:1448, 1978; R. H. Stolen. Proc. IEEE 68:1232, 1980; R. H. Stolen. IEEE J. Quant. Electron. 11:100, 1975.
29. Discussed in H. M. Gibbs, G. Khitrova, and N. Peyghamberian, eds. Nonlinear Photonics. Berlin: Springer-Verlag, 1990.
30. Reviewed in G. I. Stegeman, R. Zanoni, N. Finlayson, E. M. Wright and C. T. Seaton. J. Lightwave Techn. 6:953, 1988; G. I. Stegeman and A. Miller. In: J. Midwinter, ed. Photonic Switching, Vol I. Orlando, FL: Academic Press, 1992, p. 81.
31. Summarized in G. I. Stegeman. SPIE Proceedings on Nonlinear Optical Properties of Advanced Materials 1852:75, 1993.
32. A. D. Ellis and D. M. Spirit. Electron. Lett. 29:2115, 1993; S. Kawanishi, H. Takara, M. Saruwatari, and T. Kitoh. Electron. Lett. 29:1714, 1993.
33. For example, M. J. Adams, D. A. O. Davies, M. C. Tatham, and M. A. Fisher. Opt. Quant. Electron. 27:1, 1995.
34. For example, S. H. Park, J. F. Mohrange, A. D. Jeffrey, R. A. Morgan, A. Chavez-Pirson, H. M. Gibbs, S. W. Koch, N. Peyghamberian, M. Derstine, A. C. Gossard, J. H. English, and W. Wiegmann. Appl. Phys. Lett. 52:1201, 1988.

35. For example, G. R. Olbright and N. Peyghamberian. Solid State Comm. 58:337, 1986.
36. For example, B. Lawrence, M. Cha, J. U. Kang, W. Torruellas, G. I. Stegeman, G. Baker, J. Meth, and S. Etemad. Electron. Lett. 30:447, 1994.
37. G. I. Stegeman. J. Nonlinear Optics 15:469, 1996.
38. F. A. Hopf and G. I. Stegeman. Advanced Classical Electrodynamics Vol. II: Nonlinear Optics. New York: Wiley 1986.
39. Extensive material listing by S. Singh. In: M. J. Webber, ed. CRC Handbook of Laser Science and Technology, Supplement 2: Optical Materials. Ann Arbor, MI: CRC Press, 1995, pp. 147–266.
40. M. Yamada and K. Kishima. Electron. Lett. 27:823, 1991; H. Ito, C. Takyu, and H. Inaba. Electron. Lett. 27:1221, 1991.
41. V. Pruneri, R. Koch, P. G. Kazansky, W. A. Clarkson, P. ST. J. Russell, and D. C. Hanna. Opt. Lett. 20:2375, 1995.
42. D. Eger, M. Oron, M. Katz, and A. Zussman. Appl. Phys. Lett. 64:3208, 1994.
43. S. J. B. Yoo, C. Caneau, A. Rajhel, R. Bhat, and M. A. Koza. Technical Digest of CLEO '97 (Opt. Soc. Am., Washington, DC: 1997, p. 59.
44. S. Yilmaz, W. Wirges, W. Brinker, S. Bauer-Gogonea, S. Bauer, M. Jager, G. I. Stegeman, M. Ahlheim, M. Stahelin, F. Lehr, M. Diemeer, and M. C. Flipse. J. Opt. Soc. Am. B 15:781, 1998.
45. T. Doumuki, H. Tamada, and M. Saitoh. Appl. Phys. Lett. 65:2519, 1994.
46. For example, G. Tohmon, J. Ohya, K. Yamamoto, and T. Taniuchi. IEEE Photonics Techn. Lett. 2:629, 1990.
47. For example, A. Harada, Y. Okazaki, K. Kamiyama, and S. Umegaki. Appl. Phys. Lett. 59:1535, 1991.
48. N. Bloembergen. Nonlinear Optics. Reading, MA: Benjamin, 1965.
49. P. Bravetti, A. Fiore, V. Berger, E. Rosencher, J. Nagle, and O. Gauthier-Lafaye. Opt. Lett. 23:331, 1998.
50. T. Pliska, D. Fluck, P. Gunter, E. Gini, H. Melchior, L. Beckers, and C. Buchal. Appl. Phys. Lett. 72:2364, 1998.
51. H. Ito and H. Inaba. Opt. Lett. 2:139, 1978.
52. For example, K. Hiyata, T. Sugawara, and M. Koshiba. IEEE J. Quant. Electron. 26:123, 1990.
53. R. Normandin and G. I. Stegeman. Opt. Lett. 4:58, 1979; R. Normandin, P. J. Vella, and G. I. Stegeman. Appl. Phys. Lett. 38:759, 1981.
54. R. Lodenkamper, M. L. Bortz, M. M. Fejer, K. Bacher, and J. S. Harris. Opt. Lett. 18:1798, 1993.
55. A. Otomo, S. Mittler-Neher, C. Bosshard, G. I. Stegeman, W. H. G. Horsthuis, and G. R. Möhlmann. Appl. Phys. Lett. 63:3405, 1993; A. Otomo, G. I. Stegeman, W. H. G. Horsthuis, and G. R. Mohlmann. Appl. Phys. Lett. 68:3683, 1996.
56. R. Normandin, S. Letourneau, F. Chatenoud, and R. L. Williams. IEEE J. Quant. Electron. 27:1520, 1991.
57. A. Otomo, G. I. Stegeman, W. Horsthuis, and G. Möhlmann. In: G. A. Lindsay and K. D. Singer, eds. ACS tutorial series #601. Washington, DC: American Chemical Society, 1994, p. 469.
58. R. Normandin and G. I. Stegeman. Appl. Phys. Lett. 40:759, 1982.

59. M. H. Chou, M. A. Arbore, and M. M. Fejer. Opt. Lett. 21:794, 1996.
60. Reviewed in H. Suche and W. Sohler. Optoelectronics—Devices and Technologies 4:1, 1989.
61. S. B. Alexander et al. J. Lightwave Technol. 11:714, 1993; C. A. Brackett, A. S. Acampora, J. Sweitzer, G. Tangonan, M. T. Smith, W. Lennon, K. C. Wang, and R. H. Hobbs. J. Lightwave Technol. 11:736, 1993.
62. Reviewed in H. M. Gibbs. Optical Bistability: Controlling Light with Light. Orlando, FL: Academic Press, 1985.
63. P. LiKamWa, A. Miller, J. S. Roberts, and P. N. Robson. Appl. Phys. Lett. 58: 2055, 1991; A. Cavailles, D. A. B. Miller, J. E. Cunningham, P. LiKamWa, and A. Miller. IEEE J. Quant. Electron. 28:2486, 1992.
64. G. I. Stegeman, C. T. Seaton, C. N. Ironside, T. J. Cullen, and A. C. Walker. Appl. Phys. Lett. 50:1035, 1987.
65. J. U. Kang, G. I. Stegeman, D. C. Hutchings, J. S. Aitchison, and A. Villeneuve. IEEE J. Quant. Electron. 33:341, 1997.
66. J. Mark and A. Mecozzi. J. Opt. Soc. Am. B 8:1803, 1996.
67. R. Takahashi, Y. Kawamura, and H. Iwamura. Appl. Phys. Lett. 68:153, 1996.
68. R. Jin, C. L. Chuang, H. M. Gibbs, S. W. Koch, J. N. Polky, and G. A. Pubanz. Appl. Phys. Lett. 53:1791, 1988; P. LiKamWa, A. Miller, C. B. Park, J. S. Roberts, and P. N. Robson. Appl. Phys. Lett. 57:1846, 1990; S. Nakamura, K. Tajima, N. Hamao, and Y. Sugimoto. Appl. Phys. Lett. 62:925, 1993; C. Coriasso, D. Campi, C. Cacciatore, L. Faustini, C. Rigo, and A. Stano. Opt. Lett. 23:183, 1998.
69. S. Nakamura, K. Tajima, and Y. Sugimoto. Appl. Phys. Lett. 66:2457, 1995; 67: 2445, 1995.
70. A. Lattes, H. A. Haus, J. F. Leonberger, and E. P. Ippen. J. Quant. Electron. 19: 1718, 1983.
71. M. Sheik-Bahae, D. C. Hutchings, D. J. Hagan, and E. W. VanStryland. IEEE J. Quant. Electron. 27:1296, 1991.
72. K. K. Anderson. Ph.D. dissertation, M.I.T., 1989; M. J. Lagasse, K. K. Anderson, C. A. Wang, H. A. Haus, and J. G. Fujimoto. Appl. Phys. Lett. 56:417, 1990.
73. R. Jin, J. P. Sokoloff, P. A. Harten, C. L. Chuang, S. G. Lee, M. Warren, H. M. Gibbs, N. Peyghambarian, J. N. Polky, and G. A. Pubanz. Appl. Phys. Lett. 56: 993, 1990.
74. Summarized in J. S. Aitchison, A. Villeneuve, and G. I. Stegeman. J. Nonlinear Opt. Phys. Materials 4:871, 1995.
75. K. Al-hemyari, J. S. Aitchison, C. N. Ironside, G. T. Kennedy, R. S. Grant, and W. Sibbett. Electron. Lett. 28:1090, 1992; J. S. Aitchison, A. Villeneuve, and G. I. Stegeman. Opt. Lett. 18:1153, 1993; C. C. Yang, A. Villeneuve, G. I. Stegeman, C.-H. Lin, and H.-H. Lin. Opt. Lett. 18:1487, 1993; P. A. Snow, I. E. Day, I. H. White, R. V. Penty, H. K. Tsang, R. S. Grant, Z., Su, W. Sibbett, J. B. D. Soole, H. P. Leblanc, A. S. Gozdz, N. C. Andreadakis, and C. Caneau. Electron. Lett. 28: 2346, 1992; K. Al-hemyari, A. Villeneuve, J. U. Kang, J. S. Aitchison, C. N. Ironside, and G. I. Stegeman. Technical Digest of CLEO '94 (Opt. Soc. Am., Washington, DC, 1994, p. 56.
76. A. Villeneuve, P. Mamyshev, J. U. Kang, G. I. Stegeman, J. S. Aitchison, and C. N. Ironside. IEEE J. Quant. Electron. 31:2165, 1995.

77. C. T. Hultgren and E. P. Ippen. Appl. Phys. Lett. 59:635, 1991; C. T. Hultgren, D. J. Dougherty, and E. P. Ippen. Appl. Phys. Lett. 61:2767, 1992; K. L. Hall, A. M. Darwish, E. P. Ippen, U. Koren, and G. Rayborn. Appl. Phys. Lett. 62:1320, 1993.
78. R. S. Grant and W. Sibbett. Appl. Phys. Lett. 58:1119, 1991; M. A. Fisher, H. Wickes, G. T. Kennedy, R. S. Grant, and W. Sibbett. Electron. Lett. 29:1185, 1993.
79. R. J. Manning, D. A. O. Davies, D. Cotter, and J. K. Lucek. Electron. Lett. 30: 787, 1994; R. J. Manning and G. Sherlock. Electron. Lett. 31:307, 1995.
80. D. A. O. Davies, M. A. Fisher, D. J. Elton, S. D. Perrin, M. J. Adams, G. T. Kennedy, R. S. Grant, P. D. Roberts, and W. Sibbett. Electron. Lett. 29:1710, 1993; S. G. Lee, B. P. McGinnis, R. Jin, J. Yumoto, G. Khitrova, H. M. Gibbs, R. Binder, S. W. Koch, and N. Peyghamberian. Appl. Phys. Lett. 64:454, 1994; M.-S. Lin, D.-W. Huang, C. C. Yang, M. Hong, and Y.-K. Chen. Appl. Phys. Lett. 67:2114, 1995.
81. A. Ehrhardt, M. Eiselt, G. Grosskopf, L. Küller, R. Ludwig, W. Pieper, R. Schnabel, and H. G. Weber. J. Lightwave Technol. 11:1287, 1993.
82. J. P. Sokoloff, P. R. Prucnal, I. Glesk, and M. Kane. IEEE Phot. Techn. Lett. 5: 787, 1993; J. P. Sokoloff, I. Glesk, P. R. Prucnall, and R. K. Boncek. IEEE Phot. Technol. Lett. 6:98, 1994; M. Eiselt, W. Pieper, and H. G. Weber. Electron. Lett. 29:1167, 1993; M. Eiselt, W. Pieper, and H. G. Weber. IEEE J. Lightwave Technol. 13:2099, 1995.
83. K. I. Kang, T. G. Chang, I. Glesk, P. R. Prucnall, and R. K. Boncek. Appl. Phys. Lett. 67:605, 1995.
84. E. Jahn, N. Agrawal, M. Arbert, H.-J. Ehrke, D. Franke, R. Ludwig, W. Pieper, H. G. Weber, and C. M. Weinert. Electron. Lett. 31:1857, 1995.
85. A more sophisticated version was reported by, for example, F. Ratovelomanana, N. Vodjdani, A. Enard, G. Glastre, D. Rondi, R. Blondeau, C. Joregensen, T. Durhuus, B. Mikkelsen, K. E. Stubkjaer, A. Jourdan, and G. Soulage. IEEE Photon. Technol. Lett. 7:992, 1995.
86. M. C. Tatham, G. Sherlock, and L. D. Westbrook. IEEE Photon. Technol. Lett. 5: 1303, 1993.
87. J. Zhou, N. Park, K. J. Vahala, M. A. Newkirk, and B. I. Miller. Electron. Lett. 30:859, 1994; W. Shieh, E. Park, and A. E. Willner. IEEE Photon. Technol. Lett. 8:524, 1996; A. D'Ottavi, F. Martelli, P. Spano, A. Mecozzi, S. Scotti, R. Dall'Ara, J. Eckner, and G. Guekos. Appl. Phys. Lett. 68:2186, 1996; T. Durhuus, B. Mikkelsen, C. Joergensen, S. L. Danielsen, and K. E. Stubkjaer. IEEE J. Lightwave Technol. 14:942, 1996.
88. R. J. Manning and D. A. O. Davies. Opt. Lett. 19:889, 1994.
89. Y. Ueno, S. Nakamura, T. Tajima, and S. Kitamura. IEEE Photon. Technol. Lett. 10:346, 1998.
90. M. Jinno, J. B. Schlager, and D. L. Franzen. Electron. Lett. 30:1489, 1994; S. Dubovitsky, and W. H. Steier. J. Lightwave Technol. 14:1020, 1996; D. Nesset and M. C. Tatham. Electron. Lett. 31:896, 1995; R. Fortenberry, A. J. Lowery, W. L. Ha, and R. S. Tucker. Electron. Lett. 27:1305, 1991; M. Eiselt, G. Grosskopf, R. Ludwig, W. Pieper, and H. G. Weber. Electron. Lett. 28:1438, 1992; D. Patrick and R. J. Manning. Electron. Lett. 30:151, 1994; J. Yao, P. Barnsley, N. Walker, and M. O'Mahony. Electron. Lett. 29:1053, 1993.

91. M. Segev. J. Optic. Quant. Electron. 30:503 (1998).
92. A. Barthelemy, S. Maneuf and C. Froehly. Optic. Commun. 55:201, 1985; S. Maneuf and F. Reynaud. Optic. Commun. 66:325, 1988; J. S. Aitchison, A. M. Weiner, Y. Silberberg, M. K. Oliver, J. L. Jackel, D. E. Laird, E. M. Vogel, and P. W. E. Smith. Optic. Lett. 15:1990.
93. J. S. Aitchison, K. Al-hemyari, C. N. Ironside, R. S. Grant, and W. Sibbett. Electron. Lett. 28:1879, 1992; G. Khitrova, H. M. Gibbs, Y. Kawamura, H. Iwamura, T. Ikegami, J. E. Sipe, and L. Ming. Phys. Rev. Lett. 70:920, 1993.
94. U. Bartuch, U. Peschel, Th. Gabler, R. Waldhaus, and H.-H. Horhold. Optic. Comm. 134:49, 1997.
95. R. Schiek, Y. Baek, and G. I. Stegeman. Phys. Rev. A 53:1138, 1996.
96. G. I. Stegeman, P. Mamyshev, W. Torruellas, A. Villeneuve, and J. S. Aitchison. SPIE 2481:270, 1995.
97. J. U. Kang, G. I. Stegeman, G. Hamilton, and J. S. Aitchison. Appl. Phys. Lett. 70:1363, 1997; L. Friedrich, J. S. Aitchison, P. Millar, and G. I. Stegeman. Optic. Lett. 23:1438 (1998).
98. M. L. Sundheimer, Ch. Bosshard, E. W. VanStryland, G. I. Stegeman, and J. D. Bierlein. Optic. Lett. 18:1397, 1993; R. Schiek, M. L. Sundheimer, D. Y. Kim, Y. Baek, G. I. Stegeman, H. Suche, and W. Sohler. Optic. Lett. 19:1949, 1994.
99. G. Assanto, G. I. Stegeman, M. Sheik-Bahae, and E. W. VanStryland. Appl. Phys. Lett. 62:1323, 1993; R. Schiek. Optic. Quant. Electron. 26:415, 1994; A. De Rossi, C. Conti, and G. Assanto. Optic. Quant. Electron. 29:53, 1997; G. Assanto. In: F. Kajzar and R. Reinisch, eds. Beam Shaping and Control with Nonlinear Optics. New York: Plenum Press, 1997, pp. 341–374.
100. R. Schiek, Y. Baek, G. Krijnen, G. I. Stegeman, I. Baumann, and W. Sohler. Optic. Lett. 21:940, 1996; Y. Baek, R. Schiek, G. Krijnen, G. I. Stegeman, I. Baumann, and W. Sohler. Appl. Phys. Lett. 68:2055, 1996; Y. Baek, R. Schiek, G. I. Stegeman, and G. Assanto. Appl. Phys. Lett. 72:3405, 1998; I. Yokohama, M. Asobe, A. Yokoo, H. Itoh, and T. Kaino. J. Opt. Soc. Am. B 14:3368, 1997.
101. K. Shinozaki, Y. Miyamoto, H. Okayama, T. Kamijoh, and T. Nonaka. Appl. Phys. Lett. 58:1934, 1991; L. E. Myers, G. D. Miller, R. C. Eckardt, M. M. Fejer, R. L. Byer, and W. R. Bosenberg. Optic. Lett. 20:52, 1995.

12

Design and Simulation Tools for Integrated Optics

M. R. Amersfoort

BBV Design
Enschede, The Netherlands

J. Bos

BBV Software
Enschede, The Netherlands

X. J. M. Leijtens

Delft University of Technology
Delft, The Netherlands

H. J. van Weerden

Twente University of Technology
Enschede, The Netherlands

1 INTRODUCTION

The possibility to understand and model the propagation of light in waveguides is of paramount importance for the design of integrated optic devices. Only for very simple cases can analytical expressions be found to describe the propagation characteristics. Therefore, the design of integrated optic devices relies heavily on numerical simulation tools. In the last decade, significant progress has been made in the development and implementation of these tools, thus enabling a shorter and more cost-effective design cycle.

In this chapter we will discuss several of the most common simulation and design tools for integrated optics, most of which are now commercially available. Almost all of the commercial modeling tools available today rely on the beam propagation method (BPM). With the BPM, the field is propagated in a stepwise manner through slices of a known waveguide structure. In this chapter we will discuss the background, some simulation examples, and also the limitations of the BPM.

Alternatively, the light propagation can be described in terms of eigenmodes. Each eigenmode is characterized by its field distribution and propagation constant. We will discuss the background and application of numerical eigenmode solvers.

Although BPM and eigenmode analysis are clearly different simulation concepts, it should be mentioned that, generally speaking, they are used together. For example, a BPM simulator requires a start field as the initial condition and overlap fields to determine the (guided) power in the output waveguides. In both cases an eigenmode of the input or output waveguide is the most natural choice. The other way round, a BPM may actually be used for the calculation of waveguide eigenmodes [Chen94].

The simulation tools just mentioned deal primarily with passive integrated optic components. In addition, there is a need to model, simulate, and optimize active components, such as switches, modulators, and amplifiers, that are based on the perturbation to the optical field due to an externally applied (thermal, electrical, or optical) field interaction. As an example we will present a tool to model the signal amplification in Er-doped waveguides.

Another aspect of integrated optic design tools, whose importance is often underestimated, is the mask layout utility. Due to the unique layout requirements for waveguide devices in terms of smoothness and positioning accuracy, general electronic mask layout tools are not really suitable for integrated optic design. We will describe the characteristics of a mask layout tool that takes into account the connectivity requirements of waveguide devices and allows for the use of parametrized primitives to facilitate the design of complex devices.

As the level of complexity of integrated optic devices continues to increase, simulation of the entire integrated optic circuit becomes unpractical using current (BPM-based) modeling tools. An S-matrix-based modeling tool will be presented for the design and modeling of these highly complex devices. The transmission characteristics of each subcomponent of the circuit is described in terms of an S-matrix. The entire circuit transmission performance then simply follows from a matrix multiplication.

2 MODE SOLVERS

A powerful and natural concept to describe the propagation of light in waveguides is based on the calculation of the guided waves (or eigenmodes). Each guided wave is uniquely characterized by its field distributions (electrical and magnetic) and propagation constant as a function of the optical frequency. The ability to calculate, understand, and control the modal properties is therefore of great importance for integrated optic design and simulation.

Especially for components that contain long, longitudinally uniform sections, an eigenmode-based analysis may well be more efficient and accurate than BPM. The bidirectional eigenmode propagation (BEP) method takes full advantage of this aspect [Szte93, Will95].

Unfortunately the modal properties of most waveguide structures cannot be described analytically. Therefore, numerical techniques are required for this type of calculation. In this section we consider longitudinally uniform dielectric waveguides that consist of optically linear media and are concerned with finding the numerical solutions for the guided waves supported by such waveguides. Depending on the type of waveguide and operation region, i.e., far from or near cutoff, the general vector-wave equation can be reduced to the often-used semi-vectorial one without significant loss of accuracy.

2.1 Full Vector-Wave Equations

Throughout this section, we will assume that the field of the wave varies as exp $[i(\omega t - \beta z)]$, where t is the time, z is propagation direction, ω is the angular frequency, and β is the propagation constant. The following vector-wave equations can be derived from Maxwell's (source-free) equations for the magnetic field **H** and the electric field **E**, respectively:

$$\nabla \times ([\varepsilon]^{-1}\nabla \times \vec{H}) - k_0^2[\mu]\vec{H} = 0 \tag{1}$$

$$\nabla \times ([\mu]^{-1}\nabla \times \vec{H}) - k_0^2[\varepsilon]\vec{H} = 0 \tag{2}$$

Here k_0 is the free-space wave number and $[\varepsilon]$ and $[\mu]$ are the relative permittivity and permeability tensors. For general anisotropic waveguides, either one of these two equations must be solved. For nonmagnetic materials, which normally constitute an optical waveguide, $[\mu] = 1$. This implies that both the tangential and the normal components of the magnetic fields are continuous at material interfaces (jumps in ε). For the electric fields only, the tangential components are continuous; therefore the full **H** equation is more convenient in the light of boundary condition enforcement at material interfaces.

For isotropic waveguides with $[\varepsilon] = \varepsilon = n^2$, where n is the refractive index, this equation can be written as

$$\nabla^2\vec{H} - k_0^2 n^2\vec{H} = -\nabla \ln n^2 \times \nabla \times \vec{H} \tag{3}$$

In homogeneous regions, this equation is simply the Helmholtz equation, since $\nabla \ln n^2 = 0$. Because $\nabla \ln n^2$ cannot be defined along material interfaces, its effect should be reduced to the appropriate boundary conditions. An eigenvalue problem is obtained by substituting $\partial/\partial z = -i\beta$:

$$\nabla_t^2 \vec{H}_t + (k_0^2 n^2 - \beta^2)\, \vec{H}_t = -\nabla_t \ln n^2 \times \nabla_t \times \vec{H}_t; \qquad \nabla^2 = \nabla_t^2 - \beta^2 \tag{4}$$

where $\mathbf{H}_t$ denotes the transverse magnetic field. After this equation has been solved for $\mathbf{H}_t$, the other field components can be derived from them via Maxwell's equations.

2.2 Semivectorial Approximation

Equation (4) consists of two coupled equations. If the coupling terms are ignored, the following simpler equations are obtained:

$$\nabla_t H_x + (k_0^2 - \beta^2) H_x = \frac{\partial\ H_x \partial \ln n^2}{\partial_y \partial_y} \tag{5}$$

$$\nabla_t H_y + (k_0^2 - \beta^2) H_y = \frac{\partial H_y}{\partial_x} \frac{\partial \ln n^2}{\partial_x} \tag{6}$$

These equations are the scalar-wave equations including polarization effects. The propagation constants following from these two equations are in general different and represent the two polarization modes. By neglecting the coupling terms, one assumes the orthogonal component to be negligibly small, which is accurate for the following three classes of waveguides [Chia94]:

Weakly guiding waveguides
Waveguides with elongated slablike cross sections
Rectangular-core waveguides operated in the far-from-cutoff region.

In the case of weakly guiding waveguides the index difference between the guiding region and the substrate or cladding is small (on the order 0.1% to a few percent).

Although the index difference maintains total internal reflection, the medium is virtually homogeneous as far as polarization effects are concerned. Since Eqs. (5) and (6) are exact for slab waveguides [Snyd83, p. 596], where the orthogonal component is exactly zero, it will be clear that for slablike cross sections the orthogonal component will be small.

For waveguides operated in the far-from-cutoff region, the power of the field is confined in the core; therefore the field derivatives at the core–cladding interface, and thus the coupling between the two transverse components, will be small.

In order to get some idea of the magnitude of the error introduced by using the reduced vector-wave equation instead of the full vector-wave equation, we examined three different waveguides, two of which fall in the classes stated in

Table 1 Comparison of Three Different Waveguides

Problem	FV (FD)	FV (FMM)	RV (FD)	EIM
A	3.413127	3.413133	3.413143	3.414262
B	1.990074	1.990076	1.990078	1.990080
C	1.759147	1.759167	1.760595	1.766321

Full vector-wave (FV), reduced vector-wave (RV) equation, effective index method (EIM) results for three different problems. (A) Slablike ridge waveguide; cladding air, substrate $n = 3.4$, 0.5-μm-thick guiding layer of $n = 3.44$, $\lambda = 1.15$ μm; (B) rectangular-core waveguide 5×4 μm^2 of $n = 2$ in cladding of $n = 1.5$, $\lambda = 1.32$ μm; (C) rectangular-core waveguide 1×0.5 μm^2 of $n = 2$ in cladding of $n = 1.5$, $\lambda = 1.32$ μm.

Table 1 (A and B) and one that does not (C). Table 1 shows the effective index β/k_0 of the fundamental TE mode resulting from the full vector-wave equation and the reduced vector-wave equation. The full vector-wave equations are solved by a finite difference (FD) method based on a discretization [Luss94] and the film-mode match method (FMM) [Sudb94]. The reduced vector-wave equations were also solved using the FD method. It is clearly shown that for the waveguides that fall in the classes mentioned in Table 1, A and B (>99% of the power confined in the core), using the reduced vector-wave equations instead of the full vector-wave equation is justified.

For waveguide B, the well-known effective index method (EIM), which solves the reduced wave-equation approximately, calculates the effective index with remarkable precision. However, in the less confined case of waveguide C (~70% of the power confined in the core), the difference is noticeable, i.e., two orders larger than the difference between the two vector implementations.

2.3 Other Accuracy Aspects

Apart from the choice of the form of the wave equation, aspects like waveguide description for complex and/or patently nonrectangular waveguides and used boundary conditions can influence the accuracy of the mode fields and propagation constants to a large extent. Since these aspects are method dependent we will restrict ourselves to the finite difference method. An extensive overview of methods has been given in the literature [Chia94].

2.4 Waveguide Description

An advantage of the finite difference method is its straightforward applicability to both step-index and graded-index waveguides. Especially in combination with nonuniform meshes, complex waveguide structures can be described well. This

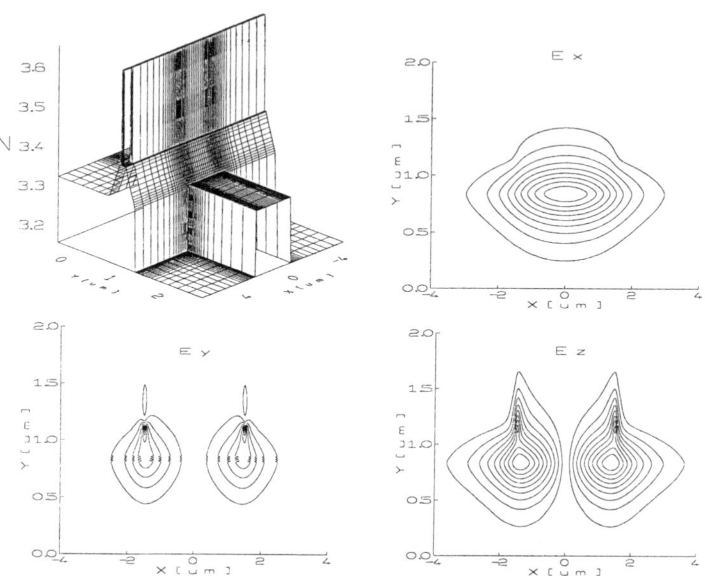

Figure 1 Index distribution and calculated electric field components of the fundamental TE mode of a ridge waveguide. Notice the discontinuities in the $E\ y$ field component at the quantum well horizontal interfaces.

is demonstrated in Fig. 1, which shows the calculated electric field components of an AlGaAs ridge waveguide that incorporates step-index as well as graded-index layers.

On the other hand, it has been shown that the transverse electric fields diverge at sharp dielectric corners [Sudb92,Ande78]. The singularity is a physical property of vector modes. For nonrectangular waveguides, however, one normally applies a staircase approximation to describe the boundary, which introduces nonphysical corners that are clearly unwanted.

2.5 Finite Computational Domain

For modes near cutoff, the mode fields extend far into the cladding. When magnetic or electric walls are used to terminate the calculation domain, one should make sure these artificial walls are placed far enough away from the core. Thus, more nodal points are needed for a given accuracy. This inefficiency can be partly overcome by the use of nonuniform meshes. More elegant ways to model the open space have been given in the literature [Had91,Vass97].

3 BEAM PROPAGATION METHODS

Most of the (commercial) integrated optic modeling tools available today rely on the beam propagation method. Probably the most important reason is that this method can be implemented rather straightforwardly to a large class of waveguide structures. In contrast to an eigenmode description, BPM calculations are based on propagation of the entire optical field. This field is then propagated in a stepwise manner through slices of the waveguide structure. Initial BPM implementations, the Fourier transform BPM (FTBPM), were based on the propagation of plane waves through a homogeneous space, with a phase correction after each propagation step to account for the refractive index inhomogeneities of the structure. However, due to their pertubative nature, FTBPMs have difficulties dealing with high-refractive-index contrast waveguides. Furthermore, they are less efficient than their finite difference counterparts and cannot deal with small details or discriminate between TE and TM polarization (for a detailed discussion see, for example, [Krij92,Hoek97]).

Nowadays most BPM implementations are based on finite difference schemes, which are superior in all the aspects mentioned to FTBPMs. In the following we will briefly outline the underlying model of the finite difference BPM (FDBPM). For the sake of simplicity we will restrict ourselves to the scalar TE approximation. The basic idea of this method is to eliminate the fast phase variations in the propagation direction due to the background homogeneous refractive index, by writing the total field electric field $E_y(x,z)$ as

$$E_y(x,z) = \psi(x,z)\exp(-in_0k_0z) \tag{7}$$

where $\psi(x,z)$ is the slowly varying field, k_0 is the free-space wave number, and n_0 is the reference index. Substitution of this expression into the scalar wave equation leads to

$$\frac{\partial^2\psi}{\partial z^2} - 2in_0k_0\frac{\partial\psi}{\partial z} - (n_0k_0)^2\,\psi + \frac{\partial^2\psi}{\partial x^2} + [n(x,z)k_0]^2\,\psi = 0 \tag{8}$$

A proper choice of the reference index is important in order to minimize the variations of the slowly varying field, and it has a direct impact on the accuracy of the results obtained [Hoek92]. Note that for propagation of an eigenmode with its effective index equal to the reference index, the z-dependence of the slowly varying field is completely eliminated.

A first approximation is obtained by making the following assumption:

$$\left|\frac{\partial^2\psi}{\partial z^2}\right| \ll \left|2n_0k_0\frac{\partial\psi}{\partial z}\right| \tag{9}$$

which is known as the *slowly varying envelope approximation* (SVEA). This leads to

$$2in_0k_0\frac{\partial\psi}{\partial z} = k_0^2[n(x,z)^2 - n_0^2]\psi + \frac{\partial^2\psi}{\partial x^2} \tag{10}$$

An expression for a propagation algorithm can be obtained by a (Crank–Nicholsen) integration combined with a discretization Δx in the transverse direction and Δz in the propagation direction. This leads to the following equation:

$$\begin{aligned}\psi_{p+1}^{s+1} + \alpha_p^+\psi + \psi_{p-1}^{s+1} &= -\psi_{p+1}^s - \alpha_p^-\,\psi_p^s - \psi_{p-1}^s\,\alpha_p^\pm \\ &= -2 \pm \frac{4ik_0n_0(\Delta x)^2}{\Delta z} + k_0^2(\Delta x)^2[n^2(x,z) - n_0^2]\end{aligned} \tag{11}$$

which relates the field at $z = (s + 1) * \Delta z$ to the field at $z = s * \Delta z$. This equation can be written in the form of a tri-diagonal matrix equation that can be solved efficiently by forward and backward substitution techniques [Pres89].

One of the appealing features of the FDBPM is that the calculation time increases only linearly with the number of points in the x-direction. Moreover, it can easily deal with small details and relatively high refractive index contrast. Also the implementation of the boundary conditions is more straightforward than for the FTBPM. Transparent boundary conditions can be implemented by assuming that the field can be approximated locally at the boundaries as a plane wave.

Nevertheless it should be recognized that the FDBPM has inherent accuracy limitations due to the SVEA approximation. By Eqs. (7) and (9) it is implicitly stated that the field propagates predominantly in one direction (the paraxial approximation), which indeed is often the case in integrated optics. For waveguides at large angles, phase and amplitude errors of the transmitted field will occur. The latter is obviously more easily observed and therefore it is conveniently used in benchmark tests to compare different BPM algorithms [Nolt95]. This is illustrated in Fig. 2, which shows the transmitted power of a 1000-μm-long single-mode waveguide versus the angle with respect to the BPM propagation direction. At an angle of about 10°, significant power loss can be observed (for the FD0 propagation algorithm). For waveguides at a large angle, the error can be minimized by choosing the reference index close to the apparent effective index of the field. Unfortunately this trick cannot be used when waveguides with significantly different directions are involved, and it is therefore of rather limited practical use. Another way to minimize the errors involved is by using a wide-angle propagation algorithm (such as the FD2 algorithm in Fig. 2), of which a discussion is considered to be beyond the scope of this section. As can be noticed in Fig. 2, the maximum allowable propagation angle can be extended significantly in this case. However, even a higher-order propagation algorithm has its limitations, so

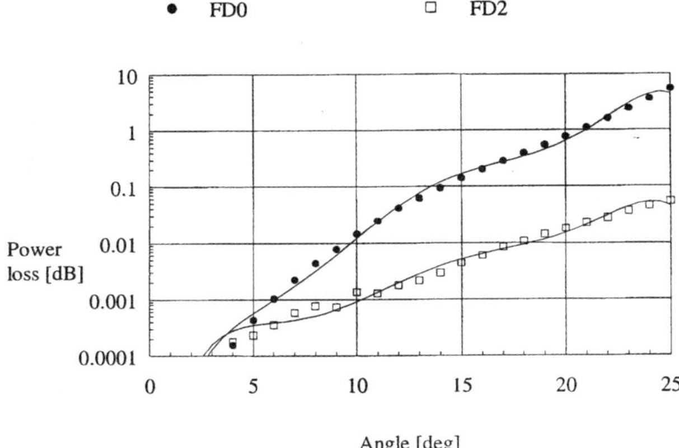

Figure 2 Apparent power loss of a straight waveguide as a function of its angle with respect to the direction of the BPM propagation. FD0 = 0th-order algorithm, FD2 = second-order algorithm (Note: the algorithm is described in Hoek 93. We did the calculation using his algorithm.) (N_{core} = 1.505, N_{clad} = 1.500, W = 6 μm, λ = 1.56 μm). [Hoek 93].

at some point one should wonder if an eigenmode-based analysis would not be a more appropriate choice.

The analysis just presented has been based on a 2D semivectorial (TE) approximation, which is used mostly in combination with the effective index method. Nevertheless, for some specific problems, a 3D BPM analysis may be required, at the price of increased simulation times. Scalar, semivectorial, and full vectorial implementations have been reported [Yama91,Xu94].

As an example, Fig. 3 shows the geometry of a polarization splitter based on an asymmetric Y-junction. In this splitter the cores of the two waveguides are identical, but they are rotated 90° with respect to each other. For these asymmetrical Y-junctions the fundamental waveguide mode has a tendency to couple into the waveguide with the highest refractive index (provided that the junction is adiabatic). For TE polarization the right waveguide (in Fig. 3) has a higher refractive index than the left waveguide. However, for TM polarized light this is exactly the opposite. Therefore this device will act as a polarization splitter. It has been simulated using a semivectorial split-step 3D BPM algorithm [Krij97]. The polarization-dependent splitting of the device can be clearly observed.

4 MODELING RARE-EARTH-DOPED WAVEGUIDES

Driven by the successful development of the erbium-doped fiber amplifier (EDFA), rare-earth-doped devices have become key components in optical tele-

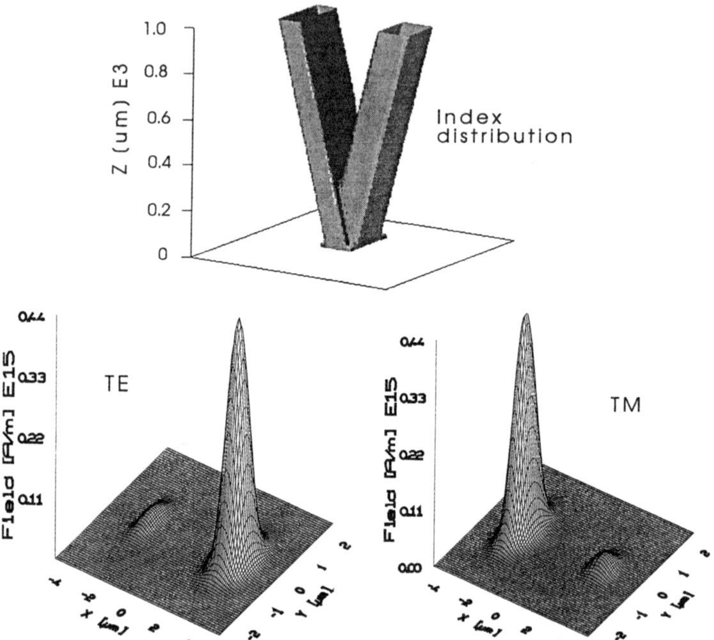

Figure 3 3D BPM simulation of a polarization splitter based an asymmetrical Y-junction.

communications. They include fiber amplifiers for the second as well as the third telecom window [Bjar93], integrated optical amplifiers for on-chip loss compensation in complex integrated optical circuits [Barb97,Baum96], as well as various laser configurations utilizing both fiber and integrated optical amplifiers as the gain medium [Baum96,Zhan96].

The design of rare-earth-doped amplifiers and lasers requires sophisticated modeling tools, allowing optimization and performance analysis of the device prior to manufacturing. These modeling tools should at least incorporate: (1) all relevant properties of the amplifying rare-earth-doped material, (2) all relevant properties of the amplifying waveguide, e.g., dopant and modal intensity distributions, and (3) the propagation of optical power through the device.

In this section, the construction of such a model, including these three key elements, will be illustrated using as an example a model for an erbium-doped waveguide amplifier (EDWA) [Hors94]. It will be shown that the model, developed at the university of Twente, successfully predicts the performance of an $Er^{3+}:Y_2O_3$ integrated optical amplifier [vanW97].

4.1 A Model for Erbium-Doped Waveguide Amplifiers

In the following, a model suitable for calculating EDWA gain will be presented [vanW97]. We will restrict ourselves to a specific, somewhat simplified amplifier configuration and focus primarily on the key elements in the model. However, as will be shown at the end of this section, the simplifications made can easily be removed with some minor modifications without affecting the essence of the model.

The EDWA considered here consists of a z-invariant planar waveguide in which one of the waveguide materials is doped with erbium. We assume copropagation of pump beam and signal beam. Although different configurations are possible, e.g., tapered amplifiers or counter-propagating pump and signal beams, this is the configuration most generally used in EDWAs. Furthermore, we assume monochromatic pump and signal sources, a pump wavelength of 1.48 μm, and negligible amplified spontaneous emission (ASE).

To explain the principle of operation for this type of EDWA, let us consider the example structure of Fig. 4. The amplifying waveguide consists of an erbium-doped Y_2O_3 core layer sandwiched between two SiO_2 cladding layers. Lateral confinement is obtained by etching a shallow ridge in the Y_2O_3 film. Optical amplification is obtained by simultaneously launching a pump beam and a signal beam in the waveguide. On propagation, the pump power is absorbed by the

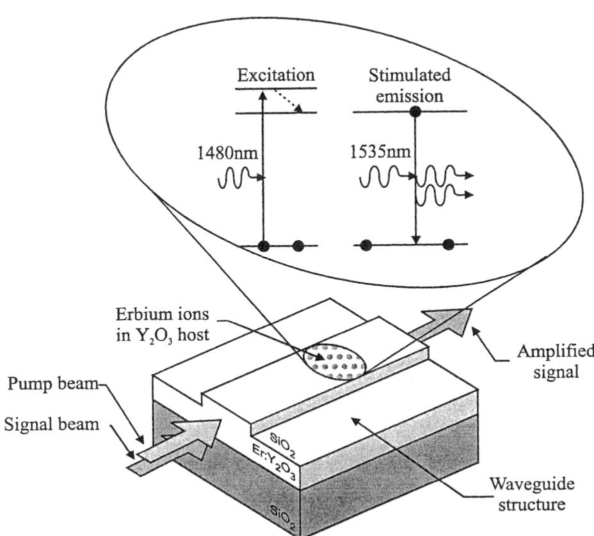

Figure 4 $Er^{3+}:Y_2O_3$ integrated optical amplifier.

erbium ions in the doped film, which are excited to a higher energy state. If the pump power is sufficiently high, a population inversion is created and the copropagating signal beam is amplified by stimulated emission.

The model describing this type of amplifier consists of three parts. Starting with the properties of the erbium-doped material, i.e., the energy level scheme and relevant transitions, we derive rate equations to obtain the population densities of the erbium energy states as a function of pump and signal intensity and dopant concentration. Next, we include the transverse distributions of pump and signal intensity and erbium concentration to obtain modal gain coefficients as a function of pump and signal power. Finally, we apply a discrete propagation algorithm to calculate the evolution of pump and signal power through the amplifier.

4.2 Properties of the Erbium-Doped Material

Basically, the amplifying properties of erbium-doped materials can be modeled as a quasi-three-level-laser material [vanW97]. However, accurate modeling generally requires incorporation of several additional effects, two of which have been included in the model:

1. *Excited-state absorption (ESA).* Once an erbium ion is excited by absorption of a pump or signal photon, it may be excited to a higher energy state by absorption of another pump or signal photon, thus reducing amplifier efficiency.
2. *Up-conversion (UC).* When two excited erbium ions are in close proximity—i.e., the erbium concentration is high—energy transfer can occur between the ions. One of the ions may transfer its energy to the other and return to the ground state, while the other ion is excited to a higher energy state. Like ESA, this process significantly reduces amplifier efficiency and is specifically important for modeling EDWAs due to the high erbium concentrations used.

Although some materials may require additional effects to be included, such as clustering and concentration quenching, we will restrict ourselves here to ESA and UC. Note, however, that incorporation of such mechanisms in our model is in most cases quite straightforward.

Including ESA and UC, the relevant erbium energy levels and transitions are shown in Fig. 5. Here, E_0–E_3 denote the (lower four) energy levels involved. The transitions labelled P_{ij}, S_{ij}, and R_{ij} denote, respectively, the pump-induced, signal-induced, and spontaneous transitions between initial and final levels i and j.* Up-conversion transitions are labeled $U_{ij,kl}$, with initial and final levels i, j,

* It should be noted that the erbium energy levels are not discrete, but consist of a number of thermally coupled sublevels. From a modeling point of view, however, the energy levels can be considered as one, introducing wavelength-dependent pump- and signal-induced rates P_{ij} and S_{ij} to account for the distribution of ions over the sublevels.

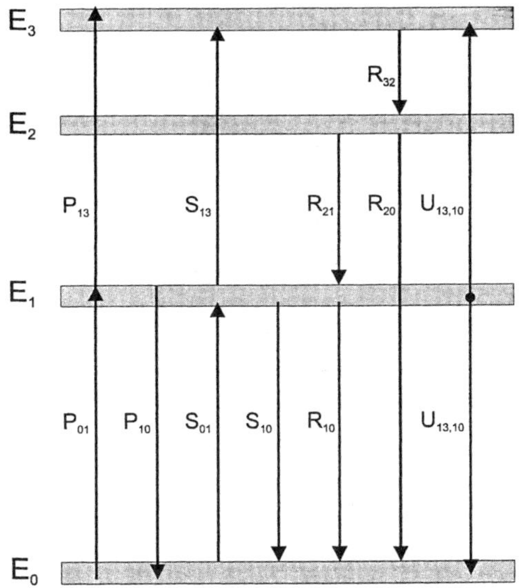

Figure 5 Er^{3+} energy level scheme and relevant transitions.

and *k*l respectively. From Fig. 5, rate Eq. (11) can be derived, describing the time dependence of the population densities n_0–n_3 of levels E_0–E_3:

$$\frac{dn_3}{dt} = (P_{13} + S_{13})n_1 + U_{13,10}n_1^2 - R_{32}n_3$$

$$\frac{dn_2}{dt} = -(R_{20} + R_{21})n_2 + R_{32}n_3$$

$$\frac{dn_1}{dt} = (P_{01} + S_{01})n_0 - (P_{10} + S_{10} + R_{10} + P_{13} + S_{13})n_1 - 2U_{13,10}n_1^2 + R_{21}n_2 \tag{12}$$

$$\frac{dn_0}{dt} = -(P_{01} + S_{01})n_0 + (P_{10} + S_{10} + R_{10})n_1 + U_{13,10}n_1^2 + R_{20}n_2$$

In steady state, the time derivatives are zero. Assuming fast relaxation from E_3 to E_2, i.e., $n_3 \approx 0$, and using the equality $n_0 + n_1 + n_2 + n_3 = n_{\mathrm{Er}}$, where n_{Er} is the total erbium concentration, the system of equations can be analytically solved for n_0–n_3. Writing P_{ij} and S_{ij} as given in Eqs. (13), with σ_{ij} the absorption ($i < j$) or emission ($i > j$) cross section of the transition, the solution of the rate equations gives the population densities n_0–n_3 as a function of pump intensity I_p, signal intensity I_s, and dopant concentration n_{Er}. It should be noted that it is generally not possible to solve the system of rate equations analytically, especially if multiple up-conversion terms play a role. In that case, the system of equations has to be solved numerically:

$$P_{ij} = \sigma_{ij}^{p} \frac{\lambda_p}{hc} I_p$$

$$S_{ij} = \sigma_{ij}^{s} \frac{\lambda_s}{hc} I_s \qquad (13)$$

4.3 Waveguide Properties

The next step in our model is to include waveguide properties. This involves calculation of the transverse distributions of pump and signal intensity and erbium concentration to obtain the distribution of population densities across the amplifying waveguide. From these, modal gain coefficients can be calculated as a function of pump and signal power. The intensity distributions are obtained from a modal analysis of the amplifying waveguide. Here, the amplifying medium is considered as a perturbation to a lossless waveguide, and standard mode solvers are used to calculate the modal intensity distributions. The erbium distribution is closely linked to the waveguide configuration, and it can usually be obtained from simple formulae. For instance, in the case of our example (Fig. 4), the erbium is uniformly distributed throughout the Y_2O_3 film.

Once the intensity and dopant distributions are known, modal gain coefficients can be obtained by calculating the overlap between the intensity and gain distributions. The result, including the modal background attenuation α_p and α_s, is shown in Eqs. (14). Here, P_{ij} and S_{ij} are related to the (position-dependent) intensity distributions through Eqs. (13), and the plane of integration A is perpendicular to the propagation direction. Note that the denominator of Eqs. (14) cannot be solved analytically and has to be evaluated numerically:

$$G_p = \frac{\iint [(P_{10} - P_{13}) n_1(I_p, I_s, n_{\mathrm{Er}}) - P_{01} n_0(I_p, I_s, n_{\mathrm{Er}})] dA}{\iint I_p dA} - \frac{\alpha_p}{10 \log(e)}$$

$$G_s = \frac{\iint [(S_{10} - S_{13}) n_1(I_p, I_s, n_{\mathrm{Er}}) - S_{01} n_0(I_p, I_s, n_{\mathrm{Er}})] dA}{\iint I_s \, dA} - \frac{\alpha_s}{10 \log(e)} \qquad (14)$$

4.4. Propagation Algorithm

The final step in our model is to include the evolution of pump and signal power along the amplifier. For that purpose we use a discrete propagation algorithm as given by Eqs. (15). Knowing the pump and signal power at the amplifier input ($z = 0$), we can use Eqs. (15) to calculate the output power after a short propagation distance Δz. By repeatedly applying Eqs. (15), evaluating Eqs. (14) at each step, we can calculate the signal output power and, consequently, the amplifier gain:

$$\begin{aligned} P_p(z + \Delta z) &= P_p(z)\exp[G_p(z)\,\Delta z] \\ P_s(z + \Delta z) &= P_s(z)\exp[G_s\,(z)\,\Delta z] \end{aligned} \tag{15}$$

4.5 Example Calculation

To illustrate the possibilities of the EDWA model, Fig. 6 shows the signal gain as a function of pump power for the example amplifer of Fig. 4. The model calculation was made using experimentally determined parameters [vanW97]. It can be seen that the experimental measurement points agree excellently with calculations.

4.6 Generalization

In the previous sections, a typical EDWA model based on the three previously mentioned key elements has been described. This model was derived based on a number of simplifications, which can in most cases easily be removed. For instance, by modifying the rate equations, a different pump wavelength or additional effects like clustering can be accommodated. Polychromatic light sources can be included by modeling them as a superposition of monochromatic light

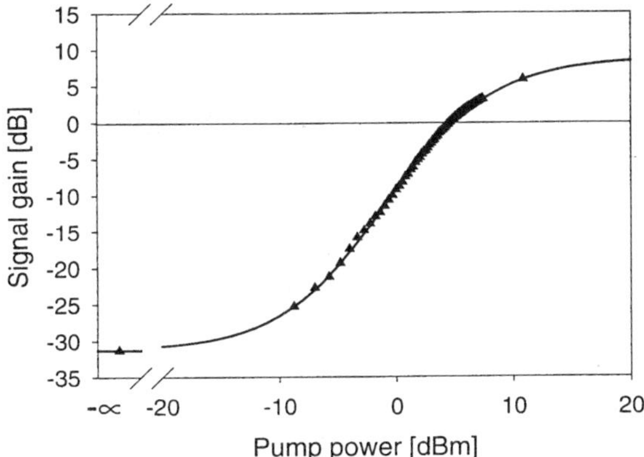

Figure 6 Er^{3+}:Y_2O_3 amplifier gain as a function of pump power (solid line: model calculation, triangles: measurement points).

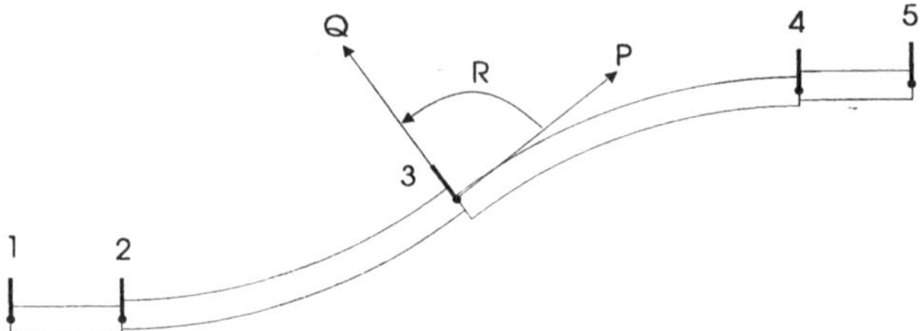

Figure 7 Graphic representation of the connectivity between different waveguide elements by using "planes" with a local coordinate system *PQR*.

sources. By including backward propagation and appropriately modifying the boundary conditions of the propagation model, counter-propagating pump and signal beams as well as ASE can be modeled. Finally, z-variant waveguides can be analyzed by varying the waveguide geometry in the direction of propagation. By including modal overlap integrals in Eqs. (15), even nonadiabatically varying structures can be analyzed.

5 MASK DESIGN TOOLS

One aspect that often gets little attention when discussing design tools for integrated optics but that is of great practical importance is the mask layout capability. As the complexity of photonic integrated circuits (PICs) continues to grow, design tools that can handle this complexity become more and more indispensable.

One of the most important and unique aspects of the layout of integrated optic components is the critical connectivity between the different waveguide sections. When certain waveguide sections have been connected, it is important that the connection remains unmodified, even when the dimensions of other parts of the circuit change. Such a connection scheme has been implemented in the integrated BPM/mask layout tool Prometheus.* As with nodes in electronic design, different waveguide elements are connected on "planes." Specifying the input and output plane number of each element determines in which way the different elements are connected. A relative coordinate system (see Fig. 7) is applied in order to allow for relative positioning between the elements (such as the offset between a straight and a curved waveguide section to account for the modal shift in the curved waveguide).

Because the layout of many integrated optic devices can be described by a limited number of mathematical expressions, ideally one would like to have the

* By BBV Software, Hengelosestraat 705, 7521 PA, Enschede, The Netherlands.

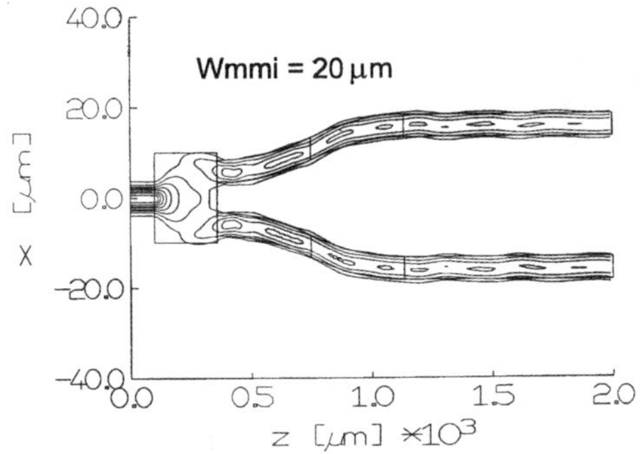

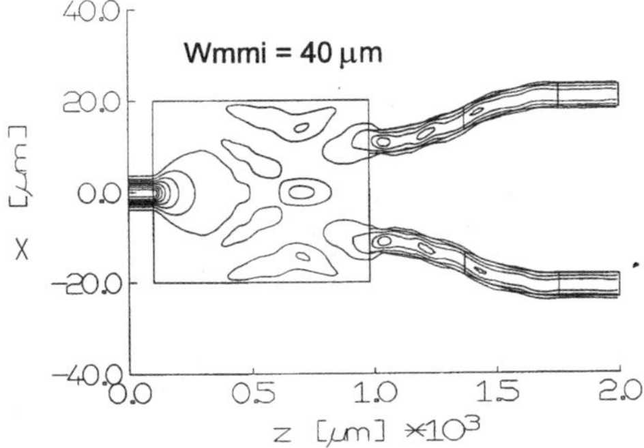

Figure 8 Parametrized design of a 1 × 2 multimode interference splitter in Prometheus. As the MMI width is increase from 20 to 40 µm, the other dimensions scale automatically according to the implemented design rules.

possibility to incorporate the design dimensions in terms of these expressions, rather than with actual numbers. The major advantage of such an approach is that the structures can be parametrized to reflect their physical operation principle and can be modified within a matter of seconds. Figure 8 shows how the design of a 1 × 2 multimode interference (MMI) splitter has been implemented in Prometheus using this approach. The required MMI section length of such a splitter simply follows from the beat length of the first two even modes of this section. As the width of the MMI section (or any other waveguide parameter) is modified, the length gets automatically updated, accounting for the different length required for the self-imaging effect [Sold95].

In order to cope with higher levels of complexity in a time-efficient manner, hierarchical design tools will become more and more important. When a subcom-

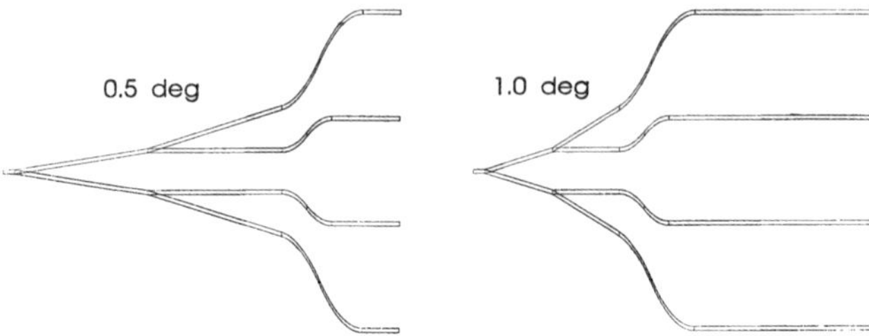

Figure 9 Layout of a 1 × 4 splitter using a 1 × 2 splitter as the building block. By proper parametrization of the design, the branching angle of the basic building block can easily be modified without affecting the integrity of the design.

ponent of a PIC has been designed, it should be possible to have the ability to use it as a basic building block for more complex devices. For example, when a 1 × 2 splitter has been designed and optimized, one would like to reuse this building block to create a 1 × 4 splitter, as is illustrated in Fig. 9. Nevertheless, sometimes it may still be required to have easy access to some of the building block variables on the higher level. This is shown in the example of the completely parametrized 1 × 4 splitter. Here the branching angle of the fundamental 1 × 2 splitter is increased from 0.5° to 1.0°. Due to the parametrization, the entire layout is automatically updated while keeping the output waveguide spacing and die length fixed.

In the previous example, the hierarchical design features are still limited to the layout level only. For even more complex devices, it is also desirable to have hierarchical simulation capabilities. An example of an *S*-matrix-based modeling tool that supports hierarchical simulations will be discussed in the next section.

6 *S*-MATRIX-BASED SIMULATOR

As described in the previous sections, most computer-aided design (CAD) tools for integrated optics are based on a layout description of the circuit and use BPM simulation techniques to carry out the simulation. For circuits with a higher level of complexity, this approach becomes impractical. In this section an alternative CAD tool, developed at the Delft University of Technology, will be presented. This tool for the design of photonic integrated circuits uses an approach also employed in design systems for electronic and microwave circuits: The circuit is designed on a symbolic level, and the simulation or generation of the mask layout is performed from that level. Both approaches are shown schematically

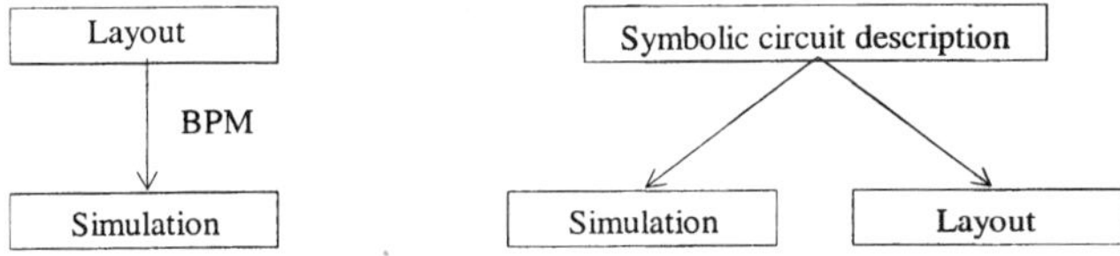

Figure 10 BPM simulation from physical description (left). Simulation and layout generation from a symbolic description (right).

in Fig. 10. The advantage of this method is that complex circuits, possibly containing loops, can be simulated; the circuit can be structured better and the best-suited simulation method can be chosen for each component or subcircuit. For example, a fast mode solver can be used for straight waveguides and a BPM simulation for a Y-junction.

6.1 Description of Optical Components

In the symbolic description, a photonic integrated circuit is composed of elementary optical components. In the simplest case these components are straight and curved waveguides. The coupling between the components takes place via guided modes and radiation fields. For most components, coupling through radiation fields is small and the optical components can be considered as individual units connected to each other at well-defined ports.

An ideal monomode waveguide is a two-port component. The waveguide properties are characterized by the propagation constant of the fundamental mode, β, and the response of the components is described by the 2×2 scattering matrix S as shown in Table 2, with ℓ the length of the waveguide. For a curved waveguide, $S_{12} = S_{21} = \exp(-i\beta_\phi \phi)$, with β_ϕ the angular propagation constant and ϕ the sector angle of the curved waveguide. Since there are no reflections in both

Table 2 Graphical Representation and S-Matrix Description of Monomode Waveguide Components

	Straight waveguide	Curved waveguide	Waveguide junction
Graphical representation	W= L=	Ang= R= W=	O= W1= W2= R2=
S-matrix	$\begin{pmatrix} 0 & e^{-j\beta\ell} \\ e^{-j\beta\ell} & 0 \end{pmatrix}$	$\begin{pmatrix} 0 & e^{-j\beta_\phi\phi} \\ e^{-j\beta_\phi\phi} & 0 \end{pmatrix}$	$\begin{pmatrix} 0 & \int U_1^* U_2 \\ \int U_1^* U_2 & 0 \end{pmatrix}$

straight and curved waveguides, the diagonal matrix elements, S_{11} and S_{22}, are zero.

The junction between two monomode waveguides is described by a 2×2 scattering matrix (see Table 2). The matrix elements are given by the overlap integral of the modal fields U_1 and U_2 in each of the waveguides $S_{12}^* = S_{21} = \int U_1 U_2^*$. Because reflections in optical chips are small for many applications, S_{11} and S_{22} are set to zero, although inclusion of reflections is straightforward.

In general, the response of an N-port component is described by its $N \times N$ S-matrix [vanD91,Leij95,Leij96]. The number of ports N is given by $N = \sum_{k=1}^{n} m_k$, where n is the number of physical ports for the component and m_n is the number of (guided) optical modes considered for input n.

In order to obtain the S-matrix elements, any mode solver calculating the modal propagation constants and fields can be used. A beam propagation method is used for components where radiation fields play an important role. However, at the junction with other components, only the guided modes are considered. Example components are (linear) tapers and Y-junctions.

The waveguide structure for the (sub-)circuit is defined by a symbol that specifies the type of waveguide (strip, rib, buried, etc.), the material parameters, and the geometry. The refractive indices of the materials in the waveguide stack can be calculated according to a model to take into account the material dispersion. This is particularly important when designing waveguide filters and routers.

6.2 Modeling of PHASAR Demultiplexer

The PHASAR demultiplexer is a combination of two star couplers connected by an array of waveguides acting as a grating [Smit96]. Figure 11 shows the symbolic representation of a 6×6 PHASAR demultiplexer. Design parameters such as the central wavelength and channel spacing are specified, and the first step in the simulation is the construction of the geometry of the component, following the description in the literature [Smit91]. Next, the simulation of the component with this geometry is started by calculating the diffraction of the input field in the star coupler, using the Rayleigh–Sommerfeld model [Born93].

The coupling coefficients of the diffracted field with each waveguide of the array are calculated. These coefficients are corrected for coupling between adjacent array waveguides by a normalization procedure using the supermodes of the waveguide array, which are constructed following the description in the literature [Chia91].

The propagation in each waveguide in the PHASAR is calculated taking into account the coupling loss at the junctions and the radiation loss in the curved waveguides. As an option, the phase transfer of each waveguide in the array can be changed randomly by a small amount with a normal distribution, in order to simulate the phase incoherence in the array due to imperfections in the fabrication

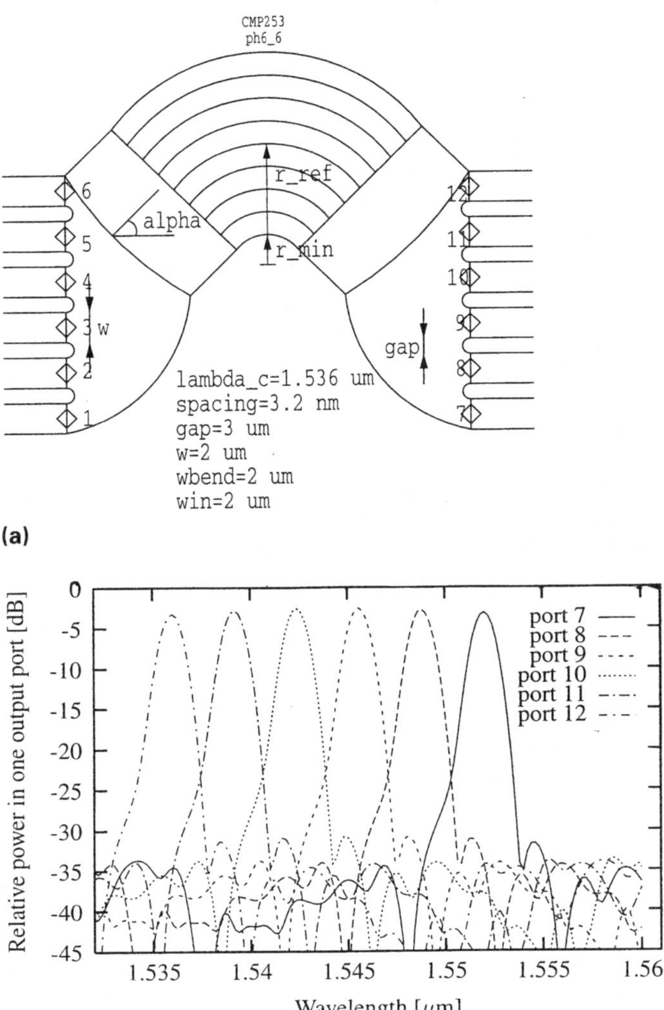

Figure 11 (a) Symbolic representation of the PHASAR. (b) The wavelength response of the PHASAR used for add–drop application, when the signal is launched in input 1.

process. In this way, a realistic crosstalk level for the PHASAR can be simulated as well.

6.3 Four-Channel Add–Drop Simulation and Comparison with Measurement

The symbolic representation of the four-channel add–drop multiplexer (ADM) is shown in Fig. 12. It consists of a combination of straight and curved waveguides, junctions, crossings, switches, and one 6 × 6 PHASAR. Switches are based on a Mach–Zehnder interferometer, and each switch symbol represents the full circuit. The 6 × 6 PHASAR has a central wavelength of 1.536 nm and a channel spacing of 3.2 nm (400 GHz). A phase noise with a spread $\sigma = 8°$ results in a realistic crosstalk level of about −30 dB. The simulated response of the PHASAR is shown in Fig. 11(b). When four channels with wavelengths $\lambda_1 = 1.5360$ μm, $\lambda_2 = 1.5392$ μm, $\lambda_3 = 1.5424$ μm, and $\lambda_4 = 1.5456$ μm are launched into the lower port (Input) of the PHASAR, they are spatially separated by the PHASAR and routed to the switches, S1–S4, respectively. Each wavelength can be switched to the drop port (D1–D4) of the switch or back to the PHASAR, where the signals are multiplexed into one output port.

The simulated response of the ADM when the signals at λ_1, λ_2, and λ_4 loop back to the PHASAR is shown in Fig. 13(a). The signal going through switch S3 is routed to the drop port D3. Three peaks (λ_1, λ_2, λ_4) are visible, and there is a fourth peak originating from the direct signal from input to output. A signal with the correct wavelength (λ_3) has been added at port A3 (dashed line in the figure). The estimated losses for the signals that are looped from input to output are between 5.9 and 7.2 dB. The peak at λ_3 has a larger width because the signal

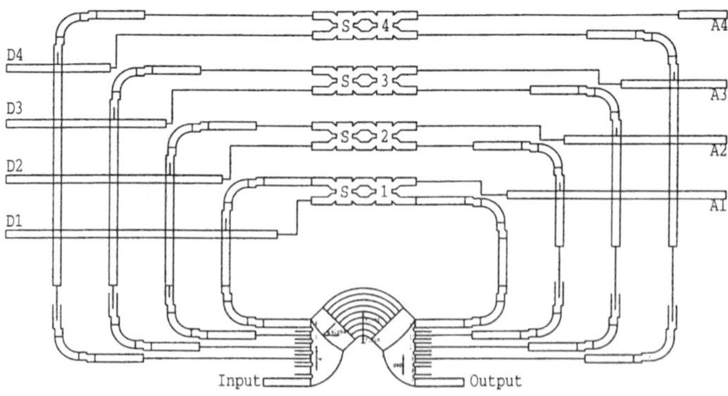

Figure 12 Symbolic representation of the four-channel ADM circuit.

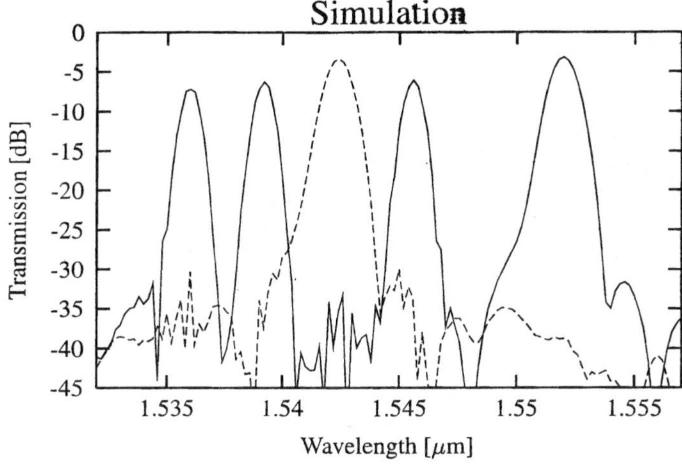

(a)

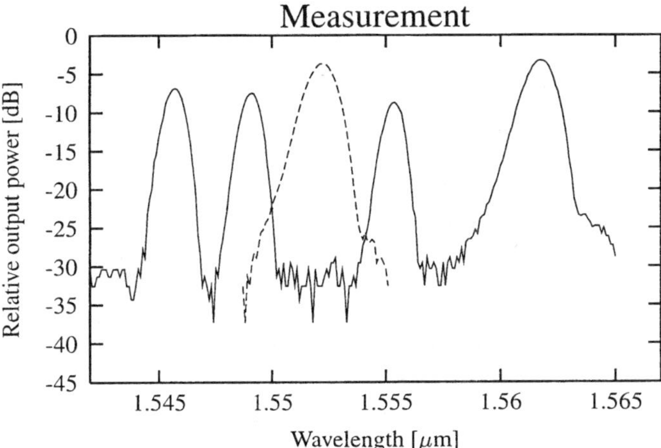

(b)

Figure 13 (a) Simulation of four-channel ADM with switch 3 in the bar state. The added signal is shown with a dashed line. (b) Measurement of four-channel ADM with switch 3 in the bar state. The added signal is shown with a dashed line.

is filtered only once by the PHASAR. Losses for this signal are also less, for the same reason.

An InP-based four-channel ADM with a design identical to the simulated ADM has been realized [Vree97]. The measured response is presented in Fig. 13(b). The shapes of the simulated and measured curves are in good agreement with each other. The difference between the designed and fabricated waveguide structures has caused a shift in the central wavelength of approximately 9 nm. In case the switches are in the cross state, the measured losses are between 7 and 9.1 dB. The simulated losses are less (5.9 dB−−7.2 dB) because the simulation doesn't take into account the transmission losses in the waveguides, which are measured to be 1.5 dB/cm. Since the length of the loop is approximately 1.2 cm, this accounts for the difference.

The simulated loss of the waveguide crossings is 0.1 dB per crossing. The measured value is 0.3 dB per crossing. This accounts for the difference in relative peak height, since the signals at λ_1, λ_2, λ_3, and λ_4 pass 1, 3, 5, and 7 crossings, respectively. Figure 13(b) shows a residual signal for about 30 dB below the original signal. This is in good agreement with the simulated value. The origin of this residual signal is the extinction ratio of the switch and the phase errors of the PHASAR.

7 CONCLUSIONS

In the previous sections we have described some of the most frequently used design and simulation tools for integrated optics. Obviously this oveview cannot be complete within the scope of this chapter. It should be recognized that, in order to model the many different integrated optic devices that are being investigated and/or developed today, a wide variety of design and simulation tools is required. This issue is amplified by the many different waveguide technologies currently used. Nonetheless, in the last couple of years a significant number of (commercial) design and simulation tools have become available to researchers and product developers. The continuing development of these tools will definitely benefit the maturing of the field of integrated optics.

REFERENCES

1. J. B. Andersen and V. V. Solodukov. Field behavior near a dielectric wedge. IEEE Trans. Ant. Prop. 26:598–602, 1978.
2. D. Barbier et al. Amplifying four-wavelength combiner, based on erbium/ytterbium-doped waveguide amplifiers and integrated splitters. IEEE photon. Technol. Lett. 9: 315–317, 1997.
3. I. Baumann et al. Er-doped integrated optical devices in $LiNbO_3$. IEEE J. Sel. Top. Quant. Electron. 2:355–366, 1996.

4. A. Bjarklev. Optical Fiber Amplifiers: Design and System Applications. Artech House, Boston, 1993.
5. M. Born and E. Wolf. Principles of Optics, 6th ed. New York: Pergamon Press, 1993.
6. J. C. Chen and S. Junglin. Computation of higher-order waveguide modes by imaginary-distance beam propagation method. Opt. Quantum Electron. 26:199–205, 1994.
7. K. S. Chiang. Effective index method for the analysis of optical waveguide couplers and arrays: an asymptotic theory. J. Lightwave Technol. 9:62–72, 1991.
8. K. S. Chiang. Review of numerical and approximate methods for the modal analysis of general optical dielectric waveguides. Opt. Quant. Electron. 26:113–134, 1994.
9. G. R. Hadley. Transparent boundary condition for beam propagation. Opt. Lett. 16: 624–626, 1991.
10. H. Hoekstra et al. On the accuracy of the finite difference method for applications in beam propagation techniques. Opt. Comm. 94:506–508, 1992.
11. H. Hoekstra et al. New formulation of the beam propagation method based on the slowly varying envelope approximation. Opt. Comm. 97:301–303, 1993.
12. H. Hoekstra. On beam propagation methods for modelling in integrated optics. Opt. Quant. Electron. 29:157–171, 1997.
13. F. Horst et al. Design of 1480-nm diode-pumped Er^{3+}-doped integrated optical amplifiers. Opt. Quant. Electron. 26:285–299, 1994.
14. G. Krijnen. All-optical switching in nonlinear integrated optic devices Ph.D. dissertation, University of Twente, The Netherlands, 1992.
15. Personal communications.
16. X. J. M. Leijtens et al. CAD tool for integrated optics. Proc. 7th Conf. on Integr. Opt. (ECIO '95), Delft, the Netherlands, April 3–6, 1995, pp. 463–466.
17. X. J. M. Leijtens et al. S-matrix oriented CAD-tool for simulating complex integrated optical circuits. J. Sel. Top. Quant. Electron. 2:257–262, 1996.
18. P. Lusse et al. Analysis of vectorial mode fields in optical waveguides by a new finite difference method. J. Lightwave Technol. 12:487–494, 1994.
19. H. P. Nolting et al. Resulting of benchmark tests for different numerical BPM algorithms. J. Lightwave Technol. 13:216–224, 1995.
20. W. H. Press et al. Numerical Recipes in Pascal. New York: Cambridge University Press, 1989, Chap. 2.6.
21. M. K. Smit. Integrated optics in silicon-based aluminum oxide. Ph.D. dissertation, Delft University of Technology, the Netherlands, 1991.
22. M. Smit and C. van Dam. PHASAR-based WDM-devices: principles, design and applications. J. Sel. Top. Quant. Electron. 2:236–250, 1996.
23. A. W. Snyder and J. D. Love. Optical Waveguide Theory. London: Chapman and Hall, 1983.
24. L. B. Soldano and E. C. M. Pennings. Optical multi-mode interference devices based on self-imaging: principles and applications. J. Lightwave Technol. 13:615–627, 1995.
25. A. Sv. Sudbo. Why are accurate computations of mode fields of rectangular dieletric waveguides difficult? J. Lightwave Technol. 10:418–419, 1992.
26. A. Sv. Sudbo. Improved formulation of the film mode matching method for mode field calculations in dielectric waveguides. Pure Appl. Opt. 3:381–388, 1994.

27. G. Sztefka and H. P. Nolting. Bidirectional eigenmode propagation for large refractive index steps. IEEE Photon. Technol. Lett. 5:554–557, 1993.
28. C. van Dam et al. Optical chip design with a microwave CAD-system. Proc. 10th Eur. Conf. on Circuit Theory and Design, Vol. 3, pp. 1316–1323, New York, Sept. 2–6, 1991.
29. H. J. van Weerden et al. Low-threshold amplification at 1.5 μm in Er:Y2O3 IO-amplifiers. Proc. 8th European Conference on Integrated Optics, ECIO '97, 1997, pp. 169–172.
30. C. Vassallo and J. M. van der Keur. Comparison of a few transparent boundary conditions for finite-difference optical mode-solvers. J. Lightwave Techn. 15: 2, pp. 397–402, 1997.
31. C. Vreeburg et al. First InP-based reconfigurable integrated add–drop multiplexer. IEEE Photon. Technol. Lett. 9:188–190, 1997.
32. J. Willems et al. The bidirectional mode expansion method for two-dimensional waveguides: the TM case. Opt. Quant. Electron. 27:995–1007, 1995.
33. C. L. Xu et al. An unconditionally stable vectorial beam propagation method for 3-D structures. IEEE Phot. Technol. Lett. 6:549–5551, 1994.
34. J. Yamauchi et al. Beam-propagation analysis of optical fibers by alternating direction implicit method. Electron. Lett. 27:1663–1665, 1991.

Index